T0203009

Foundations of Engineering Mechanics

Series Editors

Vladimir I. Babitsky, School of Mechanical, Electrical and Manufacturing Engineering, Loughborough University, Loughborough, Leicestershire, UK

Jens Wittenburg, Karlsruhe, Germany

The series "Foundations of Engineering Mechanics" includes scientific monographs and graduate-level textbooks on relevant and modern topics of application-oriented mechanics. In particular, the aim of the series is to present selected works of Russian and Eastern European scientists, so far not published in Western countries, drawing from the large pool of experience from major technological research projects. By contributing to the long tradition of enrichment of Western science and teaching by Eastern sources, the volumes of the series address to scientists, institutional and industrial researchers, lecturers and graduate students.

More information about this series at http://www.springer.com/series/3582

Nikolai Nikolaevich Polyakhov ·
Mikhail Petrovich Yushkov ·
Sergey Andreevich Zegzhda
Authors

Petr Evgenievich Tovstik
Editor

Rational and Applied Mechanics

Volume 1. Complete General Course
for Students of Engineering

 Springer

Authors
Nikolai Nikolaevich Polyakhov
Faculty of Mathematics and Mechanics
Saint Petersburg State University
St. Petersburg, Russia

Mikhail Petrovich Yushkov🅭
Faculty of Mathematics and Mechanics
Saint Petersburg State University
St. Petersburg, Russia

Sergey Andreevich Zegzhda
Faculty of Mathematics and Mechanics
Saint Petersburg State University
St. Petersburg, Russia

Editor
Petr Evgenievich Tovstik
Faculty of Mathematics and Mechanics
St. Petersburg State University
St. Petersburg, Russia

Translated by
O. S. Bukashkina
Saint Petersburg State University
St. Petersburg, Russia

A. R. Alimov
Moscow State University
Moscow, Russia

G. A. Sinilshchikova
Saint Petersburg State University
St. Petersburg, Russia

E. L. Belkind
Saint Petersburg State University
St. Petersburg, Russia

ISSN 1612-1384   ISSN 1860-6237 (electronic)
Foundations of Engineering Mechanics
ISBN 978-3-030-64063-7   ISBN 978-3-030-64061-3   (eBook)
https://doi.org/10.1007/978-3-030-64061-3

This Springer imprint is published by the registered company Springer Nature Switzerland AG
The registered company address is: Gewerbestrasse 11, 6330 Cham, Switzerland

Appraisal by Series Editor Vladimir Babitsky

This fundamental and thoroughly prepared course in rational and applied mechanics reflects the high level of teaching in this discipline by professors of the St. Petersburg State University. The main feature of the course proposed is a wide discussion covering, along with the classical problems of mechanics, new areas of study borne by developing applications. This provides rich material for the formation of numerous special courses as well. Although the main content of the course is designed for post-graduate students, many sections can be studied at the undergraduate level as well. In terms of the amount of material presented, the course has no analogues in the English literature. Once published the books will find an important place in the libraries of the major world universities.

Foreword

The present book "Rational and Applied Mechanics", which is written for classical universities and published in two volumes comprising the fourth Russian edition, was motivated by Prof. Nikolai Polyakhov (1906–1987), who was a well-known scientist in mechanics, doctor of sciences in engineering, and honored worker of science of the Russian Soviet Federative Socialist Republic.

After graduating from Moscow State University in 1929, Nikolai Polyakhov worked at Zhukovskii Central Aerohydrodynamic Institute under the supervision of S. A. Chaplygin. By the task of Prof. Chaplygin, N. N. Polyakhov and V. P. Vetchinkin developed the theory of propellers, which was published in the book "Theory and calculations of air screw-propeller". This book is still considered actual.

In 1933, N. N. Polyakhov moved to Leningrad (now St. Petersburg) and became a teacher at the chair of hydrodynamics at Leningrad Polytechnic Institute (now Peter the Great St. Petersburg Polytechnic University). In 1953, at the suggestion of Academician V. I. Smirnov, Nikolai Polyakhov headed the Chair of Analytical Mechanics (which was later called the Chair of Theoretical and Applied Mechanics) at the Department of Mathematics and Mechanics of Leningrad University. In this connection, he had to teach the new and fairly extensive course entitled "Theoretical mechanics". For a long time, he was busy in developing and continuously improving this course, incorporating not only methodological enhancements, but also embellishing the course with new scientific results, which is of course natural for fundamental university courses. Special mention should be made of the cycle of Polyakhov's papers on equations of nonholonomic mechanics and variational principles in mechanics. In particular, in these papers, he first introduced the generalized Hamilton operator (which should appropriately be called the Polyakhov operator), which proved to be instrumental in describing geometrically the reaction of perfect nonholonomic constraints.

In 1975 (22 years after he started working on the course!), Nikolai Polyakhov invited his students S. A. Zegzhda and M. P. Yushkov, who gave lectures on the same subject at other divisions of the faculty, to work jointly on a university textbook on theoretical mechanics. After ten years of work, this book was published by the Leningrad University Press and in 1987 was awarded the first premium of the university. In 2000 and 2012, the second and third editions of the textbook were published.

After Nikolai Polyakhov passed away, S. A. Zegzhda and M. P. Yushkov with their pupils continued their work in analytical mechanics. These results were published in three books (written jointly with Sh. Kh. Soltakhanov); of these three books the book "Motion equations of nonholonomic systems and variational principles in mechanics. New class of control problems" was translated into Chinese, and the book "Nonholonomic mechanics. Theory and applications" was published in English by Springer. This cycle of works on nonholonomic mechanics was awarded in 2011 a prize for science from St. Petersburg State University.

Following Nikolai Polyakhov, S. A. Zegzhda and M. P. Yushkov started working on a more complete textbook comprising the new results from the above three books. It is important that new chair staff came to work on the new edition of the book—they prepared chapters reflecting their scientific interest and the content of the special courses given by them. These chapters were included in the second volume of the book, whereas the first one contains the basic extended course in rational mechanics for university students specializing in mathematics and mechanics. In this regard, the authors decided to give a new name to the book: "Theoretical and applied mechanics", which happened to agree with their chair name.

A special role in the preparation of the book was played by P. E. Tovstik, who heads the chair since 1977, after Nikolai Polyakhov was moved to the chair of hydromechanics. Professor Tovstik is not only Editor-in-Chief of the first three editions of the textbook but also an author of the majority of chapters.

There is no doubt that the textbook by N. N. Polyakhov, P. E. Tovstik, S. A. Zegzhda, and M. P. Yushkov "Theoretical and applied mechanics" should not only meet the requirements of students and postgraduate students at classical universities, but will also be helpful for specialists in theoretical, analytical, and applied mathematics.

St. Petersburg, Russia   N. F. Morozov
September 2019   Head of the Mechanics
                                        Division of the Department of Mathematics
                                and Mechanics of the St. Petersburg State University
                                    Academician of the Russian Academy of Sciences

Introduction

This two-volume course on Rational and Applied Mechanics is the fourth revised and substantially expanded edition of the book "Theoretical mechanics"[1], which was first published by the Leningrad University Press in 1985. Our exposition is based on lecture courses given by the authors over many years in the Faculty of Mathematics and Mechanics at St. Petersburg University. This book is aimed at students and researchers from classical universities, which explains the wide range of questions considered in it. The modified name of the book "Theoretical and applied mechanics" corresponds to the name of the chair of affiliation of the authors, which in their judgment is more appropriate for the purposes of this edition. However, the English translation of the textbook will be published under the title "Rational and Applied Mechanics" according to the comment of the "Springer" Publishing advisor.

Please note that, although the textbook is published in two volumes with the chapters numbered sequentially throughout, each volume is independent and may be used separately.

The book contains three parts. Part I is dedicated to kinematics, and the second one, with general problems of theoretical mechanics. These two parts comprise the first volume of the book. The second volume incorporates Part III of the course, which is concerned with certain special applied problems in theoretical mechanics.

Part I has three chapters. In **Chapter** 1, much attention is paid to co- and contravariant components of the velocity and acceleration vectors, which in the sequel proves important in the study of a number of questions in dynamics.

The methodological coherence in the exposition of our lecture course was achieved by the repeated application of curvilinear coordinates. However, we were

---

[1]The textbook `Theoretical mechanics' by *S. V. Bolotin, A. V. Karapetyan, E. I. Kugushev, D. V. Treshchev* (Moscow, Izd. Akademiya, 2010) rightly remarks at p. 3 that "The term 'theoretical mechanics' is fairly standard, but is highly misleading, giving the impression that the rest of mechanics is practical, whereas in effect it incorporates, for example, the mechanics of continuous medium, and the statistical, quantum, and relativistic mechanics. The term 'classical mechanics' is a much better guide, but it is also not quite satisfactory, because by its opposite one usually means the quantum mechanics".

mainly focused on the simplicity of exposition, and so the appropriate mathematical tools were introduced only when required.

In **Chapter** 2, the derivation of the expression for the vector of instantaneous angular velocity involves the chain rule for differentiation of a multivariate composite function. In the study of the projections of the angular velocity onto fixed coordinate axes it is shown that they can be looked upon as quasi-velocities, because they cannot be considered as the derivatives of some new coordinates, which control the orientation of a rigid body in the space. The concept of the rotation tensor is also introduced in Chap. 2.

In **Chapter** 3, we first focus our attention on the composite motion of a point, and then present the theory of composition of motions of a rigid body.

Part II contains nine chapters. **Chapter** 4 includes general theorems, a detailed exposition of the oscillation of a particle, the relative motion, and the motion of a particle in a central force field. The Lagrange equations of the second kind are derived on the basis of results obtained in the study of the point kinematics in curvilinear coordinates. Later, we shall extend in a natural way these equations, as well as the canonical equations obtained in the same chapter, to the case of general mechanical systems.

In **Chapter** 5, we introduce the concept of a representative point in the sense of Herz and give the equations of its motion both in Cartesian and curvilinear coordinates in the form of the Lagrange equations of the second kind. The introduction of a representative point provides a methodologically unified approach to the construction of the dynamics of a point and of a system of points.

Chapter 6 is the central part of the book. It contains a number of new results. In this chapter, the motion of a material point and of the point representing a system is considered on the basis of ideal constraints (that is, the constraints the reaction force of which has the smallest magnitude). For a single material point lying on a surface, the minimal reaction is directed toward the normal vector to this surface. In general, the ideality of constraints means that the generalized reaction forces corresponding to free coordinates are zero. It is shown that the Lagrange equations of the second kind and formulas for generalized reaction forces can be obtained as a result of a linear transformation of the Lagrange equations of the first kind.

Thus, the theory of constrained motion is built without recourse to the d'Alembert–Lagrange principle, which is presented in detail in Chap. 9 on variational principles of mechanics.

A constrained motion subject to linear or nonlinear nonholonomic constraints is examined by introducing a transformation in the velocity space and subsequent application of the same logical scheme as was used in the case of holonomic systems.

Appel's equations are obtained by scalar multiplication of the principal Newton vector equation by coordinate vectors, as expressed in terms of the derivatives of the acceleration vector. As a corollary, we show that the motion equations form a necessary condition for the minimality of the Gauss constraint.

In changing from the motion of finite-point mechanical systems to general mechanical systems, the Lagrange equations of the second kind are written as a single vector equality in the tangent space. This equality, which has the form of Newton's second law, enables one to study the constrained motion of general mechanical systems under holonomic or nonholonomic constraints by the same methods as the constrained motion of a single point in arbitrary curvilinear coordinates.

Our general approach to the constrained motion is capable of considering the constraint equations as equations of some program of motion, and constraint reaction forces are looked upon as control forces.

In the next-to-last section of this chapter, we show that the principal forms of motion equations of nonholonomic systems (Chaplygin's equations, Voronets–Hamel's equations, Hamel–Novoselov's equations, Poincaré–Chetaev's equations, Udwadia–Kalaba's equations) can be obtained from Maggi's equations, which hold for nonholonomic constraints of any form (and in particular, for nonlinear constraints).

The last section is concerned with motion control using parameter-depending constraints.

Chapter 6 gives a solution to a number of problems, which illustrate the convenience of the Lagrange equations of the second kind for holonomic systems and Maggi's equations for nonholonomic systems.

In **Chapter** 7, we use the linearization of the contravariant form of the motion equations in curvilinear coordinates to derive the equations of small oscillations; this is a methodological bond between this chapter and the previous ones. The principal (normal) coordinates of a system are introduced after the completion of the study of small oscillations under the absence of resistance forces. We also discuss the minimax properties of natural frequencies.

Chapter 8 involves various approaches to the study of properties of the inertia tensor, which enables one to dwell in more detail on its physical content. In our brief presentation, the static equations are derived based on the analysis of the dynamic equations; we give a more detailed account of this part of theoretical mechanics in Chap. 10. The motion of a rigid body about a fixed axis and about a fixed point is studied by classical methods. At the end of the chapter, we present a new special form of motion equations of a rigid body, which proves instrumental in the study of the dynamics of a system of rigid bodies. These differential equations will be used in the second volume of the chapter dealing with the dynamics of a loaded Stewart platform.

In **Chapter** 9, the d'Alembert–Lagrange, Suslov–Jourdain and Gauss differential variational principles are derived from the corresponding scalar equations of motion, as written for the tangent space. We discuss the Chetaev-type constraints and the relation of the generalized d'Alembert–Lagrange principle with the Suslov–Jourdain principle. The differential variational principles obtained in this chapter are used to derive the principal forms of the motion equations for constrained mechanical systems. Differential variational principles are traditionally used in the derivation of motion equations of holonomic systems. This approach is illustrated with an example of motion of the Novoselov regulator.

The Hamilton–Ostrogradsii and Lagrange integral variational principles, which reflect the extremal properties of the curves along which the motion takes place subject to potential forces, are derived from the Hamilton principle of variable action. We show that the Hamilton–Jacobi equation can also be derived from this principle.

In comparing the differential and integral principles, we point out that in essence the different definitions of the variation are introduced in them.

Chapter 10 gives a brief account of the statics of a rigid body and of systems of rigid bodies. We give two definitions of the equivalence of systems of forces and prove that they are equivalent. Examples are given explaining the composition of equilibrium equations of mechanical systems by various methods. Fundamentals of analytical statics are presented briefly. We consider the problem of finding the centres of mass of rigid bodies, the problem on equilibrium of trusses, and the problem of a flexible inextensible string. We describe the properties of the friction forces, solve the Euler problem on the equilibrium of a string winding about a cylinder.

In **Chapter** 11, in the study of general problems of the theory of integration of mechanical equations, we employ the tight relation between the Hamilton action function and the principle of variable action. In deriving the optico-mechanical analogy, the Maupertuis—Lagrange principle, which characterizes the motion of a point, and the Fermat principle, which describes the propagation of light, are written in dimensionless variables. It proves possible to relate the optico-mechanical analogy with the Schrödinger equation, for which the Hermite polynomials are known to serve as eigenfunctions in the case of a harmonic oscillator.

Elements of the special relativity theory are outlined in **Chapter** 12. In kinematic relations, we consider the four-dimensional quadratic Poincaré form and discuss the composite motion of a point. The derivation of the dynamic equations depends on the generalized Newton law, we also prove the work-energy principle and write down the Lagrange equations of second kind.

The second volume consists of **Part III "Dynamics. Some Applied Problems of Rational Mechanics"**, which deals with special practical aspects of theoretical mechanics. In this part we consider problems of stability of motion, nonlinear oscillation, dynamics and statics of the Stewart platform, motion of mechanical systems subject to random forces, elements of the control theory, relations between the nonholonomic mechanics and the control theory, oscillations and balance of rotor systems, physical impact theory, dynamics of a thin rod, flight dynamics. The material of these chapters is based on several special courses delivered at the Chair of Theoretical and Applied Mechanics at St. Petersburg University.

In the first volume of the textbook the continuous numeration is introduced in all the twelve chapters. The double numeration for formulas is accepted within each chapter, the first digit indicates the number of the section. Figures and examples are numbered separately in each chapter.

We give a large number of examples with solutions,[2] advantage was given to problems requiring an analytical solution.

Each chapter title is followed by the names of the authors.

The work on the textbook was made in parallel with its translation into English. The first part "Kinematics" was translated by O. S. Bukashkina, the chapters 4, 5, 7, 8, 11 of the part II were translated by G. A. Sinilshchikova, the chapters 6, 9, 10, 12 were translated by A. R. Alimov. Fragments from the papers and monographs of the authors published in English are presented in the text. The English translation was edited by A. R. Alimov, E. L. Belkind, and G. A. Sinilshchikova. The very idea of translation of the treatise into English was suggested as long ago as in 1987 by Prof. J. Papastavridis (USA).

The authors wish to acknowledge their great indebtedness to their colleagues O. S. Bukashkina, D. N. Ivanov, G. A. Kuteeva, N. V. Naumova, F. F. Rodyukov, V. I. Sergeeva, G. A. Sinilshchikova, A. L. Smirnov, K. K. Tverev, L. A. Venatovskaya, as well as to students and postgraduate students at our department: S. N. Bur'yan, A. P. Deriglazov, V. V. Dodonov, K. M. Fazlyeva, V. E. Kondrenkina, D. G. Korytnikov, E. A. Kosyakov, A. S. Kozlova, I. A. Kulakovskii, D. B. Kulizhnikov, N. R. Kurbanov, D. D. Kvaratskhelia, A. S. Maksimov, G. A. Nesterchuk, D. Yu. Nikitin, A. A. Pashkina, V. I. Petrova, O. I. Ritenman, E. A. Shatrov, V. A. Shelkovina, T. S. Shugailo, L. A. Sobolev, P. P. Stepanova, A. V. Zelinskaya for their great help in the preparation of the book.

We owe very particular thanks to Prof. P. E. Tovstik, Doctor of Physics and Mathematics, Holder of the Chair of Theoretical and Applied Mechanics at St. Petersburg University, Laureate of Russian State Prize, Honoured Scientist of Russia for his continuous attention to the genesis of the book. His continuous advice helped to improve the presentation of the book. Besides, the first part contains a lot of important contributions made by Peter Tovstik, and he is one of the co-authors of the second volume. Peter Tovstik also undertook the formidable task of being Editor-in-Chief of the three editions of our book.

We will be grateful to the readers, who will take the trouble of bringing errors in the text to our attention.

Sergey Andreevich Zegzhda
Mikhail Petrovich Yushkov
yushkovmp@mail.ru

---

[2]Statements many examples were taken from the well-known problem book by I.V. Meshcherskii, which ran through many editions (see, for example, *I.V. Meshcherskii*, Collection of problems in theoretical mechanics (36th edition, Moscow, Nauka, 1986) or *I.V. Meshcherskii*, Problems in theoretical mechanics (49th edition, St. Petersburg, Lan', 2008.).

Contents

Part II DYNAMICS. GENERAL ASPECTS OF RATIONAL MECHANICS. FUNDAMENTALS OF ANALYTICAL MECHANICS

Part I KINEMATICS

Geometry deals with objects such as points, lines, surfaces, volumes, it studies the length of segments, the curvature of curves, the area of surfaces, and other elements connected with the measurement. All of such quantities are space-metric characteristics of geometric objects.

Properties of geometric objects are independent of the physical properties of the material of models, which might be constructed for such objects. Geometric constructions never employ the notion of time and never consider, for example, notions such as the point motion, which implies the change of its position in space with time.

Kinematics deals with the motion of bodies. To put it otherwise, kinematics is concerned with space-time characteristics of motion such as velocity, acceleration, and some others, which will be considered in detail later.

The dimensions of various kinematic characteristics are known to be combinations of units of measurement of space and time. For example, the dimension of velocity is the length unit divided by the time unit, and the acceleration dimension is the length unit divided by the squared time unit. Thus, quantities such as the mass or others controlling the physical properties of a material cannot have the dimension of kinematic characteristics of motion. Consequently, kinematics of point is sometimes called the geometry of four-dimensional space, where the time plays the role of the fourth dimension.

The set of points, in which the motion of every point depends on the position and the motion of the remaining ones, is called a *mechanical system*. If the distance between the points of a system is constant, then such a system is called unchangeable or an *absolutely rigid body*. Kinematics is, therefore, subdivided into the kinematics of points and the kinematics of solids. If a body represents a continuously distributed set of points (a continuum) and its form changes with time, then it is subsumed under the kinematics of continuous deformed medium (for example, the kinematics of a liquid body).

Chapter 1
Point Kinematics

N. N. Polyakhov, M. P. Yushkov⊙, and S. A. Zegzhda

In this chapter, expressions for the velocity and acceleration are derived for certain general cases of point motion. The formulas obtained can be further generalized to the case of the motion of mechanical systems, whose position is described by a finite number of independent parameters. When considering curvilinear coordinates, some notions from dynamics are introduced in order to show how the differential geometry machinery can be obtained in a natural way from problems of mechanics.

1 Point Velocity and Acceleration in Cartesian Coordinates

The position of a point in the Cartesian coordinates $Oxyz$ is given by three numbers x, y, z, which can be considered as projections of the radius vector $\mathbf{r}$:

$$\mathbf{r} = x\,\mathbf{i} + y\,\mathbf{j} + z\,\mathbf{k}, \tag{1.1}$$

where $\mathbf{i}, \mathbf{j}, \mathbf{k}$ are the unit vectors of the Cartesian coordinate system. Note that these vectors can be written as

$$\mathbf{i} = \frac{\partial \mathbf{r}}{\partial x}, \quad \mathbf{j} = \frac{\partial \mathbf{r}}{\partial y}, \quad \mathbf{k} = \frac{\partial \mathbf{r}}{\partial z}. \tag{1.2}$$

If the point position changes with time, that is, its coordinates are functions of time

$$x = f_1(t), \quad y = f_2(t), \quad z = f_3(t), \tag{1.3}$$

then the *law of point motion* is said to be given in Cartesian coordinates.

© Springer Nature Switzerland AG 2021
N. N. Polyakhov et al., *Rational and Applied Mechanics*,
Foundations of Engineering Mechanics,
https://doi.org/10.1007/978-3-030-64061-3_1

Fig. 1 Trajectory of a point

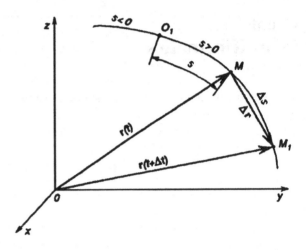

According to the very essence of the motion, the functions $f_k(t)$, $k = 1, 2, 3$, must be single-valued, because at a moment under consideration a point can occupy only a definite position. Such functions will be assumed to be twice differentiable. Equations (1.3) can be considered as a parametrical form of equations of a spatial curve (the *trajectory of a point*).

Suppose a point moves from position M to position M_1 in time Δt (see Fig. 1). The vector $\Delta \mathbf{r}$, joining M with M_1, is called the *displacement vector*. The relation $\mathbf{v}^* = \Delta \mathbf{r}/\Delta t$ is called the *mean velocity* in time Δt. The *true point velocity* at a given time is the limit (if it exists) to which the relation $\Delta \mathbf{r}/\Delta t$ tends as $\Delta t \to 0$; that is,

$$\mathbf{v} = \lim_{\Delta t \to 0} \frac{\Delta \mathbf{r}}{\Delta t}.$$

This limit will be called the *derivative* of the vector $\mathbf{r}$ in t; we denote it by $d\mathbf{r}/dt$ or $\dot{\mathbf{r}}$:

$$\mathbf{v} = \frac{d\mathbf{r}}{dt} = \dot{\mathbf{r}}. \tag{1.4}$$

Thus, the true point velocity at a given moment is the first time derivative of the radius vector at this time.

From (1.1), it follows that the radius vector $\mathbf{r}$ is a function of the variables x, y, z, which in turn are functions of time according to (1.3). That is why the derivative $\dot{\mathbf{r}}$ must be calculated according to the rule

$$\frac{d\mathbf{r}}{dt} = \frac{\partial \mathbf{r}}{\partial x} \frac{dx}{dt} + \frac{\partial \mathbf{r}}{\partial y} \frac{dy}{dt} + \frac{\partial \mathbf{r}}{\partial z} \frac{dz}{dt}.$$

Hence and from expressions (1.2), it follows that

$$\mathbf{v} = \dot{x}\,\mathbf{i} + \dot{y}\,\mathbf{j} + \dot{z}\,\mathbf{k} = v_x\,\mathbf{i} + v_y\,\mathbf{j} + v_z\,\mathbf{k}. \tag{1.5}$$

The values

$$v_x = \dot{x} = \mathbf{v}\cdot\mathbf{i}, \qquad v_y = \dot{y} = \mathbf{v}\cdot\mathbf{j}, \qquad v_z = \dot{z} = \mathbf{v}\cdot\mathbf{k}$$

are the projections of the vector $\mathbf{v}$ on the axes x, y, z, respectively. They can be also calculated by the formulas

$$v_x = v\cos\alpha_x, \qquad v_y = v\cos\alpha_y, \qquad v_z = v\cos\alpha_z,$$
$$v = \sqrt{v_x^2 + v_y^2 + v_z^2} = \sqrt{\dot{x}^2 + \dot{y}^2 + \dot{z}^2}.$$

Here, v is the absolute value of the velocity vector and $\alpha_x, \alpha_y, \alpha_z$ are the angles formed by this vector and the positive directions of coordinate axes x, y, z, respectively.

The vector $\mathbf{w}$, which is the derivative of the velocity vector $\mathbf{v}$ in time, is called the *acceleration of a point* at a present moment. According to expressions (1.4) and (1.5), we have

$$\mathbf{w} = \dot{\mathbf{v}} = \ddot{\mathbf{r}} = \ddot{x}\,\mathbf{i} + \ddot{y}\,\mathbf{j} + \ddot{z}\,\mathbf{k}. \tag{1.6}$$

This implies that the absolute value of the acceleration vector w and the angles $\beta_x, \beta_y, \beta_z$ formed by this vector with positive directions of coordinate axes can found from the formulas

$$w = \sqrt{\ddot{x}^2 + \ddot{y}^2 + \ddot{z}^2},$$
$$\cos\beta_x = \frac{\ddot{x}}{w}, \qquad \cos\beta_y = \frac{\ddot{y}}{w}, \qquad \cos\beta_z = \frac{\ddot{z}}{w}.$$

Example 1 The point motion is given in the form

$$x = a\cos\omega t, \qquad y = b\sin\omega t, \qquad z = 0,$$

where a, b, ω are constant values. Find the trajectory, velocity, and acceleration.

To find the trajectory we exclude the time from the motion equations. We have

$$x/a = \cos\omega t, \qquad y/b = \sin\omega t,$$

which gives $(x/a)^2 + (y/b)^2 = 1$ (an ellipse).

To find the velocity and acceleration we calculate the derivatives:

$$\dot{x} = -a\omega\sin\omega t, \qquad\qquad \dot{y} = b\omega\cos\omega t,$$
$$\ddot{x} = -a\omega^2\cos\omega t = -\omega^2 x, \qquad\qquad \ddot{y} = -b\omega^2\sin\omega t = -\omega^2 y.$$

Consequently

$$v = \sqrt{\dot{x}^2 + \dot{y}^2} = \omega \sqrt{\frac{a^2}{b^2} y^2 + \frac{b^2}{a^2} x^2},$$

$$\mathbf{w} = -\omega^2 \mathbf{r}, \qquad w = \omega^2 r, \qquad r = \sqrt{x^2 + y^2}.$$

We note that with $a = b = R$ the point moves along a circle with the velocity $v = \omega R$ and acceleration $w = \omega^2 R$. The velocity vector is tangent to the circle and the acceleration vector is directed toward its the center.

Further, in the study of a system dynamics, we shall introduce the representation point, which lies in the $3n$-dimensional Euclidean space (n is the number of system points). In order to easily apply the results obtained in this chapter to kinematics of the representation point, it is expedient to use the single symbol with corresponding indexes for values homogeneous in a sense. Taking this into account, we suppose

$$x_1 = x, \qquad x_2 = y, \qquad x_3 = z.$$

Hence, the principal relations of this subsection will take the following form

$$\mathbf{r} = \sum_{k=1}^{3} x_k \mathbf{i}_k, \qquad \mathbf{v} = \sum_{k=1}^{3} \dot{x}_k \mathbf{i}_k, \qquad \mathbf{w} = \sum_{k=1}^{3} \ddot{x}_k \mathbf{i}_k,$$

$$r = \sqrt{\sum_{k=1}^{3} x_k^2}, \qquad v = \sqrt{\sum_{k=1}^{3} \dot{x}_k^2}, \qquad w = \sqrt{\sum_{k=1}^{3} \ddot{x}_k^2},$$

$$\cos \alpha_k = \frac{\dot{x}_k}{v}, \qquad \cos \beta_k = \frac{\ddot{x}_k}{w}, \qquad k = 1, 2, 3. \tag{1.7}$$

These formulas correspond to the representation point if the upper limit in the sum in k is supposed to be $3n$.

2 Decomposition of the Point Velocity and Acceleration Along the Frenet Natural Trihedral Axes

Formulas from the previous subsection, which express the point velocity and the acceleration vectors as the law of its motion in the Cartesian coordinates (1.3), are fairly simple. This simplicity is explained by the fact that the Cartesian coordinate axes are at right angles to each other and constant in direction. A disadvantage of this simple expressions is that they do now show how the vectors $\mathbf{v}$ and $\mathbf{w}$ are related with the form of the curve along which the point is moving. Moreover they also depend on the position of the given curve in the space $Oxyz$.

It is clear that the point velocity and acceleration depend, first of all, on how the path made by the point depends on time and, secondly, on the form of the trajectory along which the point is moving. In order to describe the first factor, we shall consider a certain point O_1 on the trajectory as the origin of the arc coordinate s (see Fig. 1). It is also necessary to specify the positive direction of this coordinate, defining in a unique fashion the point M on the trajectory. Let us assume that s is equal to the length of the $O_1 M$ arc when $s > 0$. The point trajectory (i.e., the vector function $\mathbf{r}(s)$) and the law of motion along it $s = f(t)$ is called the *natural way of motion*. The velocity vector $\mathbf{v}$ according to this way of describing the motion is obtained by the formula

$$\mathbf{v} = \lim_{\Delta t \to 0} \frac{\Delta \mathbf{r}}{\Delta t} = \lim_{\Delta s \to 0} \frac{\Delta \mathbf{r}}{\Delta s} \lim_{\Delta t \to 0} \frac{\Delta s}{\Delta t} = \frac{d\mathbf{r}}{ds} \frac{ds}{dt} = \dot{s}\,\boldsymbol{\tau}. \tag{2.1}$$

The vector $\Delta \mathbf{r}/\Delta s$ in the limit with $\Delta s \to 0$ is tangential to the curve at the point M in the arc coordinate (in an increasing direction). Since the values $|\Delta \mathbf{r}|$ and $|\Delta s|$ are infinitesimal, the vector

$$\boldsymbol{\tau} = \lim_{\Delta s \to 0} \frac{\Delta \mathbf{r}}{\Delta s} = \frac{d\mathbf{r}}{ds}$$

is a *unit tangential vector*. Hence

$$d\mathbf{r} = \boldsymbol{\tau}\,ds$$

and since on the other hand

$$d\mathbf{r} = dx\,\mathbf{i} + dy\,\mathbf{j} + dz\,\mathbf{k},$$

we have

$$ds^2 = (d\mathbf{r})^2 = (dx)^2 + (dy)^2 + (dz)^2 = \sum_{k=1}^{3}(dx_k)^2. \tag{2.2}$$

Now we study the acceleration vector. In accordance with definition (1.6) and formula (2.1) we have

$$\mathbf{w} = \frac{d\mathbf{v}}{dt} = \ddot{s}\,\boldsymbol{\tau} + \dot{s}\,\frac{d\boldsymbol{\tau}}{ds}\frac{ds}{dt} = \ddot{s}\,\boldsymbol{\tau} + \dot{s}^2\,\frac{d\boldsymbol{\tau}}{ds}. \tag{2.3}$$

The vector $\boldsymbol{\tau}$ is a unit vector, that is

$$\boldsymbol{\tau}^2 = 1.$$

Differentiating this expression in s, we get

$$\frac{d}{ds}(\boldsymbol{\tau}^2) = 2\boldsymbol{\tau} \cdot \frac{d\boldsymbol{\tau}}{ds} = 0.$$

Fig. 2 The osculating plane

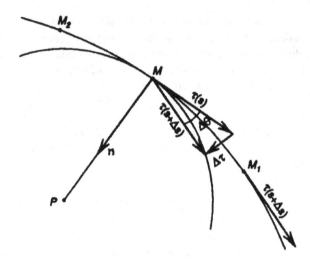

This implies that the vector $d\tau/ds$ is either zero or is perpendicular to the unit vector τ. By definition, this vector is equal to

$$\frac{d\tau}{ds} = \lim_{\Delta s \to 0} \frac{\tau(s + \Delta s) - \tau(s)}{\Delta s} = \lim_{\Delta s \to 0} \frac{\Delta \tau}{\Delta s}. \tag{2.4}$$

According to Fig. 2, the vector $d\mathbf{r}/ds$ is nonzero, the vector $\Delta\tau/\Delta s$ lies in the plane passing through the vector $\tau(s)$ and the vector $\tau(s + \Delta s)$ carried to the point M. The limit position of this plane when $\Delta s \to 0$ is called the *osculating plane*. The vector $d\tau/ds$ lies in the plane and is directed to the side of curve concavity. The unit vector at this direction $\mathbf{n}$ is called the *the principal normal vector* to the curve at a point under consideration. Accordingly, the vector $d\tau/ds$ can be represented in the form

$$\frac{d\tau}{ds} = K\mathbf{n}, \tag{2.5}$$

where K is the absolute value of the vector $d\tau/ds$.

We note that the vector $\mathbf{n}$ can be introduced by formula (2.5) if $K \neq 0$. With $K = 0$ the vector $\tau(s)$ is constant near the point under consideration to the second order of smallness in Δs. This means that the trajectory in the vicinity of the point can be approximated by the segment of straight line.

We now study in detail the case when $K \neq 0$. One introduces the *contingence angle* $\Delta\theta$ between the vectors $\tau(s)$ and $K\tau(s + \Delta s)$ (see Fig. 2) and assumes that the infinitesimal magnitudes Δs and $\Delta\theta$ have the same sign.

Introducing the angle $\Delta\theta$ makes it possible to represent the limit (2.4) as a product of two limits:

$$\frac{d\tau}{ds} = \lim_{\Delta\theta \to 0} \frac{\Delta\tau}{\Delta\theta} \lim_{\Delta s \to 0} \frac{\Delta\theta}{\Delta s}. \tag{2.6}$$

From the triangle constructed with the help of the vectors $\boldsymbol{\tau}(s)$, $\boldsymbol{\tau}(s + \Delta s)$, $\Delta \boldsymbol{\tau}$, we can see that

$$|\Delta \boldsymbol{\tau}| = 2 \sin \frac{|\Delta \theta|}{2}.$$

Therefore

$$\lim_{\Delta \theta \to 0} \frac{|\Delta \boldsymbol{\tau}|}{|\Delta \theta|} = 1.$$

Hence and from expressions (2.5) and (2.6) it follows that

$$\lim_{\Delta \theta \to 0} \frac{\Delta \boldsymbol{\tau}}{\Delta \theta} = \mathbf{n}, \qquad \lim_{\Delta s \to 0} \frac{\Delta \theta}{\Delta s} = K.$$

Let us construct a circle with radius $\rho = 1/K$ lying in the osculating plane. We assume that its center lies at the point P on the ray, which is defined by the vector $\mathbf{n}$ (in Fig. 2 the distance $MP = \rho$). This circle is called the *osculating circle* or the *circle of curvature*, the quantity K is called the *curvature* and ρ is the *radius of curvature*. The contents of these terms explain the other geometrical definition of the osculating circle. In the vicinity of the point M, we shall take two points M_1 and M_2 close to each other on the curve and cross these three points by a circle. In the limit when $M_1 \to M$ and $M_2 \to M$, this circle transforms into the osculating circle. Thus, with $K \neq 0$ the motion trajectory can be approximated by an arc of the osculating circle in the vicinity of the point under consideration.

Substituting expression (2.5) into formula (2.3) and taking into account that $K = 1/\rho$, we get

$$\mathbf{w} = \ddot{s} \boldsymbol{\tau} + \frac{\dot{s}^2}{\rho} \mathbf{n}. \tag{2.7}$$

In particular, if a point moves along a circle, then the osculating circle coincides at all points of the circle with the trajectory itself. Therefore, in formula (2.7) ρ is constant and equal to the radius of the circle.

We note that, for the motion of a point along a circle, formula (2.7) can be deduced in a more simple way. The motion under consideration can be given in the following form:

$$\mathbf{r} = \rho \cos \psi(t) \mathbf{i} + \rho \sin \psi(t) \mathbf{j}. \tag{2.8}$$

Calculating the velocity and acceleration, we get

$$\mathbf{v} = -\rho \dot{\psi} \sin \psi \, \mathbf{i} + \rho \dot{\psi} \cos \psi \, \mathbf{j},$$

$$\mathbf{w} = -\rho \ddot{\psi} \sin \psi \, \mathbf{i} + \rho \ddot{\psi} \cos \psi \, \mathbf{j} - \rho \dot{\psi}^2 (\cos \psi \, \mathbf{i} + \sin \psi \, \mathbf{j}).$$

These expressions can be written in the form (2.1) and (2.7), respectively, if we assume

$$s = \rho\psi, \quad \boldsymbol{\tau} = -\sin\psi\,\mathbf{i} + \cos\psi\,\mathbf{j}, \quad \mathbf{n} = -\cos\psi\,\mathbf{i} - \sin\psi\,\mathbf{j} = -\mathbf{r}/\rho.$$

Thus, formulas (2.1) and (2.7), which were deduced here in a general case of curvilinear point motion, can be considered as a natural generalization of formulas that are valid for a point motion along a circle. This generalization became possible because of the special choice of the unit vectors $\boldsymbol{\tau}$ and $\mathbf{n}$. The unit vectors $\mathbf{i}$, $\mathbf{j}$, $\mathbf{k}$ of the Cartesian coordinate axes are such that $\mathbf{k} = \mathbf{i} \times \mathbf{j}$. Taking this into account, we assume that $\mathbf{b} = \boldsymbol{\tau} \times \mathbf{n}$. The vector $\mathbf{b}$ is called a *binormal vector*.

The vectors $\boldsymbol{\tau}$, $\mathbf{n}$, $\mathbf{b}$ form the *natural system of axes*. From the above it follows that a plane osculating to the curve crosses the vectors $\boldsymbol{\tau}$ and $\mathbf{n}$; the *normal plane*, containing all normal vectors to the trajectory, crosses the vectors $\mathbf{n}$ and $\mathbf{b}$, and the vectors $\boldsymbol{\tau}$ and $\mathbf{b}$ define the *rectifying plane*. Thus the unit vectors $\boldsymbol{\tau}$, $\mathbf{n}$, $\mathbf{b}$ form the so-called *Frenet natural trihedral*. This trihedral is called *natural*, because the velocity and acceleration vectors have their natural representation when decomposed in the unit vectors $\boldsymbol{\tau}$, $\mathbf{n}$, $\mathbf{b}$. Note that the Frenet trihedral moves together with the point under consideration—this is why sometimes it is called *movable*.

From expression (2.7) it follows that the acceleration vector projections to the axes of the natural trihedral are as follows:

$$w_\tau = \ddot{s}, \quad w_n = \dot{s}^2/\rho = v^2/\rho, \quad w_b = 0. \tag{2.9}$$

The vector $w_\tau\boldsymbol{\tau}$ is called the *tangential component* of the acceleration vector, $w_n\mathbf{n}$ is its *normal component*. The acceleration vector projection to the binormal is zero. This means that the acceleration vector always lies in the osculating plane and points to the trajectory concavity. From the above, it follows that the tangential component of the acceleration vector appears from a change in the magnitude of the velocity vector and the normal one is consequent to a change in the direction of this vector.

Example 2 We conclude this subsection with the question of how the unit vectors $\boldsymbol{\tau}$ and $\mathbf{n}$, and values w_τ and w_n can be found if the motion is given in the Cartesian coordinates. Formulas (2.1) and (1.7) imply that

$$\boldsymbol{\tau} = \frac{\mathbf{v}}{\dot{s}} = \pm\frac{1}{v}\sum_{k=1}^{3}\dot{x}_k\,\mathbf{i}_k, \tag{2.10}$$

besides, we have the positive sign when at a given point the vector of point velocity is directed toward the increasing arc coordinate s and the negative sign occurs in the opposite case.

Using expressions (1.7), (2.7), (2.9), (2.10), we get

$$w_\tau = \mathbf{w} \cdot \boldsymbol{\tau} = \pm\frac{1}{v}\sum_{k=1}^{3}\dot{x}_k\,\ddot{x}_k, \; w_n = \sqrt{w^2 - w_\tau^2}, \; \mathbf{n} = \frac{\mathbf{w} - w_\tau\boldsymbol{\tau}}{w_n},$$

$$K = \frac{1}{\rho} = \frac{w_n}{v^2} = \frac{\sqrt{w^2 - w_\tau^2}}{v^2}.$$

Here, the quantities v and w and the vector $\mathbf{w}$ are considered to be given in the form (1.7).

We note that formulas written out here will correspond to a point motion in the $3n$-dimensional Euclidean space if the upper limit of summation index k is equal to $3n$.

3 Velocity of a Point in Cylindrical Coordinates

The choice of some or other way of describing the point motion depends on the character of problem under consideration. For example, if it is necessary to explore the point motion along a circle, then it is worth considering the radius vector $\mathbf{r}$ in the form (2.8). Let us generalize this formula, supposing that the distance ρ from the point to the z-axis can be variable. The quantities ρ and ψ specifying in a unique manner the position of a point on the plane are called the *polar coordinates*.

When the spatial motion takes place, the point radius vector $\mathbf{r}$ is considered to be given in the form

$$\mathbf{r} = \rho \cos\psi\, \mathbf{i} + \rho \sin\psi\, \mathbf{j} + z\,\mathbf{k}\,.$$

The quantities ρ, ψ, z are called the *cylindrical coordinates* (see Fig. 3). It is supposed that they are defined as functions of time.

Using the rule of differentiation of a composite function in finding the velocity vector, we get

$$\mathbf{v} = \frac{\partial \mathbf{r}}{\partial \rho}\,\dot{\rho} + \frac{\partial \mathbf{r}}{\partial \psi}\,\dot{\psi} + \frac{\partial \mathbf{r}}{\partial z}\,\dot{z}\,. \tag{3.1}$$

Fig. 3 Cylindrical coordinates

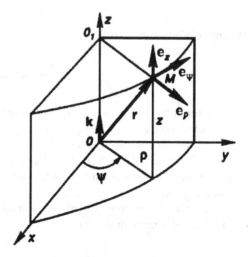

The vectors

$$\mathbf{e}_\rho = \frac{\partial \mathbf{r}}{\partial \rho} = \cos \psi \, \mathbf{i} + \sin \psi \, \mathbf{j},$$

$$\mathbf{e}_\psi = \frac{\partial \mathbf{r}}{\partial \psi} = -\rho \sin \psi \, \mathbf{i} + \rho \cos \psi \, \mathbf{j},$$

$$\mathbf{e}_z = \frac{\partial \mathbf{r}}{\partial z} = \mathbf{k}$$

form a basis of the cylindrical coordinate system (see Fig. 3). They are orthogonal to each other, besides, the vectors $\mathbf{e}_\rho$ and $\mathbf{e}_z$ are unit, but $|\mathbf{e}_\psi| = \rho$.

Let us introduce the orthonormal basis

$$\mathbf{e}_\rho^0 = \mathbf{e}_\rho, \qquad \mathbf{e}_\psi^0 = \mathbf{e}_\psi/\rho, \qquad \mathbf{e}_z^0 = \mathbf{k}. \tag{3.2}$$

In accordance with (3.1), decomposition of the velocity vector into components of this basis reads as

$$\mathbf{v} = \dot{\rho} \, \mathbf{e}_\rho + \rho \dot{\psi} \, \mathbf{e}_\psi^0 + \dot{z} \, \mathbf{k}.$$

Hence

$$v^2 = \dot{\rho}^2 + \rho^2 \dot{\psi}^2 + \dot{z}^2. \tag{3.3}$$

The unit vectors $\mathbf{e}_\rho$, $\mathbf{e}_\psi^0$, $\mathbf{k}$ form a right-hand system, and hence,

$$\mathbf{e}_\psi^0 = \mathbf{k} \times \mathbf{e}_\rho.$$

Therefore,

$$\frac{\partial \mathbf{r}}{\partial \psi} = \rho \, \mathbf{e}_\psi^0 = \mathbf{k} \times \rho \, \mathbf{e}_\rho.$$

Taking in account that $\rho \, \mathbf{e}_\rho = \mathbf{r} - z \, \mathbf{k}$, we get

$$\frac{\partial \mathbf{r}}{\partial \psi} = \mathbf{k} \times \mathbf{r}. \tag{3.4}$$

This formula gives a simple and useful representation for the partial derivative of the radius vector $\mathbf{r}$ with respect to the angle ψ. It shows that in order to find this derivative it is necessary to specify the unit vector $\mathbf{k}$. The unit vector $\mathbf{k}$ lies on the same axis, around which the radius vector $\mathbf{r}$ rotates with increasing angle ψ. We note that an observer watching from the end of the vector $\mathbf{k}$ sees rotating running counterclockwise.

Multiplying expression (3.4) by $\dot{\psi}$, we obtain the point velocity with fixed values ρ and z. Let us represent it in the form

$$\mathbf{v} = \boldsymbol{\omega} \times \mathbf{r}, \tag{3.5}$$

where $\omega = \dot{\psi}\mathbf{k}$. The coordinates ρ, z are constant if the quantities ρ, ψ, z are cylindrical coordinates of an arbitrary point of a solid body rotating around the stationary axis z. Different angles ψ correspond to different points. Nevertheless, the velocity of changing the angle ψ, that is *angular velocity* $\omega = \dot{\psi}$, is the same for all points.

Formula (3.5) implies that in the case of the rotation of a solid about a stationary axis the velocity of an arbitrary point of the body and therefore, the whole field of its velocities, depend on the vector ω (the *angular velocity vector*). The vector ω simultaneously defines the axis of rotation and the angular velocity. An observer, watching from the end of it, sees rotation, running counterclockwise.

As was shown in Sect. 2, the introduction of polar coordinates gives an opportunity to get such expressions for the velocity and acceleration of a point when moving along a circle, which permits generalization to the case of an arbitrary curvilinear motion of the point. Formula (3.5) from this subsection, which was obtained for the kinematics of a solid body, was deduced using the polar coordinates. It is essential that expression (3.5) has the same form in an arbitrary motion of a solid body with one stationary point. This can be shown easier when using the expression for velocity in spherical coordinates.

4   Point Velocity in Spherical Coordinates

The *spherical coordinates* r, ψ, φ have the following geometric sense (see Fig. 4):

(a) r is the length of the radius vector $\mathbf{r}$ of the point M;
(b) φ is the angle formed by this radius vector and the plane Oxy, considered as the origin plane, $-\pi/2 \leqslant \varphi \leqslant \pi/2$;
(c) ψ is the angle formed by the projection ρ of the radius vector $\mathbf{r}$ to the plane Oxy and the x-axis, $0 \leqslant \psi \leqslant 2\pi$.

Fig. 4 Spherical coordinates

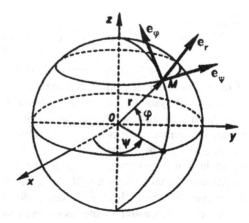

It is clear from Fig. 4 that the Cartesian coordinates are connected with the spherical ones by the formulas

$$x = \rho \cos\psi = r \cos\varphi \cos\psi,$$
$$y = \rho \sin\psi = r \cos\varphi \sin\psi,$$
$$z = r \sin\varphi.$$

In analogy with expression (3.1) the velocity vector $\mathbf{v}$ should be calculated by the formula

$$\mathbf{v} = \frac{\partial \mathbf{r}}{\partial \psi}\,\dot{\psi} + \frac{\partial \mathbf{r}}{\partial \varphi}\,\dot{\varphi} + \frac{\partial \mathbf{r}}{\partial r}\,\dot{r} = \dot{\psi}\,\mathbf{e}_\psi + \dot{\varphi}\mathbf{e}_\varphi + \dot{r}\,\mathbf{e}_r. \tag{4.1}$$

By the definition of partial derivatives, each of these summands represents the velocity of the point M moving along the curve along which it should move from the position given when the other coordinates are fixed. This curve is called the *coordinate line* corresponding to the coordinate changing during movement along it. For the coordinate ψ, we have a circle of radius ρ. The velocity $\dot{\psi}\,\mathbf{e}_\psi$ of motion along it is $\rho\,\dot{\psi}$ in modulus. Therefore, vector $\mathbf{e}_\psi$ is directed tangentially to the given circle and its absolute value is ρ. In analogy with the above, the vector $\mathbf{e}_\varphi$ can shown to be directed tangentially to the circle of radius r and $|\mathbf{e}_\varphi| = r$. The coordinate line corresponding to the coordinate r is a ray emanating from the origin and crossing the point M. Therefore, the vector $\mathbf{e}_r$ is directed along this ray and $|\mathbf{e}_r| = 1$. Figure 4 shows that the vectors $\mathbf{e}_\psi$, $\mathbf{e}_\varphi$, $\mathbf{e}_r$ are orthogonal.

Consider the following orthonormal basis of unit vectors:

$$\mathbf{e}_\psi^0 = \mathbf{e}_\psi/\rho, \qquad \mathbf{e}_\varphi^0 = \mathbf{e}_\varphi/r, \qquad \mathbf{e}_r^0 = \mathbf{e}_r, \qquad \rho = r\cos\varphi; \tag{4.2}$$

this basis allows one to represent the vector $\mathbf{v}$ in the form

$$\mathbf{v} = \rho\dot{\psi}\,\mathbf{e}_\psi^0 + r\dot{\varphi}\,\mathbf{e}_\varphi^0 + \dot{r}\,\mathbf{e}_r^0.$$

Hence

$$v^2 = r^2\cos^2\varphi\,\dot{\psi}^2 + r^2\dot{\varphi}^2 + \dot{r}^2. \tag{4.3}$$

Consider a solid body, which has one stationary point coinciding with the origin of $Oxyz$ coordinate system. Let us define the position of an arbitrary point of the body in $Oxyz$ system by spherical coordinates ψ, φ, r. By assumption the body is considered to be an absolute rigid one, that is why the coordinate r is constant, but ψ and φ coordinates will change during motion. Operator (3.4) is applied both to the angles ψ and φ. The angle ψ in the spherical coordinates is analogous in meaning to the angle ψ in cylindrical coordinates. That is why in formula (3.4) in the spherical coordinates the unit vector $\mathbf{k}$ will also coincide with the unit vector in z-direction. When changing the angle φ the radius vector $\mathbf{r}$ rotates around the axis, the positive direction of rotation about which is given by the unit vector $(-\mathbf{e}_\psi^0)$. Therefore,

$$\frac{\partial \mathbf{r}}{\partial \varphi} = -\mathbf{e}_\psi^0 \times \mathbf{r}. \tag{4.4}$$

It follows from (4.1), (3.4), and (4.4) that the velocity of an arbitrary point of a solid, possessing a stationary point, can be represented in the form

$$\mathbf{v} = \boldsymbol{\omega} \times \mathbf{r}, \tag{4.5}$$

where $\boldsymbol{\omega} = \dot{\psi}\,\mathbf{k} - \dot{\varphi}\,\mathbf{e}_\psi^0$.

Thus, formula (3.5) can be extended in case of an arbitrary rotation of a solid body about a stationary point.

Formula (4.5) was deduced by Euler and bears his name. The vector $\boldsymbol{\omega}$ from it is called the *instantaneous angular velocity*. Here, this vector is expressed in spherical coordinates of the end of vector $\mathbf{r}$, the velocity of which is to be found. This suggests that the vector $\boldsymbol{\omega}$ depends on the vector $\mathbf{r}$, while earlier, in formula (3.5), it was independent of the latter one. Indeed, the vector $\boldsymbol{\omega}$ is not related to the choice of the vector $\mathbf{r}$. A disadvantage of this easy derivation of formula (4.5) is that it does not allow one to obtain a constructive representation of the vector $\boldsymbol{\omega}$. In kinematics of solids, another derivation of *Euler's formula* deprived of this disadvantage will be given.

5 Arbitrary Curvilinear Point Coordinates. Fundamental Basis

The easiest way of giving the position of a particle in the $Oxyz$-system is to define it in the Cartesian coordinates. But if, for example, the motion of a particle over a cylindrical or a spherical surface is considered, then from the two previous subsections it follows that it is worth using the cylindrical and spherical coordinate system, respectively. Putting it in another way, the choice of a coordinate system depends on the character of a problem itself. In order to have an opportunity to solve different problems, it is necessary to have a rather general theory. For this purpose, one should start studying the particle kinematics in arbitrary curvilinear coordinates. A mathematical apparatus developed for that will be used in what follows mainly in dynamics of systems, and in particular, in dynamics of the representation point. Taking this into account, in what follows in the coordinate notation a single character with indexes will be used. Components of all vectors will be denoted by one character with indexes. It permits to reduce a transition from a typical point to a representative one to the substitution of index changing limits.

So, let us suppose that the position of a particle is defined in curvilinear coordinates, which involve three quantities q^1, q^2, q^3 and in terms of which the Cartesian coordinates of a particle can be uniquely expressed

$$x_k = F_k(q^1, q^2, q^3), \quad k = 1, 2, 3, \quad \text{or} \quad \mathbf{r} = \mathbf{r}(q^1, q^2, q^3),$$

Fig. 5 Coordinate surfaces
and the fundamental basis

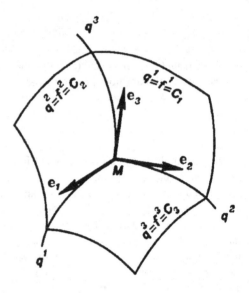

in so doing these equations are assumed to be solvable for q^1, q^2, q^3, their solutions being single-valued. Thus

$$q^k = f^k(x_1, x_2, x_3), \qquad k = 1, 2, 3.$$

It is seen directly from these equalities that making some coordinate q^k to be constant leads us to a surface equation; this surface is called the *coordinate surface* to the coordinate q^k.

Intersection of two coordinate surfaces gives a *coordinate line*, along which only one coordinate changes. For example, the intersection of the surfaces $q^1 = f^1(x_1, x_2, x_3) = C_1$, $q^2 = f^2(x_1, x_2, x_3) = C_2$ gives a coordinate line along along which the coordinate q^3 changes.

The intersection of the three coordinate surfaces (see Fig. 5) gives the point M whose position needs to be determined. If we draw the tangent lines to the coordinate lines in the direction of increasing of the quantities q^1, q^2, q^3, we get the *curvilinear coordinate axes*. The directions of these axes can form both orthogonal system (for instance, spherical or cylindric coordinate system) and non-orthogonal one (*an oblique frame*). For defining motion in curvilinear coordinates, the quantities q^1, q^2, q^3 should be given as time functions.

Let us consider the differential of the vector function $\mathbf{r}(q^1, q^2, q^3)$. In accordance with the definition of the differential of a function of several variables

$$d\mathbf{r} = \sum_{i=1}^{3} \frac{\partial \mathbf{r}}{\partial q^i} \, dq^i .$$

The vector $d\mathbf{r}$ is called an *elementary displacement*. Denoting $\partial\mathbf{r}/\partial q^i = \mathbf{e}_i$, we get

$$d\mathbf{r} = \sum_{i=1}^{3} dq^i\, \mathbf{e}_i\,.$$

From here on in order to reduce the notation, we shall use the repeated index convention for summation. In accordance with this, the previous formula should be rewritten in the form

$$d\mathbf{r} = dq^i\, \mathbf{e}_i\,. \tag{5.1}$$

The index over which summation is performed is called *dummy*. The name is explained by the fact that this index has no relation with the value obtained as a result of summation—it says nothing about it.

The vectors $\mathbf{e}_i$ directed tangentially to the coordinate lines form the *fundamental basis* of the curvilinear system under consideration (see Fig. 5). The absolute values of basis vectors may be different from 1, but in this case one may always put $\mathbf{e}_i = H_i\mathbf{e}_i^0$, where $H_i = |\mathbf{e}_i|$, $|\mathbf{e}_i^0| = 1$. The basis $\{\mathbf{e}_i^0\}$ is called *normalized*, the values H_i are called the *Lamé coefficients*.

One should note that the vectors $\mathbf{e}_i$ of the fundamental basis are vectors functions of a point in space—this is why the basis $\{\mathbf{e}_i\}$ is called *local*.

We note that in order to represent an elementary displacement $d\mathbf{r}$, which by its nature should be defined as a set of quantities dq^i in the form (5.1), a fundamental basis should be introduced. Put another way, introducing an elementary displacement vector means the simultaneous introduction of a fundamental basis. The relation of the elementary displacement vector with the fundamental basis is of general nature and will be used further in application to mechanical systems.

6 Elementary Work. Reciprocal Basis

The concept of an elementary work is directly related to that of an elementary displacement $d\mathbf{r}$. It is one of the fundamental notions in theoretical mechanics. By definition, the scalar production $\mathbf{F} \cdot d\mathbf{r}$ is called an *elementary work* δA done by a force $\mathbf{F}$ associated with the displacement $d\mathbf{r}$. If the force $\mathbf{F}$ is given by its projections to the Cartesian coordinate axes (that is, in the form $\mathbf{F} = X\,\mathbf{i} + Y\,\mathbf{j} + Z\,\mathbf{k}$), then

$$\delta A = X\,dx + Y\,dy + Z\,dz\,.$$

We note that the use of δ instead of the total differential d in this formula is related to the fact that the differential trinomial $X\,dx + Y\,dy + Z\,dz$ is not always a total differential of some function.

The elementary work depends linearly on components of the vector $d\mathbf{r}$ (a *linear form* of the vector $d\mathbf{r}$). The value of δA is independent of a system in which the vector $d\mathbf{r}$ is given. In other words, the elementary work is an invariant linear form of the vector $d\mathbf{r}$. Therefore, if in curvilinear coordinates $\{q^i\}$ the vector $d\mathbf{r}$ is given in its natural form (5.1), then the elementary work δA is a linear function of the components dq^i of the vector $d\mathbf{r}$, that is, it can be written as follows:

$$\delta A = Q_i \, dq^i \,. \tag{6.1}$$

Let us find the representation of the coefficients Q_i of this linear form. In accordance with the definition of the elementary work we have

$$\delta A = \mathbf{F} \cdot d\mathbf{r} = \mathbf{F} \cdot \mathbf{e}_i \, dq^i \,,$$

and hence

$$Q_i = \mathbf{F} \cdot \mathbf{e}_i = H_i F \cos(\widehat{\mathbf{F}, \mathbf{e}_i}) \,, \qquad i = 1, 2, 3. \tag{6.2}$$

Thus, the quantities Q_i are proportional to the orthogonal projections of the force $\mathbf{F}$ to the directions of the coordinate axes. Clearly, the force $\mathbf{F}$, *qua* a vector, is uniquely defined by these three orthogonal projections. The natural question arises whether it is possible to represent the vector $\mathbf{F}$ using the quantities Q_1, Q_2, Q_3 in the form

$$\mathbf{F} = Q_j \mathbf{e}^j \,, \tag{6.3}$$

where $\mathbf{e}^1$, $\mathbf{e}^2$, $\mathbf{e}^3$ are three unknown vectors. Substituting the expression (6.3) into (6.2), we get

$$Q_i = Q_j \mathbf{e}^j \cdot \mathbf{e}_i \,.$$

The required vectors $\mathbf{e}^j$ should be such that

$$\mathbf{e}_i \cdot \mathbf{e}^j = \delta_i^j = \begin{cases} 1, & i = j, \\ 0, & i \neq j. \end{cases} \tag{6.4}$$

Formulas (6.4) allow one to construct three new vectors $\mathbf{e}^1$, $\mathbf{e}^2$, $\mathbf{e}^3$ from the fundamental basis $\{\mathbf{e}_i\}$. The quantities δ_i^j are known as *Kronecker deltas*; the three constructed vectors $\mathbf{e}^j$ form the *reciprocal* or the *dual basis*.

Let us consider this basis in detail. We take, for example, the vector $\mathbf{e}^1$. It is orthogonal to the vectors $\mathbf{e}_2$ and $\mathbf{e}_3$, which lie in the plane tangent to the coordinate surface corresponding to coordinate q^1. Therefore, the vector $\mathbf{e}^1$ is directed along the normal to the surface. By analogy, the vectors $\mathbf{e}^2$ and $\mathbf{e}^3$ are directed along normals to the surfaces $f^2(x_1, x_2, x_3) = C_2$ and $f^3(x_1, x_2, x_3) = C_3$, respectively (see Fig. 5).

We take an elementary displacement in the form $dq^1 \mathbf{e}_1$; that is, we put $dq^2 = dq^3 = 0$. Under this displacement, the work will be done only by the component $Q_1 \mathbf{e}^1$ of the force $\mathbf{F}$, because the two other components $Q_2 \mathbf{e}^2$ and $Q_3 \mathbf{e}^3$ are orthogonal to the displacement $dq^1 \mathbf{e}_1$.

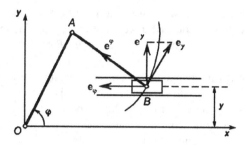

Fig. 6 Crank-and-slider mechanism

Thus, introducing the reciprocal basis allows one to represent the vector of the force as a sum of such components, each of which executes work only on one elementary displacement, as given by the differential of the corresponding coordinate.

Let us explain the expediency of introducing the reciprocal basis by considering the following example.[1] A simplest mechanism for the transformation of rotation into true rectilinear motion consists of the crank OA, the connecting rod AB and the guide B (see Fig. 6). The crank OA, rotating around the point O, moves the guide B by means of the connecting rod AB, the guide B sliding along rails parallel to the Ox-axis. It is supposed that the connecting rod is connected at points A and B with the crank and with the guide by means of cylindrical joints, respectively. A motion of the connecting rod along the rails results in the force of dry friction, which can be approximately considered as being proportional to the normal pressure of the guide on the rails. Further, it will be shown that in order to take into account the effect of rails on the motion of this mechanism, it is necessary to imaginatively remove the rails and replace their action with the forces of friction and normal pressure. In the absence of rails, the point B gets a possibility of moving along both Ox- and Oy-axes. Taking into account the specifics of this problem one should consider the angle of rotation of the crank φ and the projection y of the point B as curvilinear coordinates, defining uniquely the position of the point B. The coordinate line, corresponding to the coordinate φ, is a straight line, parallel to the Ox-axis ($y = \mathrm{const}$). The basis vector $\mathbf{e}_\varphi$ is directed to the side opposite to the direction of the Ox-axis, because the positive rotation is supposed to be the rotation of the crank OA running counterclockwise (see Fig. 6). With fixed coordinate φ the connecting rod AB rotates about the fixed point A. Therefore, the coordinate line corresponding to the coordinate y is the circle of radius AB with center at the point A. The basis vector $\mathbf{e}_y$ is tangent to this circle. We introduce the reciprocal basis as follows (Fig. 6):

$$\mathbf{e}^\varphi \cdot \mathbf{e}_\varphi = 1, \qquad \mathbf{e}^\varphi \cdot \mathbf{e}_y = 0, \qquad \mathbf{e}^y \cdot \mathbf{e}_y = 1, \qquad \mathbf{e}^y \cdot \mathbf{e}_\varphi = 0.$$

In order to satisfy the condition $y = \mathrm{const}$ when changing the angle φ it is enough to apply a force directed along the vector $\mathbf{e}^y$ at the point B. Moreover, as this force

---

[1] This example relates both to kinematics and dynamics, that is why it is useful to turn back to it after examining Chap. 6 in the Part II. On a first reading, this example can be omitted without as far as an understanding of the rest of the book is concerned.

produces no work, it would not have any effect on the law of changing the angle φ with $y = $ const (if it were not for the dry friction).

So, let us suppose that the point B of this mechanism, starting from a certain time instant, becomes free. It is required to find the force to be applied to this point in order to change the angle φ according to the law $\varphi = f(t)$. Further (in Chap. 6 of Part II) it will be shown that for this purpose it is enough to apply a force directed along the vector $\mathbf{e}^\varphi$. In this respect it can be said that the force $Q_\varphi \mathbf{e}^\varphi$ controls the φ-coordinate motion, while the force $Q_y \mathbf{e}^y$ controls the y-coordinate motion.

Expression (6.1) for elementary work is fairly general. The action of all forces applied to a mechanical system is characterized by the work done by them when this system changes from one position to another one. That is why expression (6.1) is the most general way of describing a force; it leads to a decomposition of the force in the vector projections of the reciprocal basis.

Thus, if one is concerned with a mechanical system, whose position is given by coordinates q^i, then the introduction for it the elementary displacement as an object of vector structure means the introduction of the fundamental basis, while the introduction of the concept of a force implies the introduction of the reciprocal basis.

7 Co- and Contravariant Components of the Velocity Vector. The Convention of "Raising and Lowering Indices"

The components of a vector $\mathbf{a}$ with respect to the fundamental (reciprocal) basis will be given superscripts (subscripts, respectively); that is, we shall write

$$\mathbf{a} = a^i \mathbf{e}_i, \qquad \mathbf{a} = a_j \mathbf{e}^j.$$

From expressions (6.4) it follows that

$$a^k = \mathbf{a} \cdot \mathbf{e}^k, \qquad a_k = \mathbf{a} \cdot \mathbf{e}_k. \tag{7.1}$$

We introduce the new curvilinear coordinates q^i_* and the vectors $\mathbf{e}^*_j = \partial \mathbf{r} / \partial q^j_*$ of the new fundamental basis. Taking into account the relations

$$q^j_* = q^j_*(q^1, q^2, q^3), \qquad q^i = q^i(q^1_*, q^2_*, q^3_*),$$

between the new and old coordinates, we get

$$\mathbf{e}^*_j = \frac{\partial \mathbf{r}}{\partial q^i} \frac{\partial q^i}{\partial q^j_*} = \frac{\partial q^i}{\partial q^j_*} \mathbf{e}_i, \qquad \mathbf{e}_i = \frac{\partial \mathbf{r}}{\partial q^j_*} \frac{\partial q^j_*}{\partial q^i} = \frac{\partial q^j_*}{\partial q^i} \mathbf{e}^*_j. \tag{7.2}$$

Hence, using (7.1), we find that

$$a_j^* = \mathbf{a} \cdot \mathbf{e}_j^* = \frac{\partial q^i}{\partial q_*^j} a_i, \qquad a_i = \mathbf{a} \cdot \mathbf{e}_i = \frac{\partial q_*^j}{\partial q^i} a_j^*. \tag{7.3}$$

From these formulas one can write down the following sequence of equalities:

$$\mathbf{a} = a_i \mathbf{e}^i = \frac{\partial q_*^j}{\partial q^i} a_j^* \mathbf{e}^i = a_j^* \mathbf{e}_*^j = \frac{\partial q^i}{\partial q_*^j} a_i \mathbf{e}_*^j,$$

which implies the following relations

$$\mathbf{e}_*^j = \frac{\partial q_*^j}{\partial q^i} \mathbf{e}^i, \qquad \mathbf{e}^i = \frac{\partial q^i}{\partial q_*^j} \mathbf{e}_*^j.$$

As a result, we have

$$a_*^j = \mathbf{a} \cdot \mathbf{e}_*^j = \frac{\partial q_*^j}{\partial q^i} a^i, \qquad a^i = \mathbf{a} \cdot \mathbf{e}^i = \frac{\partial q^i}{\partial q_*^j} a_*^j. \tag{7.4}$$

Comparing expressions (7.2) and (7.3) we see that the components a_j^*, a_i of the vector $\mathbf{a}$ in terms of the new and old reciprocal bases transform in the same way as the vectors of fundamental bases. That is why they are called *covariant components*. The quantities a_*^j and a^i are called *contravariant components* of a vector $\mathbf{a}$, because formulas (7.4) for change from the old a^i to the new a_*^j components involve the same coefficients, which in transformations (7.2) were applied in the converse change from the new to the old basis.

Let us consider the velocity vector of a particle. By definition (1.4) and expression (5.1) we have

$$\mathbf{v} = \dot{q}^i \mathbf{e}_i = v^i \mathbf{e}_i. \tag{7.5}$$

Thus, the derivatives $\dot{q}^i$ are contravariant components of the velocity vector of a particle. They are called *generalized velocities*.

Expression (7.5) shows that

$$\mathbf{e}^i = \frac{\partial \mathbf{v}}{\partial \dot{q}^i} = \frac{\partial \mathbf{r}}{\partial q^i}, \tag{7.6}$$

and hence

$$v_i = \mathbf{v} \cdot \mathbf{e}_i = \mathbf{v} \cdot \frac{\partial \mathbf{v}}{\partial \dot{q}^i} = \frac{\partial (v^2/2)}{\partial \dot{q}^i}. \tag{7.7}$$

The quantity $T = mv^2/2$, where m is the mass of a particle, is called its *kinetic energy*; the vector $\mathbf{p} = m\mathbf{v}$ is called the *particle momentum*. Formula (7.7) tells us that the momentum can be found from the expression for the kinetic energy; as a result it will be decomposed into the reciprocal basis. So, we have

$$\mathbf{p} = \frac{\partial T}{\partial \dot{q}^i} \, \mathbf{e}^i \, .$$

It is essential that this expression, as well as expression (6.1) for elementary work, can be extended to the case of mechanical systems. We shall also briefly mention that using these extensions one can write down the law of motion of any mechanical system, whose position of which is uniquely given by setting the parameters q^i in the form given by I. Newton for a particle; that is,

$$\frac{d\mathbf{p}}{dt} = \mathbf{F}, \qquad \frac{d}{dt}\left(\frac{\partial T}{\partial \dot{q}^i} \, \mathbf{e}^i\right) = Q_i \mathbf{e}^i \, . \tag{7.8}$$

It is worth pointing out that formulating Newton's second law in a generalized form is directly connected with introducing the covariant components of the momentum and force vectors. This question will be considered in detail in Chap. 6 in Part II.

Let us go back to formula (7.7). Expression (7.5) implies that

$$v^2 = \dot{q}^i \mathbf{e}_i \cdot \dot{q}^j \mathbf{e}_j = g_{ij} \dot{q}^i \dot{q}^j \, , \tag{7.9}$$

where

$$g_{ij} = \mathbf{e}_i \cdot \mathbf{e}_j \, . \tag{7.10}$$

Taking into account that $v^2 = ds^2/(dt)^2$ we get

$$d\mathbf{r}^2 = ds^2 = g_{ij} \, dq^i \, dq^j \, . \tag{7.11}$$

The coefficients g_{ij} of this invariant quadric form of the components dq^i of the elementary displacement vector $d\mathbf{r}$ are called the *metric coefficients*, because from them one can calculate the square of its length using the components of the vector $d\mathbf{r}$. We note that the quantities g_{ij} can be determined from the expression for the kinetic energy, since by (7.9) it can be written in the form

$$T = \frac{m}{2} g_{ij} \dot{q}^i \dot{q}^j \, . \tag{7.12}$$

From (7.1) and (7.10) it follows that the setting metric coefficients is equivalent to setting vectors $\{\mathbf{e}_i\}$ of the fundamental basis in their covariant form, that is, in the from

$$\mathbf{e}_i = g_{ij} \, \mathbf{e}^j \, . \tag{7.13}$$

Hence from (7.1) we see that the covariant and contravariant components of the vector **a** are related as

$$a_i = g_{ij}a^j . \tag{7.14}$$

In particular, for the velocity vector components we have

$$v_i = g_{ij}\dot{q}^j .$$

It is known that the determinant $|g_{ij}|$ formed by the coefficients of the positive-definite quadric form (7.11) is not zero, and hence system (7.14) is solvable for a^j, so that

$$a^j = g^{jk}a_k , \tag{7.15}$$

where g^{jk} are entries of the matrix inverse to that with entries g_{ij}. Using the expression (7.1), relation (7.15) can be written as

$$\mathbf{a} \cdot \mathbf{e}^j = g^{jk}\mathbf{a} \cdot \mathbf{e}_k .$$

Hence

$$\mathbf{a} \cdot (\mathbf{e}^j - g^{jk}\mathbf{e}_k) = 0 .$$

Since this expression holds for any vector $\mathbf{a}$, we have

$$\mathbf{e}^j = g^{jk}\mathbf{e}_k . \tag{7.16}$$

Formulas (7.13)–(7.16) allow one to change from quantities with upper indices to quantities with lower ones, and vice versa. That is why it is usually said that they define the *convention of "raising and lowering indices"*.

8 Co- and Contravariant Components of the Acceleration Vector. The Lagrange Operator

In the previous subsection, it was shown that the vectors of the fundamental basis and the velocity vector can be found from the expression for kinetic energy. We shall show that this observation also applies to the acceleration vector.

Using (7.1) and (7.6) we obtain

$$w_i = \mathbf{w} \cdot \mathbf{e}_i = \frac{d\mathbf{v}}{dt} \cdot \frac{\partial \mathbf{v}}{\partial \dot{q}^i} = \frac{d}{dt}\left(\mathbf{v} \cdot \frac{\partial \mathbf{v}}{\partial \dot{q}^i}\right) - \mathbf{v} \cdot \frac{d}{dt}\frac{\partial \mathbf{v}}{\partial \dot{q}^i} . \tag{8.1}$$

Next, from expressions (7.5) and (7.6), we see that

$$\frac{d}{dt}\frac{\partial \mathbf{v}}{\partial \dot{q}^i} = \frac{d}{dt}\frac{\partial \mathbf{r}}{\partial q^i} = \frac{\partial^2 \mathbf{r}}{\partial q^i \partial q^j}\dot{q}^j = \frac{\partial^2 \mathbf{r}}{\partial q^j \partial q^i}\dot{q}^j = \frac{\partial \mathbf{v}}{\partial q^i} .$$

As a result, expression (8.1) can be written in the form

$$w_i = \frac{d}{dt} \frac{\partial(v^2/2)}{\partial \dot{q}^i} - \frac{\partial(v^2/2)}{\partial q^i}. \tag{8.2}$$

Hence

$$m w_i = \frac{d}{dt} \frac{\partial T}{\partial \dot{q}^i} - \frac{\partial T}{\partial q^i}. \tag{8.3}$$

Thus, the covariant components of the acceleration vector can be obtained using the expression for kinetic energy. This exclusively important representation was derived by Lagrange. With the help of the Lagrange operator

$$L_i = \frac{d}{dt} \frac{\partial}{\partial \dot{q}^i} - \frac{\partial}{\partial q^i}$$

expression (8.3) assumes the form

$$m w_i = L_i(T).$$

Substituting the expression (7.9) into formula (8.2), we get

$$w_i = g_{ij} \ddot{q}^j + \frac{\partial g_{ij}}{\partial q^k} \dot{q}^j \dot{q}^k - \frac{1}{2} \frac{\partial g_{jk}}{\partial q^i} \dot{q}^j \dot{q}^k. \tag{8.4}$$

In the sum

$$\frac{\partial g_{ij}}{\partial q^k} \dot{q}^j \dot{q}^k \tag{8.5}$$

the indexes j and k are dummy (the summation is taken over them). Dummy indices can be of any sort, in particular, they can be replaced swapped, and thus expression (8.5) can be written as

$$\frac{\partial g_{ik}}{\partial q^j} \dot{q}^k \dot{q}^j,$$

and so this double sum may be represented as

$$\frac{\partial g_{ij}}{\partial q^k} \dot{q}^j \dot{q}^k = \frac{1}{2} \left(\frac{\partial g_{ij}}{\partial q^k} + \frac{\partial g_{ik}}{\partial q^j} \right) \dot{q}^j \dot{q}^k.$$

Substituting this relationship into expression (8.4), we get

$$w_i = g_{ij} \ddot{q}^j + \Gamma_{i,jk} \dot{q}^j \dot{q}^k, \tag{8.6}$$

where

$$\Gamma_{i,jk} = \frac{1}{2} \left(\frac{\partial g_{ij}}{\partial q^k} + \frac{\partial g_{ik}}{\partial q^j} - \frac{\partial g_{jk}}{\partial q^i} \right) . \tag{8.7}$$

The quantities $\Gamma_{i,jk}$ are called *Christoffel symbols of the first kind*. In view of (7.10) they can also be written as

$$\Gamma_{i,jk} = \frac{1}{2} \left(\mathbf{e}_i \cdot \frac{\partial \mathbf{e}_j}{\partial q^k} + \mathbf{e}_j \cdot \frac{\partial \mathbf{e}_i}{\partial q^k} + \mathbf{e}_i \cdot \frac{\partial \mathbf{e}_k}{\partial q^j} + \mathbf{e}_k \cdot \frac{\partial \mathbf{e}_i}{\partial q^j} - \mathbf{e}_j \cdot \frac{\partial \mathbf{e}_k}{\partial q^i} - \mathbf{e}_k \cdot \frac{\partial \mathbf{e}_j}{\partial q^i} \right) .$$

In the expressions $\partial \mathbf{e}_j / \partial q^k$ the indices j and k commute, inasmuch as

$$\frac{\partial \mathbf{e}_j}{\partial q^k} = \frac{\partial^2 \mathbf{r}}{\partial q^j \partial q^k} = \frac{\partial^2 \mathbf{r}}{\partial q^k \partial q^j} = \frac{\partial \mathbf{e}_k}{\partial q^j} ,$$

and hence,

$$\Gamma_{i,jk} = \frac{\partial \mathbf{e}_j}{\partial q^k} \cdot \mathbf{e}_i . \tag{8.8}$$

This reveals the geometric sense of the Christoffel symbols of the first kind: they are the covariant components of the vector $\partial \mathbf{e}_j / \partial q^k$.

From (7.7) it follows that the first summand on the right of (8.2) is equal to the derivative of the velocity vector covariant component in time. This makes it possible to represent the quantity w_i also in the form

$$w_i = \dot{v}_i - \frac{1}{2} \frac{\partial g_{jk}}{\partial q^i} \dot{q}^j \dot{q}^k .$$

Using again the fact that the summation indices in double sums of kind (8.2) are interchangeable, we get

$$w_i = \dot{v}_i - \Gamma_{j,ik} \dot{q}^j \dot{q}^k .$$

Next, $\dot{q}^j = g^{jl} v_l$, and hence,

$$w_i = \dot{v}_i - \Gamma_{ik}^l v_l \dot{q}^k . \tag{8.9}$$

Here, $\Gamma_{ik}^l = g^{lj} \Gamma_{j,ik}$ are the *Christoffel symbols of the second kind*. According to (8.8) and (7.16) they can be also represented as

$$\Gamma_{ik}^l = \frac{\partial \mathbf{e}_i}{\partial q^k} \cdot \mathbf{e}^l . \tag{8.10}$$

Therefore, the Christoffel symbols of the second kind are the contravariant components of the vector $\partial \mathbf{e}_i / \partial q^k$.

It should be noted that if one knows the expression for the kinetic energy, then the metric tensor $\{g_{ij}\}$ is known, using which the symbols $\Gamma_{i,jk}$ and $\Gamma_{ik}^l = g^{lj} \Gamma_{j,ik}$

can be found by formulas (8.7). Such a way of introducing the Christoffel symbols using the kinetic energy can be extended to the case of a mechanical system.

The general formula (8.9) allows one to find the rule suitable for calculating the derivative of any vector $\mathbf{a}$ in time, when represented in the form of $a_i \mathbf{e}^i$ in the case, when the reciprocal basis $\{\mathbf{e}^i\}$ moves together with the particle. Indeed, in formula (8.9), the summands proportional to $\dot{q}^k$ characterize the effect of the reciprocal basis mobility on the acceleration vector. This is effected by using the Christoffel symbols of the second kind. Therefore, setting $\mathbf{b} = \dot{\mathbf{a}}$, we have in accordance with expression (8.9)

$$b_i = \dot{\mathbf{a}} \cdot \mathbf{e}_i = \dot{a}_i - \Gamma_{ik}^l a_l \dot{q}^k \, . \tag{8.11}$$

This formula can also be checked directly. Indeed, by the definition of the derivative,

$$\dot{\mathbf{a}} = \dot{a}_j \mathbf{e}^j + a_j \dot{\mathbf{e}}^j \, .$$

Hence

$$\dot{\mathbf{a}} \cdot \mathbf{e}_i = \dot{a}_i + a_j \dot{\mathbf{e}}^j \cdot \mathbf{e}_i \, . \tag{8.12}$$

Differentiating (6.4) in time, this gives

$$\dot{\mathbf{e}}_i \cdot \mathbf{e}^j + \mathbf{e}_i \cdot \dot{\mathbf{e}}^j = 0 \, ,$$

and so

$$\dot{\mathbf{e}}^j \cdot \mathbf{e}_i = -\mathbf{e}^j \cdot \dot{\mathbf{e}}_i = -\mathbf{e}^j \cdot \frac{\partial \mathbf{e}_i}{\partial q^k} \dot{q}^k \, .$$

Substituting this expression into (8.12) and replacing the dummy index j by l, we have

$$\dot{\mathbf{a}} \cdot \mathbf{e}_i = \dot{a}_i - \frac{\partial \mathbf{e}_i}{\partial q^k} \cdot \mathbf{e}^l a_l \dot{q}^k \, .$$

In view of (8.10) this formula can be written in the form (8.11), the result required.

Now let us consider the contravariant components of the acceleration vector. Using the law of index raising, we get

$$w^i = g^{il} w_l \, .$$

Hence, using (8.6),

$$w^i = g^{il} g_{lj} \ddot{q}^j + g^{il} \Gamma_{l,jk} \dot{q}^j \dot{q}^k \, .$$

As $g^{il} \Gamma_{l,jk} = \Gamma_{jk}^i$ and since, as will be shown later

$$g^{il} g_{lj} \ddot{q}^j = \ddot{q}^i \, , \tag{8.13}$$

we finally get

$$w^i = \ddot{q}^i + \Gamma^i_{jk}\, \dot{q}^j \dot{q}^k \,. \tag{8.14}$$

contravariant components of the acceleration vector Relation (8.13) follows from the fact that the Kronecker deltas when using formulas (7.13) and (7.16) can be written as

$$\delta^i_j = \mathbf{e}^i \cdot \mathbf{e}_j = g^{il}\mathbf{e}_l \cdot g_{jk}\mathbf{e}^k = g^{il}g_{lj} \,.$$

We note that both for the velocity and acceleration vectors, the dimensions of the co- and contravariant components may in general differ from the dimensions of the velocity (m/s) and the acceleration (m/s^2), respectively. This is explained by the fact that the coordinates q^i can be measured not in meters in the SI system. In this case, it is convenient to take into consideration the unit vectors $\mathbf{e}_i/|\mathbf{e}_i|$ and $\mathbf{e}^i/|\mathbf{e}^i|$.

For example, the orthogonal projections of the acceleration vector to these unit vectors have the dimension (m/s^2); they can be calculated by the formulas

$$\widetilde{w}_i = \frac{\mathbf{w} \cdot \mathbf{e}_i}{|\mathbf{e}_i|} = \frac{w_i}{H_i}\,, \qquad \widetilde{w}^i = \frac{\mathbf{w} \cdot \mathbf{e}^i}{|\mathbf{e}^i|} = \frac{w^i}{|\mathbf{e}^i|}\,. \tag{8.15}$$

The above quantities $\widetilde{w}_i$ and $\widetilde{w}^i$ are called the *physical components of the acceleration vector*.

In particular, if a coordinate system is orthogonal, then the orthonormal fundamental and reciprocal bases coincide—this is why in this case $\widetilde{w}_i = \widetilde{w}^i$. The cylindrical and spherical coordinate systems are orthogonal. In the cylindrical coordinate system we have by (3.2), (3.3), (8.2) and (8.15)

$$\widetilde{w}_\rho = w_\rho = \frac{d}{dt}\frac{\partial(v^2/2)}{\partial\dot{\rho}} - \frac{\partial(v^2/2)}{\partial\rho} = \ddot{\rho} - \rho\dot{\psi}^2\,,$$

$$\widetilde{w}_\psi = \frac{w_\psi}{\rho} = \frac{1}{\rho}\left[\frac{d}{dt}\frac{\partial(v^2/2)}{\partial\dot{\psi}} - \frac{\partial(v^2/2)}{\partial\psi}\right] = \rho\ddot{\psi} + 2\dot{\rho}\dot{\psi}\,,$$

$$\widetilde{w}_z = w_z = \frac{d}{dt}\frac{\partial(v^2/2)}{\partial\dot{z}} - \frac{\partial(v^2/2)}{\partial z} = \ddot{z}\,.$$

Similar calculations for the spherical lead us to the following results:

$$\widetilde{w}_r = \ddot{r} - r\cos^2\varphi \cdot \dot{\psi}^2 - r\dot{\varphi}^2\,,$$

$$\widetilde{w}_\psi = r\cos\varphi \cdot \ddot{\psi} + 2\cos\varphi \cdot \dot{r}\dot{\psi} - 2r\sin\varphi \cdot \dot{\varphi}\dot{\psi}\,,$$

$$\widetilde{w}_\varphi = r\ddot{\varphi} + 2\dot{r}\dot{\varphi} + r\dot{\psi}^2\cos\varphi\sin\varphi\,,$$

formulas (4.2) and (4.3) being useful.

The advantage of using the Lagrange operator should be pointed out—it permits one to derive these formulas in a comparatively easy way.

Chapter 2
Kinematics of the Rigid Solid

N. N. Polyakhov, M. P. Yushkov⊙, and S. A. Zegzhda

In this chapter, we propose a way of describing the motion of a rigid body, derive formulas for calculation of the velocity and acceleration of any point of the body in a general case of its motion. Important particular cases of the motion of a rigid body are considered (translational motion, rotation about a stationary axis, rotation about a stationary point, planar motion). The concept of the rotation tensor is introduced.

1 Coordinates of a Rigid Body. The Euler Angles

We recall that a body is called *perfectly rigid* if the distance between its points does not change.

Let us introduce an orthogonal solid coordinate system $Ox_1x_2x_3$ with the unit vectors $\mathbf{i}_1$, $\mathbf{i}_2$, $\mathbf{i}_3$ and which is supposed to be rigidly fixed in the body. We shall also consider an orthogonal reference system $O_1\xi_1\xi_2\xi_3$ with the unit vectors $\boldsymbol{\varepsilon}_1$, $\boldsymbol{\varepsilon}_2$, $\boldsymbol{\varepsilon}_3$ with respect to which we shall study the motion of a rigid body. The system $Ox_1x_2x_3$ is usually called *moving*, the system $O_1\xi_1\xi_2\xi_3$ is called *fixed* (see Fig. 1). The position of any point M of a rigid body with respect to the movable and stationary coordinate systems is defined by setting the radius vectors $\mathbf{r}$ and $\boldsymbol{\rho}$, respectively. These vectors are related as follows:

$$\boldsymbol{\rho} = \boldsymbol{\rho}_0 + \mathbf{r}. \tag{1.1}$$

Here, $\boldsymbol{\rho}_o$ is the radius vector in the movable system of the origin in a rigid body. This point is called a *pole*.

We shall expand the vectors $\boldsymbol{\rho}$ and $\boldsymbol{\rho}_o$ in the stationary (fixed) basis $\{\boldsymbol{\varepsilon}_s\}$, and expand the vector $\mathbf{r}$ in the movable basis $\{\mathbf{i}_k\}$. Here, (1.1) can be written in the form

$$\boldsymbol{\rho} = \xi_s \boldsymbol{\varepsilon}_s = \xi_{0s} \boldsymbol{\varepsilon}_s + x_k \mathbf{i}_k. \tag{1.2}$$

© Springer Nature Switzerland AG 2021
N. N. Polyakhov et al., *Rational and Applied Mechanics*,
Foundations of Engineering Mechanics,
https://doi.org/10.1007/978-3-030-64061-3_2

Fig. 1 Motion of a rigid
body

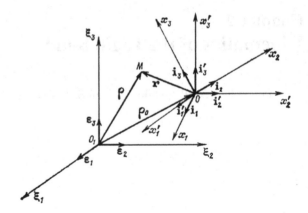

It follows that in order to find the coordinates of any point in a rigid body in the fixed
system using the coordinates of this point x_1, x_2, x_3 with respect to the movable
coordinate system $Ox_1x_2x_3$, it is necessary to find the coordinates ξ_{01}, ξ_{02}, ξ_{03} of the
pole and find the position of the movable basis $\{i_k\}$ with respect to the fixed one $\{\varepsilon_s\}$.
Let us now consider the orientation of the movable coordinate system with respect
to the fixed one.

At any fixed time the axes x_1, x_2, x_3 make with the axes ξ_1, ξ_2, ξ_3 the angles,
whose the direction cosines $\alpha_{sk} = \varepsilon_s \cdot i_k$ can be conveniently put in the tabulated
form as follows:

	i_1	i_2	i_3
$\varepsilon_1 = i'_1$	α_{11}	α_{12}	α_{13}
$\varepsilon_2 = i'_2$	α_{21}	α_{22}	α_{23}
$\varepsilon_3 = i'_3$	α_{31}	α_{32}	α_{33}

The rows of this table contain the components of the basis vectors ε_s of the stationary
axes (see Fig. 1) in the movable basis $\{i_k\}$, and the columns are formed from the
components of the basis vectors i_k in the basis $\{\varepsilon_s\}$. Thus,

$$\varepsilon_s = i'_s = \alpha_{sk} i_k, \qquad i_k = \alpha_{sk} \varepsilon_s. \qquad (1.3)$$

We introduce the new coordinate system $O'x'_1x'_2x'_3$, whose axes are parallel to the
axes ξ_1, ξ_2, ξ_3 and the origin O' coincides with the origin O. We also consider the
transformation of the coordinates when changing from one system to another.

The basis vectors i'_s of the new coordinate system coincide with the vectors ε_s.
Using (1.3), we have, for an arbitrary vector r,

$$r = x_k i_k = x_k \alpha_{sk} i'_s = x'_s i'_s = x'_s \alpha_{sk} i_k. \qquad (1.4)$$

From equalities (1.4) we have the following formulas for changing the coordinates:

$$x'_s = \alpha_{sk}x_k\,, \qquad x_k = \alpha_{sk}x'_s\,, \qquad s = 1, 2, 3, \qquad k = 1, 2, 3,$$

which can be conveniently written in the matrix form:

$$x' = \begin{pmatrix} x'_1 \\ x'_2 \\ x'_3 \end{pmatrix} = \begin{pmatrix} \alpha_{11} & \alpha_{12} & \alpha_{13} \\ \alpha_{21} & \alpha_{22} & \alpha_{23} \\ \alpha_{31} & \alpha_{32} & \alpha_{33} \end{pmatrix} \begin{pmatrix} x_1 \\ x_2 \\ x_3 \end{pmatrix} = Ax\,, \tag{1.5}$$

$$x = \begin{pmatrix} x_1 \\ x_2 \\ x_3 \end{pmatrix} = \begin{pmatrix} \alpha_{11} & \alpha_{21} & \alpha_{31} \\ \alpha_{12} & \alpha_{22} & \alpha_{32} \\ \alpha_{13} & \alpha_{23} & \alpha_{33} \end{pmatrix} \begin{pmatrix} x'_1 \\ x'_2 \\ x'_3 \end{pmatrix} = A^*x' = A^{-1}x'\,.$$

We note that the matrix A of direction cosines possesses the following remarkable property: the inverse matrix A^{-1} agrees with the transposed matrix A^*, which is obtained by interchanging the rows and columns.

Since the coordinate systems are orthogonal by the assumption and since the basis vectors are of length one, the following orthonormality conditions should be satisfied:

$$\varepsilon_i \cdot \varepsilon_j = \alpha_{ik}\alpha_{jk} = \begin{cases} 1, & i = j\,, \\ 0, & i \neq j\,. \end{cases}$$

In total, there are six such conditions, since indexes i and j vary from 1 to 3, and so of the nine cosines only three are independent. Instead of these three independent cosines, one can choose three arbitrary independent parameters can. Usually, as such parameters take the three *Euler angles*, in terms of which all α_{ik} can be expressed (see Fig. 2). These angles are introduced as follows.

Let us consider the line Ox''_1 (the *nodal line*) of intersection of the plane Ox_1x_2 and the plane $Ox'_1x'_2$. The direction of the basis vector $\mathbf{i}''_1$ of this axis is chosen to the side, watching from which the rotation of axis Ox'_3 to axis Ox_3 at the least angle denoted by θ is seen running counterclockwise. Let ψ denote the angle between Ox'_1 and Ox''_1 axes, and let φ denote the angle between the Ox''_1 and Ox_1-axis. The positive reference directions of these angles are shown in Fig. 2. The angles ψ, θ, φ are called the *precession angle*, the *nutation angle*, and the *angle of proper rotation* correspondingly. These names originated in astronomy.

It should be noted that the Euler angles are widely useful in the gyroscope theory.[1] Nevertheless, in flight dynamics other systems proved to be more preferable (see Chap. "Flight Dynamics" of the second volume).

When $\theta = 0$ and $\theta = \pi$ the movable plane Ox_1x_2 coincides with the fixed one $Ox'_1x'_2$, the angles ψ and φ turn out to be undefined. Nevertheless, their sum agrees with the angle between the axes Ox'_1 and Ox_1. In the vicinity of these two special values of the nutation angle certain calculation difficulties appear. That is why in the study of motions for which the Ox_3 and Ox'_3-axes can be collinear other char-

---

[1] See, for example, the book by *A. Ishlinsky*. Mechanics of gyroscope systems. Moscow: USSR Academy of Sciences Edition. 1963 [in Russian].

Fig. 2 The Euler angles

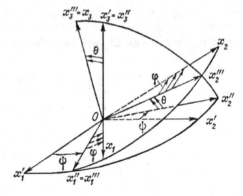

Fig. 3 The x'_k-coordinates
versus x''_k-coordinates

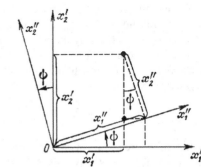

acteristics of the position of a rigid body with a stationary point are usually used, in
particular, the Rodrigues or Cayley–Klein parameters.

To superimpose the system $Ox'_1x'_2x'_3$ with the system $Ox_1x_2x_3$ it suffices:

1. to rotate the system $Ox'_1x'_2x'_3$ around the axis Ox'_3 at the angle ψ, thereby obtaining
   the system $Ox''_1x''_2x''_3$ with $x''_3 = x'_3$;
2. to rotate the system $Ox''_1x''_2x''_3$ around the axis Ox''_1 at the angle θ, thereby obtaining the system $Ox'''_1x'''_2x'''_3$ with $x'''_1 = x''_1$;
3. to rotate the system $Ox'''_1x'''_2x'''_3$ around the axis $Ox'''_3 = Ox_3$ at the angle φ,
   thereby obtaining the system $Ox_1x_2x_3$.

An arbitrary vector **r** can be given in four coordinate systems:

$$\mathbf{r} = x_k\mathbf{i}_k = x'_k\mathbf{i}'_k = x''_k\mathbf{i}''_k = x'''_k\mathbf{i}'''_k \,.$$

Figure 3 shows that the coordinates x'_k and x''_k are related as follows:

$$x'_1 = x''_1 \cos\psi - x''_2 \sin\psi + 0\,,$$

$$x'_2 = x''_1 \sin\psi + x''_2 \cos\psi + 0\,,$$

Fig. 4 The x_k''-coordinates versus x_k'''-coordinates

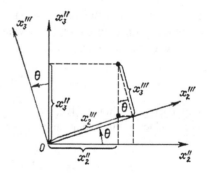

$$x_3' = 0 + 0 + x_3'',$$

or, in the matrix form,

$$x' = A_3(\psi) x''. \tag{1.6}$$

The matrix

$$A_3(\psi) = \begin{pmatrix} \cos\psi & -\sin\psi & 0 \\ \sin\psi & \cos\psi & 0 \\ 0 & 0 & 1 \end{pmatrix} \tag{1.7}$$

describes the rotation around the third axis at an angle ψ.

The transition from the system $Ox_1''x_2''x_3''$ to the system $Ox_1'''x_2'''x_3'''$ is effected by rotation around the first coordinate axis at the angle θ. The formulas of coordinates transformation, as it is seen from Fig. 4, are as follows:

$$x_1'' = x_1''' + 0 + 0,$$

$$x_2'' = 0 + x_2''' \cos\theta - x_3''' \sin\theta,$$

$$x_3'' = 0 + x_2''' \sin\theta + x_3''' \cos\theta,$$

or, in the matrix form,

$$x'' = A_1(\theta) x''', \tag{1.8}$$

where

$$A_1(\theta) = \begin{pmatrix} 1 & 0 & 0 \\ 0 & \cos\theta & -\sin\theta \\ 0 & \sin\theta & \cos\theta \end{pmatrix}.$$

The system $Ox_1'''x_2'''x_3'''$ is converted to the system $Ox_1x_2x_3$ by rotating at the angle φ around the third axis, and hence,

$$x''' = A_3(\varphi) x, \tag{1.9}$$

as this takes place, the matrix A_3 has the form of matrix (1.7).

Substituting relation (1.8), in which x''' is represented in the form (1.9), into relation (1.6), we obtain

$$x' = A_3(\psi) A_1(\theta) A_3(\varphi) x . \tag{1.10}$$

Comparing (1.5) and (1.10), we find that the required matrix A with components α_{sk} is a product of three rotation matrices

$$A(\psi, \theta, \varphi) = A_3(\psi) A_1(\theta) A_3(\varphi) =$$

$$= \begin{pmatrix} \cos\psi & -\sin\psi & 0 \\ \sin\psi & \cos\psi & 0 \\ 0 & 0 & 1 \end{pmatrix} \begin{pmatrix} 1 & 0 & 0 \\ 0 & \cos\theta & -\sin\theta \\ 0 & \sin\theta & \cos\theta \end{pmatrix} \begin{pmatrix} \cos\varphi & -\sin\varphi & 0 \\ \sin\varphi & \cos\varphi & 0 \\ 0 & 0 & 1 \end{pmatrix} =$$

$$= \begin{pmatrix} \begin{matrix}(\cos\psi\cos\varphi- \\ -\sin\psi\cos\theta\sin\varphi)\end{matrix} & \begin{matrix}(-\cos\psi\sin\varphi- \\ -\sin\psi\cos\theta\cos\varphi)\end{matrix} & \sin\psi\sin\theta \\[2mm] \begin{matrix}(\sin\psi\cos\varphi+ \\ +\cos\psi\cos\theta\sin\varphi)\end{matrix} & \begin{matrix}(-\sin\psi\sin\varphi+ \\ +\cos\psi\cos\theta\cos\varphi)\end{matrix} & -\cos\psi\sin\theta \\[2mm] \sin\theta\sin\varphi & \sin\theta\cos\varphi & \cos\theta \end{pmatrix} . \tag{1.11}$$

Let us go back to expression (1.2). Using relations (1.3) we can write (1.2) as follows:

$$\boldsymbol{\rho} = \boldsymbol{\rho}_0 + \mathbf{r} = \xi_s \boldsymbol{\varepsilon}_s = [\xi_{0s} + x_k \alpha_{sk}(\psi, \theta, \varphi)] \boldsymbol{\varepsilon}_s . \tag{1.12}$$

This vector equality can be represented in the matrix form

$$\xi = \xi_0 + A(\psi, \theta, \varphi) x .$$

Thus, in order to find the position of any given point of a e rigid body with respect to the frame $O_1\xi_1\xi_2\xi_3$, it is necessary to set the six quantities: ξ_{01}, ξ_{02}, ξ_{03}, ψ, θ, φ. That is why these quantities are called the *coordinates of a rigid body (solid)*, which uniquely define its position with respect to the fixed coordinate system.

2 Velocities and Accelerations of Points of a Rigid Body in the Case of General Motion

As it follows from formula (1.12), in order to define the law of motion of an arbitrary point of a rigid body, it is necessary to set the pole coordinates ξ_{01}, ξ_{02}, ξ_{03} and the Euler angles ψ, θ, φ as functions of time. By definition the velocity $\mathbf{v}$ of a point M

of a solid agrees with the time derivative of the radius vector $\boldsymbol{\rho}$ (see Fig. 1). Using expression (1.12) and the law of differentiation of a composite function of several variables to calculate this derivative, we find that

$$\mathbf{v} = \frac{d\boldsymbol{\rho}}{dt} = \frac{\partial\boldsymbol{\rho}}{\partial\xi_{01}}\dot{\xi}_{01} + \frac{\partial\boldsymbol{\rho}}{\partial\xi_{02}}\dot{\xi}_{02} + \frac{\partial\boldsymbol{\rho}}{\partial\xi_{03}}\dot{\xi}_{03} + \frac{\partial\boldsymbol{\rho}}{\partial\psi}\dot{\psi} + \frac{\partial\boldsymbol{\rho}}{\partial\theta}\dot{\theta} + \frac{\partial\boldsymbol{\rho}}{\partial\varphi}\dot{\varphi}.$$

Since

$$\frac{\partial\boldsymbol{\rho}}{\partial\xi_{0s}} = \boldsymbol{\varepsilon}_s, \qquad \frac{\partial\boldsymbol{\rho}}{\partial\psi} = \frac{\partial\mathbf{r}}{\partial\psi}, \qquad \frac{\partial\boldsymbol{\rho}}{\partial\theta} = \frac{\partial\mathbf{r}}{\partial\theta}, \qquad \frac{\partial\boldsymbol{\rho}}{\partial\varphi} = \frac{\partial\mathbf{r}}{\partial\varphi},$$

we have

$$\mathbf{v} = \dot{\xi}_{01}\boldsymbol{\varepsilon}_1 + \dot{\xi}_{02}\boldsymbol{\varepsilon}_2 + \dot{\xi}_{03}\boldsymbol{\varepsilon}_3 + \frac{\partial\mathbf{r}}{\partial\psi}\dot{\psi} + \frac{\partial\mathbf{r}}{\partial\theta}\dot{\theta} + \frac{\partial\mathbf{r}}{\partial\varphi}\dot{\varphi}. \tag{2.1}$$

Any such summand can be considered as the velocity of a given point in the particular case when only one chosen coordinate varies and all the rest have constant values. Suppose, for example, that only one coordinate ξ_{01} changes. In this case, the rigid body moves along the $O_1\xi_1$-axis (see Fig. 1) and all its points will the same velocity $\dot{\xi}_{01}\boldsymbol{\varepsilon}_1$, which is equal to the velocity of the poles. Now suppose that only the angle ψ changes; that is, the pole is fixed and the angles θ and φ are constant. Under these assumptions, the body rotates around the stationary axis Ox_3' (see Figs. 1 and 2). According to Sect. 3 of Chap. 1, we have in this case

$$\frac{\partial\mathbf{r}}{\partial\psi} = \mathbf{k} \times \mathbf{r}, \tag{2.2}$$

where $\mathbf{k}$ is the basis vector of that axis around which the body rotates as the angle ψ varies (that is, around the Ox_3'-axis). Therefore, $\mathbf{k} = \mathbf{i}_3'$. Analogously it can be shown that (see Fig. 2)

$$\frac{\partial\mathbf{r}}{\partial\theta} = \mathbf{i}_1'' \times \mathbf{r}, \qquad \frac{\partial\mathbf{r}}{\partial\varphi} = \mathbf{i}_3 \times \mathbf{r}. \tag{2.3}$$

Expressions (2.1)–(2.3) imply that with the fixed pole we get

$$\mathbf{v} = \frac{d\mathbf{r}}{dt} = \dot{\psi}\mathbf{i}_3' \times \mathbf{r} + \dot{\theta}\mathbf{i}_1'' \times \mathbf{r} + \dot{\varphi}\mathbf{i}_3 \times \mathbf{r} = \boldsymbol{\omega} \times \mathbf{r}, \tag{2.4}$$

where

$$\boldsymbol{\omega} = \dot{\psi}\mathbf{i}_3' + \dot{\theta}\mathbf{i}_1'' + \dot{\varphi}\mathbf{i}_3. \tag{2.5}$$

The above derivation of the Euler formula (2.4) shows that the vector $\boldsymbol{\omega}$ is independent of the choice of the vector $\mathbf{r}$. In (2.4), the vectors $\mathbf{v}$ and $\mathbf{r}$ are invariant with respect to the choice of a coordinate system. As a result, the vector $\boldsymbol{\omega}$ is also independent of the coordinates that are taken for describing the rotation of a rigid body. We shall show that this vector is also independent of the choice of a pole.

The sum of the first three summands in expression (2.1) is equal to the pole velocity $\mathbf{v}_0 = d\boldsymbol{\rho}_0/dt$. Therefore, in the general case we have

$$\mathbf{v} = \mathbf{v}_0 + \boldsymbol{\omega} \times \mathbf{r}. \tag{2.6}$$

Let us take two different poles O_1 and O_2. Then velocity of an arbitrary point M can be written in the form

$$\mathbf{v} = \mathbf{v}_{10} + \boldsymbol{\omega}_1 \times \mathbf{r}_1, \tag{2.7}$$

and in the form

$$\mathbf{v} = \mathbf{v}_{20} + \boldsymbol{\omega}_2 \times \mathbf{r}_2. \tag{2.8}$$

We note that even though the angular velocities $\boldsymbol{\omega}_1$ and $\boldsymbol{\omega}_2$ in these expressions are assumed to be different, they are independent of the choice of a point M.

Taking into account that the vectors $\mathbf{r}_1$ and $\mathbf{r}_2$ are related as

$$\mathbf{r}_1 = \overrightarrow{O_1 O_2} + \mathbf{r}_2,$$

we have

$$\mathbf{v} = \mathbf{v}_{10} + \boldsymbol{\omega}_1 \times (\overrightarrow{O_1 O_2} + \mathbf{r}_2) = \mathbf{v}_{20} + \boldsymbol{\omega}_2 \times \mathbf{r}_2. \tag{2.9}$$

By expression (2.7) the velocity of the second pole can be written as

$$\mathbf{v}_{20} = \mathbf{v}_{10} + \boldsymbol{\omega}_1 \times \overrightarrow{O_1 O_2}, \tag{2.10}$$

and hence, comparing formulas (2.8) and (2.9), this gives

$$(\boldsymbol{\omega}_1 - \boldsymbol{\omega}_2) \times \mathbf{r}_2 = 0.$$

This equality should be satisfied for any vector $\mathbf{r}_2$. This is possible only in the case when

$$\boldsymbol{\omega}_1 = \boldsymbol{\omega}_2 = \boldsymbol{\omega},$$

the result required. Thus, the angular velocity is invariant with respect to the choice of a pole.

The expression (2.10) in which one should put $\boldsymbol{\omega}_1 = \boldsymbol{\omega}$ shows that in transiting from one pole to another one there changes only the component of the pole velocity vector that is perpendicular to vector $\boldsymbol{\omega}$. In other words, the scalar product $\mathbf{v}_0 \cdot \boldsymbol{\omega}$ is also invariant, inasmuch as

$$(\boldsymbol{\omega} \times \overrightarrow{O_1 O_2}) \cdot \boldsymbol{\omega} = 0,$$

with arbitrary vector $\overrightarrow{O_1 O_2}$.

We write the velocity of the pole O as

$$\mathbf{v}_0 = \mathbf{v}_0^* + \mathbf{v}_0^{**} \,,$$

where $\mathbf{v}_0^{**} \cdot \boldsymbol{\omega} = 0$ and $\mathbf{v}_0^* = \lambda \boldsymbol{\omega}$ is an invariant vector. The pole O', whose velocity is $\mathbf{v}_0^*$, can be found using (2.6) from the equation

$$\overrightarrow{OO'} \times \boldsymbol{\omega} = \mathbf{v}_0 - \mathbf{v}_0^* \,. \tag{2.11}$$

This is a vector equation of the straight line that contains the sought-for points O'.

Thus, in the general case of motion of a rigid body with $\boldsymbol{\omega} \neq 0$ there exists a straight line in the body on which points have the same velocity $\mathbf{v}_0^*$ in the fixed moment of time. In particular, this velocity may well be zero.

Now let us find the acceleration of the point M. Using (2.6) and (2.4), this gives

$$\mathbf{w} = \frac{d\mathbf{v}}{dt} = \mathbf{w}_0 + \frac{d\boldsymbol{\omega}}{dt} \times \mathbf{r} + \boldsymbol{\omega} \times (\boldsymbol{\omega} \times \mathbf{r}) \,, \tag{2.12}$$

where $\mathbf{w}_0 = d\mathbf{v}_0/dt$ is the pole acceleration. Usually, here one introduces the vector $\boldsymbol{\varepsilon} = d\boldsymbol{\omega}/dt$, which is known as the *angular acceleration*. Formulae (2.12) represent *the Rivals theorem*.

3   Elemental Types of the Motion of a Rigid Body

Translational motion. Assume that any straight line within the body (and hence, the axes Ox_1, Ox_2, Ox_3) moves parallel to itself at any time of the motion. In this case, the vector $\boldsymbol{\omega}$ is identically equal to zero, and hence by (2.6) and (2.12) the velocities and accelerations of all points at any given time are the same:

$$\mathbf{v} = \mathbf{v}_0 \,, \qquad \mathbf{w} = \mathbf{w}_0 \,. \tag{3.1}$$

Such type of motion is called *translational*.

Rotation about a fixed axis. Suppose the pole velocity is zero and that the direction of vector $\boldsymbol{\omega}$ is stationary and coincides with that of the Ox_3-axis. Hence

$$\boldsymbol{\omega} = \dot{\varphi}\mathbf{i}_3 \,, \qquad \boldsymbol{\varepsilon} = \dot{\boldsymbol{\omega}} = \ddot{\varphi}\mathbf{i}_3 \,. \tag{3.2}$$

The trajectories of points are circles, whose centers lie on the Ox_3-axis (see Fig. 5). It is convenient to represent the radius vector $\mathbf{r}$ of a point as

$$\mathbf{r} = x_3\mathbf{i}_3 + R\,\mathbf{R}^0 \,, \qquad R = \sqrt{x_1^2 + x_2^2} \,, \qquad |\mathbf{R}^0| = 1 \,, \tag{3.3}$$

where R is the radius of the circle along which the point M moves.

The unit vector

Fig. 5 Rotation of a rigid
body about a fixed axis

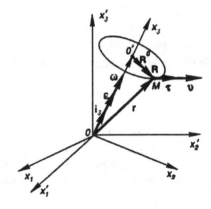

$$\tau = \mathbf{i}_3 \times \mathbf{R}^0 \qquad (3.4)$$

is directed tangentially to the circle.

In view relations (3.2)–(3.4) and since $\mathbf{v}_0 = \mathbf{w}_0 = 0$ the general formulas (2.6)
and (2.12) can be written in the form

$$\mathbf{v} = \dot{\varphi} R \tau \,, \qquad\qquad\qquad v = |\mathbf{v}| = |\omega| R \,,$$

$$\mathbf{w} = \ddot{\varphi} R \tau + \dot{\varphi}^2 R (\mathbf{i}_3 \times \tau) = \ddot{\varphi} R \tau - \dot{\varphi}^2 R \mathbf{R}^0 \,, \qquad w = |\mathbf{w}| = R \sqrt{\ddot{\varphi}^2 + \dot{\varphi}^4} \,.$$
$$(3.5)$$

The velocity is directed tangentially to the circle and is equal to $|\omega| R$.

The acceleration vector consists of two summands. The first summand $\ddot{\varphi} R \tau = \mathbf{w}_\tau$ is called the *tangential acceleration* and the quantity $\ddot{\varphi}$ is called the *angular acceleration*. The second summand $-\dot{\varphi}^2 R \mathbf{R}^0 = -\dot{\varphi}^2 \mathbf{R} = \mathbf{w}_n$, $\mathbf{R} = \overrightarrow{O'M}$, is directed to the center of the circle and is called the *centripetal acceleration*.

Screw motion. Assume that the direction of the pole velocity $\mathbf{v}_0$ and the vector ω coincide and that this direction does not change. Such a motion is called a *screw motion*. The angular velocity ω will be assumed to be given in the form (3.2). We shall also assume that the pole motion is such that

$$\mathbf{v}_0 = v_{03} \mathbf{i}_3 \,, \qquad \mathbf{w}_0 = w_{03} \mathbf{i}_3 \,.$$

These vectors are orthogonal, respectively, to the velocity and acceleration vectors, as given by (3.5), and hence the sums (2.6) and (2.12) can be easily found.

In the previous section, it was shown that in the case of general motion of a rigid body at any fixed time moment there is an axis in the body at points of which the velocities are the same and parallel to the vector ω. Hence, at a fixed time the solid has the same field of velocities as in the case of a screw motion. That is why this motion is called an *instantaneous screw motion*, the axis corresponding to such a motion is called an *instantaneous screw axis*.

We note that in the general case of motion the vectors $d\omega/dt$ and $\mathbf{w}_0$ are not collinear, respectively, to the vectors ω and $\mathbf{v}_0$. Hence, in the general case of motion the field of accelerations of a rigid body is not so simple as in the general case of a screw motion.

The loci of instantaneous screw axes in the fixed and movable spaces are called, respectively, the *fixed* and *moving (movable) axodes (axoids)*. These axodes are ruled surfaces. According to the distribution of velocities of a rigid body described in the previous subsection, the general case of motion of a rigid body can be interpreted as a rolling about the common generator of a moving axode over the fixed one with the angular velocity ω and simultaneous slipping along the instantaneous screw axis with the velocity $\mathbf{v}_0^*$.

4 Rotation of a Rigid Body About a Fixed Point

From formulas (2.6) and (2.12) for the velocities and accelerations of a rigid body in the general case of its motion and from formulas (3.1) in case of translational motion we can conclude that the motion of a free rigid body occurs in a way as if it was moved translationally together with the pole rotated simultaneously about the pole as a fixed point. In other words, the motion of a free rigid body splits into two motions, which can be studied independently from the kinematic point of view. Therefore, the study of the rotation about a fixed point of a rigid body is important not only in connection with problems arising in examining the rotor and gyroscope operations, but also in the study of all types of motion of a the rigid body, which are different from the translational motion.

From (2.6) and (2.12), it follows that the velocities and accelerations of a rigid body when it rotates about a fixed point can be calculated by the formulas

$$\mathbf{v} = \omega \times \mathbf{r}, \qquad \mathbf{w} = \frac{d\omega}{dt} \times \mathbf{r} + \omega \times (\omega \times \mathbf{r}). \qquad (4.1)$$

Expressions (4.1) show that the motion of a rigid body under consideration is determined by specifying the vector function $\omega = \omega(t)$. Translational motion is also defined by one vector function $\mathbf{v}_0 = \mathbf{v}_0(t)$. From this point of view these two motions do not differ. The principal difference is manifested especially clearly in composing the differential equations of motion of a rigid body. This difference is also manifested when defining the motion law from the functions $\omega(t)$ and $\mathbf{v}_0(t)$. Let us consider this question. .

In the case of a translational motion, the motion law from the function $\mathbf{v}_0(t)$, as given in the form

$$\mathbf{v}_0(t) = \dot{\xi}_{0s}(t)\boldsymbol{\varepsilon}_s \, ,$$

is derived by direct integration of the functions $\dot{\xi}_{0s}(t)$:

$$\xi_{0s}(t) = \xi_{0s}(t_0) + \int_{t_0}^{t} \dot{\xi}_{0s}(t')\, dt', \qquad s = 1, 2, 3.$$

We now suppose that the instantaneous angular velocity $\boldsymbol{\omega}(t)$ is known,

$$\boldsymbol{\omega}(t) = \omega_s'(t)\,\boldsymbol{\varepsilon}_s = \omega_s'(t)\mathbf{i}_s', \tag{4.2}$$

and it is required to find the Euler angles as time functions.

The vector $\boldsymbol{\omega}(t)$ is expressed in terms of the derivatives of the Euler angles by formula (2.5). The vectors $\mathbf{i}_1''$ and $\mathbf{i}_3$, which appear in this formula, can be expanded in the basis $\{\mathbf{i}_s'\}$. Figure 3 shows that the unit vector $\mathbf{i}_1''$ can be written as

$$\mathbf{i}_1'' = \cos\psi\,\mathbf{i}_1' + \sin\psi\,\mathbf{i}_2'. \tag{4.3}$$

Using formulas (1.3) and (1.11), we discover that

$$\mathbf{i}_3 = \alpha_{s3}\mathbf{i}_s' = \sin\psi\sin\theta\,\mathbf{i}_1' - \cos\psi\sin\theta\,\mathbf{i}_2' + \cos\theta\,\mathbf{i}_3'. \tag{4.4}$$

Substituting expressions (4.3) and (4.4) into formula (2.5), we get

$$\boldsymbol{\omega} = \dot{\psi}\mathbf{i}_3' + \dot{\theta}(\cos\psi\,\mathbf{i}_1' + \sin\psi\,\mathbf{i}_2') + \dot{\varphi}(\sin\psi\sin\theta\,\mathbf{i}_1' - \cos\psi\sin\theta\,\mathbf{i}_2' + \cos\theta\,\mathbf{i}_3').$$

Hence and from formula (4.2) it follows that

$$\begin{aligned}
\omega_1' &= \dot{\varphi}\sin\psi\sin\theta + \dot{\theta}\cos\psi, \\
\omega_2' &= -\dot{\varphi}\cos\psi\sin\theta + \dot{\theta}\sin\psi, \\
\omega_3' &= \dot{\varphi}\cos\theta + \dot{\psi}.
\end{aligned} \tag{4.5}$$

This system of differential equations for the unknowns functions $\varphi(t)$, $\psi(t)$, $\theta(t)$ with arbitrary $\omega_1'(t)$, $\omega_2'(t)$, $\omega_3'(t)$ can be integrated only numerically. We note that with $\theta = 0$ and $\theta = \pi$ the derivatives $\dot{\varphi}$, $\dot{\psi}$, $\dot{\theta}$ cannot be uniquely expressed from the quantities ω_1', ω_2', ω_3'. This means that the case when during its motion a rigid body begins to approach the position, where the angle θ is 0 or π involves a special consideration.

Let us again get back to expressions (4.5) and write them in the differential form

$$\begin{aligned}
\omega_1'\, dt &= \sin\psi\sin\theta\, d\varphi + \cos\psi\, d\theta, \\
\omega_2'\, dt &= -\cos\psi\sin\theta\, d\varphi + \sin\psi\, d\theta, \\
\omega_3'\, dt &= \cos\theta\, d\varphi + d\psi.
\end{aligned}$$

We claim that the quantities $\omega_s'\, dt$ cannot be considered as the differentials of new angles q^σ ($\sigma = 1, 2, 3$), which uniquely related with the Euler angles. Indeed, if for example, the quantity $\omega'_1 \, dt$ is the total differential of the function $q^1(\varphi, \psi, \theta)$, then the expression

$$dq^1 = \frac{\partial q^1}{\partial \varphi} \, d\varphi + \frac{\partial q^1}{\partial \psi} \, d\psi + \frac{\partial q^1}{\partial \theta} \, d\theta$$

implies that

$$\frac{\partial^2 q^1}{\partial \varphi \, \partial \psi} = \frac{\partial^2 q^1}{\partial \psi \, \partial \varphi}, \qquad \frac{\partial^2 q^1}{\partial \theta \, \partial \varphi} = \frac{\partial^2 q^1}{\partial \varphi \, \partial \theta}, \qquad \frac{\partial^2 q^1}{\partial \theta \, \partial \psi} = \frac{\partial^2 q^1}{\partial \psi \, \partial \theta}.$$

The coefficients of the differentials $d\varphi$, $d\psi$, $d\theta$ in the expression for $\omega'_1 \, dt$ fail to have these properties, inasmuch as

$$\frac{\partial}{\partial \psi}(\sin \psi \sin \theta) \neq \frac{\partial}{\partial \varphi}(0), \qquad \frac{\partial}{\partial \theta}(\sin \psi \sin \theta) \neq \frac{\partial}{\partial \varphi}(\cos \psi),$$

$$\frac{\partial}{\partial \theta}(0) \neq \frac{\partial}{\partial \psi}(\sin \psi).$$

Thus, the quantity $\omega'_1 \, dt$, as was claimed, is not the differential of some new angle q^1. This is the principal difference between the projections of the instantaneous angular velocity ω to the fixed axes and the projections of the velocity $\mathbf{v}_0$ of the pole to the same axes. This also results in some difficulties when changing from the vector ω to the law of motion.

In Part II it will be shown that in the differential equations describing the above case of the motion of a rigid body it is expedient to consider as unknowns the projections of the vector ω to axes x_1, x_2, x_3, which are rigidly connected with the body. Denoting these projections, respectively, by ω_1, ω_2, ω_3, we get

$$\omega = \omega_k \mathbf{i}_k \,.$$

Hence and from (2.5) it follows that in order to express the quantities ω_k in terms of the Euler angles, the vectors $\mathbf{i}'_3$ and $\mathbf{i}''_1$ should be expanded in the basis $\{\mathbf{i}_k\}$. From (1.3) and (1.11) it follows that

$$\mathbf{i}'_3 = \alpha_{3k} \mathbf{i}_k = \sin \theta \sin \varphi \mathbf{i}_1 + \sin \theta \cos \varphi \mathbf{i}_2 + \cos \theta \mathbf{i}_3 \,.$$

It is seen from Fig. 6 that the unit vector $\mathbf{i}''_1$ can be written as

$$\mathbf{i}''_1 = \cos \varphi \mathbf{i}_1 - \sin \varphi \mathbf{i}_2 \,.$$

Substituting these expressions into formula (2.5), we get

Fig. 6 Expansion of the
vectors $\mathbf{i}'_3$ and $\mathbf{i}''_1$

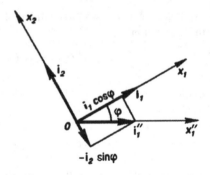

$$\omega_1 = \dot{\psi}\sin\theta\sin\varphi + \dot{\theta}\cos\varphi,$$
$$\omega_2 = \dot{\psi}\sin\theta\cos\varphi - \dot{\theta}\sin\varphi, \qquad (4.6)$$
$$\omega_3 = \dot{\psi}\cos\theta + \dot{\varphi}.$$

We find the projections of the velocity vector of an arbitrary point of the body to axes of the fixed coordinate system by using formula (4.1) and the representation of the vector product in terms of the determinant

$$\mathbf{v} = \boldsymbol{\omega} \times \mathbf{r} = \begin{vmatrix} \mathbf{i}_1 & \mathbf{i}_2 & \mathbf{i}_3 \\ \omega_1 & \omega_2 & \omega_3 \\ x_1 & x_2 & x_3 \end{vmatrix}.$$

Hence
$$v_1 = \omega_2 x_3 - \omega_3 x_2,$$
$$v_2 = \omega_3 x_1 - \omega_1 x_3, \qquad (Euler's\ formulas)$$
$$v_3 = \omega_1 x_2 - \omega_2 x_1.$$

The points lying on the instantaneous rotation axis (that is, the points for which $\mathbf{r} = \lambda\boldsymbol{\omega}$) have zero velocity. Therefore, the equation of instantaneous axis takes the following form for the movable and fixed coordinate systems, respectively,

$$\frac{x_1}{\omega_1} = \frac{x_2}{\omega_2} = \frac{x_3}{\omega_3} = \lambda, \qquad \frac{x'_1}{\omega'_1} = \frac{x'_2}{\omega'_2} = \frac{x'_3}{\omega'_3} = \lambda, \qquad -\infty < \lambda < \infty.$$

Suppose that the angular velocity is known as a function of time. In this case, these equations can be looked upon as a parametric representation of moving and fixed axodes (axoids), where time t is a parameter. If a rigid body rotates about a fixed point, the axoids appear to be conical surfaces with vertices at the origin. In this case, the motion of a rigid body can be considered as a roll without slipping of a movable axode over the fixed one.

Rotation tensor.[2] To conclude this section, we explain the above formula using the concept of a tensor. The above third-order matrix A with entries α_{ks} is called the matrix of the *rotation tensor*. Using this matrix, one is capable of presenting the material of this subsection in the matrix-vector language. We consider the column vectors (1.5) $x = (x_1, x_2, x_3)^T$ and $x' = (x'_1, x'_2, x'_3)^T$, as composed from the components of the radius vector of a point in the old and new systems of coordinates. Now (1.5) can be written as follows:

$$x' = A \cdot x, \quad x = A^T \cdot x' = A^{-1} \cdot x'. \tag{4.7}$$

A number of properties of the matrix A, which follow from the relations $\alpha_{ks} = i_k \cdot i'_s$, was formulated above: the sum of squared entries in each row and in each column of the matrix A is 1, the rows and columns are pairwise orthogonal. The inverse matrix A agrees with the transposed matrix:

$$A^T = A^{-1}. \tag{4.8}$$

From (4.8) it follows that the determinant of A is 1. Indeed,

$$1 = \mathrm{Det}[E] = \mathrm{Det}[A \cdot A^{-1}] = \mathrm{Det}[A] \cdot \mathrm{Det}[A^T] = (\mathrm{Det}[A])^2, \tag{4.9}$$

where E is the identity matrix. Besides, from (4.9) we get $\mathrm{Det}[A] = \pm 1$. However, it should be assumed that $\mathrm{Det}[A] = 1$, because in ab initio with $x = x'$ we have $A = E$, $\mathrm{Det}[A] = 1$, and the quantity $\mathrm{Det}[A]$ varies continuously in the course of motion.

If $x \neq x'$, then the new position of the body can be obtained from the old one by rotating about some axis by some angle. The position of the axis can be determined from the condition that it is fixed: $A \cdot x = x$. This linear system has nonzero solution if one of the eigenvalues of the matrix A is zero. This follows from the chain of equalities:

$$A - E = A - A \cdot A^{-1} = A \cdot (E - A^T) \quad \rightarrow \quad \mathrm{Det}[A - E] = \mathrm{Det}[A] \cdot \mathrm{Det}[E - A].$$

Rotations of a body form a non-abelian group. Two successive rotations with matrices A_1 and A_2 produce the rotation with the matrix $A = A_2 \cdot A_1$. This observation was used in the derivation of relations (1.11).

Let us find the time derivative of the rotation tensor. Assuming the vector x to be constant in the first equality in (4.7), we differentiate this equality in time: $v' = \dot{A} \cdot x$. On the other hand, by the Euler formula $v' = \omega \times x'$. We write down this formula in the matrix form

---

[2] The subsection "Rotation tensor" is written by P.E. Tovstik.

$$v' = S \cdot x', \qquad S = \begin{pmatrix} 0 & -\omega_3 & \omega_2 \\ \omega_3 & 0 & -\omega_1 \\ -\omega_2 & \omega_1 & 0 \end{pmatrix}. \qquad (4.10)$$

The skew-symmetric matrix S, which is related with the angular velocity ω, is called the *spin matrix*. We also have

$$v' = \dot{A} \cdot x = \dot{A} \cdot A^T \cdot x'. \qquad (4.11)$$

Comparing expressions (4.10) and (4.11) with arbitrary x, we see that $S = \dot{A} \cdot A^T$, and so

$$\dot{A} = S \cdot A.$$

This equality is called the *Poisson equation*. Given the angular velocity $\omega(t)$, from this equation via numerical integration one can find the components of the rotation tensor at any time. It is advantageous in comparison with the system of equation (4.6) in that it has no singular points. As was already pointed out, Eqs. (4.6) have singular points with $\theta = 0, \pi$.

5 Planar Motion

The *planar motion* of a rigid body is a motion in which the distance between points of this body and a given plane π is fixed. As a result, the velocity vectors of the points of a rigid body are parallel to the plane π and the angular velocity vector is perpendicular to it, because only in this case the second summand in formula (2.6) is always a vector that is parallel to the plane π (Fig. 7).

We choose a coordinate system $Oxyz$ connected with the body so that the z-axis be perpendicular to the plane π. Using the representation of the vector product in terms of determinants, formula (2.6) can be written in the form

$$\mathbf{v} = \mathbf{v}_0 + \begin{vmatrix} \mathbf{i} & \mathbf{j} & \mathbf{k} \\ 0 & 0 & \omega_z \\ x & y & z \end{vmatrix}.$$

Hence

$$v_x = v_{0x} - \omega_z y, \qquad v_y = v_{0y} + \omega_z x, \qquad v_z = 0.$$

From these formulas, it is seen that in every cross section of the body by the plane $z = $ const the velocity distribution is the same. That is why in the sequel we shall speak only about the cross-section of the body by the plane $z = 0$. Note that if the point O^* with zero velocity is taken as a pole, then the velocity distribution will be such as if the body would be rotating at this instant about the axis passing through the point O^* (Fig. 7). This point is called the *instantaneous velocity center*.

Fig. 7 Planar motion of a
rigid body

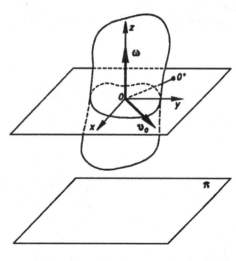

Fig. 8 The instantaneous
velocity center in the fixed
coordinates

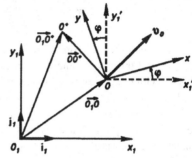

Its coordinates are obtained from Eq. (2.11) on putting $\mathbf{v}_0^* = 0$ and $O' = O^*$ in it.
Let $\overrightarrow{OO^*} = x^*\mathbf{i} + y^*\mathbf{j}$, then Eq. (2.11) is written in the form

$$\begin{vmatrix} \mathbf{i} & \mathbf{j} & \mathbf{k} \\ x^* & y^* & 0 \\ 0 & 0 & \omega_z \end{vmatrix} = v_{0x}\mathbf{i} + v_{0y}\mathbf{j}\,,$$

where

$$x^* = -\frac{v_{0y}}{\omega_z}\,, \qquad y^* = \frac{v_{0x}}{\omega_z}\,. \tag{5.1}$$

Let us find the position of the instantaneous velocity center with respect to the
fixed system $O_1 x_1 y_1$ (Fig. 8). We have

$$\overrightarrow{O_1 O^*} = x_1^*\mathbf{i}_1 + y_1^*\mathbf{j}_1 = \overrightarrow{O_1 O} + \overrightarrow{OO^*} = x_{10}\mathbf{i}_1 + y_{10}\mathbf{j}_1 + \overrightarrow{OO^*}\,.$$

The projections of the vector $\overrightarrow{OO^*}$ to the fixed axes are obtained from the equation

Fig. 9 Finding the
instantaneous velocity center

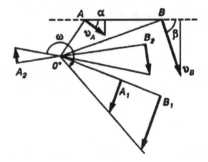

$$\begin{vmatrix} \mathbf{i}_1 & \mathbf{j}_1 & \mathbf{k}_1 \\ x_1'^* & y_1'^* & 0 \\ 0 & 0 & \omega_z \end{vmatrix} = \dot{x}_{10}\mathbf{i}_1 + \dot{y}_{10}\mathbf{j}_1 \,.$$

Thus,

$$x_1^* = x_{10} - \frac{\dot{y}_{10}}{\omega_z}, \qquad y_1^* = y_{10} + \frac{\dot{x}_{10}}{\omega_z}\,. \tag{5.2}$$

If we mark the position of the instantaneous velocity center on the moving plane $O_1x_1y_1$ and the fixed plane Oxy at different time instants, then we obtain the two curves, which are known as the *(body) moving* and *space (fixed) centroids*, respectively. Their equations can be found from systems (5.1) and (5.2).

Let us consider two arbitrary points A and B. Assume that the direction and the velocity of the point A are given. Regarding the point B, it is known along which line its velocity is directed. Consider the perpendiculars to the directions of the velocities at the points A and B (Fig. 9). If the velocity direction at the point A is not perpendicular to the segment AB and is not parallel to the velocity direction at the point B, then these perpendiculars will intersect at the instantaneous velocity center. The modulus of ω_z will be found as the ratio v_A/O^*A. If the velocity directions at the points A and B are parallel to the segment AB, then $O^*A = \infty$ and $\omega_z = 0$. It means that at a given time the body motion is translational and the velocity of an arbitrary point M is equal to the velocity of the point A.

When solving plane motion problems it is usually convenient to use the fact that the projections of the velocity of the points A and B to the direction of the interval AB are equal. Indeed, taking the point A as a pole, we see from formula (2.6) that the velocity of the point B is as follows:

$$\mathbf{v}_B = \mathbf{v}_A + \boldsymbol{\omega} \times \overrightarrow{AB}\,. \tag{5.3}$$

Multiplying both parts of the equality by the unit vector $\overrightarrow{AB}/AB$ and taking into account that the vector $\boldsymbol{\omega} \times \overrightarrow{AB}$ is perpendicular to $\overrightarrow{AB}$, we obtain

$$(\mathbf{v}_A \cdot \overrightarrow{AB})/AB = (\mathbf{v}_B \cdot \overrightarrow{AB})/AB\,,$$

Fig. 10 Acceleration of a
point in plane motion

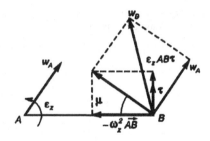

that is, $v_A \cos \alpha = v_B \cos \beta$ (see Fig. 9)

Now let us consider the case when the velocities of the points A_1 and B_1 are
perpendicular to the interval $A_1 B_1$ and have the same direction (Fig. 9). The position
of instantaneous velocity center and the modulus of the angular velocity are found
from the proportion

$$\frac{v_{A_1}}{O^* A_1} = \frac{v_{B_1}}{O^* B_1} = \frac{v_{B_1} - v_{A_1}}{A_1 B_1} = |\omega_z| .$$

If the velocities of the points A_2 and B_2 are perpendicular to the interval $A_2 B_2$ and
have opposite direction, then the instantaneous velocity center lies on the interval
$A_2 B_2$, and besides,

$$\frac{v_{A_2}}{O^* A_2} = \frac{v_{B_2}}{O^* B_2} = \frac{v_{A_2} + v_{B_2}}{A_2 B_2} = |\omega_z| .$$

The rotation direction is found directly from the directions of the velocities of the
points A_2 and B_2.

The general formula (1.12) shows that the acceleration of an arbitrary point M
during a plane-and-parallel motion of a rigid body is composed of the pole acceleration and the acceleration of the rotational motion about the pole. Let us take an
arbitrary point A for a pole. Hence, in accordance with formulas (2.12) and (3.5) the
accelerations of the points A and B are related as follows:

$$\mathbf{w}_B = \mathbf{w}_A + \varepsilon_z A B \, \boldsymbol{\tau} - \omega_z^2 \overrightarrow{AB} , \tag{5.4}$$

here $\varepsilon_z = \dot{\omega}_z = \ddot{\varphi}$ is the angular acceleration and $\boldsymbol{\tau}$ is the unit vector perpendicular
to the interval $\overrightarrow{AB}$. If the beginning of the vector $\boldsymbol{\tau}$ is translated to the point B, then
it will be oriented as counterclockwise rotation of the interval AB about the point A
(Fig. 10).

From expression (5.4) it follows that the point B may agree with such a point Q
for which $\mathbf{w}_Q = 0$. The position of the point Q, which is called the *instantaneous
acceleration center*, can be found from the equation

$$\mathbf{w}_A = \omega_z^2 \overrightarrow{AQ} - \varepsilon_z A Q \boldsymbol{\tau} , \tag{5.5}$$

Fig. 11 Finding the
instantaneous acceleration
center

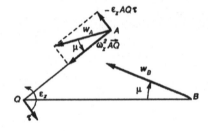

in which the quantities $\mathbf{w}_A$, ε_z and ω_z^2 are given.

Considering Eq. (5.5) a an expansion of the vector $\mathbf{w}_A$ in two orthogonal directions (Fig. 11), we find that

$$|\overrightarrow{AQ}| = \frac{w_A}{\sqrt{\varepsilon_z^2 + \omega_z^4}}.$$

The direction of the vector $\overrightarrow{AQ}$ is found by rotating the vector $\mathbf{w}_A$ about the point A at the angle $\mu = \arctan(\varepsilon/\omega_z^2)$, $(-\pi/2 < \mu < \pi/2)$, where it is assumed as usual that a counterclockwise rotation is positive (Fig. 11).

The acceleration of an arbitrary point B is proportional to the distance from it to the instantaneous acceleration center Q and is equal to $QB\sqrt{\varepsilon_z^2 + \omega_z^4}$. The angle between the vector $\mathbf{w}_B$ and the vector $\overrightarrow{BQ}$ is μ in view of the sign of ε_z (Fig. 11).

Thus, the field of accelerations in a plane-parallel motion structurally corresponds to the field of accelerations when a rigid body rotates about a fixed axis. The difference is that the axis along which the acceleration is equal to zero takes a new position at every instant of time, both in the rigid body and in the space. In general, this axis does not agree with the axis on which the points have zero velocities at a fixed instant of time. The instantaneous velocity center O^* and the instantaneous acceleration center Q represent in general two different points, which coincide only in the case of a rotation of a body about a fixed axis.

Chapter 3
Composite Motion

N. N. Polyakhov, M. P. Yushkov ⓘ, and S. A. Zegzhda

We shall first study the composite motion of a particle. It will be shown that the description of the absolute motion of a particle can be related to that of simpler motions such as the relative motion and a number of bulk (transport) ones. The results obtained for the composite motion of a particle are applied to examine the composition of motion of a rigid body. We shall consider the composition of translational motions of a rigid body and rotational ones, including a couple of rotations.

1 Composite Motion of a Particle

Assume that we are given two coordinate systems are given: the fixed $O_1\xi_1\xi_2\xi_3$-system and the movable $Ox_1x_2x_3$-system (see Fig. 1 of Chap. 2). In this case, the motion of a particle M in the $Ox_1x_2x_3$-system is called the *relative motion* and in the $O_1\xi_1\xi_2\xi_3$-system the motion is called the *absolute* one. The absolute motion of a particle M is sometimes called the *composite motion*, because it can be looked upon as that resulting from the particle motion with respect to the moving space $Ox_1x_2x_3$ and from the motion of this space together with the particle with respect to the $O_1\xi_1\xi_2\xi_3$-system.

If at a fixed time we imaginarily remove the relative motion of the particle M, then it will still move in the $O_1\xi_1\xi_2\xi_3$-system, because of the motion of the $Ox_1x_2x_3$-system itself. This imaginary motion of the particle M at a given time instant is called the *bulk (transport) motion*. This motion coincides with that of a point in the moving system with which coincides with the particle M at a fixed instant of time. As a result, the velocity and acceleration of a particle under its bulk motion can be calculated by formulas (2.6), (2.12) of Chap. 2. Thus, we have

$$\mathbf{v}_e = \mathbf{v}_0 + \boldsymbol{\omega} \times \mathbf{r}, \tag{1.1}$$

© Springer Nature Switzerland AG 2021

N. N. Polyakhov et al., *Rational and Applied Mechanics*,

Foundations of Engineering Mechanics,

https://doi.org/10.1007/978-3-030-64061-3_3

$$\mathbf{w}_e = \mathbf{w}_0 + \frac{d\boldsymbol{\omega}}{dt} \times \mathbf{r} + \boldsymbol{\omega} \times (\boldsymbol{\omega} \times \mathbf{r}). \tag{1.2}$$

These vectors have with the index e (from the French word *entraîner*, meaning to transport, carry with it).

When calculating the particle velocity and acceleration both in its relative and absolute motions, we shall use the representation of these motions in the Cartesian coordinates. Thus, in accordance with Fig. 1 of Chap. 2 and formulas (1.5), (1.6) of Chap. 2, we have

$$\mathbf{v}_r = \dot{x}_1 \mathbf{i}_1 + \dot{x}_2 \mathbf{i}_2 + \dot{x}_3 \mathbf{i}_3, \tag{1.3}$$

$$\mathbf{w}_r = \ddot{x}_1 \mathbf{i}_1 + \ddot{x}_2 \mathbf{i}_2 + \ddot{x}_3 \mathbf{i}_3, \quad \mathbf{v}_a = \dot{\xi}_1 \mathbf{i}_1' + \dot{\xi}_2 \mathbf{i}_2' + \dot{\xi}_3 \mathbf{i}_3', \quad \mathbf{w}_a = \ddot{\xi}_1 \mathbf{i}_1' + \ddot{\xi}_2 \mathbf{i}_2' + \ddot{\xi}_3 \mathbf{i}_3'. \tag{1.4}$$

The indices r and a originate from French words *relative* and *absolute*.

The question arises consider as to how the vectors $\mathbf{v}_a$, $\mathbf{v}_r$, and $\mathbf{v}_e$ are related with each other. By the definition of the velocity in the $O_1\xi_1\xi_2\xi_3$-system we have

$$\mathbf{v}_a = \dot{\boldsymbol{\rho}} = \frac{d}{dt}(\boldsymbol{\rho}_0 + \mathbf{r}) = \mathbf{v}_0 + \frac{d\mathbf{r}}{dt}. \tag{1.5}$$

The vector $\mathbf{r}$ varies in time due because its coordinates change in the moving coordinate system and because of the motion of the moving coordinate system with respect to the fixed one. Therefore,

$$\frac{d\mathbf{r}}{dt} = \frac{d}{dt}(x_k \mathbf{i}_k) = \dot{x}_k \mathbf{i}_k + x_k \frac{d\mathbf{i}_k}{dt}.$$

Since by the Euler formula (2.4) of Chap. 2

$$\frac{d\mathbf{i}_k}{dt} = \boldsymbol{\omega} \times \mathbf{i}_k,$$

we have

$$x_k \frac{d\mathbf{i}_k}{dt} = \boldsymbol{\omega} \times x_k \mathbf{i}_k = \boldsymbol{\omega} \times \mathbf{r}.$$

Thus,

$$\frac{d\mathbf{r}}{dt} = \dot{x}_k \mathbf{i}_k + \boldsymbol{\omega} \times \mathbf{r}. \tag{1.6}$$

This formula can be applied not only to the vector $\mathbf{r}$, but also to any vector $\mathbf{a}$ given in two coordinate systems, one of which moves in a defined way with respect to another system. As applied to the vector $\mathbf{a}$ formula (1.6) can be written in the form

$$\frac{d\mathbf{a}}{dt} = \frac{d^*\mathbf{a}}{dt} + \boldsymbol{\omega} \times \mathbf{a}. \tag{1.7}$$

Here ω is the instantaneous angular velocity of the moving system, da/dt is the total (or absolute) derivative in time; that is, the derivative in the fixed system. Respectively, d^*a/dt is the local (or relative) time derivative; that is, the derivative in the moving system.

From formulas (1.5), (1.6), (1.3), and (1.1), it follows that

$$\mathbf{v}_a = \mathbf{v}_r + \mathbf{v}_e \,, \tag{1.8}$$

that is, the absolute velocity of a particle is equal to the sum of the vectors of the relative and bulk (transport) velocities.

Let us now consider the *absolute acceleration* $\mathbf{w}_a$. By definition it agrees with the total derivative of the absolute velocity vector $\mathbf{v}_a$. In calculating this derivative we shall write (1.8) in the form

$$\mathbf{v}_a = \mathbf{v}_0 + \omega \times \mathbf{r} + \mathbf{v}_r \,. \tag{1.9}$$

Hence, taking into account formula (1.7),

$$\mathbf{w}_a = \frac{d\mathbf{v}_a}{dt} = \mathbf{w}_0 + \frac{d\omega}{dt} \times \mathbf{r} + \omega \times \left(\frac{d^*\mathbf{r}}{dt} + \omega \times \mathbf{r}\right) + \frac{d^*\mathbf{v}_r}{dt} + \omega \times \mathbf{v}_r \,.$$

Since

$$\frac{d^*\mathbf{r}}{dt} = \mathbf{v}_r \,, \qquad \frac{d^*\mathbf{v}_r}{dt} = \mathbf{w}_r$$

by (1.3) and (1.4), we have

$$\mathbf{w}_a = \mathbf{w}_0 + \frac{d\omega}{dt} \times \mathbf{r} + \omega \times (\omega \times \mathbf{r}) + 2\omega \times \mathbf{v}_r + \mathbf{w}_r \,.$$

Taking into account formula (1.2), this gives

$$\mathbf{w}_a = \mathbf{w}_r + \mathbf{w}_e + \mathbf{w}_c \,, \qquad \mathbf{w}_c = 2\omega \times \mathbf{v}_r \,. \tag{1.10}$$

Thus, the *absolute acceleration* is composed of the *relative acceleration* $\mathbf{w}_r$, the *bulk (transport) acceleration* $\mathbf{w}_e$, and the *additional acceleration* $\mathbf{w}_c$. The latter is also called the *Coriolis acceleration*.

Equality (1.8) expresses the *velocity addition theorem*, and relationship (1.10), the *acceleration addition theorem* for the composite motion of a particle.

Example 1 *Expansion of the velocity and acceleration of a point in the axes of the polar coordinates.* Let us consider a planar motion of a point M, which is given in the polar coordinates by the motion equations

$$\rho = \rho(t) \,, \quad \psi = \psi(t) \,.$$

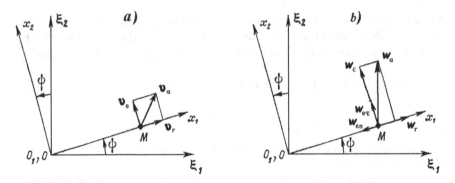

Fig. 1 Projections of the velocity and acceleration of the point onto the polar coordinates

The polar coordinates defines the principal basis $\{e_\rho, e_\psi\}$. Let us use the theory of composite motion of a point to find the projections of the velocity **v** and the acceleration **w** of a point on the axis onto the axes of the polar coordinates.

In solving practical problems on the composite motion of a point, the $Ox_1x_2x_3$-frame is usually related with some solid body on which the point is moving; the motion of this solid with respect to the $O_1\xi_1\xi_2\xi_3$-frame is considered given. In our setting, one can imagine a thin tube along which a point M rotates according to the law $\rho = \rho(t)$. With this tube, we shall associate the Ox_1x_2-frame, which will rotate in the $O_1\xi_1\xi_2$-frame along the $O_1\xi_3$-axis according to the law $\psi = \psi(t)$ (see Fig. 1).

In this case, the relative velocity is $v_r = \dot{\rho}$, its vector being directed along the Ox_1-axis. In turn, the transferred velocity of the point M agrees with the velocity of the point on the tube with which the moving point M coincides at a given time. If the tube rotates according to the law $\psi = \psi(t)$, then this velocity is $v_e = \dot{\psi}\rho$, its vector being orthogonal to the Ox_1-axis (see Fig. 1a).

Projecting formula (1.8) onto the polar axes, this gives

$$pr_{e_\rho}\mathbf{v} = \dot{\rho}, \quad pr_{e_\psi}\mathbf{v} = \dot{\psi}\rho,$$
$$v \equiv |\mathbf{v}| = \sqrt{\dot{\rho}^2 + \dot{\psi}^2\rho^2}. \tag{1.11}$$

In applying formula (1.10) one should take into account that the accelerations $\mathbf{w}_r$ and $\mathbf{w}_e$ may have both the tangential and the normal components. In our setting, since the relative motion is linear $w_{rn} = 0$, we have

$$w_r \equiv w_{r\tau} = \ddot{\rho}.$$

In contrast to this, the transferred acceleration has both components; moreover (because the transferred motion is characterized by the rotation of the tube according to the law $\psi = \psi(t)$), we have

$$w_{e\tau} = \ddot{\psi}\rho\,, \quad w_{en} = \dot{\psi}^2\rho\,.$$

The vectors $\mathbf{w}_{e\tau}$ and $\mathbf{w}_{en}$ are shown in Fig. 1b.

The motion of the Ox_1x_2-frame with respect to the $O_1\xi_1\xi_2$-frame is the solution of the first frame about the $O_1\xi_3$-axis with the angular velocity $\omega = \dot{\psi}\,\mathbf{k}_3$. Hence from the formula $\mathbf{w}_c = 2\boldsymbol{\omega} \times \mathbf{v}_r$ we have

$$w_c = 2\dot{\psi}\dot{\rho}\,,$$

the vector $\mathbf{w}_c$ having the same direction as the vector $\mathbf{w}_{e\tau}$. Projecting the vector

$$\mathbf{w}_a = \mathbf{w}_r + \mathbf{w}_{e\tau} + \mathbf{w}_{en} + \mathbf{w}_c$$

onto the polar axis, we see that (see Fig. 1b):

$$pr_{\mathbf{e}_\rho}\mathbf{w} = \ddot{\rho} - \dot{\psi}^2\rho\,, \quad pr_{\mathbf{e}_\psi}\mathbf{w} = \ddot{\psi}\rho + 2\dot{\psi}\dot{\rho}\,,$$

$$w \equiv |\mathbf{w}| = \sqrt{(\ddot{\rho} - \dot{\psi}^2\rho)^2 + (\ddot{\psi}\rho + 2\dot{\psi}\dot{\rho})^2}\,. \tag{1.12}$$

We note that in Fig. 1 the directions of the vectors correspond to the case when

$$\dot{\rho} > 0\,, \quad \ddot{\rho} > 0\,, \quad \dot{\psi} > 0\,, \quad \ddot{\psi} > 0\,.$$

Formulas (1.11) and (1.12) coincide with the formulas obtained in Sects. 3, 8 of Chap. 1 for the cylindrical system of coordinates, of which the polar system of coordinate is a particular case.

It is worth pointing out that with the help of the theory of composite motion of a point it proved possible to represent any plant motion of a point as the superposition of two simplest motions: the rectilinear motion and the rotation of a point along a circle.

2 Particle Velocity Under Bulk Motions

Assume that a point M moves in the system $O_1x_1^{(1)}x_2^{(1)}x_3^{(1)}$ moving with respect to the system $O_2x_1^{(2)}x_2^{(2)}x_3^{(2)}$, which in turn moves with respect to the system $O_3x_1^{(3)}x_2^{(3)}x_3^{(3)}$, and so on. How will the velocity of the particle M be expressed with respect to the (absolute) system $O_{n+1}x_1^{(n+1)}x_2^{(n+1)}x_3^{(n+1)}$ in terms of the relative velocity $\mathbf{v}_r^{(1)}$ in the system $O_1x_1^{(1)}x_2^{(1)}x_3^{(1)}$ and the bulk velocities, whose number is n ?

On the basis of formula (1.9), which expresses the absolute velocity of a particle in terms of its relative and bulk velocities in the case of one moving system, we have, for the particle velocity in the system $O_2x_1^{(2)}x_2^{(2)}x_3^{(2)}$,

Fig. 2 Point velocity with
several bulk motions

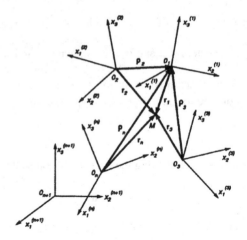

$$\mathbf{v}_r^{(2)} = \mathbf{v}_1 + \boldsymbol{\omega}_1 \times \mathbf{r}_1 + \mathbf{v}_r^{(1)}, \qquad \mathbf{r}_1 = \overrightarrow{O_1 M}$$

(see Fig. 2), where $\mathbf{v}_1$ is the velocity of the origin O_1 in the system $O_2 x_1^{(2)} x_2^{(2)} x_3^{(2)}$, and $\boldsymbol{\omega}_1$ is the angular velocity of rotation of the axes $x_1^{(1)}, x_2^{(1)}, x_3^{(1)}$ with respect to the axes $x_1^{(2)}, x_2^{(2)}, x_3^{(2)}$.

Similarly, the expression for the velocity of particle M in the system $O_3 x_1^{(3)} x_2^{(3)} x_3^{(3)}$ can be written in the form

$$\mathbf{v}_r^{(3)} = \mathbf{v}_2 + \boldsymbol{\omega}_2 \times \mathbf{r}_2 + \mathbf{v}_r^{(2)}, \qquad \mathbf{r}_2 = \overrightarrow{O_2 M},$$

where $\mathbf{v}_2$ is the velocity of the origin O_2 in the system $O_3 x_1^{(3)} x_2^{(3)} x_3^{(3)}$, and $\boldsymbol{\omega}_2$ is the angular velocity of rotation of the axes $x_1^{(2)}, x_2^{(2)}, x_3^{(2)}$ with respect to the axes $x_1^{(3)}, x_2^{(3)}, x_3^{(3)}$.

The resulting general rule is as follows:

$$\mathbf{v}_a = \mathbf{v}_r^{(n+1)} = \mathbf{v}_n + \boldsymbol{\omega}_n \times \mathbf{r}_n + \mathbf{v}_r^{(n)}, \qquad \mathbf{r}_n = \overrightarrow{O_n M}.$$

Expressing in succession $\mathbf{v}_r^{(k)}$ in terms of $\mathbf{v}^{(k-1)}$, $k = n, n-1, \ldots, 2$, we obtain

$$\mathbf{v}_a = \sum_{\nu=1}^{n} (\mathbf{v}_\nu + \boldsymbol{\omega}_\nu \times \mathbf{r}_\nu) + \mathbf{v}_r^{(1)}. \tag{2.1}$$

Taking into account that $\mathbf{r}_\nu = \boldsymbol{\rho}_\nu + \mathbf{r}_1$, where $\boldsymbol{\rho}_\nu$ is the radius vector the origin O_ν to the origin O_1 (Fig. 2), expression (2.1) can be represented in the form

$$\mathbf{v}_a = \mathbf{V} + \boldsymbol{\Omega} \times \mathbf{r}_1 + \mathbf{v}_r^{(1)}, \tag{2.2}$$

where

$$\Omega = \sum_{\nu=1}^{n} \omega_\nu, \qquad V = \sum_{\nu=1}^{n} (\mathbf{v}_\nu + \omega_\nu \times \rho_\nu), \qquad \rho_\nu = \overrightarrow{O_\nu O_1}.$$

The vectors Ω and V are by no means related with the relative motion of a particle in the system $O_1 x_1^{(1)} x_2^{(1)} x_3^{(1)}$, and hence they characterize the resulting bulk motion.

Formula (2.2), which is completely identical to formula (1.9), expresses the absolute velocity of the particle M. The system $O_1 x_1^{(1)} x_2^{(1)} x_3^{(1)}$ rotates about the origin O_1 with instantaneous angular velocity Ω, the origin O_1 moving with velocity V.

3 Composition of Motions of a Rigid Body

Assume that we are given n moving coordinate systems $O_k x_1^{(k)} x_2^{(k)} x_3^{(k)}$ (Fig. 2). It can be assumed that a rigid body is considered in all such systems. The motion of kth body relative to the $(k + 1)$st body is supposed to be known: it is represented by the velocity of the translational motion of the body $\mathbf{v}_k$ together with the pole O_k and the angular velocity ω_k of the rotation about this pole. It is required to find what the type of motion of the first body relative to the $(n + 1)$st body (that is, with respect to the system $O_{n+1} x_1^{(n+1)} x_2^{(n+1)} x_3^{(n+1)}$).

We shall employ the results of the previous subsection. The velocity $\mathbf{v}$ of an arbitrary point M of the first body relative to the $(n + 1)$st body is found by formula (2.2), in which the velocity $\mathbf{v}_r^{(1)}$ of the same point in the system $O_1 x_1^{(1)} x_2^{(1)} x_3^{(1)}$ is equal to zero. As a result, we have

$$\mathbf{v} = V + \Omega \times \mathbf{r}_1, \qquad \mathbf{r}_1 = \overrightarrow{O_1 M}, \tag{3.1}$$

where

$$V = \sum_{k=1}^{n} (\mathbf{v}_k + \omega_k \times \rho_k), \qquad \Omega = \sum_{k=1}^{n} \omega_k, \qquad \rho_k = \overrightarrow{O_k O_1}. \tag{3.2}$$

Thus, the velocity of the point M is that as if the first body move translationally with the velocity V rotating about the pole O_1 with the angular velocity Ω.

The coordinate system $O_k x_1^{(k)} x_2^{(k)} x_3^{(k)}$, which is connected with the k-th body, is chosen arbitrarily. With each body we connect a new coordinate system $O_k' x_1^{(k)'} x_2^{(k)'} x_3^{(k)'}$. The angular velocity of the rigid body, as was shown in Chap. 2, does not depend on the choice of a pole. Therefore, the angular velocity ω_k' of the system $O_k' x_1^{(k)'} x_2^{(k)'} x_3^{(k)'}$ relative to the system $O_{k+1}' x_1^{(k+1)'} x_2^{(k+1)'} x_3^{(k+1)'}$ is equal to the angular velocity ω_k. The velocity $\mathbf{v}$ of the point M relative to the $(n + 1)$-st body is invariant to the choice of the coordinate system. That is why with this choice of the new axes we get the same velocity, which by analogy with expressions (3.1) and (3.2) and taking into account that $\omega_k' = \omega_k$ can be written in the form

$$\mathbf{v} = \mathbf{V}' + \boldsymbol{\Omega} \times \mathbf{r}_1', \qquad \mathbf{r}_1' = \overrightarrow{O_1'M}, \tag{3.3}$$

$$\mathbf{V}' = \sum_{k=1}^{n} (\mathbf{v}_k' + \boldsymbol{\omega}_k \times \boldsymbol{\rho}_k'), \qquad \boldsymbol{\Omega} = \sum_{k=1}^{n} \boldsymbol{\omega}_k, \qquad \boldsymbol{\rho}_k' = \overrightarrow{O_k'O_1'}. \tag{3.4}$$

Here, $\mathbf{v}_k'$ is the velocity of a new pole O_k' relative to the $(k+1)$st body. With the velocity of the previous pole $\mathbf{v}_k$ it is related as follows:

$$\mathbf{v}_k' = \mathbf{v}_k + \boldsymbol{\omega}_k \times \overrightarrow{O_k O_k'}. \tag{3.5}$$

Comparing expression (3.3) with expression (3.1) and taking into account that

$$\mathbf{r}_1 = \overrightarrow{O_1 O_1'} + \mathbf{r}_1',$$

we get

$$\mathbf{V}' = \mathbf{V} + \boldsymbol{\Omega} \times \overrightarrow{O_1 O_1'}. \tag{3.6}$$

This formula is analogous to (3.5).

Thus, the angular velocity $\boldsymbol{\Omega}$ is invariant with respect to the choice of the coordinate system; the translational velocity when changing to the new pole O_1' is calculated by formula (3.6).

Let us consider some particular cases.

Composition of translational motions. Assume that all bodies move translationally relative to each other (that is, $\boldsymbol{\omega}_k = 0$ for all k). Then the resulting motion is also translational; by formulas (3.1), (3.2) its velocity reads as

$$\mathbf{v} = \sum_{k=1}^{n} \mathbf{v}_k.$$

Composition of rotations about intersecting axes. Assume that all $\mathbf{v}_k = 0$ and all $\boldsymbol{\omega}_k$ passing through the points O_k intersect at one point O. It is convenient to locate the origins of all moving coordinate systems at this point on putting $O_k' = O$, $k = \overline{1, n}$. Then (see Fig. 3)

$$\overrightarrow{O_k O_k'} = \lambda_k \boldsymbol{\omega}_k, \qquad \boldsymbol{\rho}_k' = \overrightarrow{O_k'O_1'} = 0.$$

Hence, by formulas (3.5) and (3.4)

$$\mathbf{v}_k' = \mathbf{v}_k = 0, \qquad \mathbf{V}' = \sum_{k=1}^{n} \boldsymbol{\omega}_k \times \boldsymbol{\rho}_k' = 0, \qquad k = \overline{1, n}.$$

Fig. 3 Composition of
rotations about intersecting
axes

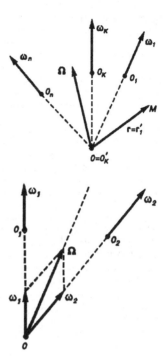

Fig. 4 Composition of
rotations about two
intersecting axes

According to formula (3.3) the velocity of an arbitrary point can be represented in
the form

$$\mathbf{v} = \mathbf{\Omega} \times \mathbf{r}, \qquad \mathbf{r} = \overrightarrow{OM}.$$

Thus, under a simultaneous rotation of a rigid body about the instantaneous axes,
which intersect at one point, we have the resulting rotation with the instantaneous
angular velocity $\mathbf{\Omega}$, which agrees with the vector sum of the instantaneous angular
velocities $\boldsymbol{\omega}_k$. The direction of the resulting instantaneous axis is that of the vector $\mathbf{\Omega}$,
the point of intersection O is a fixed point. The simplest example of this composition
of rotations is the case of a couple of rotations about axes that intersect at one point
(Fig. 4).

The case of a couple of rotations. Let $\mathbf{\Omega} = 0$. Then

$$\mathbf{v} = \mathbf{V} = \sum_{k=1}^{n} (\mathbf{v}_k + \boldsymbol{\omega}_k \times \boldsymbol{\rho}_k), \qquad \boldsymbol{\rho}_k = \overrightarrow{O_k O_1},$$

i.e., the velocities of all body points are equal; as a result, the body moves translationally.

Of special interest is the case of *couple of rotations*—here we mean a set of two
vectors $\boldsymbol{\omega}_1$ and $\boldsymbol{\omega}_2$, having parallel action lines and such that $\boldsymbol{\omega}_2 = -\boldsymbol{\omega}_1$ (Fig. 5).

Fig. 5 Couple of rotations

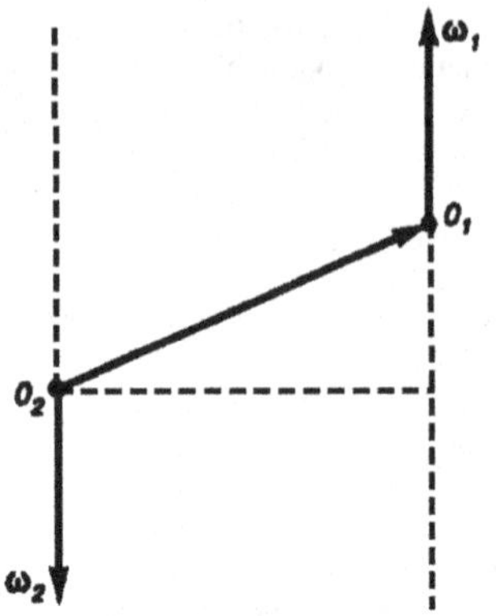

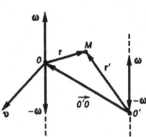

Fig. 6 Equivalence of a couple of rotations to the translatory motion

Assuming for simplicity that $\mathbf{v}_1 = \mathbf{v}_2 = 0$ we get

$$\Omega = \omega_1 + \omega_2 = 0,$$

$$\mathbf{v} = \mathbf{V} = \omega_2 \times \overrightarrow{O_2 O_1} = \omega_1 \times \overrightarrow{O_1 O_2} = \overrightarrow{O_2 O_1} \times \omega_1.$$

The vector product $\overrightarrow{O_2 O_1} \times \omega_1 = \overrightarrow{O_1 O_2} \times \omega_2$ is called the *moment of a couple of rotations*. Thus, a couple of rotations produces a translational motion with the velocity equal to the moment of the couple. This velocity is equal to ωp, where $\omega = |\omega_1|$, and p is the shortest distance between the action lines of the vectors ω_1 and ω_2 (the *arm of the couple*,), and it is perpendicular to the plane of the couple of rotations according to the right-hand screw rule.

The equivalence between the couple of rotations and the translational motion allows us to state that the body rotation with angular velocity ω about an instantaneous axis passing through the point O' is equivalent to the rotation with the same angular velocity about the axis passing through the point O parallel to the first one and is also equivalent to the translational motion with the velocity $\mathbf{v} = \omega \times \overrightarrow{O'O}$. This result follows directly from Fig. 6. Indeed, if one locates the system of vectors ω, $-\omega$ equivalent to zero at the point O , then the vector $\omega(O')$ is equivalent to $\omega(O)$ and to the couple $\{\omega(O'), -\omega(O)\}$ with the moment $\mathbf{v} = \overrightarrow{OO'} \times \omega = \omega \times \overrightarrow{O'O}$.

This can be also verified by considering the velocity of an arbitrary point M. During the body rotation about the axis passing through the point O', the velocity of point M is as follows:

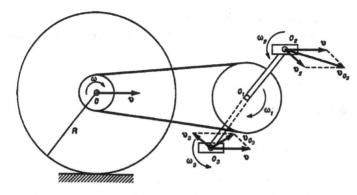

Fig. 7 Motion of bike pedals

$$\mathbf{v}_M = \boldsymbol{\omega} \times \mathbf{r}', \qquad \mathbf{r}' = \overrightarrow{O'M}.$$

Writing $\mathbf{r}'$ as $\mathbf{r}' = \overrightarrow{O'O} + \mathbf{r}, \ \ \mathbf{r} = \overrightarrow{OM}$, we get

$$\mathbf{v}_M = \boldsymbol{\omega} \times \overrightarrow{O'O} + \boldsymbol{\omega} \times \mathbf{r} = \mathbf{v} + \boldsymbol{\omega} \times \mathbf{r},$$

which corresponds to the transition to rotation about a parallel axis owing to introducing the couple of rotation.

Example 2 Let us consider the motion of a bicycle pedal from the point of view of the motion composition. Let a bicycle frame moves with velocity $\mathbf{v}$ (see Fig. 7), the center O of the back bicycle wheel moves with the same velocity. If a wheel of radius R moves without slipping, then the instantaneous velocity center is located at its lowest point—this is why it rotates clockwise with angular velocity $\omega = v/R$. Under a motion without free travel the angular velocity ω appears due to the rotation of cranks $O_1 O_2$ and $O_1 O_3$ with he angular velocity $\omega_1 = \omega_z/z_1$, where z and z_1 are, respectively, the numbers of teeth of the little and big-toothed wheels on the O- and O_1-axes. The angular velocity ω_1 corresponds to the rotation of cranks clockwise. Relative to cranks, the cyclist rotates pedals counterclockwise with angular velocities $\omega_2 = \omega_1$ and $\omega_3 = \omega_1$. Thus, relative to the framework, every pedal takes part in a couple of rotations: in the bulk motion of crank rotation about the point O_1 with angular velocity ω_1 and in the relative motion of rotation about the point O_2 with angular velocity $\omega_2 = -\omega_1$ (or about the point O_3 with angular velocity $\omega_3 = -\omega_1$). As was shown above, the couple of rotations is equivalent to the translational motion with velocity $\mathbf{v}_k = \overrightarrow{O_1 O_k} \times \omega_k$. Here, $k = 2$ for the upper pedal, and $k = 3$ for the lower one. As a result, relative to the fixed coordinate system the composition of two translational motions takes place for pedals: the bulk motion together with the framework with velocity $\mathbf{v}$ and the relative motion with respect to the framework with velocity $\mathbf{v}_k, \ k = 2, 3$. In composition of translational motions, the velocities are added vectorially and the results of these additions are shown in Fig. 7. That is why at a given time all the points of the upper pedal move relative to the fixed coordinate system with velocities $\mathbf{v}_{0_2} = \mathbf{v} + \mathbf{v}_2$ and all the points of the lower pedal move with velocities $\mathbf{v}_{0_3} = \mathbf{v} + \mathbf{v}_3$.

4 Kinematic Screw

Let us go back to formula (3.1)[1]:

$$\mathbf{v} = \mathbf{V} + \mathbf{\Omega} \times \mathbf{r}, \qquad \mathbf{r} = \overrightarrow{OM}. \tag{4.1}$$

Assume that nonzero vectors $\mathbf{V}$ and $\mathbf{\Omega}$ form angle α at the point O (Fig. 8). We decompose the vector $\mathbf{V}$ into the vector $\mathbf{V}^*$, directed along the action line $\mathbf{\Omega}$, and the vector $\mathbf{V} - \mathbf{V}^*$ perpendicular to $\mathbf{\Omega}$. The vector $\mathbf{V}^*$ can be represented as

$$\mathbf{V}^* = \frac{\mathbf{\Omega}(\mathbf{V} \cdot \mathbf{\Omega})}{\Omega^2} = \frac{V \cos \alpha \, \mathbf{\Omega}}{\Omega} = \frac{\Omega_k \mathbf{i}_k (V_\nu \Omega_\nu)}{\Omega^2}. \tag{4.2}$$

As was shown above, the vector $\mathbf{\Omega}$ does not change when passing to a new pole O' and the vector $\mathbf{\Omega} \times \overrightarrow{OO'}$, which is perpendicular to the vector $\mathbf{\Omega}$ (and hence, to the vector $\mathbf{V}^*$) is added to the vector $\mathbf{V}$ (see formula (3.6)). This implies that the vector $\mathbf{V}^*$ (as well as the vector $\mathbf{\Omega}$) does not depend on the choice of a pole. When transforming to the new pole only the vector $\mathbf{V} - \mathbf{V}^*$ changes, and besides

$$\mathbf{V}' - \mathbf{V}^* = \mathbf{V} - \mathbf{V}^* + \mathbf{\Omega} \times \overrightarrow{OO'}.$$

By choosing the point O', it is always possible to satisfy the equality $\mathbf{V}' = \mathbf{V}^*$. In the vector form, the equation of the line along which $\mathbf{V}' = \mathbf{V}^*$ is as follows:

$$\overrightarrow{OO'} \times \mathbf{\Omega} = \mathbf{V} - \mathbf{V}^*. \tag{4.3}$$

If the pole O' lies on this line, then formula (3.3) takes the form

$$\mathbf{v} = \mathbf{V}^* + \mathbf{\Omega} \times \mathbf{r}', \qquad \mathbf{r}' = \overrightarrow{O'M}.$$

Hence, the distribution of velocities in the body is such as if rotates with angular velocity $\mathbf{\Omega}$ about the above straight line and simultaneously moves translationally along it with velocity $\mathbf{V}^*$. Recall that such a motion of a rigid body is called an *instantaneous screw motion*.

Thus, in the most general case, the set of some translational motions and rotations is equivalent to the velocity instantaneous distribution at a fixed moment of time, this

---

[1] In this section the index "1" is omitted for the simplicity of all notation for the first body.

Fig. 8 Construction of the
instantaneous screw axis

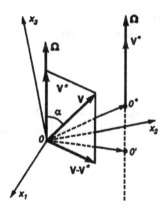

screw motion being characterized by vectors V^* and Ω, directed along the instantaneous screw axis. The set of vectors V^* and Ω, lying on the same straight line, forms the so-called *kinematic screw* .

The *instantaneous screw axis equation* in the canonical form is as follows:

$$\frac{x_1 - x_1^*}{\Omega_1} = \frac{x_2 - x_2^*}{\Omega_2} = \frac{x_3 - x_3^*}{\Omega_3} = \lambda, \quad -\infty < \lambda < +\infty.$$

Here, x_1, x_2, x_3 are the coordinates of the current point O', and x_1^*, x_2^*, x_3^* are the coordinates of that point O^* on the straight line for which

$$\Omega \cdot \overrightarrow{OO^*} = 0. \tag{4.4}$$

From equation (4.3), we have

$$\Omega \times (\overrightarrow{OO^*} \times \Omega) = \Omega \times (V - V^*) = \Omega \times V.$$

Writing the double vector product in the form

$$a \times (b \times c) = b(a \cdot c) - c(a \cdot b),$$

and taking into account relation (4.4) we see that the coordinates of the given point O^* can be derived by the formula

$$\overrightarrow{OO^*} = \frac{\Omega \times V}{\Omega^2} = x_k^* i_k = \frac{1}{\Omega^2} \begin{vmatrix} i_1 & i_2 & i_3 \\ \Omega_1 & \Omega_2 & \Omega_3 \\ V_1 & V_2 & V_3 \end{vmatrix}.$$

The equation of instantaneous axis is frequently useful not in the canonical form, but in another form, which is derived from the definition of the instantaneous screw axis as a locus of the rigid body points with velocities $\mathbf{V}^*$ at a given time. Therefore, according to the general formula (4.1), the unknown radius vector $\mathbf{r}$ of a point belonging to the screw axis satisfies the equation

$$\mathbf{V}^* = \mathbf{V} + \mathbf{\Omega} \times \mathbf{r}.$$

Taking into account the representation of the vector $\mathbf{V}^*$ in form (4.2) in the coordinate system connected with the body, we have

$$p\,\Omega_k \mathbf{i}_k = V_k \mathbf{i}_k + \begin{vmatrix} \mathbf{i}_1 & \mathbf{i}_2 & \mathbf{i}_3 \\ \Omega_1 & \Omega_2 & \Omega_3 \\ x_1 & x_2 & x_3 \end{vmatrix}, \tag{4.5}$$

where $p = V \cos \alpha / \Omega$ is a parameter of kinematic screw.

From the vector equality (4.5) it follows that the straight lines, which are projections of the screw axis onto coordinate planes, are given in these planes by the equations

$$V_1 + \Omega_2 x_3 - \Omega_3 x_2 = \Omega_1 p,$$
$$V_2 + \Omega_3 x_1 - \Omega_1 x_3 = \Omega_2 p, \tag{4.6}$$
$$V_3 + \Omega_1 x_2 - \Omega_2 x_1 = \Omega_3 p.$$

Each of these three equations defines a plane in the space parallel to the corresponding coordinate axis. Since any line in the space is the intersection of two planes, one of the equations can be neglected. If the straight line is parallel to one of the coordinate axes, then its projection onto the coordinate plane perpendicular to this axis degenerates to a point, the corresponding equation of system (4.6) becoming the identity. In the case when the straight line is perpendicular to one of the axes, two of the planes (4.6) merge into one plane, which is perpendicular to this axis. In all other cases, the screw axis can be looked upon as the intersection of any two planes (4.6).

The system of equations (4.6) is usually written as the sequence of equalities

$$\frac{V_1 + \Omega_2 x_3 - \Omega_3 x_2}{\Omega_1} = \frac{V_2 + \Omega_3 x_1 - \Omega_1 x_3}{\Omega_2} = \frac{V_3 + \Omega_1 x_2 - \Omega_2 x_1}{\Omega_3} = p. \tag{4.7}$$

If we omit the parameter p, then the first three equalities give the general equations of the three planes. All the three planes intersect at the screw axis.

It is clear that if the vector $\mathbf{V}$ is perpendicular to the vector $\mathbf{\Omega}$, then $\mathbf{V}^* = 0$ and at all points of the instantaneous axis (4.4) we get only the vector $\mathbf{\Omega}$. This shows that the velocities of the rigid body points are such as if the body rotates about the instantaneous axis with angular velocity $\mathbf{\Omega}$ at a fixed moment of time. A set of vectors ω_k distributed in one plane and also a set of parallel vectors ω_k serve as an example of the above systems with $\mathbf{\Omega} \neq 0$. This directly follows from the general formula (3.2).

Indeed, if $\boldsymbol{\omega}_k$ lie in one plane, then the vectors $\boldsymbol{\rho}_k$ also lie in this plane, but the vectors $\boldsymbol{\omega}_k \times \boldsymbol{\rho}_k$ and $\mathbf{V} = \sum\limits_{k=1}^{n} \boldsymbol{\omega}_k \times \boldsymbol{\rho}_k$ are perpendicular to it. Thus, the planar system of vectors $\boldsymbol{\omega}_k$ is equivalent to one rotation with angular velocity $\boldsymbol{\Omega}$ about the axis $O'O^*$.

If all $\boldsymbol{\omega}_k$ are parallel to each other, then we can write

$$\boldsymbol{\omega}_k = \omega_k \boldsymbol{\omega}^0 ,$$

which shows that the vectors $\boldsymbol{\omega}_k \times \boldsymbol{\rho}_k = \omega_k(\boldsymbol{\omega}^0 \times \boldsymbol{\rho}_k)$ are perpendicular to $\boldsymbol{\omega}^0$, and hence, the vector $\mathbf{V} = \sum\limits_{k=1}^{n} \boldsymbol{\omega}_k \times \boldsymbol{\rho}_k$ is perpendicular to the vector $\boldsymbol{\Omega} = \Omega\boldsymbol{\omega}^0$.

Part II DYNAMICS. GENERAL ASPECTS OF RATIONAL MECHANICS. FUNDAMENTALS OF ANALYTICAL MECHANICS

The part of mechanics that studies the motion of masses (bodies) on the basis of physical postulates with due regard to the causes changing the motion (forces) and the mass of the moving body is called *dynamics*, of which the theory of equilibrium—the *statics*—is a specific case.

In dynamics, new concepts come into being: the mass, the momentum, the force. Newton defines the *mass* as the quantity of the matter. Being an atomist, he considered the quantity of a substance to be proportional to the number of infissionable homogeneous atoms and, therefore, proportional to their volume. On the basis of experiments with pendula, he showed that the mass of a body is proportional to its weight, and consequently, can be measured with the help of the latter.

The second basic concept of dynamics is the notion of *momentum* of a body or its impulse, by which we mean the vector quantity p = mv, where m is the mass of the body; v is the velocity of its translational motion. As applied to this case, we agree to speak of a particle of mass m. For the general case of a body motion, we shall determine the impulse by the formula

$$\mathbf{p} = \int_{\tau} \mathbf{v}\, dm = m\mathbf{v}^*$$

where $\mathbf{v}^*$ is the quantity found by the mean-value theorem. The point with the position vector, which we will call the *center of mass*, corresponds to this quantity.

The third basic concept of dynamics is the notion of *force*. According to Newton, an applied force is the action exerted on the body in order to change its state of rest or that of a uniform and rectilinear motion. Thus, a force is estimated by its action that is exhibited in changing the motion of the body. In this case, the particle introduced above is meant by the body. Below, in the cases when we assume the rotational motion of the body be absent, the meaning of the terms "particle" and "body" will be the same.

Along with the dynamic effect of a force, we can consider the force as the action causing the change in the shape of the body or its deformation. On the basis of this principle, a force is measured by a spring balance (dynamometer).

The introduced definitions enabled Newton to postulate the following laws of motion.

The first law "Every body preserves in its state of rest, or of uniform motion in a right line, unless it is compelled to change that state by forces impressed thereon".[1]

The second law pooling experimental data is formulated as follows: "A change in motion is ever proportional to the motive force impressed; and is made in the direction of the right line, in which that force is impressed".[2] By a change in motion one should understand the derivative dp/dt. Thus, Newton's second law should be written in the form

$$\frac{d(m\mathbf{v})}{dt} = \mathbf{F}. \tag{1}$$

Note that in cases when the particle moves with velocities close to that of light the mass m entering this expression is viewed as a variable quantity.[3] Note that in the classical mechanics, the mass of a moving material point is considered constant, Eq. (1) taking the form

$$m\frac{d\mathbf{v}}{dt} = \mathbf{F}. \tag{2}$$

We shall call the quantity m the inertial mass.

In theoretical mechanics, there exists the branch "Dynamics of a particle of variable mass" in which Newton's second law is applied not to a single particle, but to a system of particles. As a result, it turns out that in the right-hand side of equation (2) an extra component containing dm/dt emerges, the mass m on the left-hand side being its instantaneous value at the given instant.[4]

Simple experiments make it possible to measure both the acceleration and the force acting on the moving body. The force is measured by a dynamometer, and the acceleration is calculated from the change in the distance covered depending on the time. Since the mass of a homogeneous body can be computed from its volume, the correctness of relation (2) can be verified experimentally.

---

[1] *I. Newton*. The Mathematical Principles of Natural Philosophy. Benjamin Motte. 1729. Vol. I. P. 19.

[2] Ibidem, p. 19.

[3] Analyzing the notation of the fundamental law of dynamics in the form (1), the well-known physicist A. Sommerfeld remarks, that "the mass is not always constant, for example, it is not constant in the relativity theory in which Newton's formulation of the law (Eq. (2) has proved almost prophetic" (see: *A. Sommerfeld*. Mechanics, Moscow, 1947, p. 10 [in Russian]).

[4] This issue will be studied in detail in Chap. "Flight Dynamics" of the second volume.

Following the principles of "inductivism", Newton considered that "in experimental philosophy we are to look upon propositions collected by general induction from phenomena as accurately or very nearly true, notwithstanding any contrary hypotheses that may be imagined, till such as other phenomena occur, by which they may either be made more accurate, or liable to exceptions".[5] Physics, statements derived from occurring phenomena with the help of adjustment in spite of the possibility of assumptions opposite to them should be regarded as correct, either precisely or approximately, until such phenomena come to light by which they will be refined further or subjected to exclusion. Proceeding from these principles, he introduced relation (1) as the axiom, or the law of motion of a body in general. This law makes it possible, having measured two out of the three quantities entering it, to determine the third one from Eq. (2). Since it is easier to measure directly velocities and accelerations, Eq. (2) can be used for determining the mass or the force. So, for instance, proceeding from Keplerian laws based on observations, Newton established that the acceleration of a planet in its orbital motion is invertially proportional to its squared separation from the Sun. Multiplying the obtained acceleration by the mass of the planet, he deduced the *law of gravitation*.

Kirchhoff went even further and suggested that the force should be viewed, by definition, as the derivative of impulse. Such an approach proved to be convenient when constructing a more general mechanics similar, however, in the form of Newtonian mechanics.

Let us finally discuss in what reference frame Newton's second law containing the derivative of momentum (of impulse) with respect to the time was formulated. Consider a method of measuring the time.

Everyone perfectly knows what is a time interval (for example, a second measured by a watch). To ensure the precision of its measurement, equal time intervals should correspond to equal sections of the face. To put it another way, time is required to pass uniformly. However, it is impossible to guarantee that the watch indicates a uniformly passing time. It can only be assumed. Such an assumption makes it possible to mathematize the time by putting a timeline in correspondence with it. Then the difference of coordinates of two points in this line or, as they say, of two instants of time will correspond to the time interval Δt. If, in addition, we require that the time interval Δt be independent of the fact whether or not the watch itself moves and also of the masses of bodies surrounding it, then we shall get an abstraction that Newton called the *true mathematical or absolute time*. According to Newton, "Absolute, True, and Mathematical Time, of itself, and from its own nature flows equably without regard to any thing external, and by another name is called Duration".

"Relative, Apparent, and Common Time is some sensible and external (whether accurate or unequable) measure of Duration by the means of motion, which is commonly used instead of the True time; such as an Hour, a Day, a Month, a Year".[6]

---

[5] *I. Newton*. The Mathematical Principles of Natural Philosophy. Vol. II, p. 205.

[6] Ibidem, Vol. I, p. 9.

In everyday life, we perfectly know what are the dimensions of a body. Assume that this body moves relative to some reference frame that, in turn, may also move relative to another reference frame, etc. In classical mechanics, the dimensions of a body are independent of the manner in which the reference frames move, neither do they depend on the masses surrounding the given body.

According to Newton, a moving body is a bounded movable part of some absolute space: "Absolute Space, in its own nature, without regard to any thing external remains always similar and immovable.

Relative Space is some moveable dimension or measure of the absolute spaces; which our senses determine, by its position to bodies, and which is vulgarly taken for immovable space...".[7]

Thus, according to Newton, the absolute time and the absolute space are notions on the basis of which two invariant quantities can be introduced: the time interval and the length of segment. And after the choice of the reference frame of displacements and of the units of time and length, these quantities are expressed in terms of the differences of time instants and of the coordinates of segment ends. In accordance with the basic postulates of differential and integral calculus, the notions of time interval and segment length make it possible to define the velocity and acceleration at a given instant of time as well as the distance covered.

Newton considers the absolute and relative motions. The absolute motion is originated and altered by forces applied to the body itself. The relative or seeming one is originated by the motion of the reference frame. Hence, the second law of dynamics is formulated relative to the reference frame $Ox_1x_2x_3$ in which the forces applied to the body exist and can be measured. However, we should especially emphasize that the indicated coordinate system is not unique.

Let us introduce a new coordinate system $O'x_1'x_2'x_3'$, the axes of which are parallel to the axes x_1, x_2, x_3. Suppose that this system moves relative to the initial one uniformly and translationally with a velocity $v = (v_1, v_2, v_3)$. Then, the coordinates of a moving particle in both systems turn out to be related to one another via the transformation

$$x_i' = x_i - v_i (t - t_0), \ i = 1, 2, 3, \ t' - t_0' = t - t_0, \tag{3}$$

which is called the *Galilei transformation*. From (3) we see that the accelerations $\mathbf{w}$ and $\mathbf{w}'$ in both systems are equal; i. e., $\mathbf{w} = \mathbf{w}'$. On the basis of this, we conclude that Newton's second law in the new coordinate system is of the same form as in the initial one, to put it another way, Newton's equation is invariant under the group of transformations given by formulas (3). Obviously, there exists an infinite number of such systems. We shall call them the *inertial systems*.

---

[7] Ibidem. Vol. I, p. 9.

Classical mechanics rests on the assumption that the source of a force acting on the body is always another body. Thus, according to Newton, a force is brought about by the interaction of at least two bodies.

The Third law (on interaction) is formulated as follows: "To every Action there is always opposed an equal Reaction: or the mutual actions of two bodies upon each other are always equal and directed to contrary parts".[8]

In a note to the indicated laws, Newton considers a joint action of forces. He asserts: "A body subject to aggregate forces describes the diagonal of a parallelogram during the same time its sides describe it when subject to separate forces".[9] This statement was entitled the *law of the parallelogram of forces*. We give some examples. If the forces are constant, then under the zero initial conditions for rectilinear displacements, we have

$$\mathbf{r}_1 = F_1 t^2 / (2m) , \ r_2 = F_2 t^2 / (2m) ,$$

and hence, $r = r_1 + r_2 = (F_1 + F_2) \, t^2 / (2m)$. As the displacements are added by the rule of parallelogram, then so are the forces F_1 and F_2. It is seen from the indicated formula that the resultant displacement is executed along the diagonal of the parallelogram built on the force vectors.

The form of the right-hand side of Newton equation (2) is usually specified by examining the force in its static display. So, for instance, according to the Hooke law, it can be ascertained for a statically stretched spring that the force of its tension is proportional to the extension of the latter. In accordance with this, in a dynamical problem of oscillations of a heavy particle suspended by a spring, we can consider that the elasticity force is also proportional to the deformation.

Similarly, if the resistance force, acting from the flow to the body located in it, is tested under static conditions, then in a certain range of velocities (approximately from 10 to 100 m/s) it turns out to be proportional to the squared velocity. On the basis of the foregoing, we can also solve the dynamical problem of motion of a body in the medium with resistance proportional to the squared velocity.

One should note that experimental studies of bodies moving with acceleration in a resisting medium make it possible to detect, along with the mentioned forces proportional to the squared velocity, a force proportional to the acceleration. We can explain the emergence of this force relying on the Newton laws. So, a body moving with acceleration in the fluid imparts accelerations different at different points to its particles. The presence of acceleration allows us to assert that particles of the fluid are acted upon by the forces from the body. By Newton's third law, a force emerges that acts on the body in question from the fluid. Detailed computations and experiments show that this force is proportional to the acceleration of the body.

---

[8] Ibidem, Vol. I, p. 20.

[9] Ibidem, Vol. I, p. 21.

Follow Newton, from integration of equation (2) for a circular mathematical pendulum one can establish the proportionality of the body's inertial mass m to its weight P, thereby proposing a method for its measurement. Indeed, the Newton equation projected on the tangent to the trajectory of the pendulum load reads

$$m \frac{dv}{dt} = -P\sin\varphi ,$$

where φ is the angle of the thread deviation from the vertical line. For a circular motion, we have $v = l\dot{\varphi}$ (l is the length of the pendulum thread), and hence for small φ we have the differential equation

$$\ddot{\varphi} + \frac{P}{ml} \varphi = 0 .$$

Integrating, we have

$$\varphi = A\cos\omega t + B\sin\omega t ,$$

where $\omega = \sqrt{P/(ml)}$, and A, B are arbitrary constants. Hence, the oscillation period of the pendulum is expressed by the formula

$$T = \frac{2\pi}{\omega} = 2\pi\sqrt{\frac{ml}{P}} . \tag{4}$$

In experiments with the pendulum the quantity l is given and the oscillation period T is determined quite accurately from observations.

Formula (4) implies that

$$\frac{P}{m} = \frac{4\pi^2 l}{T^2} \equiv k .$$

High-precision experiments show that in all cases the quantity $4\pi^2 l / T^2$ is the same. The most essential result of these experiments is the fact that the constant k having the units of acceleration is just the free-fall acceleration $g = 9.81 m/s^2$ and, consequently, the inertial mass m can be calculated in terms of the weight P by the formula

$$m = P/g .$$

We shall agree to measure the body weight P by (a spring balance) a dynamometer and regard the quantity $m_g = P/g$ as the *gravitation mass* of the body. Using the balance for measuring the gravity force P depends, in essence, on the law of gravitation for a body located on the Earth surface. This justifies the term "gravitational" or "gravitating" mass.

The equality $m = m_g$ is called the *postulate of equivalence* of the inertial and gravitational masses.

It is by no means obvious that the statically measured weight of the body divided by the constant k coincides with the inertial mass, which enters the axiomatically introduced Newton's second law as the factor of proportionality.

The postulate of equivalence of the inertial and gravitational masses is equivalent to the statement that all the bodies placed at the same point of the gravitational field gain equal accelerations. The validity of this postulate was experimentally verified by many researchers all over the world: no departure (deviation) from the indicated equality was so far detected.

In our exposition, we shall have two parts on the dynamics: in the first one (see the first volume), the general aspects of rational and analytical mechanics are presented, and in the second one (see the second volume), some special problems of rational mechanics having application significance.

Chapter 4
Particle Dynamics

N. N. Polyakhov, M. P. Yushkov⦿, and S. A. Zegzhda

In this chapter, the most widespread forms of differential equations of particle motion are presented, key theorems of particle dynamics are proved, and the conservative force field is studied. Further on, the derivation of Lagrange equations of the second kind and canonical equations for a particle makes it possible to generalize them to the motion of a system of mass points with the help of the notion of representation point. The main cases of oscillatory motion of a particle, those of motion of a particle subject to central forces as well as dynamics of the particle relative motion are analyzed in detail.

1 Differential Equations of Particle Motion in Various Coordinate Systems

According to the second Newton law

$$m\mathbf{w} = \mathbf{F} . \tag{1.1}$$

In a Cartesian coordinate system this equation can be represented as

$$m\frac{d^2 x_i}{dt^2} = X_i , \quad i = 1, 2, 3. \tag{1.2}$$

In a curvilinear coordinate system $q^1(t)$, $q^2(t)$, $q^3(t)$ for contra- and covariant components of acceleration, in accordance with formulas (8.14), (8.2) and (8.6) of Chap. 1, we have

$$w^i = \ddot{q}^i + \Gamma^i_{jk}\dot{q}^j\dot{q}^k , \quad i = 1, 2, 3, \tag{1.3}$$

$$w_i = \frac{d}{dt}\frac{\partial v^2/2}{\partial \dot{q}^i} - \frac{\partial v^2/2}{\partial q^i} = g_{ij}\ddot{q}^j + \Gamma_{i,jk}\dot{q}^j\dot{q}^k , \quad i = 1, 2, 3, \tag{1.4}$$

© Springer Nature Switzerland AG 2021
N. N. Polyakhov et al., *Rational and Applied Mechanics*,
Foundations of Engineering Mechanics,
https://doi.org/10.1007/978-3-030-64061-3_4

where Γ^i_{jk}, $\Gamma_{i,jk}$ are Christoffel symbols of the second and first kind. When decomposing the right- and left-hand sides of equation (1.1) by the fundamental basis $\mathbf{e}_i$, we should invoke formula (1.3). In doing so, we obtain the system of equations

$$m(\ddot{q}^i + \Gamma^i_{jk} \dot{q}^j \dot{q}^k) = Q^i, \quad i = 1, 2, 3,$$

where $Q^i = \mathbf{F} \cdot \mathbf{e}^i$ is the contravariant component of the vector $\mathbf{F} = Q^i \mathbf{e}_i$.

Scalar multiplication of equation (1.1) by the vectors of the reciprocal basis $\mathbf{e}^i$ that leads immediately to a system of equations resolved for the second derivatives $\ddot{q}^i$, is not always convenient, especially in the case of a skew-angular coordinate system, when the unit vectors of the fundamental and reciprocal bases do not coincide. The most tedious step in writing the equations in this form is finding the Christoffel symbols of the second kind.

The formula (1.4) makes it possible to determine $w_i = \mathbf{w} \cdot \mathbf{e}_i$ directly through the expression $v^2/2$ without preliminary calculation of the Christoffel symbols. Thus, the scalar multiplication of equation (1.1) by the vectors of the fundamental basis $\mathbf{e}_i$ leads to a system of equations, which can be written in the form

$$\frac{d}{dt} \frac{\partial T}{\partial \dot{q}^i} - \frac{\partial T}{\partial q^i} = Q_i, \quad i = 1, 2, 3, \tag{1.5}$$

where

$$Q_i = \mathbf{F} \cdot \mathbf{e}_i = \mathbf{F} \cdot \frac{\partial \mathbf{r}}{\partial q^i} = \sum_{\nu=1}^{3} X_\nu \frac{\partial x_\nu}{\partial q^i}, \quad T = \frac{mv^2}{2}.$$

The covariant component Q_i of the force $\mathbf{F} = Q^i \mathbf{e}_i$ is called the *generalized force* corresponding to the generalized coordinate q^i, and the quantity T as was already pointed out in Part I, the *kinetic energy of a particle* of mass m. Equations of motion written in the form (1.5) are called the *Lagrange equations of the second kind*.

Let us project both sides of equation (1.1) onto the unit vectors $\boldsymbol{\tau}, \mathbf{n}, \mathbf{b}$ of the natural trihedral. In this case, we obtain the *Euler equations*:

$$m\ddot{s} = F_\tau, \quad m\frac{v^2}{\rho} = F_n, \quad 0 = F_b.$$

Here, s is the arc coordinate, ρ is the radius of curvature of the trajectory. Recall that the vector $\boldsymbol{\tau}$ is directed to the side of arc coordinate s increasing.

Note that of all the presented forms of the second law of mechanics the notation of form (1.5) is most general. All of them are systems of second-order differential equations, where coordinates of the point are unknown functions. The right-hand sides of these equations contain supposedly continuous and sufficiently smooth functions of the time, coordinates, velocities, and sometimes, of accelerations. Let us consider that the differential equations are completed with initial data and all the conditions ensuring the existence and uniqueness of the solution of the system of ordinary differential equations are satisfied.

The problems of particle dynamics can be separated into two types. *The direct problem of dynamics* consists in determining the forces acting on the particle with its mass and the law of motion given. Obviously, in this case, the solution can be readily obtained by differentiation. Much more complicated proves to be the *inverse problem of dynamics*, when the law of motion of the point should be determined basing on the given functions residing in the right-hand sides of differential equations and being the projections of forces, with the help of integrating a system of differential equations, for example, (1.2). As it is a system of three second-order equations, then when solving it, one manages to determine each coordinate x_i as a function of time and six arbitrary constants:

$$x_i = x_i(t, C_1, C_2, \ldots, C_6), \quad i = 1, 2, 3. \tag{1.6}$$

Let us find $C_1, C_2, \ldots, C_6$ using the initial conditions. From equations of motion (1.6) the law of changes in the projections of the velocity of the moving particle can be easily derived:

$$\dot{x}_i = \dot{x}_i(t, C_1, C_2, \ldots, C_6), \quad i = 1, 2, 3. \tag{1.7}$$

Let us take the initial value of time $t = t_0$, for which the values of coordinates and of the projections of velocity x_{i0}, $\dot{x}_{i0}$, $i = 1, 2, 3$, are known. Then invoking (1.6) and (1.7) we can write

$$x_{i0} = x_i(t_0, C_1, C_2, \ldots, C_6), \quad \dot{x}_{i0} = \dot{x}_i(t_0, C_1, C_2, \ldots, C_6),$$
$$i = 1, 2, 3. \tag{1.8}$$

We shall assume that for the system of equations (1.8) the solvability conditions are satisfied. In this case, the values of arbitrary constants C_j, $j = 1, 2, \ldots, 6$, which are of interest to us, can be found as functions of the initial data

$$C_j = \Phi_j(t_0, x_{10}, x_{20}, x_{30}, \dot{x}_{10}, \dot{x}_{20}, \dot{x}_{30}), \quad j = 1, 2, \ldots, 6.$$

At that point, one could conclude the exposition of particle mechanics, because solving concrete problems now reduces to the integration of a system of differential equations of type (1.2), which can usually be done numerically accurately enough with the help of modern computers. However, often the complete determination of the particle motion is not required, while for an analytic investigation of the pattern of motion it is sufficient to know just a few integrals of motion.

2   General Principles of Particle Dynamics

As was already noted, the inverse problem of dynamics reduces to the integration of a system of differential equations of motion:

$$\frac{d}{dt}(m\dot{x}_i) = X_i(t, x, \dot{x}), \quad i = 1, 2, 3,$$

where by x we mean the tuple of all coordinates x_1, x_2, x_3. Similarly, $\dot{x} = (\dot{x}_1, \dot{x}_2, \dot{x}_3)$. To put it another way, the problem consists in finding the six integrals of the following normal system:

$$\frac{dp_i}{dt} = X_i(t, x, p), \quad \frac{dx_i}{dt} = \frac{p_i}{m}, \quad i = 1, 2, 3,$$

where $p_i = mv_i$ and $p = (p_1, p_2, p_3)$. The *integral* of this system is a functions $\Phi(t, x, p) = C = \text{const}$ that in motion retains constant values depending on the initial conditions. The integrals of equations of motion can sometimes be found on the basis of general theorems of dynamics that are corollaries of the fundamental equation expressing the Newton second law. Let us consider these theorems.

The principle of momentum is a corollary of the fundamental Newton law

$$\frac{d\mathbf{p}}{dt} = \mathbf{F}(t, x, p), \quad \mathbf{F} = (X_1, X_2, X_3), \tag{2.1}$$

where the vector $\mathbf{p} = m\mathbf{v}$ is the *momentum* or the *impulse*. The force $\mathbf{F}$ can depend on the time t, coordinates x_i, and velocities $\dot{x}_i$. In this case, the existence and uniqueness of the solution under the initial conditions $x_i(t_0) = x_{i0}$, $\dot{x}_i(t_0) = \dot{x}_{i0}$ takes place providing that the force $\mathbf{F}$ is continuous in the vicinity of the initial values and the Lipschitz conditions are satisfied for all the functions X_1, X_2, X_3 in all the arguments x_1, x_2, x_3, $\dot{x}_1$, $\dot{x}_2$, $\dot{x}_3$.

If the force $\mathbf{F}$ is given as a function of time, then the variables in Eq. (2.1) can be separated, and after integrating we obtain

$$\mathbf{p} - \mathbf{p}_0 = \int_{t_0}^{t} \mathbf{F}(t)\, dt \equiv \mathbf{I}(t). \tag{2.2}$$

Let us call the definite integral $\mathbf{I} = \int_{t_0}^{t} \mathbf{F}\, dt$ the *impulse of force* for the time interval $t - t_0$. Formula (2.2) means that the change of impulse for the time $t - t_0$ is equal to the corresponding impulse of force.

By virtue of the fundamental theorem of integral calculus, we have

$$\mathbf{I}(t) = \int_{t_0}^{t} \mathbf{F}(t)\, dt = \mathbf{S}(t) - \mathbf{S}(t_0)\,,$$

where $\mathbf{S}(t)$ is the primitive function. Consequently, formula (2.2) can be written in the form

$$\mathbf{p}(t) - \mathbf{S}(t) = \mathbf{p}(t_0) - \mathbf{S}(t_0) = \mathbf{C}\,.$$

This formula is the integral of impulses in the vector form. Projecting onto the axes of the Cartesian coordinate system, we have three scalar integrals:

$$p_i(t) - S_i(t) = C_i\,, \quad i = 1, 2, 3,$$

where C_i, S_i are the components of vectors $\mathbf{C}$ and $\mathbf{S}$, respectively. In particular, if $\mathbf{F} \equiv 0$, then $\mathbf{p} = \mathbf{p}_0 = \mathbf{p}(t_0)$, which expresses the *law of conservation of impulse*. In the most general case, when $\mathbf{F}$ depends on t, x_i, p_i, for calculating the generating function $\mathbf{S}$ one should know the behavior of x_i depending on t. However, if the functions $x_i(t)$ are known, then we have already solved the second problem of dynamics, and the integrals of its equations are known as well. In this case we can write

$$\mathbf{p}(t) - \mathbf{p}(t_0) = \int_{t_0}^{t} \mathbf{F}(t, x(t), p(t))dt\,. \tag{2.3}$$

This equality is the analytic expression of the *principle of linear momentum* in the integral form.

Applying the mean-value theorem from integral calculus to the integral (2.3) for $\mathbf{p}_0 = 0$, we can write

$$\mathbf{p} = \mathbf{F}^* \tau\,,$$

where $\tau = t - t_0$ and $\mathbf{F}^*$ is the mean value of the force $\mathbf{F}$ in the interval τ. If we let τ tend to zero with simultaneously increasing $\mathbf{F}^*$ so that the quantity $\mathbf{p}$ remains bounded all the time, then we arrive at the idea of an instantaneously acting force of "infinitely large" value. Such a force is called the *impact*. Integrating formula (2.3) between 0 and τ and assuming that $v_0 = 0$, we obtain

$$\mathbf{r} - \mathbf{r}_0 = \int_{0}^{\tau} \frac{\mathbf{I}}{m}\, dt = \frac{\mathbf{I}^*}{m}\tau\,,$$

where $\mathbf{I}^*$ is the mean value of the impulse in the interval τ.

Letting τ tend to zero and taking into account that in so doing $\mathbf{I}^*$ remains finite, we find that if an impact takes place, then[1]

$$\mathbf{r} - \mathbf{r}_0 = 0.$$

The angular impulse-momentum principle. Multiplying both sides of the fundamental Eq. (2.1) vectorially by $\mathbf{r}$, we find

$$\mathbf{r} \times \frac{d\mathbf{p}}{dt} = \mathbf{r} \times \mathbf{F},$$

or

$$\frac{d(\mathbf{r} \times \mathbf{p})}{dt} - \frac{d\mathbf{r}}{dt} \times \mathbf{p} = \mathbf{r} \times \mathbf{F}. \tag{2.4}$$

But $d\mathbf{r}/dt = \mathbf{v}$, $\mathbf{p} = m\mathbf{v}$, and so

$$\frac{d\mathbf{r}}{dt} \times \mathbf{p} = \mathbf{v} \times m\mathbf{v} = 0.$$

Hence formula (2.4) reduces to the form

$$\frac{d(\mathbf{r} \times \mathbf{p})}{dt} = \mathbf{r} \times \mathbf{F}.$$

The vector products $\mathbf{l} = \mathbf{r} \times \mathbf{p}$, $\mathbf{L} = \mathbf{r} \times \mathbf{F}$ are called the *moment of impulse* and the *moment of force*, respectively, about the pole O, which is the origin of the vector $\mathbf{r}$. So,

$$d\mathbf{l}/dt = \mathbf{L}. \tag{2.5}$$

This equality is the analytic expression of the *angular impulse-momentum principle*. We can see from Eq. (2.5) that if the moment of force $\mathbf{L} = 0$, then

$$\mathbf{l} = \mathbf{C}', \tag{2.6}$$

which is the *integral of moments*. This integral takes place in the trivial cases when $\mathbf{r} = 0$ or $\mathbf{F} = 0$, and also when the vector $\mathbf{r}$ is collinear to $\mathbf{F}$, i.e., when the force $\mathbf{F}$ is directed to the center O. In this case, the force is called *as central*. Multiplying (2.6) by $\mathbf{r}$, we find

$$\mathbf{r} \cdot \mathbf{l} = \mathbf{r} \cdot (\mathbf{r} \times m\mathbf{v}) = \mathbf{r} \cdot \mathbf{C}' = 0,$$

or, in the Cartesian coordinates with origin at the point O,

$$C_1' x_1 + C_2' x_2 + C_3' x_3 = 0. \tag{2.7}$$

---

[1] For more details, see the impact theory in chapter "Physical Impact Theory" of the second volume.

This is the equation of the *Laplace plane* orthogonal to the vector $\mathbf{C}'$. This plane contains the trajectory described by a particle subject to a central force.

The vector product

$$\mathbf{r} \times \Delta\mathbf{r} = 2\,\Delta\boldsymbol{\sigma}\,, \tag{2.8}$$

where $\Delta\mathbf{r}$ is the displacement of the particle, is a vector equal in magnitude to the doubled area of the triangle built on the vectors $\mathbf{r}$ and $\Delta\mathbf{r}$. Dividing equality (2.8) by the time interval Δt in which the displacement takes place, and passing to the limit, we have

$$2\frac{d\boldsymbol{\sigma}}{dt} = \mathbf{r} \times \mathbf{v}\,.$$

The vector $d\boldsymbol{\sigma}/dt$, which the limit of the ratio of vector $\Delta\boldsymbol{\sigma}$ to the corresponding time increment, is naturally called the *sectorial velocity*, the vector $\Delta\boldsymbol{\sigma}$ characterizing the increment of the area of curvilinear sector that is described by the position vector $\mathbf{r}$. It follows from the above that if the integral of moments exists, then

$$2\frac{d\boldsymbol{\sigma}}{dt} = \frac{1}{m}\mathbf{C}' = \mathbf{C}^*\,,$$

which implies that

$$\boldsymbol{\sigma} = \frac{\mathbf{C}^*}{2}t + \boldsymbol{\sigma}_0\,.$$

As in this case the motion is executed in a plane given by equation (2.7), then taking into account the expression for $\mathbf{v}$ in polar coordinates $\mathbf{v} = \dot{r}\mathbf{e}_r^0 + r\dot{\varphi}\mathbf{e}_\varphi^0$, we obtain $\mathbf{l}/m = \mathbf{r} \times \mathbf{v} = r^2\dot{\varphi}\mathbf{l}^0 = \mathbf{C}^*$, where $\mathbf{l}^0 = \mathbf{e}_r^0 \times \mathbf{e}_\varphi^0$, or, in the scalar form,

$$r^2\dot{\varphi} = C^*\,,$$

which is the *areal integral*.

The vector integral (2.6) is equivalent to the three scalar integrals

$$\mathbf{l} \cdot \mathbf{i}_1 = l_1 = (\mathbf{r} \times m\mathbf{v})_1 = m(x_2v_3 - x_3v_2) = C_1'\,,$$

$$\mathbf{l} \cdot \mathbf{i}_2 = l_2 = (\mathbf{r} \times m\mathbf{v})_2 = m(x_3v_1 - x_1v_3) = C_2'\,,$$

$$\mathbf{l} \cdot \mathbf{i}_3 = l_3 = (\mathbf{r} \times m\mathbf{v})_3 = m(x_1v_2 - x_2v_1) = C_3'\,.$$

The projections of the vector $\mathbf{l}$ onto the coordinate axes are called the *moments of impulse with respect to the axes* x_1, x_2, and x_3, respectively.

In the projections onto the coordinate axes, we have from formula (2.5)

$$\frac{dl_i}{dt} = L_i\,, \quad i = 1, 2, 3,$$

Fig. 1 Evaluating the
elementary work of force

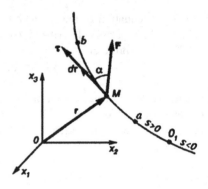

where L_i are the *moments of force vector with respect to the axes* $x_1, x_2,$ and x_3. As projections of a vector product, they are expressed by the formulas

$$L_1 = x_2 X_3 - x_3 X_2 , \quad L_2 = x_3 X_1 - x_1 X_3 , \quad L_3 = x_1 X_2 - x_2 X_1 .$$

The work-energy principle. Again, we will proceed from the fundamental equation of mechanics (2.1)

$$m \, d\mathbf{v} = \mathbf{F}(t, \mathbf{r}, \dot{\mathbf{r}}) \, dt .$$

Multiplying the both sides by $\mathbf{v}$, we obtain

$$m \, \mathbf{v} \cdot d\mathbf{v} = \mathbf{F} \cdot \mathbf{v} \, dt = \mathbf{F} \cdot d\mathbf{r} ,$$

where $d\mathbf{r}$ is the elemental displacement of the particle. This expression can also be written as follows

$$d \frac{mv^2}{2} = \mathbf{F} \cdot d\mathbf{r} = \mathbf{F} \cdot \boldsymbol{\tau} \, ds = F_\tau \, ds = F \cos \alpha \, ds , \tag{2.9}$$

where $\boldsymbol{\tau}$ is the unit vector tangent to the trajectory and directed so that the arc coordinate s is increasing; α is the angle between the vectors $\mathbf{F}$ and $\boldsymbol{\tau}$ (Fig. 1). The scalar product $\delta A = \mathbf{F} \cdot d\mathbf{r}$, as was already noted in Sect. 6 of Chap. 1 of Part I, is called the *elementary work done by the force* $\mathbf{F}$ *on the displacement* $d\mathbf{r}$.

The equality (2.9) expresses the *work-energy principle of a particle in the differential form*: the differential of the kinetic energy agrees with the elementary work done by the resultant of forces applied to the particle.

In a motion of the particle along the trajectory from point a to point b (Fig. 1) the force $\mathbf{F}$ does the work

$$A = \int_{\smile ab} \mathbf{F} \cdot d\mathbf{r} = \int_{\smile ab} F_\tau \, ds = \int_{\smile ab} F \cos \alpha \, ds .$$

As the scalar product $\mathbf{F} \cdot d\mathbf{r}$ can be represented as

$$\mathbf{F} \cdot d\mathbf{r} = X_1 dx_1 + X_2 dx_2 + X_3 dx_3 \,,$$

the expression for the work can be written as follows

$$A = \int_{\smile ab} (X_1 dx_1 + X_2 dx_2 + X_3 dx_3) \,.$$

Integrals of the form $\int_{\smile ab} X_i \, dx_i$ are *curvilinear integrals of the second kind*. Note particularly that these integrals change the sign in motion along the same path but in the opposite direction, i.e.,

$$\int_{\smile ab} X_i \, dx_i = - \int_{\smile ba} X_i \, dx_i \,.$$

Integrating expression (2.9), we obtain the *work-energy principle of a particle in the integral form*:

$$\frac{mv^2}{2} - \frac{mv_0^2}{2} = \int_{\smile ab} F \cos \alpha \, ds = \int_{\smile ab} (X_1 dx_1 + X_2 dx_2 + X_3 dx_3) \,. \qquad (2.10)$$

Here, v_0 is the velocity of the particle at the starting point a; v is the velocity at the terminal point b. Formula (2.10) indicates that the increment to kinetic energy of the particle on a path segment is equal to the work done by the force on this segment.

If the tangential component F_τ of the force is given as a function of the arc of trajectory, i.e., $F_\tau = F_\tau(s)$, then the calculation of the curvilinear integral expressing the work reduces to the computation of the ordinary definite integral

$$A = \int_{s_0}^{s_1} F_\tau(s) \, ds \,.$$

Similarly, if the law of motion of the particle $x_k = x_k(t)$ is known, then we can write

$$A = \int_{t_0}^{t_1} F_\tau(t, x(t), \dot{x}(t)) \sqrt{\dot{x}_1^2(t) + \dot{x}_2^2(t) + \dot{x}_3^2(t)} \, dt \,,$$

or

$$A = \sum_{k=1}^{3} \int_{t_0}^{t_1} X_k(t, x(t), \dot{x}(t)) \, \dot{x}_k(t) \, dt \,,$$

i.e., we again obtain ordinary integrals.

In a specific case, when the trinomial of elemental work

$$\delta A = \sum_{k=1}^{3} X_k dx_k = dU(x_1, x_2, x_3) \equiv dU(x)$$

is the total differential of some scalar function U of coordinates of the particle, the work done on the path segment is expressed as follows

$$A = \int_{a}^{b} (X_1 dx_1 + X_2 dx_2 + X_3 dx_3) = \int_{a}^{b} dU = U(b) - U(a),$$

and, therefore, formula (2.10) takes the form

$$\frac{mv^2}{2} - \frac{mv_0^2}{2} = U(b) - U(a),$$

or

$$\frac{mv^2}{2} - U(b) = \frac{mv_0^2}{2} - U(a) = \text{const}. \qquad (2.11)$$

This expression is the integral of differential equations of motion. It is quite easy to see that in the case under consideration the work A is independent of the shape of the path, but it is controlled by the difference of the values which the function U takes at the starting and terminal points of the path. The function U, which is a function of coordinates of the particle, is called the *force potential* or the *force function*. The function $\Pi = -U$ is called the *potential energy of the particle*. Using this function, the integral (2.11) assumes the form

$$T + \Pi = T_0 + \Pi_0 = \text{const}; \qquad (2.12)$$

it expresses the constancy of the sum of kinetic and potential energies. Expression (2.12) is called the *energy integral*, and the sum $T + \Pi = E$ is the *total mechanical energy of a particle*.

The above shows that the energy integral does not always exist; it exists only when the condition

$$F(s) \cos \alpha(s) \, ds = \sum_{k=1}^{3} X_k \, dx_k = dU(x_1, x_2, x_3)$$

is satisfied, which can also be written in the form

$$\sum_{k=1}^{3} X_k \, dx_k = \sum_{k=1}^{3} \frac{\partial U}{\partial x_k} \, dx_k \, .$$

Hence, we have

$$X_k = \frac{\partial U}{\partial x_k} \quad \text{or} \quad \mathbf{F} = \sum_{k=1}^{3} \frac{\partial U}{\partial x_k} \, \mathbf{i}_k \, . \tag{2.13}$$

A vector given in the form (2.13) is called the *gradient of the scalar function* $U(x)$. The symbolic operator (the *Hamiltonian operator*) ∇ is sometimes called the "nabla" operator $\nabla = \sum_{k=1}^{3} \frac{\partial}{\partial x_k} \mathbf{i}_k$. So,

$$\mathbf{F} = \sum_{k=1}^{3} \frac{\partial U}{\partial x_k} \, \mathbf{i}_k \equiv \nabla U = \text{grad } U \, .$$

3 Potential Force Field

A spatial region on which the function $U(x_1, x_2, x_3)$ is defined is called the *stationary potential field*. Assuming that such a function is twice differentiable, we can build a vector field for the force: $\mathbf{F} = \nabla U$, in so doing, the components $X_k = \partial U / \partial x_k$ are continuous. Apparently, the equation

$$U(x_1, x_2, x_3) = C = \text{const} \tag{3.1}$$

is the equation of a family of surfaces depending on the parameter C. On each surface, i.e., for fixed C, the potential $U(x)$ retains a constant value, and that is why the family (3.1) is called the *family of surfaces of equal potential* or *equipotential surfaces*. Clearly, if the vector $d\mathbf{r}$ lies in a plane tangent to this surface, then

$$\mathbf{F} \cdot d\mathbf{r} = \nabla U \cdot d\mathbf{r} = dU = 0 \, ,$$

i.e., the vector ∇U is orthogonal to the surface of equal potential at the point for which it is calculated. Let us call a curve whose tangent at each point coincides in direction with the force corresponding to this point the *force line*. Differential equations of such lines have the form

$$\frac{dx_1}{X_1} = \frac{dx_2}{X_2} = \frac{dx_3}{X_3} \, .$$

It follows from the above that the family of force lines of a potential field is orthogonal to the family of equipotential surfaces, as the vector $d\mathbf{r}$ is collinear to the vector $\mathbf{F}$ directed along the normal to a surface of equal potential.

As was already noted, if $\mathbf{F} = \nabla U(x_1, x_2, x_3)$, then the work will not depend on the shape of the path described by the point. Let us prove the inverse: if the work in a force field is path-independent, then the field is potential, the force field being an area of space to each point of which a force vector $\mathbf{F}$ corresponds.

We choose a curve connecting the points a and b in the force field and calculate the work done in moving through this segment

$$A = \int_{\smile ab} \mathbf{F} \cdot d\mathbf{r} = \int_{\smile ab} \mathbf{F} \cdot \boldsymbol{\tau}\, ds = \int_{\smile ab} F_\tau\, ds \,.$$

As the integral is path-independent by the assumption, then at a fixed point a it is a function of coordinates (x_1, x_2, x_3) of the point b:

$$A = \int_{\smile ab} F_\tau\, ds = U(\mathbf{r}) \,, \quad \mathbf{r} = x_1 \mathbf{i}_1 + x_2 \mathbf{i}_2 + x_3 \mathbf{i}_3 \,.$$

Moving the point b to the position b' with coordinates $x_k + \Delta x_k$, we have

$$A(\mathbf{r} + \Delta\mathbf{r}) = \int_{\smile ab'} F_\tau\, ds = U(\mathbf{r} + \Delta\mathbf{r}) \,.$$

Hence,

$$\int_{\smile bb'} F_\tau\, ds = U(\mathbf{r} + \Delta\mathbf{r}) - U(\mathbf{r}) \,.$$

Applying the mean-value theorem to this integral, we find

$$F_\tau(\mathbf{r} + \lambda\Delta\mathbf{r})\, \Delta s = U(\mathbf{r} + \Delta\mathbf{r}) - U(\mathbf{r}) \,, \quad 0 < \lambda < 1 \,.$$

Dividing both sides of this equality by Δs and letting Δs tend to zero, we have a derivative with respect to the direction s:

$$\mathbf{F} \cdot \boldsymbol{\tau} = F_\tau = \frac{\partial U(\mathbf{r})}{\partial s} \,.$$

This direction is characterized by the vector $\boldsymbol{\tau} = \lim_{\Delta s \to 0} (\Delta\mathbf{r}/\Delta s)$. Replacing Δs sequentially by the quantities Δx_1, Δx_2, and Δx_3, we obtain

$$X_1 = \frac{\partial U}{\partial x_1} \,, \quad X_2 = \frac{\partial U}{\partial x_2} \,, \quad X_3 = \frac{\partial U}{\partial x_3}$$

or

$$\mathbf{F} = \nabla U \,.$$

Fig. 2 Small triangle

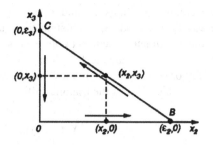

It is this expression that proves the statement formulated earlier.

From the path independence of the integral A we also see that the integral $A = \oint \mathbf{F} \cdot d\mathbf{r}$ about a closed circuit is zero if the function U is single valued.

Note that the derivative $\partial U / \partial s$ can be obtained by the formula

$$F_\tau = \mathbf{F} \cdot \boldsymbol{\tau} = \sum_{k=1}^{3} \frac{\partial U}{\partial x_k} \frac{\partial x_k}{\partial s} = \frac{\partial U}{\partial s},$$

where dx_k / ds are the direction cosines of $\boldsymbol{\tau}$.

Now let us find the conditions under which the work done about a closed circuit in a force field be zero. With this aim in mind, we first calculate the work along the contour of the triangle displayed in Fig. 2. The direction in which the contour is traversed is shown by arrows. The work along the whole contour is the sum of three integrals

$$\Delta A_{23} = \int_{\smile OB} \mathbf{F} \cdot d\mathbf{r} + \int_{\smile BC} \mathbf{F} \cdot d\mathbf{r} + \int_{\smile CO} \mathbf{F} \cdot d\mathbf{r} =$$

$$= \int_{\smile OB} X_2 \, dx_2 + \int_{\smile BC} (X_2 \, dx_2 + X_3 \, dx_3) + \int_{\smile CO} X_3 \, dx_3 =$$

$$= \int_0^{\varepsilon_2} X_2(x_2, 0) \, dx_2 + \int_{\varepsilon_2}^0 X_2(x_2, x_3) \, dx_2 + \int_0^{\varepsilon_3} X_3(x_2, x_3) \, dx_3 +$$

$$+ \int_{\varepsilon_3}^0 X_3(0, x_3) \, dx_3 = \int_0^{\varepsilon_3} (X_3(x_2, x_3) - X_3(0, x_3)) \, dx_3 -$$

$$- \int_0^{\varepsilon_2} (X_2(x_2, x_3) - X_2(x_2, 0)) \, dx_2. \tag{3.2}$$

Here, for simplicity, X_2 and X_3 are represented as functions of just two independent variables x_2 and x_3. The extracted arguments $(x_2, 0)$, (x_2, x_3), and $(0, x_3)$ are also the coordinates of current points on the lines OB, BC, and CO, respectively.

Assuming that the functions $X_2(x_2, x_3)$ and $X_3(x_2, x_3)$ can be expanded into Taylor series near the point O and considering only the linear terms, i.e., choosing the dimensions of the triangle sufficiently small, we have with fixed x_2

$$X_2(x_2, x_3) - X_2(x_2, 0) = \frac{\partial X_2}{\partial x_3} x_3, \tag{3.3}$$

and with fixed x_3

$$X_3(x_2, x_3) - X_3(0, x_3) = \frac{\partial X_3}{\partial x_2} x_2. \tag{3.4}$$

Recall that we calculate the derivatives $\partial X_2 / \partial x_3$ and $\partial X_3 / \partial x_2$ at the point O, though for simplicity it is not mentioned.

The curvilinear integral (3.2) includes the difference of values of the function X_2 at the point (x_2, x_3) lying on the line BC and at the point $(x_2, 0)$ lying in the line OB (see Fig. 2). When substituting the difference (3.3) into the integral (3.2), one should take into account that x_3 satisfies the equation of the line BC

$$\frac{x_2}{\varepsilon_2} + \frac{x_3}{\varepsilon_3} = 1. \tag{3.5}$$

Similarly, when integrating with respect to x_3 by formula (3.4), one should keep in mind that x_2 satisfies Eq. (3.5) and

$$\int_0^{\varepsilon_2} x_3 \, dx_2 = \int_0^{\varepsilon_3} x_2 \, dx_3 = \frac{1}{2} \varepsilon_2 \varepsilon_3 = \Delta S_{23}.$$

Thus, if the triangle OBC is sufficiently small, then

$$\Delta A_{23} = \left(\frac{\partial X_3}{\partial x_2} - \frac{\partial X_2}{\partial x_3} \right) \Delta S_{23} = R_1 \, \Delta S_{23}.$$

By advancing the letters, we obtain

$$\Delta A_{31} = \left(\frac{\partial X_1}{\partial x_3} - \frac{\partial X_3}{\partial x_1} \right) \Delta S_{31} = R_2 \, \Delta S_{31},$$

$$\Delta A_{12} = \left(\frac{\partial X_2}{\partial x_1} - \frac{\partial X_1}{\partial x_2} \right) \Delta S_{12} = R_3 \, \Delta S_{12}.$$

Here, R_1, R_2, and R_3 are components of the vector $\mathbf{R}$, which is called the *rotation of vector* $\mathbf{F}$. With the help of the operator ∇, it can be represented as

Fig. 3 Small tetrahedron

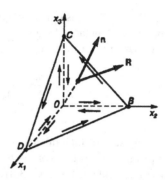

$$\mathbf{R} = \nabla \times \mathbf{F} = \mathrm{rot}\,\mathbf{F}.$$

Consider the tetrahedron $OBCD$ (Fig. 3) constructed on the platforms ΔS_{ij}. Let the area of its face BCD be ΔS. Then

$$\Delta S_{23} = \Delta S \cos \alpha_1, \quad \Delta S_{31} = \Delta S \cos \alpha_2, \quad \Delta S_{12} = \Delta S \cos \alpha_3.$$

Here, $\cos \alpha_k = \mathbf{n} \cdot \mathbf{i}_k$ are the components of the unit vector $\mathbf{n}$ of the outer (external) normal to the face BCD.

Summing the work along the contours of the triangles OBC, OCD, and ODB in the directions indicated by the arrows and assuming that the dimensions of the tetrahedron are sufficiently small, we obtain

$$\Delta A = \Delta A_{23} + \Delta A_{31} + \Delta A_{12} = (R_1 \cos \alpha_1 + R_2 \cos \alpha_2 + R_3 \cos \alpha_3)\,\Delta S =$$

$$= (\mathbf{R} \cdot \mathbf{n})\,\Delta S = R_n\,\Delta S. \tag{3.6}$$

When calculating ΔA, we traverse twice each of the edges OB, OC, and OD but in opposite directions, and therefore, the sum of work done on the sides of the indicated triangles is zero. Only the work done on the hypotenuses remains. Therefore, ΔA is the work done around the contour of the triangle BCD of the larger (major) face; this contour is traversed counterclockwise, as viewed from the end of the vector $\mathbf{n}$. The equality (3.6) is an approximate one. The smaller dimensions of the tetrahedron are, the more accurate it is. Making $\Delta S \to 0$, we obtain the following exact equality:

$$\lim_{\Delta S \to 0} \frac{\Delta A}{\Delta S} = \mathbf{R} \cdot \mathbf{n} = R_n. \tag{3.7}$$

The point O, at which the vector $\mathbf{R}$ is determined, was placed for simplicity at the origin. The final formula (3.7) holds for every point of the space. It is written in the form that bears no relation to the choice of the coordinate system, and it can be shown that it is valid not only for a triangle but also for an area ΔS of arbitrary shape.

Now let us consider a smooth surface limited by a smooth closed contour. It can be roughly replaced by a polyhedron with the faces of triangle forms of areas ΔS_k. Applying the formula (3.6) to each of these platforms and summing over k, we obtain

$$A = \sum_{k=1}^{N} \Delta A_k = \sum_{k=1}^{N} \int_{C_k} \mathbf{F_k} \cdot d\mathbf{r} = \sum_{k=1}^{N} R_{nk}\, \Delta S_k \,. \tag{3.8}$$

The sum of integrals taken along the segments common to adjacent triangles is equal to zero, as the work in this case is summed in opposite directions. Hence, the left-hand side of formula (3.8) can be transformed to a sum of integrals taken over the segments of the outer (external) contour:

$$A = \sum_{j=1}^{n} \int_{\smile a_j b_j} \mathbf{F} \cdot d\mathbf{r} = \sum_{j=1}^{n} \mathbf{F}_j^* \cdot \Delta \mathbf{r}_j \,,$$

where $\mathbf{F}_j^*$ is the value of $\mathbf{F}$ at some point of the segment $\Delta \mathbf{r}_j$. Now formula (3.8) assumes the form

$$\sum_{j=1}^{n} \mathbf{F}_j^* \cdot \Delta \mathbf{r}_j = \sum_{k=1}^{N} R_{nk}\, \Delta S_k \,.$$

Making n and N go off to infinity and assuming that $\max_k \Delta S_k \to 0$, we find

$$A = \oint_C \mathbf{F} \cdot d\mathbf{r} = \int_S \mathbf{R} \cdot \mathbf{n}\, dS = \int_S (\nabla \times \mathbf{F}) \cdot \mathbf{n}\, dS \,. \tag{3.9}$$

In vector analysis, the curvilinear integral on the left-hand side of formula (3.9) is called the *circulation of the vector* $\mathbf{F}$ *along the contour C*. The right-hand side of expression (3.9) includes the surface integral taken over any smooth surface limited by the contour C. An integral containing the projection of some vector onto the normal to the surface is called the *flux of this vector through the surface*. Thus, formula (3.9) means that in a vector field the circulation of a vector is equal to the flux of the rotation of this vector through any smooth surface resting on that contour (*Stokes's theorem*). It follows from this theorem that if $\mathbf{R} = \operatorname{rot} \mathbf{F} = 0$, then in a simply connected domain the work in a force field is independent of the path of integration and, consequently, it is a potential one.

Examples of potential forces

1. *Central force depending on the distance r.* If we place the center into the origin and denote the projection of the force $\mathbf{F}$ onto the direction $\mathbf{r}^0 = \mathbf{r}/r$ as $F(r)$, then this force can be represented as

$$\mathbf{F}(\mathbf{r}) = \frac{F(r)}{r}\,\mathbf{r}\,.$$

In this case the elementary work is

$$\mathbf{F}{\cdot}d\mathbf{r} = \frac{F(r)\,\mathbf{r}\cdot d\mathbf{r}}{r} = \frac{F(r)\,d(r^2/2)}{r} = F(r)\,dr = dU\,,$$

whence $U = \int F(r)\,dr$.

In a specific case when $F(r) = -c_1 r$ (a quasielastic force),

$$U = -c_1 r^2/2 + \text{const}\,,$$

where c_1 is the stiffness.

If $F(r) = -\gamma m M/r^2$ (attraction by the law of gravitation), then

$$U = \gamma m M/r + \text{const}\,.$$

Here, m is the mass of the particle; M is the mass of the attracting center; γ is the gravity constant.

2. *Plane central field.* Let a force $\mathbf{F}$ be a function of the distance ρ between its point of application and the x_3-axis, we assume that the direction of the force is collinear to the vector $\boldsymbol{\rho}^0 = \boldsymbol{\rho}/\rho$, i.e.,

$$\mathbf{F} = F(\rho)\,\boldsymbol{\rho}/\rho\,,\quad \boldsymbol{\rho} = x_1\mathbf{i}_1 + x_2\mathbf{i}_2\,.$$

In this case

$$\mathbf{F}\cdot d\mathbf{r} = \mathbf{F}\cdot d\boldsymbol{\rho} = F(\rho)\,d\rho = dU\,.$$

As the properties of this field are identical in all the planes perpendicular to the axis x_3, we can limit ourselves to considering just one of them, the plane Ox_1x_2. An example of a plane central field is a field, for which

$$F(\rho) = C/\rho\,,\quad U = \int F(\rho)\,d\rho = C\ln\rho + \text{const}\,,\quad C = \text{const}\,.$$

The surfaces of equal potential are circumferences, the force lines are the lines defined by the differential equation

$$\frac{dx_1}{X_1} = \frac{dx_2}{X_2}\,,\quad X_k = \frac{F(\rho)\,x_k}{\rho}\,,\quad k = 1, 2,$$

or $dx_1/x_1 = dx_2/x_2$, whence $x_2/x_1 = \text{const}$.

Such a force field can be created if we continuously distribute electric charges $de = (C/2)\,dx_3$ on the infinitely long line coinciding with the x_3-axis, and add together their actions on a unit charge located at the point with the coordinates

Fig. 4 Planar force field
with many-valued potential

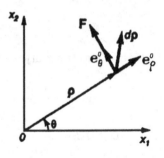

$(x_1, x_2, 0)$ taking into account Coulomb's law. It is quite easy to see that all the above potentials are single valued.

3. *Plane force field with multivalued potential.* For this case a force field is typical, in which

$$X_1 = -\frac{C \sin \theta}{\rho} = -\frac{C x_2}{\rho^2}, \qquad X_3 = 0,$$

$$X_2 = \frac{C \cos \theta}{\rho} = \frac{C x_1}{\rho^2}, \qquad |\mathbf{F}| = \frac{|C|}{\rho},$$

(3.10)

where θ is the polar angle (Fig. 4).

We find the force lines from the equation

$$\frac{dx_1}{x_2} = -\frac{dx_2}{x_1} \qquad \text{or} \qquad d(x_1^2 + x_2^2) = 0,$$

whence $x_1^2 + x_2^2 = \rho^2 = \text{const}$, i.e., we have obtained the equations of circumferences.

Due to the orthogonality conditions, the lines of equal potential are lines emanating from the origin.

Figure 4 shows that the radial component of the force $\mathbf{F}$ is equal to zero, while the transverse one can be represented by the formula

$$F_\rho = \mathbf{F} \cdot \mathbf{e}_\theta^0 = C/\rho.$$

On the basis of this, the work done by the force $\mathbf{F}$ through the displacement $d\rho$ is as follows

$$\mathbf{F} \cdot d\rho = \frac{C}{\rho} \mathbf{e}_\theta^0 \cdot (\mathbf{e}_\rho^0 \, d\rho + \mathbf{e}_\theta^0 \rho \, d\theta) = C \, d\theta = dU,$$

whence, $U = C\theta + \text{const}$, which is a multivalued function. When traversing n times the closed path about the origin, the work is expressed by the formula

$$A = 2\pi C n.$$

Let us calculate the rotation $\mathbf{R} = \nabla \times \mathbf{F}$. We have

$$R_1 = R_2 = 0, \qquad R_3 = \frac{\partial X_2}{\partial x_1} - \frac{\partial X_1}{\partial x_2} =$$

$$= \frac{C}{\rho^2} - \frac{2Cx_1^2}{\rho^4} + \frac{C}{\rho^2} - \frac{2Cx_2^2}{\rho^4} = \frac{2C}{\rho^2}\left(1 - \frac{x_1^2 + x_2^2}{\rho^2}\right) = 0.$$

This shows that the vector $\mathbf{R}$ is 0 everywhere except at the origin. The force $\mathbf{F}$ increases unboundedly as $\rho \to 0$. To put it another way, we have a singular point at the origin. If we agree not to count this point among those in the field, then the latter becomes a doubly connected potential field with multivalued potential.

The considered field can be realized physically if we imagine that a current I takes the path along the axis x_3 from $-\infty$ to $+\infty$. Now the magnetic intensity reads as

$$\mathbf{F} = \frac{C}{\rho^2}(\mathbf{e}_3^0 \times \boldsymbol{\rho}), \tag{3.11}$$

where $C = I/c$, c is the constant depending on the choice of the units, and $\rho = (x_1^2 + x_2^2)^{1/2}$. From (3.11) we have

$$X_1 = -\frac{Cx_2}{\rho^2}, \quad X_2 = \frac{Cx_1}{\rho^2}, \quad X_3 = 0,$$

which coincides with formulas (3.10).

So far, we have considered the potential field $U(x)$ that does not depend on the time t explicitly. Such a field is called *stationary*. If the function U, in addition, depends on t explicitly, then the field is called *nonstationary*. In this case

$$\mathbf{F} \cdot d\mathbf{r} = \sum_{k=1}^{3} \frac{\partial U}{\partial x_k} dx_k = dU - \frac{\partial U}{\partial t} dt,$$

and consequently,

$$A = U(t, x) - U(t_0, x_0) - \int_{t_0}^{t} \frac{\partial U}{\partial t'} dt',$$

formula (2.10) assuming the form

$$\frac{mv^2}{2} + \Pi(t, x) = \frac{mv_0^2}{2} + \Pi(t_0, x_0) + \int_{t_0}^{t} \frac{\partial \Pi(t', x)}{\partial t'} dt'.$$

To calculate the quadrature one must know the functions $x_k(t)$ (i.e., the law of motion of the particle).

Fig. 5 Radius vector defined
in the form $\mathbf{r} = \mathbf{r}(t, q)$

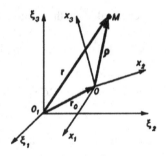

4 Derivation of the Lagrange Equations of the Second Kind in Case of Nonstationary Basis

It was established in Sect. 1 that if the position of a point is given by the curvilinear coordinates q^σ, $\sigma = 1, 2, 3$, i.e., if $\mathbf{r} = \mathbf{r}(q)$, then the equations of dynamics can be written in the form of the Lagrange equations of the second kind (1.5).

Now let us consider a more general case when the position vector $\mathbf{r}$ depends not only on q^σ but also explicitly on the time t, i.e., it is a function of the form $\mathbf{r} = \mathbf{r}(t, q)$. In particular, this is possible when the curvilinear coordinates q^σ specify the position of the particle relative to a coordinate system that moves in a certain way relative to the fixed (absolute) coordinate system $O_1\xi_1\xi_2\xi_3$ (Fig. 5). Even for fixed values of q^σ, the position vector $\mathbf{r}$ changes depending on the time due to the bulk (transport) motion of the system $Ox_1x_2x_3$.

In the case under consideration, we calculate the absolute velocity $\mathbf{v}$ by the formula

$$\mathbf{v} = \dot{\mathbf{r}} = \frac{\partial \mathbf{r}}{\partial t} + \frac{\partial \mathbf{r}}{\partial q^\sigma} \dot{q}^\sigma . \tag{4.1}$$

Introducing for brevity the following notation $t = q^0$, we can express (4.1) as

$$\mathbf{v} = \frac{\partial \mathbf{r}}{\partial q^\alpha} \dot{q}^\alpha , \quad \alpha = 0, 1, 2, 3, \quad \dot{q}^0 = 1 . \tag{4.2}$$

We emphasize that such a representation is introduced only for notational brevity, and therefore, cannot be carried over in the four-dimensional space. The coordinate vectors are still as follows

$$\mathbf{e}_\sigma(t, q) = \frac{\partial \mathbf{r}}{\partial q^\sigma} , \quad \sigma = 1, 2, 3.$$

Thus, a nonstationary basis changes not only when going from one point to another but also at every point with time.

Let us calculate the covariant component of the acceleration $\mathbf{w}$:

$$w_\sigma = \mathbf{w} \cdot \mathbf{e}_\sigma = \frac{d\mathbf{v}}{dt} \cdot \frac{\partial \mathbf{r}}{\partial q^\sigma} = \frac{d}{dt}\left(\mathbf{v} \cdot \frac{\partial \mathbf{r}}{\partial q^\sigma}\right) - \mathbf{v} \cdot \frac{d}{dt}\frac{\partial \mathbf{r}}{\partial q^\sigma}. \tag{4.3}$$

Differentiating the expression (4.1) with respect to $\dot{q}^\rho$ and then with respect to q^ρ, $\rho = 1, 2, 3$, we have

$$\frac{\partial \mathbf{v}}{\partial \dot{q}^\rho} = \frac{\partial \mathbf{r}}{\partial q^\rho}, \quad \frac{\partial \mathbf{v}}{\partial q^\rho} = \frac{\partial^2 \mathbf{r}}{\partial t \partial q^\rho} + \frac{\partial^2 \mathbf{r}}{\partial q^\sigma \partial q^\rho}\dot{q}^\sigma = \frac{d}{dt}\frac{\partial \mathbf{r}}{\partial q^\rho}.$$

Hence, the terms entering the expression (4.3) can be represented in the form

$$\mathbf{v} \cdot \frac{\partial \mathbf{r}}{\partial q^\sigma} = \mathbf{v} \cdot \frac{\partial \mathbf{v}}{\partial \dot{q}^\sigma} = \frac{1}{2}\frac{\partial v^2}{\partial \dot{q}^\sigma} = \frac{\partial T_1}{\partial \dot{q}^\sigma},$$

$$\mathbf{v} \cdot \frac{d}{dt}\frac{\partial \mathbf{r}}{\partial q^\sigma} = \mathbf{v} \cdot \frac{\partial \mathbf{v}}{\partial q^\sigma} = \frac{1}{2}\frac{\partial v^2}{\partial q^\sigma} = \frac{\partial T_1}{\partial q^\sigma},$$

therefore, for w_σ we finally obtain

$$w_\sigma = \frac{d}{dt}\frac{\partial T_1}{\partial \dot{q}^\sigma} - \frac{\partial T_1}{\partial q^\sigma}, \quad T_1 = \frac{v^2}{2}. \tag{4.4}$$

Thus, the Lagrange form of representation of the covariant component w_σ also does not change for a nonstationary basis.

The kinetic energy of the particle, in accordance with the expression (4.2), reads as

$$T = \frac{mv^2}{2} = \frac{m}{2}\frac{\partial \mathbf{r}}{\partial q^\alpha} \cdot \frac{\partial \mathbf{r}}{\partial q^\beta}\dot{q}^\alpha \dot{q}^\beta = \frac{m}{2}g_{\alpha\beta}\dot{q}^\alpha \dot{q}^\beta, \quad \alpha, \beta = 0, 1, 2, 3. \tag{4.5}$$

If in this expression we single out the terms containing $\frac{\partial \mathbf{r}}{\partial q^0} = \frac{\partial \mathbf{r}}{\partial t}$ explicitly, then we can write

$$T = T^{(2)} + T^{(1)} + T^{(0)},$$

where

$$T^{(2)} = \frac{m}{2}\left(\frac{\partial \mathbf{r}}{\partial q^\rho} \cdot \frac{\partial \mathbf{r}}{\partial q^\sigma}\right)\dot{q}^\rho \dot{q}^\sigma = \frac{m}{2}g_{\rho\sigma}\dot{q}^\rho \dot{q}^\sigma,$$

$$T^{(1)} = m\left(\frac{\partial \mathbf{r}}{\partial t} \cdot \frac{\partial \mathbf{r}}{\partial q^\sigma}\right)\dot{q}^\sigma = mg_{0\sigma}\dot{q}^\sigma,$$

$$T^{(0)} = \frac{m}{2}\left(\frac{\partial \mathbf{r}}{\partial t}\right)^2 = \frac{m}{2}g_{00}.$$

Note that in this case the metric coefficients are represented only by the quantities $g_{\rho\sigma}$ entering the expression for $T^{(2)}$.

It follows from formulas (4.4) and (4.5) that multiplying both sides of the Newton equation $m\mathbf{w} = \mathbf{F}$ by $\mathbf{e}_\sigma$, we obtain the Lagrange equations of the second kind:

$$\frac{d}{dt}\frac{\partial T}{\partial \dot{q}^\sigma} - \frac{\partial T}{\partial q^\sigma} = Q_\sigma, \quad Q_\sigma = \mathbf{F} \cdot \mathbf{e}_\sigma, \quad \sigma = 1, 2, 3. \tag{4.6}$$

If the vectors $\mathbf{F}$ and $\mathbf{e}_\sigma$ are given by their components in the Cartesian coordinates, then Q_σ assumes the form

$$Q_\sigma = X_i \frac{\partial x_i}{\partial q^\sigma}.$$

In practice, the following approach can be used to find Q_σ. Consider the vector $\delta\mathbf{r}$ defined as

$$\delta\mathbf{r} = \delta q^\sigma \, \mathbf{e}_\sigma = \frac{\partial \mathbf{r}}{\partial q^\sigma}\, \delta q^\sigma,$$

where δq^σ are the arbitrary changes of the coordinates q^σ called the *variations* of q^σ. The vector $\delta\mathbf{r}$ is the differential of the vector function $\mathbf{r}(t, q)$ calculated for fixed t. It is called the *variation of the vector function* $\mathbf{r}(t, q)$ as well as *virtual deviation* or *displacement* of the particle under consideration. Calculating the elementary work δA done by the force $\mathbf{F}$ in moving through the displacement $\delta\mathbf{r}$, we have

$$\delta A = \mathbf{F} \cdot \delta\mathbf{r} = \mathbf{F} \cdot \mathbf{e}_\sigma\, \delta q^\sigma = Q_\sigma\, \delta q^\sigma.$$

This expression is called the *virtual elementary work*. Here, each summand, for example, $Q_1 \delta q^1$, is the work done by the force $\mathbf{F}$ in moving through the displacement $\delta\mathbf{r}_1 = \delta q^1\, \mathbf{e}_1$. The latter makes it possible to consider the generalized force Q_σ as a coefficient of the variation δq^σ in the expression for the elementary work δA.

If the nonstationary force field is potential, i.e., if for any t

$$\mathbf{F} = \nabla U(t, x),$$

then the generalized force Q_σ assumes the form

$$Q_\sigma = \mathbf{F} \cdot \frac{\partial \mathbf{r}(t, q)}{\partial q^\sigma} = \nabla U(t, x) \cdot \frac{\partial \mathbf{r}(t, q)}{\partial q^\sigma} =$$

$$= \frac{\partial U(t, x)}{\partial x_k}\frac{\partial x_k}{\partial q^\sigma} = \frac{\partial U(t, q)}{\partial q^\sigma}.$$

Note that the gradient ∇U, as follows from the equalities presented, can be written as follows

$$\nabla U(t, q) = Q_\sigma \mathbf{e}^\sigma = \frac{\partial U(t, q)}{\partial q^\sigma}\, \mathbf{e}^\sigma.$$

Now let us ascertain the explicit form of the Lagrange equations of the second kind in case of a nonstationary basis. Substituting the expression (4.5) into equations (4.6), we obtain

$$m \left(g_{\sigma\tau} \ddot{q}^{\tau} + \Gamma_{\sigma,\alpha\beta} \dot{q}^{\alpha} \dot{q}^{\beta} \right) = Q_{\sigma}, \quad \sigma, \tau = 1, 2, 3, \quad \alpha, \beta = 0, 1, 2, 3, \qquad (4.7)$$

where

$$\Gamma_{\sigma,\alpha\beta} = \frac{1}{2} \left(\frac{\partial g_{\sigma\alpha}}{\partial q^{\beta}} + \frac{\partial g_{\sigma\beta}}{\partial q^{\alpha}} - \frac{\partial g_{\alpha\beta}}{\partial q^{\sigma}} \right).$$

The sums $\Gamma_{\sigma,\alpha\beta} \dot{q}^{\alpha} \dot{q}^{\beta}$ in equations (4.7) can be written as

$$\Gamma_{\sigma,\alpha\beta} \dot{q}^{\alpha} \dot{q}^{\beta} = \Gamma_{\sigma,\rho\tau} \dot{q}^{\rho} \dot{q}^{\tau} + 2 \Gamma_{\sigma,0\tau} \dot{q}^{\tau} + \Gamma_{\sigma,00}.$$

The double sum on the right of this expression is encountered in the case of a stationary basis. The sum $2 \Gamma_{\sigma,0\tau} \dot{q}^{\tau}$ and the term $\Gamma_{\sigma,00}$ emerge due to the nonstationarity of the basis. Note that only the quantities $\Gamma_{\sigma,\rho\tau}$ are the Christoffel symbols of the first kind.

5 Derivation of the Energy Integral and the Jacobi Integral From the Lagrange Equations of the Second Kind

If the forces acting on a particle have a potential that does not depend on the time explicitly, then, as was shown in Sect. 2, the equations of motion have the energy integral (2.12). Let us show that this integral can be derived directly from the Lagrange equations of the second kind on the assumption that the basis is stationary. As this takes place, $Q_{\sigma} = \partial U / \partial q^{\sigma}$ and the kinetic energy T is a homogeneous quadratic form of velocities:

$$T = \frac{m}{2} g_{\sigma\tau} \dot{q}^{\sigma} \dot{q}^{\tau},$$

and $g_{\sigma\tau}$ are functions of only the coordinates q^{σ}. Multiplying Lagrange equations of the second kind by $\dot{q}^{\sigma}$ and summing over σ, we obtain

$$\left(\frac{d}{dt} \frac{\partial T}{\partial \dot{q}^{\sigma}} \right) \dot{q}^{\sigma} - \frac{\partial T}{\partial q^{\sigma}} \dot{q}^{\sigma} = \frac{\partial U}{\partial q^{\sigma}} \dot{q}^{\sigma},$$

or

$$\frac{d}{dt} \left(\frac{\partial T}{\partial \dot{q}^{\sigma}} \dot{q}^{\sigma} \right) - \frac{\partial T}{\partial \dot{q}^{\sigma}} \ddot{q}^{\sigma} - \frac{\partial T}{\partial q^{\sigma}} \dot{q}^{\sigma} = \frac{\partial U}{\partial q^{\sigma}} \dot{q}^{\sigma}. \qquad (5.1)$$

As T and U are not supposed to depend on the time explicitly, we have

$$\frac{dT(q,\dot{q})}{dt} = \frac{\partial T}{\partial q^\sigma}\dot{q}^\sigma + \frac{\partial T}{\partial \dot{q}^\sigma}\ddot{q}^\sigma\,, \quad \frac{dU(q)}{dt} = \frac{\partial U}{\partial q^\sigma}\dot{q}^\sigma\,.$$

Now formula (5.1) can be written in the form

$$\frac{d}{dt}\left(\frac{\partial T}{\partial \dot{q}^\sigma}\dot{q}^\sigma - T - U\right) = 0\,.$$

Since $\frac{\partial T}{\partial \dot{q}^\sigma}\dot{q}^\sigma = 2T$ by Euler's theorem on homogeneous functions, we finally obtain

$$\frac{d(T - U)}{dt} = 0\,,$$

whence,

$$E \equiv T + \Pi = \text{const}\,. \tag{5.2}$$

In the general case of a nonstationary problem the following expression would take place

$$\frac{\partial T}{\partial q^\sigma}\dot{q}^\sigma + \frac{\partial T}{\partial \dot{q}^\sigma}\ddot{q}^\sigma = \frac{dT}{dt} - \frac{\partial T}{\partial t}\,,$$

where

$$T = T^{(2)} + T^{(1)} + T^{(0)}\,.$$

Consider the expression $\frac{d}{dt}\left(\frac{\partial T}{\partial \dot{q}^\sigma}\dot{q}^\sigma\right)$ from formula (5.1). Applying Euler's theorem to the functions $T^{(2)}$ and $T^{(1)}$, we obtain

$$\frac{d}{dt}\left(\frac{\partial T}{\partial \dot{q}^\sigma}\dot{q}^\sigma\right) = \frac{d}{dt}(2T^{(2)} + T^{(1)})\,.$$

As a result, formula (5.1) can be written as

$$\frac{d}{dt}(2T^{(2)} + T^{(1)}) - \frac{dT^{(2)}}{dt} - \frac{dT^{(1)}}{dt} - \frac{dT^{(0)}}{dt} +$$

$$+\frac{\partial T}{\partial t} - \frac{dU}{dt} + \frac{\partial U}{\partial t} = 0\,,$$

or

$$\frac{d(T^{(2)} - T^{(0)} - U)}{dt} = -\frac{\partial U}{\partial t} - \frac{\partial T}{\partial t}\,.$$

This shows that if the functions U and T do not contain t explicitly, then the following expression will be valid

$$T^{(2)} - T^{(0)} + \Pi = \text{const}\,, \tag{5.3}$$

that is called the *Jacobi integral*. This integral is used most often in solving special problems of celestial mechanics.

6 Hamilton's Canonical Equations

The Lagrange equations of the second kind, when written in the explicit form (4.7), constitute a system of ordinary second-order differential equations. There are three such equations. It is known that when theoretically studying the solutions behavior of differential equations, it is quite convenient to use the *equations in the normal form*. By this form we mean a system of first-order equations solved for derivatives.

To replace the system of second-order differential equations (4.7) by a system of six first-order equations, it suffices to set

$$\dot{q}^{\sigma} = \widehat{v}^{\sigma}, \quad \sigma = 1, 2, 3 \tag{6.1}$$

and write equations (4.7) in the form

$$m\left(g_{\sigma\tau}\,\dot{\widehat{v}}^{\tau} + \Gamma_{\sigma,\alpha\beta}\,\dot{q}^{\alpha}\dot{q}^{\beta}\right) = Q_{\sigma}, \quad \sigma = 1, 2, 3. \tag{6.2}$$

Solving the system (6.2) for $\dot{\widehat{v}}^{\tau}$, we obtain

$$
\begin{aligned}
\dot{\widehat{v}}^{\tau} &= \frac{g^{\tau\sigma}\,Q_{\sigma}}{m} - g^{\tau\sigma}\,\Gamma_{\sigma,\alpha\beta}\,\dot{q}^{\alpha}\dot{q}^{\beta} = \frac{Q^{\tau}}{m} - \Gamma^{\tau}_{\alpha\beta}\,\dot{q}^{\alpha}\dot{q}^{\beta} = \\
&= \frac{Q^{\tau}}{m} - \Gamma^{\tau}_{\rho\sigma}\,\widehat{v}^{\rho}\widehat{v}^{\sigma} - 2\Gamma^{\tau}_{0\sigma}\,\widehat{v}^{\sigma} - \Gamma^{\tau}_{00}.
\end{aligned}
\tag{6.3}
$$

Here, as also in Sect. 7 of Chap. 1, $g^{\tau\sigma}$ stand for the elements of the matrix inverse to the matrix with elements $g_{\sigma\tau}$. The system of equations (6.1) and (6.3) is a system of six first-order differential equations solved for derivatives.

If the forces acting on the particle have a potential, the system of equations (4.7) can be reduced to *equations in canonical form*. The Lagrange equations of the second kind in the presence of force potential can be represented as

$$\frac{d}{dt}\frac{\partial L}{\partial \dot{q}^{\sigma}} - \frac{\partial L}{\partial q} = 0, \quad \sigma = 1, 2, 3, \tag{6.4}$$

where the function $L = T + U$ is called the *kinetic potential* or the *Lagrange function*. Assuming that U depends only on the coordinates and time and, consequently, $\partial U / \partial \dot{q}^{\sigma} = 0$, we can write

$$\frac{\partial L}{\partial \dot{q}^{\sigma}} = \frac{\partial T}{\partial \dot{q}^{\sigma}}, \quad \frac{\partial L}{\partial q^{\sigma}} = \frac{\partial T}{\partial q^{\sigma}} + \frac{\partial U}{\partial q^{\sigma}},$$

therefore, the Lagrange equations of the second kind

$$\frac{d}{dt}\frac{\partial T}{\partial \dot{q}^\sigma} - \frac{\partial T}{\partial q^\sigma} = \frac{\partial U}{\partial q^\sigma}, \quad \sigma = 1, 2, 3$$

can be expressed in the form (6.4).

When reducing the differential equations of mechanics to a system of first-order equations, instead of $\widehat{v}^\sigma$ one gains the benefit from using the variables

$$p_\sigma = \frac{\partial L}{\partial \dot{q}^\sigma} = \frac{\partial T}{\partial \dot{q}^\sigma} = \frac{\partial}{\partial \dot{q}^\sigma}\left(\frac{m}{2} g_{\alpha\beta}\, \dot{q}^\alpha \dot{q}^\beta\right) = m\,(g_{\sigma\tau}\,\dot{q}^\tau + g_{\sigma 0})\,, \tag{6.5}$$

$$\sigma = 1, 2, 3.$$

Further on, we shall call the introduced quantities p_σ the *generalized momenta*. The linear system (6.5) can always be solved for $\dot{q}^\tau$, $\tau = 1, 2, 3$, as the coefficients $g_{\sigma\tau}$ are coefficients of a positive definite quadratic form. Solving this system, we obtain

$$\dot{q}^\tau = g^{\tau\sigma}\, p_\sigma/m - g^{\tau\sigma}\, g_{\sigma 0}\,, \quad \sigma, \tau = 1, 2, 3. \tag{6.6}$$

Let us introduce the function

$$H(t, q, p) = p_\tau \dot{q}^\tau - L(t, q, \dot{q})\,. \tag{6.7}$$

In the right-hand side of this equation the generalized velocities $\dot{q}^\tau$ are considered to be expressed in terms of the generalized momenta p_σ by formulas (6.6). The change from the function $L(t, q, \dot{q})$ to the function $H(t, q, p)$ by formula (6.7) is affected by the *Legendre transformation*; the function H is called the *Hamiltonian function*. Calculating the derivatives we see that

$$\frac{\partial H}{\partial q^\sigma} = p_\tau \frac{\partial \dot{q}^\tau}{\partial q^\sigma} - \frac{\partial L}{\partial q^\sigma} - \frac{\partial L}{\partial \dot{q}^\tau}\frac{\partial \dot{q}^\tau}{\partial q^\sigma}\,.$$

However, according to formulas (6.5), $\partial L/\partial \dot{q}^\tau = p_\tau$, and therefore,

$$\frac{\partial H}{\partial q^\sigma} = -\frac{\partial L}{\partial q^\sigma}, \quad \sigma = 1, 2, 3. \tag{6.8}$$

On the basis of these formulas, Lagrange equations (6.4) can be written in the form

$$\dot{p}_\sigma = -\frac{\partial H(t, q, p)}{\partial q^\sigma}\,.$$

Similarly,

$$\frac{\partial H}{\partial p_\sigma} = \dot{q}^\sigma + p_\tau \frac{\partial \dot{q}^\tau}{\partial p_\sigma} - \frac{\partial L}{\partial \dot{q}^\tau}\frac{\partial \dot{q}^\tau}{\partial p_\sigma}\,.$$

Invoking (6.5), we obtain

$$\dot{q}^\sigma = \frac{\partial H(t, q, p)}{\partial p_\sigma}.$$ (6.9)

The system of equations

$$\dot{p}_\sigma = -\frac{\partial H(t, q, p)}{\partial q^\sigma}, \quad \dot{q}^\sigma = \frac{\partial H(t, q, p)}{\partial p_\sigma}, \quad \sigma = 1, 2, 3$$ (6.10)

is called the *canonical system of equations in Hamiltonian form.*

From the system (6.10) it follows that if any of the coordinates q^ρ does not appear in the expression for the function H, then the derivative $\partial H / \partial q^\rho = 0$, and consequently, we obtain the integral

$$p_\rho = \text{const}.$$

Such coordinates are called the *cyclic coordinates*, and the integrals corresponding to them are the *cyclic integrals*. Note that cyclic coordinates do not enter the function $L(t, q, \dot{q})$ either.

The total derivative of H with respect to time is

$$\frac{dH}{dt} = \frac{\partial H}{\partial t} + \frac{\partial H}{\partial q^\tau} \dot{q}^\tau + \frac{\partial H}{\partial p_\tau} \dot{p}_\tau.$$

Replacing $\dot{q}^\tau$ and $\dot{p}_\tau$ by their values (6.10), we find

$$\frac{dH}{dt} = \frac{\partial H}{\partial t}.$$

Therefore, in the case of a stationary system, we have the integral

$$H = \text{const}.$$

It is easy to make sure that it is the energy integral. Indeed, in the case of a stationary basis the sum

$$p_\tau \dot{q}^\tau = \frac{\partial T}{\partial \dot{q}^\tau} \dot{q}^\tau$$

is equal to $2T$ by Euler's theorem on homogeneous functions, and the expression for H takes on the form

$$H = 2T - L = T - U = T + \Pi,$$

consequently, in the stationary case the Hamiltonian function is equal to the total mechanical energy of the system:

$$H = T + \Pi = E.$$

If among the forces acting on the particle, apart from the forces having a potential, there exist nonpotential generalized forces Q^*_σ, then the Lagrange equations read as

$$\frac{d}{dt}\frac{\partial L}{\partial \dot{q}^\sigma} - \frac{\partial L}{\partial q^\sigma} = Q^*_\sigma, \quad \sigma = 1, 2, 3,$$

and so

$$\dot{p}_\sigma = \frac{\partial L}{\partial q^\sigma} + Q^*_\sigma. \tag{6.11}$$

We can see from formula (6.7) defining the function H that the relations (6.8) and (6.9) do not depend on Q^*_σ, therefore, they are also valid for the case under consideration. As a result, in accordance with formulas (6.11), (6.8) and (6.9), we have

$$\dot{p}_\sigma = -\frac{\partial H}{\partial q^\sigma} + Q^*_\sigma, \quad \dot{q}^\sigma = \frac{\partial H}{\partial p_\sigma}, \quad \sigma = 1, 2, 3.$$

7 Oscillatory Motion of a Mass Point

Free oscillations of a mass point. Consider the simplest example of motion of a body of mass m moving on a smooth horizontal plane when it is attached to the end of a horizontal spring, the other end of which being stationary. For translational motion of the given body it suffices to study the motion of just one its point, for instance, the point M coinciding with the attachment point of the movable end of the spring. Suppose that the initial velocity of translational motion of the body is directed along the spring. Then the subsequent motion of the point M will be executed along a straight line with which it is convenient to connect the x-axis. Let us locate the origin O of this axis at the point M, the position of which corresponds to the undeformed spring. Obviously, $\mathbf{r} = \overrightarrow{OM} = x\,\mathbf{i}$.

Suppose that a force of c Newtons is required to extend the spring by 1 meter. In this case, we shall say that the stiffness of the spring equals c N/m. It is obvious that if the body in question is attached to such a spring, then it is acted upon by the elasticity force $F = cr$ exerted by the spring. As this force strives to return the displaced body to the origin, it can be written vectorially as

$$\mathbf{F} = -c\mathbf{r} = -cx\,\mathbf{i}.$$

Under real conditions, various resistance (damping) forces act along with the elasticity force. Consider the case, when we can assume that the resistance force $\mathbf{R}$ is proportional to the first degree of the velocity $\mathbf{v}$. Since this force is always oriented in the direction opposite to that of the velocity, it can be represented as

$$\mathbf{R} = -\mu\,\mathbf{v} = -\mu\dot{x}\,\mathbf{i},$$

where μ is the proportionaity factor.

Along with elasticity and resistance forces, the gravity force and the smooth plane normal pressure force act on the body. But, since the last two are vertical and the dry friction force is assumed to be absent, the equation of motion of a mass point along the horizontal x-axis is as follows

$$m\ddot{x} = -cx - \mu\dot{x},$$

or

$$\ddot{x} + 2n\dot{x} + \omega^2 x = 0, \tag{7.1}$$

where

$$\omega^2 = \frac{c}{m}, \qquad n = \frac{\mu}{2m}. \tag{7.2}$$

The general solution of the second-order differential equation (7.1) can be represented as

$$x = C_1 x_{\mathrm{I}} + C_2 x_{\mathrm{II}}, \tag{7.3}$$

where x_{I} and x_{II} are linearly independent solutions of the equation in hand. Depending on n and ω, the roots of the characteristic equation (7.3) can be complex ($n < \omega$), real and equal ($n = \omega$), or real and distinct from each other ($n > \omega$). For $n < \omega$ (for low resistance) particular solutions read

$$x_{\mathrm{I}} = e^{-nt} \cos \sqrt{\omega^2 - n^2}\, t, \qquad x_{\mathrm{II}} = e^{-nt} \sin \sqrt{\omega^2 - n^2}\, t, \tag{7.4}$$

therefore, the general solution (7.3) can be represented as follows

$$x = e^{-nt} \left(C_1 \cos \sqrt{\omega^2 - n^2}\, t + C_2 \sin \sqrt{\omega^2 - n^2}\, t \right). \tag{7.5}$$

It often pays to represent the general solution (7.5) differently, replacing arbitrary constants C_1 and C_2 by new constants a and α according to the formulas

$$C_1 = a \sin \alpha, \qquad C_2 = a \cos \alpha.$$

Now expression (7.5) reads as

$$x = e^{-nt} a \sin \left(\sqrt{\omega^2 - n^2}\, t + \alpha \right). \tag{7.6}$$

Clearly, the quantities a and α can be calculated in terms of C_1 and C_2:

$$a = \sqrt{C_1^2 + C_2^2}; \quad \alpha = \operatorname{arccotan} \frac{C_2}{C_1}, \quad C_1 > 0;$$

$$\alpha = \pi + \text{arccotan}\, \frac{C_2}{C_1}, \quad C_1 < 0;$$

$$\alpha = 0, \quad C_1 = 0, \quad C_2 > 0; \quad \alpha = \pi, \quad C_1 = 0, \quad C_2 < 0.$$

If the initial conditions are specified

$$x|_{t=0} = x_0, \qquad \dot{x}|_{t=0} = \dot{x}_0, \tag{7.7}$$

then the arbitrary constants can be found by the formulas

$$C_1 = x_0, \qquad C_2 = \frac{\dot{x}_0 + nx_0}{\sqrt{\omega^2 - n^2}}, \tag{7.8}$$

or

$$a = \sqrt{x_0^2 + \frac{(\dot{x}_0 + nx_0)^2}{\omega^2 - n^2}}; \quad \alpha = \text{arccotan}\, \frac{\dot{x}_0 + nx_0}{x_0\sqrt{\omega^2 - n^2}}, \quad x_0 > 0;$$

$$\alpha = \pi + \text{arccotan}\, \frac{\dot{x}_0 + nx_0}{x_0\sqrt{\omega^2 - n^2}}, \quad x_0 < 0; \tag{7.9}$$

$$\alpha = 0, \quad x_0 = 0, \quad \dot{x}_0 + nx_0 > 0; \quad \alpha = \pi, \quad x_0 = 0, \quad \dot{x}_0 + nx_0 < 0.$$

Let us consider the special case when $n = 0$; this is the case of an idealized case of no resistance to the motion at all. Then according to formulas (7.5) and (7.6), the general solution can be represented as

$$x = C_1 \cos \omega t + C_2 \sin \omega t \tag{7.10}$$

or

$$x = a \sin(\omega t + \alpha). \tag{7.11}$$

The values of C_1, C_2, and a, α, specifying the particular solution satisfying the initial data (7.7) can be calculated using (7.8) or (7.9) with $n = 0$. Clearly, the motion described by Eq. (7.10) or (7.11) is periodic with the period $T = 2\pi/\omega$. The quantities a, ω and α are called the *amplitude, the angular (cyclic) frequency* and the *initial phase*, respectively. Note that since by formula (7.2) $\omega = \sqrt{c/m}$, it follows that the frequency and, consequently, the oscillation period are independent of the initial conditions. The described phenomenon is called the *isochronism* or *tautochronism* of harmonic oscillations. It is typical for linear differential equation (7.1) when $n = 0$ and is absent when oscillations are described by nonlinear equations (for details see chapter "Nonlinear Oscillations" of the second volume). Note that the angular frequency ω is also called the *frequency of free oscillations* or the *natural frequency*.

Let us go back to the case of low resistance ($0 < n < \omega$). From the above solution (7.6) it is seen that in this case we have a periodic change of sign of the x-

coordinate of the particle under investigation. The multiplier e^{-nt} shows that the amplitude of oscillations decreases with time and so we are dealt with damped oscillations. Even though function (7.6) is aperiodic, in a case of a similar motion one usually introduces the notion of the period

$$T_1 = \frac{2\pi}{\sqrt{\omega^2 - n^2}},$$

by which we mean the time interval between two successive passings of the particle through the state of equilibrium in the same direction is meant.

The considered damped oscillations have a very interesting property that if a sequence $x_1, x_2, x_3, \ldots$ is formed of all the extreme deviations of the particle in the same direction, then it turns out that it represents a geometric progression. Indeed, these extreme deviations are observed when the velocity of the particle vanishes

$$\dot{x} = e^{-nt} a \left(\sqrt{\omega^2 - n^2} \cos \left(\sqrt{\omega^2 - n^2}\, t + \alpha \right) - n \sin \left(\sqrt{\omega^2 - n^2}\, t + \alpha \right) \right) = 0.$$

If this transcendental equation is satisfied for $t = t_i$, then its next root, which corresponds to another extreme deviation of the particle in the same direction, is equal to $t_{i+1} = t_i + T_1$. Thus the sequence $t_1, t_2, t_3, \ldots$ is an arithmetic progression with the common difference equal to the value of the above conventional period T_1 of damped oscillations. Two successive extreme deviations in the same direction are calculated by the formulas

$$x_i = e^{-nt_i} a \sin \left(\sqrt{\omega^2 - n^2}\, t_i + \alpha \right),$$

$$x_{i+1} = e^{-n(t_i + T_1)} a \sin \left(\sqrt{\omega^2 - n^2}\, (t_i + T_1) + \alpha \right) = e^{-nt_i} x_i,$$

and, therefore, their ratio is indeed the constant

$$x_{i+1}/x_i = e^{-nT_1} = D. \tag{7.12}$$

The common ratio D of this geometric progression is called the *decrement of oscillations*, and its natural logarithm $\ln D = -nT_1$, the *logarithmic decrement of oscillations*. Note that formula (7.12) enables us to use the experimentally read quantities x_i and x_{i+1} to evaluate the coefficient $\mu = 2mn$ (see formulas (7.2)).

In case of motion with high resistance ($n \geqslant \omega$), the eigenvalues of the differential equation (7.1) are real. For $n = \omega$ the general solution of equation (7.1) can be written as

$$x = e^{-nt}(C_1 + C_2 t),$$

and for $n > \omega$, it reads as

$$x = e^{-nt} \left(C_1 e^{\sqrt{n^2 - \omega^2}\, t} + C_2 e^{-\sqrt{n^2 - \omega^2}\, t} \right).$$

With the help of hyperbolic functions, the last formula can be represented as follows:

$$x = e^{-nt} \left(A \cosh \sqrt{n^2 - \omega^2}\, t + B \sinh \sqrt{n^2 - \omega^2}\, t \right)$$

or

$$x = e^{-nt} a \, \sinh \left(\sqrt{n^2 - \omega^2}\, t + \beta \right).$$

The arbitrary constants C_1, C_2, or A, B as well as α, β can be determined from the initial conditions (7.7). For $n = \omega$, for instance, we have

$$C_1 = x_0, \qquad C_2 = \dot{x}_0 + n x_0,$$

and for $n > \omega$

$$C_1 = \frac{x_0 \left(\sqrt{n^2 - \omega^2} + n \right) + \dot{x}_0}{2 \sqrt{n^2 - \omega^2}}, \qquad C_2 = \frac{x_0 \left(\sqrt{n^2 - \omega^2} - n \right) - \dot{x}_0}{2 \sqrt{n^2 - \omega^2}}.$$

It is clear from the above solutions that for $n \geqslant \omega$ motions are of explicitly nonoscillating character and hence are called *aperiodic*.

Forced oscillations of a particle. Now consider forced oscillations of a particle when it is subject, in addition to the above examined elasticity and resistance forces, to a force $\mathbf{Q}(t)$ called a *disturbing force*. We will assume that $\mathbf{Q}(t)/m = f(t)\,\mathbf{i}$. In this case, the motion of a particle is described by the following differential equation

$$\ddot{x} + 2n\dot{x} + \omega^2 x = f(t). \tag{7.13}$$

We will search for the solution of inhomogeneous differential equation (7.13) using the method of variation of (arbitrary) constants. According to this method, the solution should be sought in the form of solution (7.3) of the homogeneous equation (7.10), where the arbitrary constants are treated as unknown functions $C_1(t)$ and $C_2(t)$ of time. Differentiating equation (7.3) under this assumption, we obtain

$$\dot{x} = \dot{C}_1 x_\mathrm{I} + \dot{C}_2 x_\mathrm{II} + C_1 \dot{x}_\mathrm{I} + C_2 \dot{x}_\mathrm{II}.$$

As we have only one Eq. (7.13) for finding two unknown functions, these functions can be subjected to an additional condition. We choose this condition to be

$$\dot{C}_1 x_\mathrm{I} + \dot{C}_2 x_\mathrm{II} = 0. \tag{7.14}$$

Condition (7.14) is convenient, because if it holds, the expression for $\dot{x}$ has the same form as for the constants C_1 and C_2:

$$\dot{x} = C_1 \dot{x}_\mathrm{I} + C_2 \dot{x}_\mathrm{II}. \tag{7.15}$$

Differentiating (7.15) with respect to time again, we obtain

$$\ddot{x} = \dot{C}_1 \dot{x}_I + \dot{C}_2 \dot{x}_{II} + C_1 \ddot{x}_I + C_2 \ddot{x}_{II}. \tag{7.16}$$

Substituting (7.3), (7.15), and (7.16) into Eq. (7.13) and taking into account that solution (7.3) satisfies the corresponding homogeneous differential equation (7.1), we obtain

$$\dot{C}_1 \dot{x}_I + \dot{C}_2 \dot{x}_{II} = f(t). \tag{7.17}$$

This equation and condition (7.14) constitute a system of differential equations. Integrating this system we can find the unknown functions $C_1(t)$ and $C_2(t)$ and thereby build up the general solution of the inhomogeneous equation (7.13) in the form (7.3).

Let us elaborate a derivation of the general solution using the method of variation of arbitrary constants for the case of low resistance. For $n < \omega$, the particular solutions x_I and x_{II} have the form of (7.4), therefore, the additional condition (7.14) assumes now the form

$$\dot{C}_1 e^{-nt} \cos \sqrt{\omega^2 - n^2}\, t + \dot{C}_2 e^{-nt} \sin \sqrt{\omega^2 - n^2}\, t = 0. \tag{7.18}$$

Equation (7.17) transforms to

$$\begin{aligned}
-\dot{C}_1 e^{-nt} \left(n \cos \sqrt{\omega^2 - n^2}\, t + \sqrt{\omega^2 - n^2} \sin \sqrt{\omega^2 - n^2}\, t \right) + \\
+\dot{C}_2 e^{-nt} \left(\sqrt{\omega^2 - n^2} \cos \sqrt{\omega^2 - n^2}\, t - n \sin \sqrt{\omega^2 - n^2}\, t \right) = f(t).
\end{aligned} \tag{7.19}$$

Integrating the system of differential equations (7.18)–(7.19) we may find the unknown functions $C_1(t)$ and $C_2(t)$ of interest to us. Let us draw attention to the fact that due to holding (7.18), Eq. (7.19) can be simplified:

$$-\dot{C}_1 \sin \sqrt{\omega^2 - n^2}\, t + \dot{C}_2 \cos \sqrt{\omega^2 - n^2}\, t = \frac{f(t)\, e^{nt}}{\sqrt{\omega^2 - n^2}}.$$

Solving it simultaneously with Eq. (7.18) multiplied by e^{nt}, we obtain

$$\dot{C}_1 = -\frac{f(t)\, e^{nt}}{\sqrt{\omega^2 - n^2}} \sin \sqrt{\omega^2 - n^2}\, t, \qquad \dot{C}_2 = \frac{f(t)\, e^{nt}}{\sqrt{\omega^2 - n^2}} \cos \sqrt{\omega^2 - n^2}\, t,$$

or

$$C_1(t) = -\frac{1}{\sqrt{\omega^2 - n^2}} \int\limits_0^t f(\xi)\, e^{n\xi} \sin \sqrt{\omega^2 - n^2}\, \xi \, d\xi + D_1,$$

$$C_2(t) = \frac{1}{\sqrt{\omega^2 - n^2}} \int_0^t f(\xi) e^{n\xi} \cos \sqrt{\omega^2 - n^2}\, \xi \, d\xi + D_2 \,,$$

where D_1 and D_2 are arbitrary constants. Now, according to relations (7.3) and (7.4), the general solution of equation (7.13) for $n < \omega$ can be written as

$$x = D_1 e^{-nt} \cos \sqrt{\omega^2 - n^2}\, t + D_2 e^{-nt} \sin \sqrt{\omega^2 - n^2}\, t +$$

$$+ \frac{1}{\sqrt{\omega^2 - n^2}} \int_0^t f(\xi) e^{-n(t-\xi)} \sin \sqrt{\omega^2 - n^2}\,(t - \xi)\, d\xi \,. \tag{7.20}$$

According to formula (7.5), the first two terms here are the general solution of the corresponding homogeneous differential equation (7.1), the last term, which is called the *Duhamel integral*, is the particular solution of the inhomogeneous differential equation (7.13) of interest. Note that the particular solution, which is represented as the Duhamel integral, satisfies the zero initial conditions. Therefore, the solution of equation (7.13) satisfying the initial data (7.7) can be written as

$$x = e^{-nt} \left(x_0 \cos \sqrt{\omega^2 - n^2}\, t + \frac{\dot{x}_0 + nx_0}{\sqrt{\omega^2 - n^2}} \sin \sqrt{\omega^2 - n^2}\, t \right) +$$

$$+ \frac{1}{\sqrt{\omega^2 - n^2}} \int_0^t f(\xi) e^{-n(t-\xi)} \sin \sqrt{\omega^2 - n^2}\,(t - \xi)\, d\xi \,, \quad n < \omega \,. \tag{7.21}$$

In the case of no resistance ($n = 0$) it read as

$$x = x_0 \cos \omega t + \frac{\dot{x}_0}{\omega} \sin \omega t + \frac{1}{\omega} \int_0^t f(\xi) \sin \omega(t - \xi)\, d\xi \,, \qquad n = 0 \,. \tag{7.22}$$

Similarly, we can obtain the solutions of equation (7.13) satisfying the initial conditions (7.7) for the cases $n = \omega$ and $n > \omega$:

$$x = e^{-nt} \left(x_0 + (\dot{x}_0 + nx_0)\, t \right) + \int_0^t f(\xi) e^{-n(t-\xi)}\,(t - \xi)\, d\xi \,, \qquad n = \omega \,, \tag{7.23}$$

$$x = e^{-nt} \left(x_0 \cosh \sqrt{n^2 - \omega^2}\, t + \frac{\dot{x}_0 + nx_0}{\sqrt{n^2 - \omega^2}} \sinh \sqrt{n^2 - \omega^2}\, t \right) +$$

$$+ \frac{1}{\sqrt{n^2 - \omega^2}} \int_0^t f(\xi) e^{-n(t-\xi)} \sinh \sqrt{n^2 - \omega^2}\,(t - \xi)\, d\xi \quad n > \omega \,. \tag{7.24}$$

Let us consider the action of a harmonic disturbing force

$$\mathbf{Q}(t) = H \sin \nu t \, \mathbf{i} \,. \tag{7.25}$$

We have

$$f(t) = h \sin \nu t \,, \qquad h = H/m \,.$$

In case of no resistance ($n = 0$), the Duhamel integral in (7.22) for $\nu \neq \omega$ reduces to the tabulated form

$$\frac{h}{\omega} \int_0^t \sin \nu \xi \, \sin \omega (t - \xi) \, d\xi = \frac{h}{2\omega} \int_0^t \left\{ \cos \left[(\nu + \omega) \xi - \omega t \right] - \cos \left[(\nu - \omega) \xi + \omega t \right] \right\} d\xi$$

and therefore can be readily evaluated

$$\frac{h}{\omega} \int_0^t \sin \nu \xi \, \sin \omega (t - \xi) \, d\xi = \frac{h\nu}{\omega(\nu^2 - \omega^2)} \sin \omega t + \frac{h}{\omega^2 - \nu^2} \sin \nu t \,, \qquad \nu \neq \omega \,.$$

$$\tag{7.26}$$

For $\nu = \omega$ we have

$$\frac{h}{\omega} \int_0^t \sin \omega \xi \, \sin \omega (t - \xi) \, d\xi = \frac{h}{2\omega} \int_0^t \left[\cos(2\omega \xi - \omega t) - \cos \omega t \right] d\xi = \tag{7.27}$$

$$= \frac{h}{2\omega^2} \sin \omega t - \frac{ht}{2\omega} \cos \omega t \,.$$

The first term in formula (7.26) and the first two terms in formula (7.22) display oscillations of a particle with the natural frequency ω. These terms do not die out only in the idealized no resistance case ($n = 0$). However, in all real systems, where resistance, even if an insignificant one, is present, the mentioned terms die out with time, as we shall see below. Therefore, of special interest is the second term in formula (7.26), characterizing the so-called forced oscillations of the particle that take place with the frequency ν of the disturbing force (7.25). The amplitude $h/(\omega^2 - \nu^2)$ of these forced oscillations increases unrestrictedly when $\nu \to \omega$. The evolution of those oscillations with time when $\nu = \omega$ is characterized by the second term in formula (7.27). This phenomenon is called the *resonance*. Sometimes the component $-(ht \cos \omega t)/(2\omega)$ is called the *secular term*. This notion originates in astronomy, where similar expressions characterize the deviations in the planetary motion accumulating over for centuries.

In radio-receiving sets it is essential to tune the natural frequency ω of a circuit most sharply to the frequency of a transmitter ν, whereas in the case of mechanical vibrations the resonance usually turns out to be dangerous and often causes the collapse of structures.[2] However, nowadays quite a few machines (for example, in vibrators) depend on tuning the system to resonance. By the way, it was since ancient times that experienced bell ringers knew that while swinging the heavy clapper of the bell, one must pull the string with the frequency coinciding with the clapper natural frequency.

If $n \neq 0$, then, substituting $f(t) = h \sin \nu t$ into the formula (7.28), after quite tedious calculations we can obtain

$$
x = e^{-nt} \left(x_0 \cos \sqrt{\omega^2 - n^2}\, t + \frac{\dot{x}_0 + n x_0}{\sqrt{\omega^2 - n^2}} \sin \sqrt{\omega^2 - n^2}\, t \right) +
$$
$$
+ e^{-nt} A \left(\sin \gamma \cos \sqrt{\omega^2 - n^2}\, t + \frac{n \sin \gamma - \nu \cos \gamma}{\sqrt{\omega^2 - n^2}} \sin \sqrt{\omega^2 - n^2}\, t \right) + \quad (7.28)
$$
$$
+ A \sin(\nu t - \gamma),
$$

where

$$
A = \frac{h}{\sqrt{(\omega^2 - \nu^2)^2 + 4n^2 \nu^2}}, \qquad \tan \gamma = \frac{2n\nu}{\omega^2 - \nu^2}. \quad (7.29)
$$

It is perfectly displayed by formula (7.28) that the terms containing the multiplier e^{-nt} die out with time. A motion, for which they must be taken into account, is called the *transient process*. Upon its termination the oscillations are quite accurately characterized by the term $A \sin(\nu t - \gamma)$, describing the system steady-state forced oscillations with the disturbing force frequency. The dependence of the amplitude A on the disturbing force frequency ν is called the *amplitude-frequency characteristic* of the given system.

It is interesting to compare the static elongation A_0 under the constant force H and the amplitude A of steady-state forced oscillations under the action of the variable disturbing force (7.25) of amplitude H. The ratio $\eta = A/A_0$ is called the *dynamic coefficient*. Since

$$
A_0 = H/c = hm/c = h/\omega^2
$$

and by the first formula of (7.29) this coefficient can be represented as

---

[2] This is how A.N. Krylov describes some tragic consequences of a resonance in his book "Vessel Vibrations" [in Russian], Leningrad-Moscow: Izdat."ONTI", 1936, p. 7: "It must have been back to Napoleon wars in Spain that a military detachment was marching over a bridge, stamping energetically (perhaps, there were some important authorities standing on the bridge or behind it). It was a chain suspension bridge, and the stamping happened to be of the same beat as the bridge oscillation period, so the swings increased so much that the chains ruptured, and the bridge fell down into the river. After that accident it was forbidden in all the armies to march in step while crossing a bridge; but about thirty years ago the chain suspension bridge called Egypetsky over the Fontanka river in St. Petersburg of those days was being crossed by an elite cavalry squadron, no matter which regiment; the horses were well trained, especially for a well-disposed ceremonious march, so they were marching in a brilliant cadence, and their stamping was of just the same beat as the bridge oscillations, — the chains ruptured, the bridge collapsed into the water, about 40 people died".

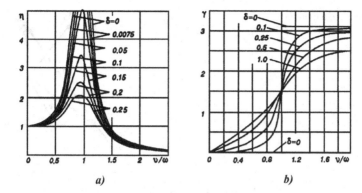

Fig. 6 The frequency response and the phase-frequency functions

$$\eta = \left[\left(1 - \frac{v^2}{\omega^2}\right)^2 + \frac{4n^2}{\omega^2}\cdot\frac{v^2}{\omega^2}\right]^{-1/2}.$$

The dependence of the dynamic coefficient values on the ratio v/ω for various values of the quantity $\delta = 2n/\omega$ is shown in Fig. 6a. The figure shows that if the disturbing force frequency is small relative to the natural frequency of system oscillations, then the dynamic coefficient is close to one. This means that in similar cases the spring elongation at any instant of time can be calculated accurately enough on the assumption that the disturbing force $H \sin vt$ acts statically. In the other extreme case, when the frequency v is high relative to the natural frequency ω, the dynamic coefficient becomes quite small. In both extreme cases the curves corresponding to various values of δ converge very close together. This means that in these cases the damping forces have little influence on the amplitude of the forced oscillations. But if the disturbing force frequency approaches the natural frequency, then the dynamic coefficient increases sharply and, as seen from Fig. 6a, its value becomes quite sensitive to changes in damping, especially when it is small. Setting the dynamic coefficient derivative with respect to v/ω equal to zero, we discover that the quantity η is at maximum when

$$\frac{v^2}{\omega^2} = 1 - \frac{2n^2}{\omega^2},$$

and

$$\eta_{max} = \frac{\omega}{2n}\left(1 - \frac{n^2}{\omega^2}\right)^{-1/2}.$$

For studying the *phase-frequency characteristics*, which is the dependence of the phase shift γ on the disturbing force frequency v, it is convenient to rewrite the second formula in (7.29) as follows:

Fig. 7 Motion of a point in
two coordinate frames

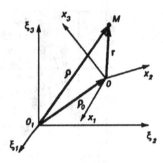

$$\tan \gamma = \frac{2(n/\omega) \cdot (v/\omega)}{1 - v^2/\omega^2} .$$

Hence, in particular, it is clear that when the resonance takes place ($v = \omega$), we have
the phase shift $\gamma = \pi/2$. The γ—v/ω dependence for various values of $\delta = 2n/\omega$
is displayed in Fig. 6b.

In Fig. 6 the limiting curves have discontinuities corresponding to the case $n = 0$
and are consistent with the results obtained above.

8   Relative Motion Dynamics of a Particle

Basic concepts. Let us study the relative motion of a particle.

Let a coordinate system $O_1\xi_1\xi_2\xi_3$ be given, with respect to which a particle M
moves with an acceleration $\mathbf{w}$ (Fig. 7). Then according to Newton second law, we
can write

$$m\mathbf{w} = \mathbf{F} , \qquad (8.1)$$

where $\mathbf{F}$ is the force acting on the particle. Choose another coordinate system
$Ox_1x_2x_3$. Assume that its motion with respect to the reference system $O_1\xi_1\xi_2\xi_3$
is known, i.e., the velocity $\mathbf{v}_0$ of the auxiliary coordinate system origin relative to the
initial system and the angular velocity $\boldsymbol{\omega}$ of its rotations about the latter are specified.
Then, as known from kinematics, the absolute velocity of the particle M with respect
to the system $O_1\xi_1\xi_2\xi_3$ is

$$\mathbf{v}_a = \mathbf{v}_e + \mathbf{v}_r , \qquad \text{where} \qquad \mathbf{v}_e = \mathbf{v}_0 + \boldsymbol{\omega} \times \mathbf{r} .$$

Similarly, by the theorem on accelerations addition (composition), the absolute
acceleration of the particle is constituted of the bulk (transport, drag), relative, and
Coriolis acceleration vectors:

$$\mathbf{w}_a = \mathbf{w}_e + \mathbf{w}_r + \mathbf{w}_C , \qquad (8.2)$$

where

$$\mathbf{w}_C = 2\boldsymbol{\omega} \times \mathbf{v}_r .\tag{8.3}$$

The acceleration $\mathbf{w}$ entering (8.1) equals $\mathbf{w}_a$. On the basis of (8.2), the Newton second law can be written in the form

$$m\mathbf{w}_r = \mathbf{F} + (-m\mathbf{w}_e) + (-m\mathbf{w}_C) ,\tag{8.4}$$

which is the equation of motion of the particle subject to the forces $\mathbf{F}$, $-m\mathbf{w}_e$, $-m\mathbf{w}_C$. Here, and throughout the book we agree to call the negative of mass times acceleration the *inertia force*.[3] Therefore, we will call the expressions

$$\mathbf{J}_e = -m\mathbf{w}_e , \qquad \mathbf{J}_C = -m\mathbf{w}_C\tag{8.5}$$

the *inertial forces resulting from the transport (bulk, drag)* and the *Coriolis accelerations*, respectively. Thus, an observer in the $Ox_1x_2x_3$-system can assert that, along with the force $\mathbf{F}$, the inertial forces $\mathbf{J}_e$ and $\mathbf{J}_C$ act on the particle. From the viewpoint of the observer in the system $Ox_1x_2x_3$ the inertial forces appear as usual forces causing additional accelerations $-\mathbf{w}_e$ and $-\mathbf{w}_C$.

In view of the notation (8.5), we rewrite equation (8.4) as

$$m\mathbf{w}_r = \mathbf{F} + \mathbf{J}_e + \mathbf{J}_C .\tag{8.6}$$

The characteristic feature of the inertial forces is that they cause equal accelerations to particles of distinct masses, i.e., they are proportional to the masses. Gravity forces also have a similar property, what makes them in a sense akin to inertial forces. Besides, inertia forces have a property, which distinguishes them from forces satisfying the Newton third law: for them no concrete bodies originating them can be pointed out directly.[4]

Consider an example that shows that in particular cases a constant gravitational field can be put in equilibrium with the action of inertial forces. Let us determine the force of a weight pressure on the floor of an elevator moving upward with a constant acceleration $\mathbf{a}$. Clearly,

$$m\mathbf{a} = \mathbf{N} + \mathbf{P} .$$

---

[3] The inertia forces issue is considered more profoundly and comprehensively in the monograph by *A.Y.Ishlinskii*. Classical Mechanics and Inertia Forces. Moscow: Izdat. "Nauka". 1987. 320 p. [in Russian].

[4] There exists an opinion in theoretical physics that inertial forces satisfy the Newton third law and are originated by the aggregation of cosmic material bodies (masses). This kind of an explanation is, in essence, an attempt to substantiate Newton first and second laws. Note that Newton himself postulates the following statement: "The Vis Infita, or Innate Force of Matter is a power of resisting, by which every body, as much as in it lies, endeavors to persevere in its present state, whether it be of rest, or of moving uniformly forward in a right line" (*I. Newton*. The Mathematical Principles of Natural Philosophy. Vol. I, p. 2.).

Hence, the reaction force that is equal to the floor pressure on the body is expressed by the formula

$$\mathbf{N} = m\mathbf{a} - \mathbf{P} = -m\,(\mathbf{g} - \mathbf{a}) = -m\mathbf{g}'\,.$$

Thus, we can think that the body is in the gravitational field with the free fall acceleration $\mathbf{g}'$.

If the elevator fell with the acceleration $\mathbf{a} = k\,\mathbf{g}$ $(k > 0)$, then we would obtain

$$\mathbf{N} = -m\mathbf{g}\,(1 - k) = -m\mathbf{g}'\,.$$

It is clear that when the elevator is falling free $(k = 1)$ $g' = 0$, and in the relative reference frame the gravity vanishes: one says that weightlessness (zero-gravity) occurs. Of course, the same thing happens to an astronaut in a satellite moving along a circular orbit. We can say that he/she rests relative to the cockpit, therefore, all the forces are balanced, i.e., the gravity force is compensated by the inertial force. One should not think, however, basing on the considered example that in the entire space any gravitational field, for instance, Newton central gravity force field can be counterbalanced by an inertial force field, originated by the motion of the reference frame unique for the entire space.

Note that if, in particular, the motion of the reference frame $Ox_1x_2x_3$ is such that $\mathbf{w}_e = \mathbf{w}_C = 0$, then equations (8.4) and (8.1) coincide identically. This indicates that inertial forces are absent in this system. Such systems are called the *inertial reference frames*.

Consequently, for determining whether the given reference frame is inertial or not, one must find out, if along with active forces originated by concrete bodies' action, it contains forces, which impart equal accelerations to any particles placed at the given point in space regardless of their masses. As was already pointed out, along with inertial forces gravity forces have the above property.

Consider the case, when the relative motion is absent $(\mathbf{v}_r = 0,\ \mathbf{w}_r = 0)$. Then in view of formulas (8.3) and (8.5), Eq. (8.6) transforms to

$$\mathbf{F} + \mathbf{J}_e = 0\,, \tag{8.7}$$

which corresponds to the *relative equilibrium of the particle*.

Specific cases. Consider a number of problems on the equilibrium and the motion of a particle with respect to the Earth axes $Ox_1x_2x_3$. We designate the system, the origin of which is located at the Earth center and the axes are directed toward the stars that appear motionless to us, as the inertial coordinate system $O_1\xi_1\xi_2\xi_3$. This coordinate system is, strictly speaking, noninertial, because precise physical experiments would reveal inertial forces in it that emerge as a cause of the Earth motion around the Sun.

First, *consider the equilibrium of a particle M suspended on a thread and resting with respect to the rotating Earth* (Fig. 8). With φ denoting the place latitude, the inertial force of bulk (drag) motion can be written as

Fig. 8 The direction of the
plumb line near the Earth
surface

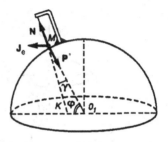

$$J_e = m\,\omega^2\,R\cos\varphi\,, \tag{8.8}$$

where R is the Earth radius, ω is the angular velocity of its rotation. In the given case
Eq. (8.7) transforms to

$$\mathbf{P}^* + \mathbf{N} + \mathbf{J}_e = 0\,, \tag{8.9}$$

where $\mathbf{N}$ is the thread tension, $\mathbf{P}^*$ is the force with which the Earth attracts the particle.
Let us represent P^* as follows:

$$P^* = m\mathbf{g}^*\,. \tag{8.10}$$

Here, the proportionality multiplier g^* accounts only for the Earth pull of gravity.

The vanishing of the vector sum (8.9) is equivalent to the closedness of the triangle
built on the vectors $\mathbf{P}^*$, $\mathbf{N}$ and $\mathbf{J}_e$. Clearly, this triangle is geometrically similar to the
triangle $O_1 K M$ displayed in Fig. 8. Therefore, invoking the sine law, we can write

$$\frac{\sin(\varphi + \gamma)}{\sin \gamma} = \frac{P^*}{J_e}\,,$$

hence, in view of (8.8) and (8.10) we obtain

$$\tan \gamma = \frac{\omega^2 R \sin 2\varphi}{2\,(g^* - \omega^2 R \cos^2 \varphi)}\,. \tag{8.11}$$

It is easy to calculate the angular velocity of the Earth $\omega = 0.0000727\,\text{s}^{-1}$. Due to
the smallness of this quantity, the angle γ, of the plumb line deviation from the
normal to the surface (Fig. 8), computed by formula (8.10) turns out small as well.
Its maximum value, as follows from relation (8.11), corresponds to $\varphi = 45°$ and
equals $11'$. Similarly, the Earth rotation about its axis can explain the changes in a
body weight depending on the place latitude: a body weight close to the equator is
almost by $1/290$ less than the weight of the same object near the poles.

Proceeding from the presented considerations, we represent the examined gravity
force $\mathbf{P}$ which is the vector sum of the forces $\mathbf{P}^*$ and $\mathbf{J}_e$, in the form $P = mg$, where g
denotes the local gravitational acceleration depending on the place latitude φ.

Consider *the free fall of a particle nearby the Earth surface* in latitude φ. For the
purpose of solving the given problem it pays to introduce a noninertial coordinate

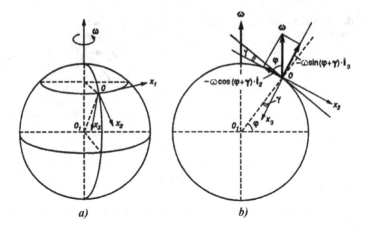

Fig. 9 Deflection of a falling body near the Earth surface

system $Ox_1x_2x_3$ (Fig. 9a). Let us locate the origin at the particle initial position. We will direct the x_3-axis downward along the plumb line. Recall that this line does not coincide with the normal to the Earth surface at the given place. The x_1-axis is directed to the east and the x_2-axis is directed to the south.

We will rewrite Eq. (8.6) as

$$m\mathbf{w}_r = \mathbf{P} + \mathbf{J}_C, \tag{8.12}$$

where $\mathbf{P} = \mathbf{P}^* + \mathbf{J}_e$. Here, as before, $\mathbf{P}^* = \mathbf{F}$ is the Earth attraction force. For projecting this equation on the coordinate axes, we shall preliminarily calculate the components of the Earth rotation angular velocity vector. As follows from Fig. 9b representing the meridional section of the Earth, we can write

$$\boldsymbol{\omega} = -\omega \cos(\varphi + \gamma)\mathbf{i}_2 - \omega \sin(\varphi + \gamma)\mathbf{i}_3. \tag{8.13}$$

Next, invoking formulas (8.3) and (8.13) from equation of relative motion (8.12) after reducing by mass m, we obtain

$$\begin{aligned}
\ddot{x}_1 &= 2\omega\left(\dot{x}_3 \cos(\varphi + \gamma) - \dot{x}_2 \sin(\varphi + \gamma)\right), \\
\ddot{x}_2 &= 2\dot{x}_1\omega \sin(\varphi + \gamma), \\
\ddot{x}_3 &= g - 2\dot{x}_1\omega \cos(\varphi + \gamma).
\end{aligned} \tag{8.14}$$

Integrating the given differential equations under zero initial conditions, we have

$$\begin{aligned}
\dot{x}_1 &= 2\omega\left(x_3 \cos(\varphi + \gamma) - x_2 \sin(\varphi + \gamma)\right), \\
\dot{x}_2 &= 2x_1\omega \sin(\varphi + \gamma), \\
\dot{x}_3 &= gt - 2x_1\omega \cos(\varphi + \gamma).
\end{aligned} \tag{8.15}$$

The subsequent integrating of the obtained system of differential equations could be continued routinely. But for revealing the basic patterns of the motion we will use the method of successive approximations. Setting in (8.15) $\omega = 0$, we obtain a familiar solution describing the particle free fall on the stationary (fixed) Earth

$$x_1 = 0, \qquad x_2 = 0, \qquad x_3 = g\,t^2/2. \tag{8.16}$$

We adopt the functions (8.16) as zero approximation and substitute into the right-hand side of system (8.15):

$$\dot{x}_1 = t^2 \omega\,g\,\cos(\varphi + \gamma),$$
$$\dot{x}_2 = 0, \tag{8.17}$$
$$\dot{x}_3 = gt.$$

Integrating the system (8.17) under the same zero initial conditions, we find the first approximation for the system under consideration (8.15):

$$x_1 = \frac{t^3}{3}\,\omega g\,\cos(\varphi + \gamma),$$
$$x_2 = 0, \tag{8.18}$$
$$x_3 = gt^2/2.$$

Substituting (8.18) into the right-hand side of system (8.15) and integrating the obtained equations under zero initial conditions, we have the second approximation in the form

$$x_1 = \frac{t^3}{3}\,\omega g\,\cos(\varphi + \gamma),$$
$$x_2 = \frac{t^4}{6}\,\omega^2 g\,\sin(\varphi + \gamma)\,\cos(\varphi + \gamma), \tag{8.19}$$
$$x_3 = \frac{gt^2}{2} - \frac{t^4}{6}\,\omega^2 g\,\cos^2(\varphi + \gamma).$$

We could go on solving with the method of successive approximations. As a result, we would manage to build up the solution of system (8.15) as a series in terms of powers of small parameter ω ($\omega = 0.0000727\,\mathrm{s}^{-1}$). But even the first approximation (8.18) shows that along with the fall along the plumb line by usual law $x_3 = g\,t^2/2$, we detect a deflection of the particle to the east that is proportional to the quantity ω. If we consider the second approximation (8.19), then in addition to some refinement to the fall along the x_3-axis, we can also reveal a deflection to the south. However, it is much less than the deflection to the east, as it is proportional to the quantity ω squared.

The Earth rotation can also explain the *Baer law*, according to which in the Northern Hemisphere the rivers wash out the right bank, and in the Southern Hemisphere the erosion occurs on the left banks. Indeed, if a river runs in the Northern Hemisphere with the velocity $\mathbf{v}_r$ with respect to the Earth, then the component of Coriolis

Fig. 10 The Baer law

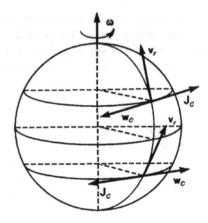

acceleration $\mathbf{w}_C = 2\boldsymbol{\omega} \times \mathbf{v}_r$ is directed in a horizontal plane perpendicular to the stream to the left, if one looks in the direction of the river current. Therefore, the water particles are acted upon by the Coriolis inertia force, oriented in the opposite direction; thus, the water will wash out the right bank of the river. For convenience, a river running in the northern direction, for instance, the Pechora (the Ob, the Yenisei) is displayed in Fig. 10. In the case, when the river is running south, like the Volga does, with a change in the direction of the vector $\mathbf{v}_r$, the $\mathbf{w}_C$, and $\mathbf{J}_C$ vectors will also reverse their directions, but the river will wash out its right bank again. For the rivers in the Southern Hemisphere the relocation of the vectors $\boldsymbol{\omega}$ and $\mathbf{v}_r$ relative to one another occurs, therefore, the vectors $\mathbf{w}_C$ and $\mathbf{J}_C$ have directions opposite to those in the Northern Hemisphere, consequently, rivers wash out left banks (see Fig. 10).

Note that in the places of abrupt changes in the river current direction the water particles inertial force will play a more essential role in the bank erosion than the Coriolis inertial forces will. Therefore, in case when a river turns right the left bank is being destroyed, and if a river turns left the right bank is being washed out.

9   Motion of a Particle Subject to Central Forces

Let a particle of mass m, the position of which with respect to the $Oxyz$-system is determined by specifying the radius vector $\mathbf{r}$, is subject to a force $\mathbf{F}$ that is collinear to the vector $\mathbf{r}$:

$$\mathbf{F} = \lambda\, \mathbf{r},$$

where $\lambda = \lambda(x, y, z, t)$ is a scalar function of the particle coordinates and the time.

If only the above force is applied to the particle in question, then according to Newton second law, the equation of its motion reads

$$m\mathbf{w} \equiv m\ddot{\mathbf{r}} = \lambda\, \mathbf{r}. \tag{9.1}$$

Fig. 11 Illustration of the area integral

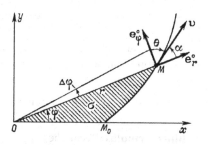

The force **F** that always acts along a straight line passing through the same point O is called the *central force*. When $\lambda > 0$ the mass m is repelled from the fixed center, and when $\lambda < 0$ it is attracted to it.

As was noted in Sect. 2 of this chapter, for a central force there always exists the integral of moments $\mathbf{l} = \mathbf{r} \times m\mathbf{v} = \mathbf{const} = \mathbf{C}$, which implies that the trajectory of motion of the particle subject to a central force is a flat curve. If $\mathbf{C} = 0$, it degenerates into a straight line, since in this case the velocity vector $\mathbf{v} = \dot{\mathbf{r}}$ is collinear to the vector $\mathbf{r}$.

For the sake of simplicity, we will assume that the vector $\mathbf{C}$ is directed along the Oz-axis. Then, we write the integral of moments as

$$\mathbf{r} \times \mathbf{v} = c\,\mathbf{k}, \quad \text{where} \quad c > 0. \tag{9.2}$$

Area integrals. On the plane Oxy, where the motion takes place, it is convenient to introduce a system of polar coordinates r, φ (Fig. 11). Now Eq. (9.2), as projected along the z-axis, reads as

$$r^2\dot{\varphi} = c = \text{const}.$$

This relation is called the *area integral*, as the derivative of the area σ of curvilinear sector OM_0M with respect to the time is calculated by the formula

$$\frac{d\sigma}{dt} = \frac{1}{2}r^2\dot{\varphi}. \tag{9.3}$$

Indeed, the increment $\Delta\sigma$ of the area σ, corresponding to the increment $\Delta\varphi$ of angle φ accurate to the second-order infinitesimal is

$$\Delta\sigma = \frac{1}{2}r^2\Delta\varphi.$$

Hence, dividing by Δt and passing to the limit we obtain expression (9.3).

If we consider the angles α and θ between the velocity vector **v** and the polar coordinate system unit vectors $\mathbf{e}_r^0$ and $\mathbf{e}_\varphi^0$, as shown in Fig. 11, then since $v\sin\alpha = v\cos\theta = r\dot{\varphi}$, $v = |\mathbf{v}|$, the constant c can be represented as

$$c = rv \sin \alpha = rv \cos \theta .$$

Thus, the area integral that is of great importance in studying a motion subject to a central force can be represented in the form of the following set of equations:

$$r^2 \dot{\varphi} = rv \sin \alpha = rv \cos \theta = 2 \frac{d\sigma}{dt} = c . \tag{9.4}$$

Binet formulas. Recall the expressions for the velocity and acceleration in a polar coordinate system

$$pr_{e_r} \mathbf{v} = \dot{r} , \qquad pr_{e_\varphi} \mathbf{v} = r\dot{\varphi} , \qquad T_1 = \frac{v^2}{2} = \frac{\dot{r}^2 + r^2 \dot{\varphi}^2}{2} , \tag{9.5}$$

$$pr_{e_r} \mathbf{w} = \frac{w_r}{H_r} = \frac{d}{dt} \frac{\partial T_1}{\partial \dot{r}} - \frac{\partial T_1}{\partial r} = \ddot{r} - r\dot{\varphi}^2 , \tag{9.6}$$

$$pr_{e_\varphi} \mathbf{w} = \frac{w_\varphi}{H_\varphi} = \frac{1}{r} \left(\frac{d}{dt} \frac{\partial T_1}{\partial \dot{\varphi}} - \frac{\partial T_1}{\partial \varphi} \right) = \frac{1}{r} \frac{d}{dt} (r^2 \dot{\varphi}) . \tag{9.7}$$

The Newton equation (9.1), as projected along the polar coordinate system axes, transforms to

$$m \, pr_{e_r} \mathbf{w} = \lambda r , \tag{9.8}$$

$$m \, pr_{e_\varphi} \mathbf{w} = 0 . \tag{9.9}$$

From equality (9.9) in view of relation (9.7), it follows that

$$r^2 \dot{\varphi} = \text{const} = c .$$

Thus, the area integral (9.4) can be obtained directly from the Lagrange equations of the second kind, without invoking the angular impulse-momentum principle.

The presence of the area integral enables us to replace the operator of differentiating with respect to the time by the operator of differentiating with respect to the variable φ:

$$\frac{d}{dt} = \frac{d\varphi}{dt} \frac{d}{d\varphi} = \frac{c}{r^2} \frac{d}{d\varphi} .$$

Denoting $u = 1/r$, we obtain

$$\frac{d}{dt} = c u^2 \frac{d}{d\varphi} .$$

This relation and the area integral enable us to represent the expressions (9.5)–(9.7) as

$$pr_{e_r} \mathbf{v} = \dot{r} = \frac{dr}{dt} = -\frac{1}{u^2} \frac{du}{dt} = -c \frac{du}{d\varphi} , \tag{9.10}$$

$$pr_{e_{\varphi}}\mathbf{v} = r\,\dot{\varphi} = c\,u\,, \tag{9.11}$$

$$pr_{e_r}\mathbf{w} = \frac{dv_r}{dt} - u^{-1}c^2u^4 =$$
$$= c\,u^2\frac{d}{d\varphi}\left(-c\frac{du}{d\varphi}\right) - c^2u^3 = -c^2u^2\left(\frac{d^2u}{d\varphi^2} + u\right), \tag{9.12}$$

$$pr_{e_{\varphi}}\mathbf{w} = 0\,.$$

From the expressions (9.10)–(9.12) we can obtain the following relations

$$v^2 = c^2\left[\left(\frac{du}{d\varphi}\right)^2 + u^2\right], \tag{9.13}$$

$$pr_{e_r}\mathbf{w} = -c^2u^2\left(\frac{d^2u}{d\varphi^2} + u\right), \tag{9.14}$$

which are the *first* and the *second Binet formulas*, respectively.

Let us emphasize that the Binet formulas make it possible to determine the velocity and acceleration at any point of the trajectory based on the trajectory equation in the polar coordinates $r = r(\varphi)$, are valid for the motion of a particle in any central force field, i.e., for any function $\lambda = \lambda(x, y, z, t)$.

Energy integral. A test of the fact that the force field is central is the presence of the area integral. But a test of the existence of the energy integral is the path-independence of work. Therefore, if the field is central, then the area integral always exists, and the energy integral does only in some cases.

In a particular case of great importance, when the magnitude of central force depends only on the distance r, i.e., when $\lambda = \lambda(r)$, the elementary work can be represented as

$$\delta A = \mathbf{F}\cdot d\mathbf{r} = F_r(r)\,dr = \lambda(r)\,r\,dr = -d\Pi\,,$$

where

$$\Pi = -\int F_r\,dr = -\int \lambda(r)\,r\,dr. \tag{9.15}$$

Consequently, when $\lambda = \lambda(r)$, the energy integral exists and the potential energy can be calculated by formula (9.15).

Motion of particle in a central potential field. Inserting expression (9.13) for the squared velocity into the energy integral $mv^2/2 + \Pi(r) = E_0 = \text{const}$, we obtain

$$\left(\frac{du}{d\varphi}\right)^2 = \frac{2}{mc^2}\left(E_0 - \Pi(u) - \frac{mc^2u^2}{2}\right).$$

Switching from u to r, we have

$$\left(\frac{dr}{d\varphi}\right)^2 = \frac{2\,r^4}{mc^2}\,P(r)\,, \tag{9.16}$$

where $P(r) = E_0 - \Pi(r) - mc^2/(2\,r^2)$.

Assume that *ab initio* $dr/d\varphi \neq 0$. Then the function $P(r)$ is positive when $r = r_0$. Being continuous, it is also positive in some neighborhood of the point $r = r_0$. The pattern of the trajectory of motion depends on the size of this neighborhood.

We highlight the following four cases, in which the equation $P(r) = 0$:

(1) has two distinct positive roots r_1 and r_2 such that $P(r) > 0, r_1 < r < r_2$;
(2) has a positive root $r_1 < r_0$ such that $P(r) > 0, r_1 < r < +\infty$;
(3) has a root $r_2 > r_0$ such that $P(r) > 0, 0 < r < r_2$;
(4) has no positive roots: $P(r) > 0, 0 < r < +\infty$.

In the first case, the motion takes place between two concentric circumferences of radii r_1 and r_2. For simplicity assume that angle φ is measured from the position in which $r = r_1$ and $dr/d\varphi = 0$. When going from this point to the circumference of a greater radius $r = r_2$ the quantity r increases as φ increases and, therefore, differential equation (9.16) in this region can be represented as

$$d\varphi = f(r)\,dr\,, \quad \text{where} \quad f(r) = \frac{c\,\sqrt{m}}{\sqrt{2}\,r^2\,\sqrt{P(r)}}\,.$$

Integrating this equation, we obtain

$$\varphi = I(r) = \int_{r_1}^{r} f(r)\,dr\,, \quad 0 \leqslant \varphi \leqslant \Delta\varphi = I(r_2)\,, \quad r_1 \leqslant r \leqslant r_2. \tag{9.17}$$

After touching the circumference of a greater radius, the quantity r decreases as φ increases and, consequently, $d\varphi = -f(r)\,dr$ and, therefore

$$\varphi = \Delta\varphi - \int_{r_2}^{r} f(r)\,dr = \Delta\varphi - \int_{r_2}^{r_1} f(r)\,dr - \int_{r_1}^{r} f(r)\,dr = 2\Delta\varphi - I(r)\,, \tag{9.18}$$

$$\Delta\varphi \leqslant \varphi \leqslant 2\,\Delta\varphi\,.$$

As angle φ increases further, arguing as before, we have

$$\varphi = 2\,\Delta\varphi + I(r)\,, \quad 2\,\Delta\varphi \leqslant \varphi \leqslant 3\,\Delta\varphi\,,$$
$$\varphi = 4\,\Delta\varphi - I(r)\,, \quad 3\,\Delta\varphi \leqslant \varphi \leqslant 4\,\Delta\varphi\,.$$

Comparing these two equalities with relations (9.17) and (9.18), we see that the function $r = r(\varphi)$ is periodic and the angle $2\,\Delta\varphi$ is its period. Thus, construction of the function $r = r(\varphi)$ reduces to solving the integral (9.17).

If the angle $2 \Delta\varphi$ corresponding to returning to the circumference where the motion started can be represented as

$$2 \Delta\varphi = 2\pi k/n,$$

where k and n are integers, then after the n-th return, the angle φ will change by $2\pi k$. This means that the particle has returned to the initial position. The trajectory of motion is in this case a closed curve. As the study of the integral (9.17) has shown, it is possible for arbitrary c only when $\Pi = -\mu/r$, $\Pi = k r^2/2$, where μ and k are some positive constants.

Note an important property of the function $r = r(\varphi)$. If we make the change $\varphi_1 = -\varphi$, i.e., we reverse the direction of angle φ's positive reading, then the assumed differential equation will retain its form, since $(dr/d\varphi)^2 = (dr/d\varphi_1)^2$. Other reasonings when constructing the function $r = r(\varphi_1)$ are identical to those just presented and, therefore, $r(\varphi_1) = r(\varphi)$, i.e.,

$$r(\varphi) = r(-\varphi). \tag{9.19}$$

Thus, the function $r = r(\varphi)$ is even and, consequently, is readily extended to the interval of negative values of the angle φ.

Next, consider the motion of a particle along the constructed trajectory. The initial data in the problem in question are the quantities φ_0, $\dot\varphi_0$, r_0, $\dot r_0$. The constants E_0 and c that determine the shape of the trajectory, are characterized by the values of $\dot\varphi_0$, r_0, $\dot r_0$. The quantity φ_0 depends on the choice of zero of angle φ. This choice is arbitrary, therefore, the numeric value of angle φ_0 is not essential in the given problem. Recall that for the convenience of constructing the curve $r = r(\varphi)$ it was assumed that the angle φ is measured from the point, where $r = r_1$ and $dr/d\varphi = 0$. The presence of the area integral indicates that in motion the quantity $\dot\varphi$ is of the same sign as $\dot\varphi_0$. We shall assume that the chosen direction of angle φ measurement is such that in motion the angle φ increases. For this approach, the initial point on the curve $r = r(\varphi)$ is determined by the magnitude of r_0 and the sign of $\dot r_0$.

We mark the points of intersection of the curve $r = r(\varphi)$ with the circumference $r = r_0$. At some of them $\dot r > 0$, while at the rest $\dot r < 0$. Any of those points, $\dot r$ of which is of the same sign as $\dot r_0$, can be treated as the initial point of the trajectory. For certainty, when $\dot r_0 > 0$ adopt the point corresponding to the smallest value of angle φ_0 as initial. Let denote this angle φ_0^*. Since $r(\varphi) = r(-\varphi)$, then $\dot r(\varphi) = -\dot r(-\varphi)$. Therefore, when $\dot r_0 < 0$, the point corresponding to the angle $\varphi = -\varphi_0^*$ can be regarded as initial.

Let us consider the second case, when $P(r_1) = 0$, $r_1 < r_0$, $P(r) > 0$, $r_1 < r < +\infty$. In this case the curve $r = r(\varphi)$, constructed by the formulas (9.17) and (9.19), has only two points of intersection with the circumference $r = r_0$ at $\varphi = \varphi_0^*$ and $\varphi = -\varphi_0^*$, respectively. If $\dot r_0 > 0$, then the particle monotonously goes to infinity. In the case $\dot r_0 < 0$ when leaving the point $(r_0, -\varphi_0^*)$ the value of r first declines as φ increases, next the particle touches the circumference $r = r_1$ and then monotonously goes to infinity.

In the third case, when $P(r_2) = 0$, $r_0 < r_2$, $P(r) > 0$, $0 < r < r_2$, it is expedient to connect the zero of angle φ with the point at which $r = r_2$ and $dr/d\varphi = 0$. Then the curve $r = r(\varphi)$ is given by the relations

$$\varphi = \int\limits_r^{r_2} f(r)\, dr\,, \quad \varphi \geqslant 0\,, \quad r \leqslant r_2\,, \quad r(\varphi) = r(-\varphi)\,, \quad \varphi \leqslant 0\,.$$

In motion when $\dot{r}_0 > 0$, the particle touches the circumference and then starts to approach the center monotonously. When $\dot{r}_0 < 0$ the approaching starts right away. In the fields, where the third case holds under some initial conditions, the particle's drop onto the center is possible.

The fourth case, when the equation $P(r) = 0$ has no positive roots, is not very common and, therefore, is not worth considering.

Motion of a particle subject to a force proportional to the distance. The simplest example of a central potential field is the field, in which the force is given in the form $\mathbf{F} = -k\,\mathbf{r}$, $\lambda = -k = \text{const}$. When $k > 0$ the particle is attracted to the fixed center, and when $k < 0$, it is repelled from it. As follows from formula (9.15), the potential energy in this case is $\Pi = k\,r^2/2$.

Consider the equation $P(r) = 0$. We have

$$E_0 - \frac{k\,r^2}{2} - \frac{mc^2}{2\,r^2} = 0\,,$$

or

$$k\,r^4 - 2\,E_0 r^2 + mc^2 = 0\,, \tag{9.20}$$

hence

$$r_{1,2}^2 = \frac{E_0 \mp \sqrt{E_0^2 - mc^2 k}}{k}\,.$$

As is known, for any a and b the following inequality holds:

$$(a + b)^2 \geqslant 4\,ab\,, \tag{9.21}$$

therefore, since

$$v^2 \geqslant r^2\dot{\varphi}^2 = c^2/r^2\,, \tag{9.22}$$

when $k > 0$, we have

$$E_0^2 = \left(\frac{mv^2}{2} + \frac{k\,r^2}{2}\right)^2 \geqslant \left(\frac{mc^2}{2\,r^2} + \frac{k\,r^2}{2}\right)^2 \geqslant mc^2 k\,.$$

It follows from the above that in the case of the particle attracted to the center ($k > 0$) the quantities r_1^2 and r_2^2 are real and positive numbers.

The trajectory of motion lying between the concentric circumferences of radii r_1 and r_2 is constructed by formula (9.17). In this case we obtain

$$\varphi = c\sqrt{m} \int_{r_1}^{r} \frac{dr}{r\sqrt{2r^2 E_0 - kr^4 - mc^2}} = \frac{c\sqrt{m}}{2} \int_{r_1^2}^{r^2} \frac{d(r^2)}{r^2\sqrt{2r^2 E_0 - kr^4 - mc^2}} .$$

It is known that

$$\int \frac{dx}{x\sqrt{a + bx + cx^2}} = \frac{1}{\sqrt{-a}} \arcsin \frac{2a + bx}{x\sqrt{b^2 - 4ac}}, \qquad a < 0, \quad b^2 - 4ac > 0,$$
$$\tag{9.23}$$

therefore,

$$\varphi = \frac{1}{2} \arcsin \frac{E_0 r^2 - mc^2}{r^2 \sqrt{E_0^2 - mc^2 k}} \bigg|_{r_1^2}^{r^2} . \tag{9.24}$$

Taking into account the properties of the roots of equation (9.20)

$$r_1^2 r_2^2 = \frac{mc^2}{k}, \qquad r_1^2 + r_2^2 = \frac{2E_0}{k}, \qquad r_2^2 - r_1^2 = \frac{2\sqrt{E_0^2 - mc^2 k}}{k},$$

we represent the expression (9.24) as

$$\varphi = \frac{1}{2} \arcsin \frac{r^2(r_1^2 + r_2^2) - 2r_1^2 r_2^2}{r^2(r_2^2 - r_1^2)} + \frac{\pi}{4} .$$

Hence, it follows that

$$-\cos 2\varphi = \frac{r^2(r_1^2 + r_2^2) - 2r_1^2 r_2^2}{r^2(r_2^2 - r_1^2)}$$

or

$$r^2(r_2^2 - r_1^2)\sin^2 \varphi - r^2(r_2^2 - r_1^2)\cos^2 \varphi = r^2(r_1^2 + r_2^2) - 2r_1^2 r_2^2,$$

which finally yields

$$\frac{r^2 \cos^2 \varphi}{r_1^2} + \frac{r^2 \sin^2 \varphi}{r_2^2} = 1 .$$

In view of the formulas $x = r\cos\varphi$, $y = r\sin\varphi$, relating the polar coordinates to Cartesian ones, we have

$$\frac{x^2}{r_1^2} + \frac{y^2}{r_2^2} = 1 . \tag{9.25}$$

Thus, in the case of attraction to the center with a force, proportional to the distance, the trajectory of motion at any initial data is an ellipse with semiaxes r_1 and r_2. When $c \to 0$ as $r_1 \to 0$, this ellipse degenerates into a straight line segment of length $2\,r_2$.

If $k < 0$, then the first root r_1^2 of Eq. (9.20), when treated as a quadratic equation for r^2, is positive and the second one r_2^2 is negative. This shows that when $k < 0$ the equation $P(r) = 0$ has one positive root $r = r_1$ such that $P(r) > 0$ for $r > r_1$. As was already shown, in this case the trajectory is a curve going to infinity. The equation of this curve is again given by formula (9.24). But now, when passing to r_1^2 and r_2^2 in this formula and the subsequent ones, one should keep in mind that in this case the quantity r_2^2 is negative.

Setting $r_1^2 = a^2, r_2^2 = -b^2$, Eq. (9.25) assumes the form

$$\frac{x^2}{a^2} - \frac{y^2}{b^2} = 1 \, .$$

Consequently, in the case of repelling from the center with a force proportional to the distance, the trajectory of motion is a hyperbola and, namely, its right arm, since $r = r_1$ at $\varphi = 0$.

Common properties of Keplerian motion. The theory of motion of material bodies in a central potential field, with which we are concerned, was created and developed basically in connection with the study of the motion of celestial bodies subjected to forces of their mutual attraction. According to the law of gravitation, any two material bodies act on one another with forces proportional to their masses and inverse proportional to their squared separation. The coefficient of this dependence is the same for all bodies. In other words, the force F of masses m_1 and m_2 interaction can be represented as

$$F = \gamma \, \frac{m_1 m_2}{r^2} \, , \tag{9.26}$$

where γ is the gravity constant, $\gamma = 6.67 \cdot 10^{-11} \ \mathrm{Nm^2/kg^2}$, r is the distance between the masses. This law is formulated for two points, or localized masses. Now apply this law to a point mass m and to a mass M, which is continuously distributed inside of a sphere. Suppose that in every spherical layer the mass distribution can be treated as uniform. It is believed that many celestial bodies have such spherically symmetrical mass distribution. Decompose the mass M into a sum of elementary masses Δm_ν, each acting on the mass m in accordance with the law (9.26). Summing up the forces applied to the mass m from the masses Δm_ν, i.e., integrating over the inside of the sphere, we obtain the force $\mathbf{F}$, which, as can be shown, reads

$$\mathbf{F} = -\frac{\mu \, m}{r^3} \, \mathbf{r} \, , \qquad \mu = \gamma M \, ,$$

where $\mathbf{r}$ is the position vector connecting the center of the sphere and the mass m, $r = |\mathbf{r}|$. The constant μ is called the *gravity parameter of attracting center*.

Thus, a central gravitational field is formed around the mass M. By formula (9.15) its potential Π reads as

$$\Pi = \mu m \int \frac{dr}{r^2} = -\frac{\mu m}{r} \,.$$

All the presented general formulas, characterizing the motion of the mass m in a central gravitational field, were established on the assumption that the field center is a fixed point. The immovability of the coordinate system $Oxyz$, the center of which coincides with the field center, enables us to introduce no inertia forces and to write the law of motion in this system in the form (9.1). In reality, the force $-\mathbf{F}$ is applied to the mass M and, consequently, the system $Oxyz$ moves translationally with some acceleration. However, if the mass of m is considerably less than than the mass of M, then the motion of the mass M can be neglected. This actually means that the action of the force $-\mathbf{F}$ on the mass M is disregarded. In other words, we consider a problem on motion of a "nonattracting" particle in the gravity field of attracting one. In celestial mechanics such problem is called the *restricted two-body problem*. Let us consider this problem in detail.

A central field, in which the electrostatic force is invertially proportional to the squared separation from the center, in accordance with Coulomb's law, is created around a charged body of spherical shape. A particle, the charge of which is of the same sign as that of the body, placed into this field is repelled from the center, while an oppositely charged particle is attracted to it. Therefore, to make possible applying the obtained results to the motion of not only celestial bodies, but also that of charged particles, we shall suppose that the constant μ characterizing the field is not only positive (attraction), but also negative (repelling).

As was already shown, the behavior of the trajectory of motion is determined by the roots of the equation $P(r) = 0$. In the given case this equation reads

$$E_0 + \frac{\mu m}{r} - \frac{mc^2}{2r^2} = 0$$

or

$$hr^2 + 2\mu r - c^2 = 0\,, \tag{9.27}$$

where

$$\frac{2E_0}{m} \equiv h = v^2 - \frac{2\mu}{r} = \text{const}\,, \tag{9.28}$$

hence

$$r_{1,2} = \frac{-\mu \pm \sqrt{\mu^2 + hc^2}}{h}\,. \tag{9.29}$$

If $h > 0$, then regardless of the sign of μ, there exists a single positive root $r_1 = (-\mu + \sqrt{\mu^2 + hc^2})/h$ such that $P(r) > 0, r > r_1$. Consequently, if $h > 0$ the trajectory extends to infinity.

If $h < 0$, which is possible only when $\mu > 0$, both roots are positive, since

$$\mu^2 \geqslant -hc^2, \quad h < 0. \tag{9.30}$$

In order to show that this inequality holds when $h < 0$, we find the expression for μ using formula (9.28):

$$\mu = \frac{r}{2}(v^2 - h).$$

In view of equations (9.21) and (9.22), we have

$$\mu^2 \geqslant \frac{r^2}{4}\left(\frac{c^2}{r^2} - h\right)^2 \geqslant -hc^2.$$

Thus, if $h < 0$ the trajectory of motion always remains a limited (finite) curve lying between the circumferences of radii r_1 and r_2.

If $h = 0$, Eq. (9.27) becomes linear:

$$2\mu r - c^2 = 0, \quad \text{hence} \quad r_1 = c^2/(2\mu),$$

and $P(r) > 0$ when $r > r_1$.

To demonstrate the geometric and physical meaning in the case $h = 0$, one should examine the values of h close to zero, and next make $h \to 0$. From formulas (9.27), (9.29) as $h \to -0$ it follows that

$$\lim_{h \to 0} r_1 = \frac{c^2}{2\mu}, \quad \lim_{h \to -0} r_2 = +\infty.$$

Thus, as $h \to -0$ the finite trajectory transforms to infinite.

For determining the trajectory equation, we consider expression (9.17). In the given case, in view of notation (9.28), we have

$$\varphi = \int_{r_1}^{r} \frac{c\,dr}{r\sqrt{hr^2 + 2\mu r - c^2}}.$$

Using formula (9.23), we obtain

$$\varphi = \arcsin \frac{\mu r - c^2}{r\sqrt{\mu^2 + hc^2}} + \frac{\pi}{2},$$

hence

$$r = \frac{c^2}{\mu + \sqrt{\mu^2 + hc^2}\cos\varphi}. \tag{9.31}$$

This equation holds for any values of the parameters c and μ other than zero. Parameter h that characterizes the total mechanical energy, as follows from the relations (9.28), (9.30), lies between the limits

$$-\mu^2/c^2 \leqslant h < +\infty, \quad \mu > 0; \quad 0 < h < +\infty, \quad \mu < 0.$$

If $\mu > 0$, then introducing the notation

$$p = c^2/\mu, \quad e = \sqrt{1 + hc^2/\mu^2}, \tag{9.32}$$

we can write

$$r = \frac{p}{1 + e \cos \varphi}. \tag{9.33}$$

The above is the equation of a conic section in polar coordinates. The quantity p is called the *focal parameter (latus rectum)* and e is the *eccentricity*.

The minimal value of the total mechanical energy for given c and μ corresponds to the value $e = 0$. The trajectory of motion in this case is a circumference of radius $r = p = c^2/\mu$. The velocity of circular motion, or *circular velocity* v_{circ}, is perpendicular to the position vector, therefore $\sqrt{\mu r} = c = v_{\text{circ}} r$, hence

$$v_{\text{circ}} = \sqrt{\mu/r}. \tag{9.34}$$

The circular velocity by the surface of a celestial body having the shape of a sphere of radius R is called the *first cosmic (orbital) velocity* v_{I} relative to this celestial body. Such velocity can be expressed through the body's free fall acceleration g by its surface. In accordance with the Newton second law $mg = \mu m/R^2$, therefore,

$$\mu = gR^2, \quad v_{\text{I}} = \sqrt{gR}.$$

For the Earth, the value of R is accepted equal to the average radius of the Earth: $R_{\text{E}} = 6371\,\text{km}$. The acceleration g $= 9.820\,\text{m/s}^2$ corresponds to this value of R_{E} and $v_{\text{I}} = 7.910\,\text{km/s}$.

It is also interesting to consider the motion of the so-called *geostationary satellite* rotating along a circular orbit in the equatorial plane of the Earth with its angular velocity $\omega_{\text{E}} = 2\pi/(24 \cdot 3600)\,\text{s}^{-1}$. Such a satellite will be situated over the same point of the Equator. Having launched several such satellites one can provide a TV- and radio contact almost for all points of the Earth, except for the not large zones near its poles. For a geostationary satellite that is at the distance R_{stat} and has a constant velocity v_{stat}, the following equality should be satisfied:

$$\frac{m\, v_{\text{stat}}^2}{R_{\text{stat}}} = \frac{\gamma\, M_{\text{E}}\, m}{R_{\text{stat}}^2}, \quad 2\pi R_{\text{stat}} = v_{\text{stat}}\, 24 \cdot 3600.$$

This implies $v_{\text{stat}} = 3.075\,\text{km/s}$, $R_{\text{stat}} = 42\,164\,\text{km}$. It is obvious that a geostationary satellite is at the distance h_{stat} from the Earth surface:

$$h_{\text{stat}} = R_{\text{stat}} - R_E = 35\,793\,\text{km}\,.$$

For the quantity h varying from $-\mu^2/c^2$ to 0 ($0 < e < 1$) the trajectory of motion is an ellipse. The major semiaxis a of this ellipse, equal to $(r_1 + r_2)/2$, as follows from the Vieta theorem on the properties of equation (9.27) roots, can be represented as

$$a = -\mu/h\,. \tag{9.35}$$

The motion along the ellipse for $h \to -0$, when $e \to 1 - 0$ and $a \to \infty$, approaches to a motion along a parabola, and it is closer to the parabola the smaller is r. The limit case motion along the parabola is remarkable in the fact that in it the velocity, called the *parabolic velocity* v_{par}, at every point on the trajectory, as follows from the energy integral (9.28), can be represented as

$$v_{\text{par}} = \sqrt{2\mu/r}\,. \tag{9.36}$$

The point on the trajectory closest to the attracting center is called the *pericenter*. In the accepted reference frame the pericentre has the coordinates ($r = r_1$, $\varphi = 0$). The parabolic velocity in the pericenter for r_1 equal to the average radius R of the celestial body in question is called the *second cosmic velocity* v_{II} relative to this body. It is the minimal velocity of a satellite near the surface of a celestial body required to escape from this body. As follows from formulas (9.34) and (9.36), the first and the second cosmic velocities are related to one another via the relation $v_{\text{II}} = \sqrt{2}\,v_{\text{I}}$. For the Earth $v_{\text{II}} = 11.19\,\text{km/s}$.

As the total mechanic energy increases further, when $0 < h < +\infty$, $1 < e < +\infty$, the trajectory of motion is a hyperbola. Then the angle φ, as follows from Eq. (9.33), lies between the limits $-\varphi_{\text{max}} < \varphi < \varphi_{\text{max}}$, where $\varphi_{\text{max}} = \arccos(-1/e)$, $\pi/2 < \varphi_{\text{max}} < \pi$.

As the trajectory goes to infinity, it approaches the hyperbola asymptotes that are specified by the angles $\varphi = \pm\varphi_{\text{max}}$. Then the velocity decreases monotonically to the value $v = v_{\infty}$, which, as seen from formula (9.28), is $v_{\infty} = \sqrt{h}$.

To demonstrate the changes in the trajectory shape depending on h in the entire range of this parameter, the curves $r = r(\varphi)$ are displayed in Fig. 12 for some typical values of h for fixed $c > 0$ and $\mu > 0$. When constructing the trajectory corresponding to the specified values of c, μ and h, it pays to use the trajectory equation in polar and Cartesian coordinates simultaneously. From the formulas

$$\cos \varphi = x/r\,, \qquad r = \sqrt{x^2 + y^2}$$

it follows that Eq. (9.33) in Cartesian coordinates reads

$$r \equiv \sqrt{x^2 + y^2} = p - ex\,, \qquad 0 \leqslant e < +\infty\,, \tag{9.37}$$

or

Fig. 12 Trajectory forms under thee action of an attracting central force

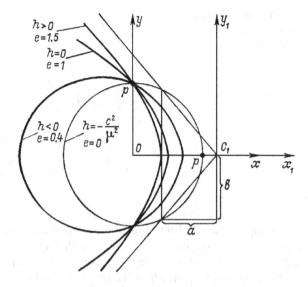

$$x^2(1 - e^2) + 2\,p\,ex + y^2 = p^2,$$

hence for $e = 1$ and $e \neq 1$ we find, respectively,

$$y^2 = 2\,p\left(\frac{p}{2} - x\right), \qquad \frac{x_1^2}{a^2} + \frac{1 - e^2}{|1 - e^2|}\frac{y_1^2}{b^2} = 1, \qquad (9.38)$$

where

$$x_1 = x - \frac{ep}{e^2 - 1}, \qquad y_1 = y, \qquad a = \frac{p}{|1 - e^2|}, \qquad b = a\sqrt{|1 - e^2|}. \qquad (9.39)$$

The hyperbola given by Eq. (9.38) has two branches. As follows from the original equation (9.37), the motion is executed along such a branch on which r decreases as x increases. The outlined property is exhibited by the left branch, the symmetry axis of which contains the center of force.

On the right branch of the hyperbola located in the semiplane $x_1 > 0$, the value of r increases with increasing x, therefore, for it the following equation is the original one

$$r \equiv \sqrt{x^2 + y^2} = ex - p, \qquad e > 1,$$

or in polar coordinates

$$r = \frac{p}{-1 + e\cos\varphi}, \qquad e > 1. \qquad (9.40)$$

Fig. 13 Trajectory forms
under thee action of
a repelling central force

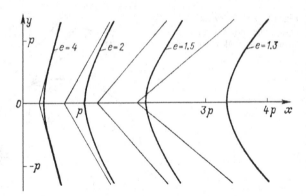

We have mentioned this equation of hyperbola right branches in the connection that Eq.(9.31) for $\mu < 0$ after introducing the notation $p = -c^2/\mu$, $e = \sqrt{1 + hc^2/\mu^2}$, $h > 0$, $e > 1$ transforms to (9.40). Thus, the unique trajectory for the particles being repelled from the center by Coulomb's law is the hyperbola right branches. Figure 13 displays the changes in the shape of these trajectories depending on e.

Trajectory equations (9.33) and (9.40) contain the parameters p and e. The quantity p can be adopted as a unit of length. In the dimensionless coordinates $\bar{x} = x/p$, $\bar{y} = y/p$, the trajectory equations depend on one dimensionless parameter e. Thus, the parameter p specifies the scope of the trajectory and the parameter e determines its shape and location in the plane $O\bar{x}\bar{y}$.

Next, consider how the motion along the constructed trajectories proceeds with time. The relation between the angle φ and the time t can be determined using the area integral (9.4). We first study the case when $\mu > 0$. Substituting the trajectory equation (9.33) into the relation $r^2\dot{\varphi} = c$, we obtain

$$c\,dt = \frac{p^2\,d\varphi}{(1 + e\cos\varphi)^2}.\qquad(9.41)$$

For simplicity, we will measure the time from the moment, when the particle is in the pericenter, i.e., when $\varphi = 0$. Integrating equation (9.41) under these initial conditions, we have

$$ct = \int\limits_0^{\varphi} r^2\,d\varphi = p^2\int\limits_0^{\varphi} \frac{d\varphi}{(1 + e\cos\varphi)^2}.\qquad(9.42)$$

The definite integral written above is expressed by elementary functions. The analytic form of this expression depends essentially on the value of e. For a motion along an ellipse, when $0 \leqslant e < 1$, using integral tables we obtain

$$ct = \frac{2p^2}{(1 - e^2)\sqrt{1 - e^2}} \arctan\left(\sqrt{\frac{1 - e}{1 + e}} \tan\frac{\varphi}{2}\right) - \frac{p^2 e \sin\varphi}{(1 - e^2)(1 + e \cos\varphi)},$$
$$0 \leqslant e < 1.$$

$$(9.43)$$

The expression found for the time t as a function of φ is hard to invert. For given t the value of φ is found as the solution of the transcendental equation (9.43). This equation is complicated and cannot be solved in closed form. Consequently, applying the relation (9.43) to the motion of celestial bodies, when it is necessary to determine with high precision, where the celestial body in question is located at the given instant, is connected with certain computational difficulties. The given problem is simplified substantially, if we invoke the parametric equation of an ellipse when solving it. In the x_1- and y_1-coordinates introduced by formulas (9.39) the Eq. (9.38) in parametric form for $0 \leqslant e < 1$ can be given by

$$x_1 = a \cos E, \qquad y_1 = b \sin E.$$

For finding out the geometric meaning of the parameter E let us plot a circumference of radius a with the center located at the origin of the system $O_1 x_1 y_1$ (Fig. 14). We drop the perpendicular $P_1 Q_1$ from a point P_1 of this circumference that is specified by the angle E to the $O_1 x_1$-axis. The point P of intersection of this perpendicular with the ellipse has the polar coordinates (r, φ). In astronomy, the angle E is called the *eccentric anomaly* and the angle φ is called the *true anomaly*. To evaluate r and φ as functions of E, we invoke the relations

$$x = r \cos\varphi = a(\cos E - e), \qquad y = r \sin\varphi = b \sin E,$$
$$0 \leqslant e < 1, \qquad a = \frac{p}{1 - e^2}, \qquad b = a\sqrt{1 - e^2} = \frac{p}{\sqrt{1 - e^2}}, \qquad (9.44)$$

which follow directly from formulas (9.39), hence

$$x^2 + y^2 = a^2(\cos E - e)^2 + a^2(1 - e^2)(1 - \cos^2 E) = a^2(1 - e \cos E)^2,$$

or

$$r = a(1 - e \cos E). \qquad (9.45)$$

Considering this relation simultaneously with the equality $r \cos\varphi = a(\cos E - e)$, we find

$$r(1 - \cos\varphi) = a(1 + e)(1 - \cos E),$$
$$r(1 + \cos\varphi) = a(1 - e)(1 + \cos E).$$

Hence, since $\tan^2(x/2) = (1 - \cos x)/(1 + \cos x)$, we have

$$\tan\frac{E}{2} = \sqrt{\frac{1 - e}{1 + e}} \tan\frac{\varphi}{2}. \qquad (9.46)$$

Fig. 14 Eccentric and true
anomalies

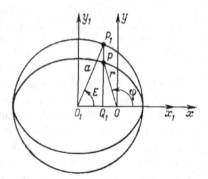

Substituting this expression into the formula (9.43) and taking account (9.33), (9.44)
and (9.32), we obtain

$$nt = E - e \sin E \,, \tag{9.47}$$

where

$$n = \sqrt{\mu} / a^{3/2} \,. \tag{9.48}$$

Next, specifying t, we come to the equation

$$E - e \sin E = N \,, \qquad N = nt \tag{9.49}$$

for the parameter E. The transcendental equation (9.49) is called *Kepler's equation.*
Having determined the value of E from it, by formulas (9.45) and (9.46) we find r
and φ. Thus, the problem is completely solved.

If a moving particle makes all the way along the ellipse, then φ and E change
by 2π. Hence, the time of a full rotation $T = 2\pi/n$, thus

$$n = 2\pi/T \,. \tag{9.50}$$

Consequently, n is the mean angular velocity of the position vector $\mathbf{r}$ rotations.
Therefore, in astronomy the quantity n is called the *mean motion.*

Formulas (9.48), (9.50) enable us to represent the constant μ as

$$\mu = a^3 n^2 = 4\pi^2 a^3 / T^2 \,. \tag{9.51}$$

Now consider the motion along a hyperbola. Solving the integral from expression
(9.42) for $e > 1$ leads us to an even more complicated relation between t and φ
than the one written for $e < 1$. Therefore, we turn immediately to the parametric
form of specifying a hyperbola. As was already noted, for $\mu > 0$ and $e > 1$ the
motion is executed along the left branch of the hyperbola (9.38). In this branch the
x_1-coordinate is negative, therefore, in parametric form the curve is given by the
equations

$$x_1 = -a \cosh H, \qquad a = \frac{p}{e^2 - 1},$$
$$y_1 = b \sinh H, \qquad b = a\sqrt{e^2 - 1},$$

hence, in accordance with formulas (9.39)

$$x = r \cos \varphi = a(e - \cosh H),$$
$$y = r \sin \varphi = b \sinh H, \tag{9.52}$$

and, consequently,

$$r = a(e \cosh H - 1). \tag{9.53}$$

Using this formula and the formula (9.52) we obtain

$$r(1 - \cos \varphi) = a(e + 1)(\cosh H - 1),$$
$$r(1 + \cos \varphi) = a(e - 1)(\cosh H + 1), \tag{9.54}$$

hence,

$$\tan \frac{\varphi}{2} = \sqrt{\frac{e + 1}{e - 1}} \tanh \frac{H}{2}. \tag{9.55}$$

For changing in the integral (9.42) from the variable φ to a new variable H it is necessary to express the quantity $d\varphi$ in terms of dH. Calculating the differentials of the right- and left-hand sides of expression (9.55) we obtain

$$\frac{d\varphi}{\cos^2(\varphi/2)} = \sqrt{\frac{e + 1}{e - 1}} \frac{dH}{\cosh^2(H/2)}.$$

In view of relations (9.54), we have

$$d\varphi = \sqrt{\frac{e + 1}{e - 1}} \frac{1 + \cos \varphi}{1 + \cosh H} dH = \frac{a\sqrt{e^2 - 1}}{r} dH. \tag{9.56}$$

Substituting expressions (9.53) and (9.56) into the integral (9.42), we find

$$ct = \int_0^\varphi r^2 \, d\varphi = a^2\sqrt{e^2 - 1} \int_0^H (e \cosh H - 1) \, dH = a^2\sqrt{e^2 - 1}(e \sinh H - H),$$

or

$$nt = e \sinh H - H, \tag{9.57}$$

where n, just as for $e < 1$, is calculated by formula (9.48).

Specifying t we come to Eq. (9.57), which is identical to Kepler's equation. Solving this equation, we find the parameter H and then determine r and φ by formulas (9.53) and (9.55).

The technique of constructing the functions $r(t)$ and $\varphi(t)$ for a hyperbolic motion under this approach corresponds completely to the technique of constructing such functions for an elliptic motion. For values of e close to the unit, this technique complicates in both cases, as then in a vicinity of the pericenter the right-hand sides of equations (9.47) and (9.57) are the differences between two quantities of close values. Therefore, the case of a parabolic motion, when $e = 1$, should be considered separately.

Parabolic motion is the limit case of elliptic or hyperbolic motions and is never realized in practice. However, in astronomy the cases of comets moving in quite extended elliptic orbits and of meteorites — in quite extended hyperbolic ones are often observed. In both cases the eccentricity of the orbit is quite close to the unit, and, consequently, the motion of the celestial body, at least near the pericenter (i.e. near the Sun), hardly differs from a parabolic motion and can, therefore, be analyzed by the formulas corresponding to this motion.

The integral (9.42) for $e = 1$ reads

$$ ct = p^2 \int\limits_0^\varphi \frac{d\varphi}{(1 + \cos\varphi)^2} = \frac{p^2}{4} \int\limits_0^\varphi \frac{d\varphi}{\cos^4(\varphi/2)} = \frac{p^2}{2}\left(\tan\frac{\varphi}{2} + \frac{1}{3}\tan^3\frac{\varphi}{2} \right) $$

or

$$ n_1 t = \tan\frac{\varphi}{2} + \frac{1}{3}\tan^3\frac{\varphi}{2}, \qquad \text{where} \qquad n_1 = \frac{2\sqrt{\mu}}{p^{3/2}}. $$

Determining the true anomaly φ corresponding to the given value of t reduces in this case to solving the cubic equation

$$ \sigma^3 + 3\sigma = 3n_1 t, \qquad \text{where} \qquad \sigma = \tan(\varphi/2). $$

Now consider the case when $\mu < 0$. Then, as was already shown, the motion is executed along the right branch of a hyperbola. In parametric form it is given by the equations

$$ x_1 = a\cosh H, \qquad p = -c^2/\mu = a\,(e^2 - 1), $$
$$ y_1 = b\sinh H, \qquad b = a\sqrt{e^2 - 1}. $$

Calculations that are completely identical to those for the hyperbolic motion for $\mu > 0$ give the following results

$$ r = a(e\cosh H + 1), \qquad \tan\frac{\varphi}{2} = \sqrt{\frac{e-1}{e+1}}\,\tanh\frac{H}{2}, $$
$$ nt = e\sinh H + H, \qquad n = \frac{\sqrt{-\mu}}{a^{3/2}}. $$

Derivation of the law of gravitation from the Kepler laws. The basic properties of planetary motion were discovered by Kepler in 1609–1619. He has formulated them in the form of the following three laws:

1. *The heliocentric motion of every planet is executed in a fixed plane passing through the center of the Sun such that the area of the sector swept by the position vector of the planet changes proportionally to the time.*

2. *The orbit of each planet is an ellipse, the Sun being located at one of its focuses.*

3. *The squared times of planets' circulations around the Sun relate as the cubes of their orbits' major semiaxes.*

These laws called the *Kepler laws* are presented here in the order specified by Kepler himself. Later it has been changed: the first law was renamed as the second and the second was called the first one. The formulations have also been slightly changed.

On the basis of the Kepler laws, the law of gravitation was discovered by Newton in 1666, which made it possible to create a consistent mathematical theory of celestial bodies' motion, the basic elements of which were considered in this paragraph.

When deriving the law of gravitation from the Kepler laws, Newton used a complicated geometric method. Under analytic approach the reasonings simplify substantially. They become most simple after invoking the above presented formulas.

It follows from the first Kepler law that the integral of areas exists. As was found out, the presence of such integral is the attribute of the fact that the force field is central. Equation (9.8), in accordance with Binet's formula (9.14), can be represented as

$$mw_r \equiv -mc^2u^2\left(\frac{d^2u}{d\varphi^2} + u\right) = -F, \qquad u = \frac{1}{r}, \tag{9.58}$$

where F is the force of attraction of a planet of mass m to the Sun.

The analytic expression for the second Kepler law is given by ellipse equation in polar coordinates (9.33). Taking into account that $u = 1/r$, we have

$$u = (1 + e\cos\varphi)/p. \tag{9.59}$$

Substituting this expression into Eq. (9.58), we obtain

$$F = \frac{mc^2}{p\,r^2}. \tag{9.60}$$

The first and second Kepler laws lead to the Kepler equation (9.47), thereby enabling us to determine, how the motion along the ellipse given by Eq. (9.59) proceeds with time. When going from Eq. (9.43) to the Kepler equation (9.47), the quantity μ was introduced in the expression for n in accordance with the notation (9.32). If we treat only the motion parameters p, c and a as given, then it follows from formulas (9.48) and (9.32) that

$$n = \frac{c}{\sqrt{p}\, a^{3/2}} \cdot$$

Taking into account that the mean motion n is related to the period T of the planetary circulation around the Sun via the relation (9.50), we obtain

$$\frac{c^2}{p} = \frac{4\pi^2 a^3}{T^2} \cdot \qquad (9.61)$$

Since, according to the third Kepler law, the ratio a^3/T^2 is constant for all planets in the solar system, then it follows from this law and the equality (9.61) that the quantity c^2/p entering the expression (9.60) is a constant. Letting μ denote it, we have

$$F = \frac{\mu m}{r^2} \cdot \qquad (9.62)$$

The solar-planet interaction force F is proportional to the mass m of the planet, and hence the constant μ is proportional to the mass M of the Sun. Consequently, it can be represented as $\mu = k^2 M$, where k is a new constant common to all planets in the solar system and called the *Gaussian gravitational constant*. The expression for the interaction force then transforms to

$$F = k^2 \frac{mM}{r^2} \cdot \qquad (9.63)$$

To make sure that the Gaussian constant k is the same not only for the Sun and the planets, but also for any two bodies, Newton applied his law of gravitation (9.63) to the Earth and the Moon.

In accordance with formula (9.61) the quantity μ for the Earth can be calculated from the period of the Moon circulations around the Earth. Inserting this expression for μ into formula (9.62), we find that the acceleration g of a body of mass m by the Earth surface, i.e., for $r = R_E = 6.371 \cdot 10^6$ m, can be represented as

$$g = \frac{\mu}{R^2} = \frac{4\pi^2 a^3}{T^2 R^2} \cdot \qquad (9.64)$$

Taking into account that for the Moon $T = 27.3\,\text{days} = 2.358 \cdot 10^6\,\text{s}$, $a = 3.844 \cdot 10^8$ m, we obtain g $= 9.93$ m/s.

The somewhat overestimated value of g is explained by the fact that the examination is carried out within a limited two-body problem. A correction providing a more accurate result will be inserted into the formula (9.64) later.

The data on the distance from the Moon to the Earth Newton used in 1666 were far from accurate, therefore, the value of g obtained by the formula (9.64) did not coincide with the value of g found from the oscillation period of a pendulum. That is why the law of gravitation discovered by him was published 20 years later after repeated calculations using more accurate data.

Fig. 15 Hohmann transfer

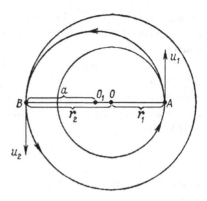

Thus, the quantity k^2 in formula (9.63), as Newton investigations showed and, subsequently, so did those of many other scientists, indeed is a universal constant. In formula (9.26), from which the consideration of Keplerian motion started, it was denoted by γ.

The transfer from one circular orbit to another one along the Hohmann ellipse. The methods of studying the celestial bodies motion worked out by Newton, Euler, d'Alembert, Lagrange, Laplace, and other scientists, are widely applied in studying the motion of artificial celestial bodies. With the development of cosmonautics a brand new factor has arisen in the problems of dynamics dealing with the motion of bodies in the gravitational field, — control of a spacecraft motion. This control is executed by means of jet propulsion. We can assume that the necessary change in the velocity vector of a spacecraft in this case occurs instantaneously in the sense that the change in the spacecraft position for the time of the jet propulsion action can be neglected (and, namely, the distance the body covers in that time is negligibly small compared to the distance from it to the attracting center). In such case it is said that an impulsive change in the velocity vector of the spacecraft occurred — *an impulsive orbital maneuver.* The quantity Δv that characterizes the maneuver, is usually called the *velocity impulse* in cosmodynamics, or simply the *impulse.* The transfer from one orbit to another, executed as a result of a k-fold impulsive change in the velocity is called the *k-impulse transfer.*

Among two-impulse transfers, the maneuver, as a result of which the spacecraft transfers from one circular orbit to another lying in the same plane is of particular importance. The velocity impulses Δv_1, Δv_2 are applied on the circumferences of radii r_1 and r_2 respectively. As a result of applying the first impulse Δv_1, the circular orbit transforms to an elliptical one. At the point of contact of this ellipse with the circumference of radius r_2, the second impulse is applied, the one that transfers the spacecraft to a new circular orbit (Fig. 15). It can be shown that such a maneuver suggested by Hohmann proves to be most expedient from the energy consumption viewpoint. Let us consider it in detail.

Let A and B denote the points of contact of the ellipse with the circumferences of radii r_1 and r_2, respectively. Let the impulses applied at those points be given by scalar quantities u_1 and u_2. When the spacecraft transfers to the orbit of greater radius, it speeds up at the points A and B ($u_1 > 0$, $u_2 > 0$), and so it slows down when $r_2 < r_1$ ($u_1 < 0$, $u_2 < 0$).

Before the first impulse is applied, the spacecraft has the angular velocity v_1, calculated by formula (9.34): $v_1 = \sqrt{\mu/r_1}$. After applying the impulse u_1 and arrival in elliptical orbit, the magnitude v of velocity at every point of this orbit can be found from the energy integral that, in accordance with the expressions (9.28), (9.35), rewrites as

$$v^2 = \frac{2\mu}{r} - \frac{\mu}{a}, \qquad a = \frac{r_1 + r_2}{2}.$$

It follows from the above that the velocities at the points A and B of the elliptical orbit are

$$v_A \sqrt{\frac{2\mu}{r_1} - \frac{2\mu}{r_1 + r_2}} = v_1 \sqrt{\frac{2\rho}{1 + \rho}},$$

$$v_B = \sqrt{\frac{2\mu}{r_2} - \frac{2\mu}{r_1 + r_2}} = v_1 \sqrt{\frac{2}{\rho(1 + \rho)}}.$$

Here, $\rho = r_2/r_1$. In addition,

$$v_A = v_1 + u_1, \qquad v_2 = v_B + u_2,$$

where $v_2 = \sqrt{\mu/r_2}$, therefore,

$$\frac{u}{v_1} = \frac{\rho - 1}{\rho} \sqrt{\frac{2\rho}{1 + \rho}} + \frac{1}{\sqrt{\rho}} - 1 \equiv f(\rho). \tag{9.65}$$

Here, $u = u_1 + u_2$ is the total impulse. The function corresponding to the expression (9.65) is displayed in Fig. 16. For $\rho = 15.58$ the function $f(\rho)$ has a maximum which equals 0.539. The limiting value of $f(\rho)$ as $\rho \to +\infty$ equals $\sqrt{2} - 1 = 0.414$. Then

$$v_A \to \sqrt{2}\, v_1, \qquad u_1 \to (\sqrt{2} - 1)\, v_1,$$

which corresponds to applying the impulse at the point A alone and the transfer to the parabolic orbit. The other limiting value $f(+0) = -\infty$ means that the energy consumption increases unrestrictedly, when striving to revolve around the attracting center directly.

As known, the Earth's orbit is nearly complanar to the orbits of all the planets apart from Pluto, therefore, Fig. 16 can be used for a rough estimation of the consumption of energy on the spacecraft transfer from the Earth orbit to orbits of other planets in the solar system. Here, the values of the function $f(\rho)$ corresponding to the specific planets are presented. Recall that the average velocity of the Earth motion around the Sun $v_1 = 29.765$ km/s.

Fig. 16 Power consumption of a spacecraft

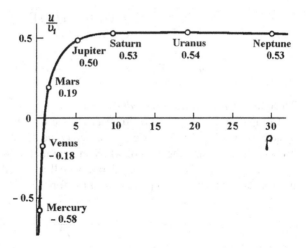

Fig. 17 Consideration of the motion of mass M in the two-body problem

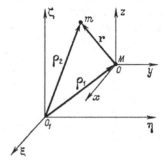

Two-body problem. When considering the motion of mass m around the mass M in a central field created by gravity forces or Coulomb's forces, the motion of mass M subject to the force applied to it from the mass m has not been taken into account so far.

Assume that in a Cartesian coordinate system $O_1\xi\eta\zeta$ the masses M and m are acted upon by the forces of their interaction only (Fig. 17). In this case, the equations of motion of the centers of mass of the bodies in question read

$$M\ddot{\boldsymbol{\rho}}_1 = -\mathbf{F}, \qquad m\ddot{\boldsymbol{\rho}}_2 = \mathbf{F},$$

where $\boldsymbol{\rho}_1$ and $\boldsymbol{\rho}_2$ are the position vectors of the centers of mass of the bodies of mass M and m, respectively, and $\mathbf{F}$ is the force applied to the mass m from the mass M. Suppose that the vector $\mathbf{F}$ can be represented as

$$\mathbf{F} = \lambda\,\mathbf{r}, \qquad \text{where} \qquad \mathbf{r} = \boldsymbol{\rho}_2 - \boldsymbol{\rho}_1.$$

Differentiating the above relation twice with respect to time and substituting the accelerations $\ddot{\boldsymbol{\rho}}_1$ and $\ddot{\boldsymbol{\rho}}_2$ in terms of the force $\mathbf{F}$ into it, we obtain

$$\ddot{\mathbf{r}} = (1/M + 1/m)\,\mathbf{F}\,,$$

or

$$m_*\ddot{\mathbf{r}} = \mathbf{F}\,, \qquad m_* = \frac{mM}{m+M} = m\left(1 + \frac{m}{M}\right)^{-1}. \tag{9.66}$$

The quantity m_* is called the *reduced mass*.

Comparing Eq. (9.66) with Eq. (9.1) we see that the translational motion of the system $Oxyz$ (Fig. 17), the origin of which coincides with the center of mass of the body of mass M, and the axes of which are parallel in motion to the axes of system $O_1\xi\eta\zeta$, is easily taken into consideration. All we need is to replace the reduced mass m by the mass m_*. In particular, if the force

$$\mathbf{F} = -\gamma\,\frac{mM}{r^3}\,\mathbf{r}\,,$$

then we can rewrite the equation of motion of the mass m in the system $Oxyz$ in accordance with the equality (9.66) as

$$\ddot{\mathbf{r}} + \frac{\mu_*\mathbf{r}}{r^3} = 0\,, \qquad \mu_* = \gamma\,(M+m)\,. \tag{9.67}$$

If the motion of the mass M is not taken into account, then the corresponding equation is

$$\ddot{\mathbf{r}} + \frac{\mu\mathbf{r}}{r^3} = 0\,, \qquad \mu = \gamma M\,.$$

It follows from comparing these equations that replacing the quantity μ by the quantity $\mu_* = \mu\,(1 + m/M)$ in all the preceding formulas, we thereby take account of the motion of the mass M subject to the force $-\mathbf{F}$.

After replacing μ by μ_*, the relation (9.51) expressing the third Kepler law transforms to

$$\frac{4\pi^2 a^3}{T^2} = \mu\left(1 + \frac{m}{M}\right). \tag{9.68}$$

This shows that the ratio a^3/T^2 is distinct for different planets and depends on their masses. However, this correction to the third Kepler law is very insignificant, since even for Jupiter having the greatest mass of all the planets the ratio is $m/M \approx 0.001$.

The indicated correction should be taken into account, when deriving the law of gravitation from the Kepler laws. However, such correction to the formula (9.51) should be implemented simultaneously with correcting equation (9.58) and the equality (9.60) by replacing the mass m by the reduced mass m_* in them. The relation (9.61) following immediately from the Kepler equation remains valid under the new approach. But in accordance with the correction (9.68), the quantity c^2/p in formula (9.60) should be denoted by $\mu\,(1 + m/M)$, where μ is a constant. But since now in formula (9.60) m is replaced by m_*, the final expressions (9.62) and (9.63) remain unchanged.

The formula (9.64) used by Newton to test the law of gravitation, in view of the correction (9.68), can be rewritten as

$$g = \frac{\mu}{R^2} = \frac{4\pi^2 a^3}{T^2 R^2 (1 + m/M)} \, .$$

As known, the ratio of the Moon mass m to the Earth mass M equals $1.23 \cdot 10^{-2}$. Correcting the calculations by formula (9.64), we obtain $g = 9.82 \, \text{m/s}^2$, which almost identically corresponds to the value of g for the average radius R_E of the Earth.

In reality, the motion of a planet in question around the Sun is affected not only by its interaction with the Sun but also with all the other bodies in the solar system. These additional effects are called *perturbations* in celestial mechanics. The motion described by the differential equation (9.67) disregarding these perturbations is called the *unperturbed Keplerian motion*. The necessity of taking into account the perturbations from other planets leads to the arising of the following problem.

Suppose that in the system $O\xi\eta\zeta$, in which the motion of n bodies is considered, we can limit ourselves to accounting for the forces of their interaction by the law of gravitation only. Analyzing the motion of those bodies in the given system is called the *n-body problem*. For $n = 2$ the problem reduces to Eq. (9.67). For $n \geqslant 3$ this problem has no solutions in closed form. This is why various qualitative and numeric methods are widely used in celestial mechanics to examine solutions of differential equations.

Chapter 5
System Dynamics

N. N. Polyakhov, M. P. Yushkov⊕, and S. A. Zegzhda

With the help of introducing the concept of a representative point the general types of equations of motion that were obtained earlier for a particle are extended in this chapter to the case of motion of a system consisting of n particles. The general dynamics theorems making it possible to get the integrable combinations of equations of motion of a system in certain cases are proved.

1 Representative Point. Its Equations of Motion

Representative point. A set of particles interacting with each other is called *a mechanical system*. The interaction of particles changes their motion comparing to that which they would have if they were isolated. As the change in motion occurs under the action of forces we shall assume that between the particles of a system there are forces which we shall call *internal forces*.

If the coordinates and velocities of particles of a system are not connected to each other with any given restriction, then the system is called *unconstrained*. As the position of each particle belonging to an unconstrained system is defined by three independent coordinates, then the system consisting of n particles is characterized by $3n$ coordinates. For instance, the solar system is an unconstrained system of particles.

The Newton law for a particle belonging to the system consisting of n particles reads as

$$m_\nu \ddot{\mathbf{r}}_\nu = \mathbf{F}_\nu, \quad \mathbf{F}_\nu = \mathbf{F}_\nu^{(i)} + \mathbf{F}_\nu^{(e)}, \quad \nu = \overline{1, n}, \tag{1.1}$$

where ν is the number of a particle; $\mathbf{r}_\nu$ is its position vector; $\mathbf{F}_\nu^{(i)}$ is the internal force which is applied to the particle and is the torque-free (pure force) resultant of all interaction forces of the particle in question and other particles of the system; $\mathbf{F}_\nu^{(e)}$ is the torque-free resultant of all external forces. Suppose that the forces $\mathbf{F}_\nu^{(i)}$ and $\mathbf{F}_\nu^{(e)}$

© Springer Nature Switzerland AG 2021
N. N. Polyakhov et al., *Rational and Applied Mechanics*,
Foundations of Engineering Mechanics,
https://doi.org/10.1007/978-3-030-64061-3_5

are the functions of t, $\mathbf{r}_1$, $\mathbf{r}_2$, $\ldots$, $\mathbf{r}_n$, $\dot{\mathbf{r}}_1$, $\dot{\mathbf{r}}_2$, $\ldots$, $\dot{\mathbf{r}}_n$. Thus, the differential equations of motion of the system are n second-order vector equations of containing n unknown vector functions $\mathbf{r}_\nu$.

Equations (1.1) in projections onto the axes of the Cartesian coordinate system can be represented as

$$m_\nu \ddot{x}_{\nu k} = X_{\nu k}(t, x, \dot{x}), \quad X_{\nu k}(t, x, \dot{x}) = X_{\nu k}^{(i)}(t, x, \dot{x}) + X_{\nu k}^{(e)}(t, x, \dot{x}),$$
$$\nu = \overline{1, n}, \quad k = 1, 2, 3, \tag{1.2}$$

where $x_{\nu k}$, $X_{\nu k}$, $X_{\nu k}^{(i)}$, $X_{\nu k}^{(e)}$ are the projections of the vectors $\mathbf{r}_\nu$, $\mathbf{F}_\nu$, $\mathbf{F}_\nu^{(i)}$, $\mathbf{F}_\nu^{(e)}$ respectively, onto the x_k-axis. Recall that here as before when writing the arguments of a multivariable function the letter without a subscript denotes all the set of variables introduced by the same symbol with subscripts.

We shall assume that the functions in the right-hand side of differential equations are continuous in the vicinity of the initial values of $x_{\nu k}(t_0)$, $\dot{x}_{\nu k}(t_0)$ and have the partial derivatives with respect to $x_{\nu k}$ and $\dot{x}_{\nu k}$ are bounded in this vicinity. According to the general theory of differential equations these conditions are sufficient for the unique solvability of system (1.2) under given initial conditions.

The solutions of the system of differential equations (1.2) are $3n$ functions of time t and $6n$ arbitrary constants

$$x_{\nu k} = x_{\nu k}(t, C_1, C_2, \ldots, C_{6n}),$$

in this case to find C_j, $j = 1, 2, \ldots, 6n$, one should employ the initial conditions

$$x_{\nu k}(t_0) = x_{\nu k}(t_0, C_1, C_2, \ldots, C_{6n}), \quad \dot{x}_{\nu k}(t_0) = \dot{x}_{\nu k}(t_0, C_1, C_2, \ldots, C_{6n}),$$

which make it possible to express C_j in terms of the initial coordinates $x_{\nu k}(t_0)$ and initial velocities $\dot{x}_{\nu k}(t_0)$.

It is convenient to write the system of $3n$ differential equations (1.2) in the form using the through (continuous) numeration of coordinates as follows. Setting $x_{\nu k} = x_\mu$, where $\mu = 3(\nu - 1) + k$, $\nu = \overline{1, n}$, $k = 1, 2, 3$, we get

$$x_{11} = x_1, \quad x_{12} = x_2, \quad x_{13} = x_3,$$

$$x_{21} = x_4, \quad x_{22} = x_5, \quad x_{23} = x_6,$$

$$\ldots\ldots\ldots\ldots\ldots\ldots\ldots\ldots\ldots\ldots\ldots\ldots$$

$$x_{n1} = x_{3n-2}, \quad x_{n2} = x_{3n-1}, \quad x_{n3} = x_{3n}.$$

The same can be also performed for the quantities $X_{\nu k}$. Therefore, the system (1.2) appears as

$$m_\mu \ddot{x}_\mu = X_\mu, \quad \mu = \overline{1, 3n}. \tag{1.3}$$

As in system (1.2) the same mass m_ν is multiplied by $\ddot{x}_{\nu k}$, $k = 1, 2, 3$, successively, and so in system (1.3) we suppose that the mass m_μ is equal to m_ν for $\mu = 3\nu - 2$, $3\nu - 1$, 3ν.

Let us introduce the $3n$-dimensional vectors $\mathbf{y} = (y_1, y_2, \ldots, y_{3n})$ and $\mathbf{Y} = (Y_1, Y_2, \ldots, Y_{3n})$ with the components

$$y_\mu = x_\mu \sqrt{\widetilde{m}_\mu}, \quad Y_\mu = \frac{X_\mu}{\sqrt{\widetilde{m}_\mu}}, \tag{1.4}$$

where

$$\widetilde{m}_\mu = \frac{m_\mu}{M}, \quad M = \sum_{\nu=1}^{n} m_\nu. \tag{1.5}$$

We agree to assume that the length of the vector $\mathbf{y}$ as

$$|\mathbf{y}| = \sqrt{\sum_{\mu=1}^{3n} y_\mu^2} = \sqrt{\sum_{\mu=1}^{3n} m_\mu x_\mu^2},$$

and its direction is characterized by means of the quantities $\cos \alpha_\mu = y_\mu / |\mathbf{y}|$. The same convention applies to the vector $\mathbf{Y}$.

If we introduce the unit vectors $\mathbf{i}_\mu$ corresponding to the $3n$-dimensional orthogonal system of Cartesian coordinates y_μ, then the vectors $\mathbf{y}$ and $\mathbf{Y}$ are represented as

$$\mathbf{y} = y_\mu \mathbf{i}_\mu, \quad \mathbf{Y} = Y_\mu \mathbf{i}_\mu.$$

On the basis of the above the system of differential equations (1.3), which has the form

$$M \ddot{y}_\mu = Y_\mu, \quad \mu = \overline{1, 3n}$$

in new coordinates, can be written in the form of a vector equation for a single point (particle) of mass M moving in the $3n$-Euclidean space. This point is called *a Hertz representative point*. The equation of its motion is

$$M \ddot{\mathbf{y}} = \mathbf{Y}, \quad \text{or} \quad M \mathbf{W} = \mathbf{Y}, \quad \text{where} \quad \mathbf{W} = \dot{\mathbf{V}} = \ddot{\mathbf{y}} = \ddot{y}_\mu \mathbf{i}_\mu. \tag{1.6}$$

Note that if we define the kinetic energy of the system as the sum of the kinetic energies of its particles, then we can easily see that this energy agrees with the energy of the representative point, which is $M V^2 / 2$, where V is the absolute magnitude of the velocity $\mathbf{V} = \sum_{\mu=1}^{3n} \dot{y}_\mu \mathbf{i}_\mu$ of the representative point. Indeed, according to the definition

$$T = \frac{1}{2} \sum_{\nu=1}^{n} m_\nu v_\nu^2 = \frac{1}{2} \sum_{\mu=1}^{3n} m_\mu \dot{x}_\mu^2 =$$

$$= \frac{M}{2} \sum_{\mu=1}^{3n} \tilde{m}_\mu \dot{x}_\mu^2 = \frac{M}{2} \sum_{\mu=1}^{3n} \dot{y}_\mu^2 = \frac{MV^2}{2}. \tag{1.7}$$

If instead of the Cartesian coordinates y_μ we introduce the curvilinear ones q^σ with the basis vectors $\mathbf{e}_\sigma = \partial \mathbf{y}(t, q)/\partial q^\sigma$, $\sigma = \overline{1, 3n}$, then the velocity vector of the representative point can be represented as

$$\mathbf{V} = \frac{\partial \mathbf{y}}{\partial t} + \frac{\partial \mathbf{y}}{\partial q^\sigma} \dot{q}^\sigma = \frac{\partial \mathbf{y}}{\partial q^\alpha} \dot{q}^\alpha, \quad \alpha = \overline{0, 3n}, \quad q^0 = t. \tag{1.8}$$

Comparing this expression with the expression (4.2) of Chap. 4 for the velocity of a particle in the case of a non-stationary basis, we note that the only difference is in the fact that the summation is done up to $3n$ but not 3. Obviously, this assertion also applies to the acceleration $\mathbf{W}$ of a representative point, comparing to that of a particle $\mathbf{w}$. This implies that the formulas characterizing the kinematics of a particle in arbitrary curvilinear coordinates can be applied when studying the motion of a representative point. General aspects of the particle dynamics in arbitrary curvilinear coordinates were studied in Sects. 4–6 of Chap. 4. The general vector equation of motion for a representative point (1.6) and its kinetic energy (1.7) are identical with the expressions for a particle. That is why all the results and formulas obtained in the indicated subsections are generalized to the case of representative point. To this point, the mass m should be replaced by the mass M of the system, and the vectors $\mathbf{r}$ and $\mathbf{F}$ should be changed by the vectors $\mathbf{y}$ and $\mathbf{Y}$, respectively, the summation over the recurring indices being performed to "$3n$".

Lagrange equations of the second kind and canonical equations. In order to write Lagrange equations of the second kind (4.6) of Chap. 4 in an explicit form

$$M \left(g_{\sigma\tau} \ddot{q}^\tau + \Gamma_{\sigma,\alpha\beta} \dot{q}^\alpha \dot{q}^\beta \right) = Q_\sigma, \quad \sigma, \tau = \overline{1, 3n}, \quad \alpha, \beta = \overline{0, 3n}, \tag{1.9}$$

one should define the kinetic energy T and compute the generalized forces Q_σ. For an unconstrained mechanical system consisting of n particles, we have

$$T = T^{(2)} + T^{(1)} + T^{(0)},$$

where

$$T^{(2)} = \frac{M}{2} g_{\rho\sigma} \dot{q}^\rho \dot{q}^\sigma, \quad T^{(1)} = M g_{0\sigma} \dot{q}^\sigma, \quad T^{(0)} = \frac{M}{2} g_{00}.$$

Formulas (1.8), (1.4), and (1.5) imply that

$$T^{(2)} = \frac{M}{2} \left(\frac{\partial \mathbf{y}}{\partial q^\rho} \cdot \frac{\partial \mathbf{y}}{\partial q^\sigma} \right) \dot{q}^\rho \dot{q}^\sigma = \frac{M}{2} \sum_{\mu=1}^{3n} \frac{\partial y_\mu}{\partial q^\rho} \frac{\partial y_\mu}{\partial q^\sigma} \dot{q}^\rho \dot{q}^\sigma =$$

$$= \sum_{\mu=1}^{3n} \frac{m_\mu}{2} \frac{\partial x_\mu}{\partial q^\rho} \frac{\partial x_\mu}{\partial q^\sigma} \dot{q}^\rho \dot{q}^\sigma = \sum_{\nu=1}^{n} \frac{m_\nu}{2} \frac{\partial \mathbf{r}_\nu}{\partial q^\rho} \cdot \frac{\partial \mathbf{r}_\nu}{\partial q^\sigma} \dot{q}^\rho \dot{q}^\sigma \, ,$$

$$T^{(1)} = M \left(\frac{\partial \mathbf{y}}{\partial t} \cdot \frac{\partial \mathbf{y}}{\partial q^\sigma} \right) \dot{q}^\sigma = \sum_{\nu=1}^{n} m_\nu \left(\frac{\partial \mathbf{r}_\nu}{\partial t} \cdot \frac{\partial \mathbf{r}_\nu}{\partial q^\sigma} \right) \dot{q}^\sigma \, ,$$

$$T^{(0)} = \frac{M}{2} \left(\frac{\partial \mathbf{y}}{\partial t} \right)^2 = \sum_{\nu=1}^{n} \frac{m_\nu}{2} \left(\frac{\partial \mathbf{r}_\nu}{\partial t} \right)^2 .$$

Consider the generalized forces. By definition we have

$$Q_\sigma = \mathbf{Y} \cdot \mathbf{e}_\sigma = \mathbf{Y} \cdot \frac{\partial \mathbf{y}}{\partial q^\sigma}, \quad \sigma = \overline{1, 3n} \, .$$

Using formulas (1.4), we get

$$Q_\sigma = \sum_{\mu=1}^{3n} Y_\mu \frac{\partial y_\mu}{\partial q^\sigma} = \sum_{\mu=1}^{3n} X_\mu \frac{\partial x_\mu}{\partial q^\sigma} = \sum_{\nu=1}^{n} \mathbf{F}_\nu \cdot \frac{\partial \mathbf{r}_\nu}{\partial q^\sigma} \, . \tag{1.10}$$

Multiplying Eqs. (1.9) by elements $g^{\sigma\rho}$ of the matrix inverse to the matrix $g_{\sigma\tau}$ and summing over σ we have

$$M \left(\ddot{q}^\rho + \Gamma^\rho_{\alpha\beta} \dot{q}^\alpha \dot{q}^\beta \right) = Q^\rho, \quad \rho = \overline{1, 3n}, \quad \alpha, \beta = \overline{0, 3n} \, . \tag{1.11}$$

Here

$$\Gamma^\rho_{\alpha\beta} = g^{\sigma\rho} \Gamma_{\sigma,\alpha\beta}, \quad Q^\rho = g^{\sigma\rho} Q_\sigma \, .$$

It follows from Eqs. (1.11) that the vectors $\mathbf{W}$ and $\mathbf{Y}$ in Eqs. (1.6) in the contravariant representation read as

$$\mathbf{W} = \left(\ddot{q}^\sigma + \Gamma^\sigma_{\alpha\beta} \dot{q}^\alpha \dot{q}^\beta \right) \mathbf{e}_\sigma, \quad \mathbf{Y} = Q^\sigma \mathbf{e}_\sigma \, . \tag{1.12}$$

We can obtain the *Jacobi integral* $T^{(2)} - T^{(0)} + \Pi = $ const, from the Lagrange equations of the second kind in the non-stationary case reasoning in the same way as in the case of one particle, and the *energy integral* $T + \Pi = $ const can be obtained in the stationary case.

By analogy, in the case of a representative point moving in the potential field the system of Lagrange equations of the second kind reduces to the system of *canonical equations*

$$\dot{p}_\sigma = -\frac{\partial H(t, q, p)}{\partial q^\sigma}, \quad \dot{q}^\sigma = \frac{\partial H(t, q, p)}{\partial p_\sigma},$$

$$\sigma = \overline{1, 3n}, \quad H = p_\sigma \dot{q}^\sigma - L, \quad L = T - \Pi.$$

Thus, introducing the representative point allows us to consider the dynamical equations of an unconstrained system (1.9) as the equations of motion of one point (particle), the mass of which is equal to the mass of the whole system and is moving in the $3n$-dimensional Euclidean space.

So far we have supposed that in a potential field the force function U depends on the coordinates and time only and the generalized forces Q_σ are expressed in terms of U by the formulas

$$Q_\sigma = \frac{\partial U}{\partial q^\sigma}, \quad \sigma = \overline{1, 3n}.$$

But there are such force fields, in which the generalized forces can be represented as

$$Q_\sigma = \frac{\partial U}{\partial q^\sigma} - \frac{d}{dt}\frac{\partial U}{\partial \dot{q}^\sigma}, \quad \sigma = \overline{1, 3n},$$

where U is the function of time, coordinates, and velocities. It is called the *generalized force potential*. Such a force field is, in particular, the field of Lorentz force with which the electromagnetic field acts on the electric charge moving in it.

Note that if the function U depends linearly on the generalized velocities $\dot{q}^\sigma$, then the generalized forces Q_σ do not depend on the acceleration.

It is obvious that in such generalized potential fields Lagrange equations of the second kind remain as before

$$\frac{d}{dt}\frac{\partial L}{\partial \dot{q}^\sigma} - \frac{\partial L}{\partial q^\sigma} = 0, \quad L = T + U.$$

Equations of motion of a representative point in projections on the tangent and the normal to the trajectory. Let us establish one more form of equations of motion of a representative point. It is natural to suppose that the velocity vector of a representative point $\mathbf{V} = \dot{\mathbf{y}}$ is directed along the tangent to the trajectory, and as in the case of one particle it can be represented as

$$\mathbf{V} = V_\tau \boldsymbol{\tau}, \tag{1.13}$$

where $\boldsymbol{\tau}$ is the unit vector of the tangential direction. Differentiating this equation we get

$$\mathbf{W} = \dot{\mathbf{V}} = \frac{dV_\tau}{dt}\boldsymbol{\tau} + V_\tau \frac{d\boldsymbol{\tau}}{dt}.$$

The vector $d\boldsymbol{\tau}/dt$ is perpendicular to $\boldsymbol{\tau}$, because $\boldsymbol{\tau} \cdot \boldsymbol{\tau} = 1$, and therefore $\boldsymbol{\tau} \cdot d\boldsymbol{\tau} = 0$, which implies that the vectors $d\boldsymbol{\tau}$ and $\boldsymbol{\tau}$ are orthogonal. As in the case of one particle, in three-dimensional space we can write

$$\frac{d\boldsymbol{\tau}}{dt} = \frac{d\boldsymbol{\tau}}{ds}\frac{ds}{dt} = \dot{s}\frac{d\boldsymbol{\tau}}{ds}, \tag{1.14}$$

by analogy with three-dimensional space, meaning by ds the quantity defined by the formula

$$ds^2 = \sum_{\mu=1}^{3n} dy_\mu^2.$$

In accordance with this expression the velocity of a representative point is represented as

$$V^2 = \left(\frac{ds}{dt}\right)^2 = \sum_{\mu=1}^{3n} \dot{y}_\mu^2, \quad \mathbf{V} = \frac{d\mathbf{y}}{dt} = \frac{ds}{dt}\frac{d\mathbf{y}}{ds}.$$

Comparing this equation with formula (1.13) we obtain

$$V_\tau = \dot{s}, \quad \boldsymbol{\tau} = \frac{d\mathbf{y}}{ds}.$$

We shall call the quantity

$$K = \left|\frac{d\boldsymbol{\tau}}{ds}\right| = \left|\frac{d^2\mathbf{y}}{ds^2}\right| = \sqrt{\sum_{\mu=1}^{3n}\left(\frac{d^2 y_\mu}{ds^2}\right)^2}.$$

the *curvature of a 3n-dimensional curve.* Now expression (1.14) reads as

$$\frac{d\boldsymbol{\tau}}{dt} = K V_\tau \mathbf{n},$$

where we agree to call the vector $\mathbf{n} = \frac{1}{K}\frac{d\boldsymbol{\tau}}{ds}$ the *vector of a principal normal to a curve.* Thus, the acceleration can be expressed by the formula

$$\mathbf{W} = \dot{V}_\tau \boldsymbol{\tau} + V^2 K \mathbf{n}, \quad \text{or} \quad \mathbf{W} = \ddot{s}\boldsymbol{\tau} + \dot{s}^2 K \mathbf{n},$$

and hence, the equation of motion of a representative point can be written as

$$M(\dot{V}_\tau \boldsymbol{\tau} + V^2 K \mathbf{n}) = Y_\tau \boldsymbol{\tau} + Y_n \mathbf{n}.$$

In particular, if $Y_\tau = 0$, then $\dot{V}_\tau = \ddot{s} = 0$, $V = \text{const}$, and the curvature of the curve is

$$K = \frac{|\mathbf{Y}|}{M}\frac{1}{V^2}.$$

Integration of the system of differential equations of motion of a representative point is a quite difficult problem, no matter what coordinate system is used for writing these equations. However, in some cases it becomes possible to obtain the first integrals of the system of differential equations of motion with the help of general theorems of dynamics. Let us consider these theorems.

2   The Principle of Linear Momentum and the Principle of Motion of the Center of Mass

We shall call the sum

$$\mathbf{K} = \sum_\nu m_\nu \mathbf{v}_\nu = \sum_\nu m_\nu \dot{\mathbf{r}}_\nu$$

the *impulse* (or the *momentum*) of a system. Using Eq. (1.1), we get

$$\frac{d}{dt} \sum_\nu m_\nu \mathbf{v}_\nu = \sum_\nu \mathbf{F}_\nu^{(i)} + \sum_\nu \mathbf{F}_\nu^{(e)} . \tag{2.1}$$

As the internal forces should be subject to the Newton third law, their geometrical sum is zero: $\sum_\nu \mathbf{F}_\nu^{(i)} = 0$. This is why formula (2.1) can be written as

$$\frac{d\mathbf{K}}{dt} = \mathbf{F}, \quad \text{or} \quad \mathbf{K} - \mathbf{K_0} = \int_{t_0}^{t} \mathbf{F} dt , \tag{2.2}$$

where $\mathbf{F} = \sum_\nu \mathbf{F}_\nu^{(e)}$ is the geometrical sum of all external forces, which is called the *resultant* of these forces.

The relation between the vectors $\mathbf{K}$ and $\mathbf{F}$, as given by Eqs. (2.2), is called the *principle of linear momentum*.

It is obvious that the sum $\sum_\nu m_\nu \mathbf{r}_\nu$ can be represented as

$$\sum_\nu m_\nu \mathbf{r}_\nu = \mathbf{r}_c \sum_\nu m_\nu = M \mathbf{r}_c ,$$

where

$$\mathbf{r}_c = \sum_\nu m_\nu \mathbf{r}_\nu / M . \tag{2.3}$$

We shall call the point with the position vector $\mathbf{r}_c$ defined by formula (2.3) the *center of mass* (or the *center of inertia*) of a system. Formula (2.3) implies that

$$M \mathbf{v}_c = \sum_\nu m_\nu \mathbf{v}_\nu = \mathbf{K} ,$$

and so Eq. (2.2) can be written in the form

$$M\frac{d\mathbf{v}_c}{dt} = \sum_{\nu} \mathbf{F}_{\nu}^{(e)}, \quad \text{or} \quad M\frac{d\mathbf{v}_c}{dt} = \mathbf{F}. \tag{2.4}$$

This relation has the form of the equation of motion for a particle with mass M subject to a force $\mathbf{F}$. In other words, when a system is moving, its center of mass is moving as a particle, whose mass is equal to that of the system and to which the force that equals the resultant $\mathbf{F}$ of all external forces is applied (the *principle of motion of the center of mass*).

Note that the principle of linear momentum and the principle of motion of the center of mass can be extended to the case of motion of a continuously distributed set of points (a continuum or a continuous medium). In this case all the above sums should be considered as integrals. In so doing, the forces of interaction between the particles of a continuous medium are considered as internal forces.

These theorems can be also applied to the mechanical systems that consist of heterogeneous parts, which are used, for example, in robotics, and as objects of animated and inanimate nature. Application of these principles to the motion of aircrafts will be used in Chap. "Flight Dynamics" of the second volume.

A particular case of a continuous medium is an absolutely rigid body. The forces providing the constancy of distances between its points can be considered as internal forces. Another approach to absolutely rigid bodies (solids) will be presented in Chap. 6 "Constrained Motion".

As was mentioned above the resultant of the internal forces does not enter the principle of motion of the center of mass (2.4). However, one should not think that the internal forces have no effect on the motion of the center of mass at all. The external forces often depend on the internal ones. For instance, the muscle tension of a sportsman influences the force with which he drives off the earth when high-jumping. The internal forces also define the force with which an electric locomotive hauls the train pushing off the rails.

3 The Angular Impulse-Momentum Principle

The angular impulse-momentum principle for the case of a fixed pole. For any point belonging to the system we can write the angular impulse-momentum principle (see formula (2.5) of Chap. 4)

$$\frac{d\mathbf{l}_{\nu}}{dt} = \mathbf{L}_{\nu}, \quad \nu = \overline{1, n},$$

or in expanded form

$$\frac{d(\mathbf{r}_\nu \times \mathbf{p}_\nu)}{dt} = \mathbf{r}_\nu \times (\mathbf{F}_\nu^{(e)} + \mathbf{F}_\nu^{(i)}), \quad \nu = \overline{1,n}, \tag{3.1}$$

where $\mathbf{F}_\nu^{(e)}$ and $\mathbf{F}_\nu^{(i)}$ denote, respectively, the resultants of all external and internal forces applied to the point under discussion, as usual when considering a system of particles.

Summing all the equalities in (3.1), we get

$$\frac{d\left(\sum_{\nu=1}^{n} \mathbf{r}_\nu \times \mathbf{p}_\nu\right)}{dt} = \sum_{\nu=1}^{n} \mathbf{r}_\nu \times \mathbf{F}_\nu^{(e)} + \sum_{\nu=1}^{n} \mathbf{r}_\nu \times \mathbf{F}_\nu^{(i)}. \tag{3.2}$$

The expression $\sum_{\nu=1}^{n} \mathbf{r}_\nu \times \mathbf{F}_\nu^{(i)}$ is called *the resultant moment of internal forces.* According to the Newton third law the forces of interaction of any two points of a system are equal in value and opposite in direction. Besides, if they are acting along one line, then the moment $\sum_{\nu=1}^{n} \mathbf{r}_\nu \times \mathbf{F}_\nu^{(i)}$ vanishes. Indeed, under this assumption for any pair of points the forces of their interaction create the moments that are equal in value and opposite in direction with respect to any pole. The forces of interaction of two points, as was already mentioned, are equal in value and opposite in direction according to the Newton third law. The lines of their action can be parallel or, in particular, they can coincide. In classical mechanics, the case is considered in which the resultant moment of internal forces is always zero. In problems of electrodynamics interactions occur to which a nonzero moment of internal forces corresponds.

Another sum $\mathbf{L}_O = \sum_{\nu=1}^{n} \mathbf{r}_\nu \times \mathbf{F}_\nu^{(e)}$ on the right of (3.2) is called the *resultant moment of external forces*, and the expression $\mathbf{l}_O = \sum_{\nu=1}^{n} \mathbf{r}_\nu \times \mathbf{p}_\nu$ is called *the resultant moment of momentum of the system* or *its angular momentum*. Hence, according to (3.2), the *angular impulse-momentum principle* can be rewritten as

$$\frac{d\mathbf{l}_O}{dt} = \mathbf{L}_O, \tag{3.3}$$

i.e., the time derivative of the angular momentum of the system with respect to the fixed pole is equal to the resultant moment of external forces.

Equation (3.3) has motion integrals in the case when $\mathbf{L}_O$ is a given time function (or a constant, in particular). In a more particular case, when $\mathbf{L}_O = 0$, we have the *law of conservation of the angular momentum*

$$\mathbf{l}_O = \mathbf{l}_O(0) = \mathbf{C},$$

that is, the angular momentum is constant in magnitude and direction when the resultant moment of the external forces is absent.

The angular impulse-momentum principle, as well as the principle of linear momentum, can be extended to an arbitrary mechanical system, including continu-

Fig. 1 Rotation of a body about a fixed axis

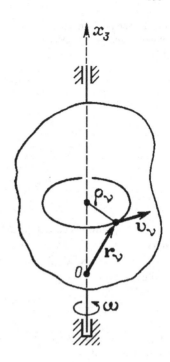

ous media. We shall consider the rotation of a solid with an angular velocity $\boldsymbol{\omega}$ about a fixed axis x_3 (Fig. 1) as an example of application of theorem (3.3). In this case, the moment of momentum of the point with number ν can be written as

$$\mathbf{l}_\nu = \mathbf{r}_\nu \times m_\nu \mathbf{v}_\nu = \mathbf{r}_\nu \times m_\nu (\boldsymbol{\omega} \times \mathbf{r}_\nu) = m_\nu [\omega r_\nu^2 - \mathbf{r}_\nu (\boldsymbol{\omega} \cdot \mathbf{r}_\nu)] \,.$$

So, its projection onto the x_3-axis of rotation (or the moment of momentum of this point with respect to the x_3-axis) is

$$l_{\nu x_3} = m_\nu \omega_{x_3} (x_{\nu 1}^2 + x_{\nu 2}^2) = m_\nu \omega_{x_3} \rho_\nu^2 \,.$$

The projection of the resultant moment of momentum of the whole system is

$$l_{O x_3} = \omega_{x_3} \sum_\nu m_\nu \rho_\nu^2 \,.$$

The expression $J_{x_3} = \sum_\nu m_\nu \rho_\nu^2$ is called the *moment of inertia of a system of particles with respect to x_3-axis*. We note that it can be represented as $J_{x_3} = M R_{x_3}^2$. The quantity R_{x_3} is called the *radius of gyration of a system with respect to x_3-axis*. If the mass is continuously distributed, the sum becomes an integral. In the case under discussion, the projection of Eq. (3.3) onto the x_3-axis assumes the form

Fig. 2 The law of moment of momentum with respect to a movable center of mass

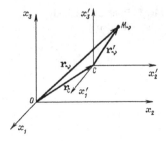

$$J_{x_3}\frac{d\omega_{x_3}}{dt} = L_{Ox_3}.\qquad(3.4)$$

This equation is called the *differential equation of rotation of a solid about a fixed axis*.

In some cases Eq. (3.4) has an integral. In particular, if $L_{Ox_3} = f(t)$, then we can write

$$J_{x_3}(\omega - \omega_0) = \int_{t_0}^{t} f(t)dt.$$

If $f(t) \equiv 0$, then we have $\omega = \omega(0) = $ const; that is, a solid rotates with constant angular velocity.

The angular impulse-momentum principle with respect to a moving center. Let us introduce a coordinate system $Cx_1'x_2'x_3'$ that is translationally moving with respect to the $Ox_1x_2x_3$-system. The origin of a moving system coincides with the center of inertia C of the system (Fig. 2). It is obvious that

$$\mathbf{r}_\nu = \mathbf{r}_c + \mathbf{r}_\nu', \quad \mathbf{v}_\nu = \mathbf{v}_c + \mathbf{v}_\nu'.\qquad(3.5)$$

Hence, the angular momentum can be represented as

$$\mathbf{l}_O = \sum_\nu (\mathbf{r}_c + \mathbf{r}_\nu') \times m_\nu(\mathbf{v}_c + \mathbf{v}_\nu').$$

Expanding the brackets, we get

$$\mathbf{l}_O = \sum_\nu \mathbf{r}_c \times m_\nu \mathbf{v}_c + \sum_\nu \mathbf{r}_c \times m_\nu \mathbf{v}_\nu' + \sum_\nu \mathbf{r}_\nu' \times m_\nu \mathbf{v}_c + \sum_\nu \mathbf{r}_\nu' \times m_\nu \mathbf{v}_\nu'.\quad(3.6)$$

Let us denote the first sum by

$$\mathbf{l}_C = \sum_\nu \mathbf{r}_c \times m_\nu \mathbf{v}_c = \mathbf{r}_c \times M\mathbf{v}_c,\qquad(3.7)$$

where $M = \sum_\nu m_\nu$ is the mass of the whole system. The second sum can be transformed as follows

$$\sum_\nu \mathbf{r}_c \times m_\nu \mathbf{v}'_\nu = \mathbf{r}_c \times \sum_\nu m_\nu \frac{d\mathbf{r}'_\nu}{dt} = \mathbf{r}_c \times \frac{d}{dt} \sum_\nu m_\nu \mathbf{r}'_\nu .$$

However, from the definition of the center of inertia (2.3) and formulas (3.5) it follows that

$$\sum_\nu m_\nu \mathbf{r}'_\nu = 0 ,$$

and, hence, the second sum in expression (3.6) is zero.

In a similar manner, the third sum also vanishes, since it can be rewritten as

$$\sum_\nu \mathbf{r}'_\nu \times m_\nu \mathbf{v}_c = \sum_\nu m_\nu \mathbf{r}'_\nu \times \mathbf{v}_c = 0 .$$

We denote by $\mathbf{l}'_C = \sum_\nu \mathbf{r}'_\nu \times m_\nu \mathbf{v}'_\nu$ the last sum in (3.6). This is the resultant moment of momentum of a system with respect to the moving coordinate system $Cx'_1x'_2x'_3$ (in what follows, for brevity the vector $\mathbf{l}'_C$ will be called the *angular momentum of a system when it moves about its center of mass*).

Now formula (3.6) assumes its final form

$$\mathbf{l}_O = \mathbf{r}_c \times M\mathbf{v}_c + \mathbf{l}'_C ,$$

that is, the angular moment of a system with respect to a fixed point O is equal to the vector sum of the angular momentum with respect to the point O of a material point of mass M lying at the center of mass C of the system and the angular moment of the system when it moves about its center of mass.

The above results imply that the angular impulse-momentum principle (3.3) can be represented as

$$\frac{d\mathbf{l}_C}{dt} + \frac{d\mathbf{l}'_C}{dt} = \mathbf{r}_c \times \mathbf{F} + \sum_\nu \mathbf{r}'_\nu \times \mathbf{F}^{(e)}_\nu , \tag{3.8}$$

where $\mathbf{F} = \sum_\nu \mathbf{F}^{(e)}_\nu$ is the resultant vector of external forces.

Multiplying the both sides of Eq. (2.4) (that is the principle of motion of the center of mass) by $\mathbf{r}_c$ we get

$$\mathbf{r}_c \times M\frac{d\mathbf{v}_c}{dt} = \mathbf{r}_c \times \mathbf{F} .$$

Taking into account expressions (3.7), this implies

$$\frac{d\mathbf{l}_C}{dt} = \mathbf{r}_c \times \mathbf{F} ,$$

since $(d\mathbf{r}_c/dt) \times M\mathbf{v}_c = 0$. So, after collecting similar terms in formula (3.8) we arrive at

$$\frac{d\mathbf{l}'_C}{dt} = \mathbf{L}'_C . \tag{3.9}$$

Fig. 3 Moment theorem
with respect to a moving
center O'

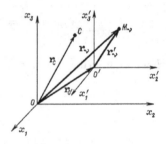

This equality is an analytical expression of *the angular impulse-momentum principle for the system moving with respect to its center of mass*. Note that the form of this theorem is the same as that of the angular impulse-momentum principle (3.3) with respect to the fixed center O.

Analogously to the differential equation of rotation of a solid about a fixed axis (3.4), in the given case of rotation of a solid about the x_3'-axis we obtain

$$J_{x_3'} \frac{d\omega_{x_3'}}{dt} = L_{Cx_3'}.$$

Consider a particular case of existence of an integral of equation (3.9). Suppose, for example, $L_{Cx_i'} = 0$. Then projecting Eq. (3.9) onto the x_i'-axis we get $l_{Cx_i'} = $ const. If points of the system do not rotate about the x_i'-axis at the initial instant, then this constant is zero.

To retain the zero value, a part of points of a system should rotate in one direction, and another part of points should rotate in another direction. This property is used by a cat, when it is falling with its back down: turning its tail and head about its longitudinal axis in one direction, it turns thereby its paws in the other direction.

Let us now consider a more involved case[1] when one takes into account the relative motion of a system of material points with respect to a system $O'x_1'x_2'x_3'$ which moves linearly and whose origin O' does not coincide with the center of mass C (Fig. 3).

The following clear relations hold:

$$\mathbf{r}_\nu = \mathbf{r}_{o'} + \mathbf{r}_\nu', \quad \mathbf{v}_\nu = \mathbf{v}_{o'} + \mathbf{v}_\nu'. \tag{3.10}$$

Consider the vectors

$$\mathbf{l}_{O'}' = \sum_\nu \mathbf{r}_\nu' \times m_\nu \mathbf{v}_\nu', \quad \mathbf{L}_{O'}' = \sum_\nu \mathbf{r}_\nu' \times \mathbf{F}_\nu. \tag{3.11}$$

Hence, using the second Newton law

$$m_\nu \mathbf{w}_\nu = \mathbf{F}_\nu$$

---

[1] The theorem that follows and its proof is taken from P.E. Tovstik's lectures.

and formulas (3.10), (3.11), we have

$$\frac{d\mathbf{l}'_{O'}}{dt} = \sum_\nu \mathbf{r}'_\nu \times m_\nu \frac{d\mathbf{v}'_\nu}{dt} = \sum_\nu \mathbf{r}'_\nu \times m_\nu \left(\frac{d\mathbf{v}_\nu}{dt} - \frac{d\mathbf{v}_{o'}}{dt} \right) =$$

$$= \sum_\nu \mathbf{r}'_\nu \times \mathbf{F}_\nu - \sum_\nu \mathbf{r}'_\nu \times m_\nu \mathbf{w}_{o'} = \mathbf{L}_{O'} - \overrightarrow{O'C} \times M\mathbf{w}_{o'} .$$

It follows that

$$\frac{d\mathbf{l}'_{O'}}{dt} = \mathbf{L}_{O'} + \overrightarrow{CO'} \times M\mathbf{w}_{o'} . \tag{3.12}$$

Formula (3.12) is an analytic expression of the *law of moment of momentum of a system when moving with respect to a pole O'* which is different from the center of mass C. With $C = O'$ formula (3.12) assumes the form (3.9).

4 The Work-Energy Principle of a System

As was already mentioned, the kinetic energy of a system is the sum of all elementary masses included in it: $T = \frac{1}{2} \sum_{\nu=1}^n m_\nu v_\nu^2$. According to the equation of motion of an isolated (separated) point of a system

$$m_\nu \frac{d\mathbf{v}_\nu}{dt} = \mathbf{F}_\nu^{(e)} + \mathbf{F}_\nu^{(i)} , \quad \nu = \overline{1,n} , \tag{4.1}$$

variation of its squared velocity occurs as a result of both the action of the pure force resultant $\mathbf{F}_\nu^{(e)}$ of all external forces and the action of the pure force resultant $\mathbf{F}_\nu^{(i)}$ of all internal forces applied to this point.

Multiplying scalarly equality (4.1) by an elementary displacement of the point distinguished (separated) $d\mathbf{r}_\nu = \mathbf{v}_\nu dt$, we get

$$m_\nu \mathbf{v}_\nu \cdot d\mathbf{v}_\nu = \mathbf{F}_\nu^{(e)} \cdot d\mathbf{r}_\nu + \mathbf{F}_\nu^{(i)} \cdot d\mathbf{r}_\nu$$

or

$$d \left(\frac{m_\nu v_\nu^2}{2} \right) = \mathbf{F}_\nu^{(e)} \cdot d\mathbf{r}_\nu + \mathbf{F}_\nu^{(i)} \cdot d\mathbf{r}_\nu . \tag{4.2}$$

Summing up the right- and left-hand parts of equalities (4.2), we have

$$dT = \delta A^{(e)} + \delta A^{(i)} , \tag{4.3}$$

where

$$\delta A^{(e)} = \sum_\nu \mathbf{F}_\nu^{(e)} \cdot d\mathbf{r}_\nu, \quad \delta A^{(i)} = \sum_\nu \mathbf{F}_\nu^{(i)} \cdot d\mathbf{r}_\nu$$

is the elementary work done by external and internal forces, respectively. Equality (4.3) expresses the *work-energy principle for a system in the differential form*.

It is worth pointing out the fact that while the resultant vector and the resultant moment of all internal forces are both zero, the sum of works done by internal forces, generally speaking, does not equal zero. If the particles of a system move relative to each other, then the distance between them changes, which leads to appearance of the work of internal forces.

In motion of a system every point with number ν moves from the position $M_{0\nu}$ to the position M_{ν}. Accordingly, the total work should be written as

$$A = A^{(e)} + A^{(i)} = \sum_{\nu} \int_{\smile M_{0\nu} M_{\nu}} \mathbf{F}_{\nu}^{(e)} \cdot d\mathbf{r}_{\nu} + \sum_{\nu} \int_{\smile M_{0\nu} M_{\nu}} \mathbf{F}_{\nu}^{(i)} \cdot d\mathbf{r}_{\nu} . \qquad (4.4)$$

On the basis of the above, integrating relation (4.3), we get

$$T - T_0 = A . \qquad (4.5)$$

Here T_0 is the kinetic energy of the system at time t_0.

Equality (4.5) is the *work-energy principle for a system in the integral form*: the increment of kinetic energy is equal to the sum of all external and internal forces applied to the system.

We should note that the curvilinear integrals (4.4), by means of which the work A is expressed, generally depend not only on the initial and terminal positions $M_{0\nu}$ and M_{ν} of particles of the system, but also on the shape of the trajectory corresponding to the true motion.

The equation of motion of a representative point (1.6) in $3n$-dimensional space coincides in its form with the Newton equation for a single point in three-dimensional space. This is why, as in the case of a single point we get

$$d\left(\frac{MV^2}{2}\right) = \mathbf{Y} \cdot d\mathbf{y} .$$

The left-hand part of this equality is the differential of the kinetic energy of the system, since its kinetic energy is equal to the kinetic energy of a representative point. Let us now show that the right-hand part of this equality corresponds to the elementary work of all forces applied to the points of the system. Indeed, formulas (1.4) imply that

$$\mathbf{Y} \cdot d\mathbf{y} = \sum_{\mu=1}^{3n} Y_\mu dy_\mu = \sum_{\mu=1}^{3n} \frac{X_\mu}{\sqrt{\widetilde{m}_\mu}} \sqrt{\widetilde{m}_\mu}\, dx_\mu = \sum_{\nu=1}^{n} \mathbf{F}_\nu \cdot d\mathbf{r}_\nu =$$

$$= \sum_{\nu=1}^{n} \mathbf{F}_\nu^{(e)} \cdot d\mathbf{r}_\nu + \sum_{\nu=1}^{n} \mathbf{F}_\nu^{(i)} \cdot d\mathbf{r}_\nu = \delta A^{(e)} + \delta A^{(i)} = \delta A .$$

Thus, as a result of introducing of a representative point into the equations of motion we can rewrite the work-energy principle for the system in the following form

$$\frac{MV^2}{2} - \frac{MV_0^2}{2} = \int_{\smile ab} \mathbf{Y} \cdot d\mathbf{y},$$

which coincides with formula (2.10) of Chap. 4 obtained for dynamics of one particle.

As an example, consider the case when the magnitude of the interaction forces of points of the system (that is, the internal forces) depends only on the distance between the corresponding points, the forces themselves being directed along the line connecting the points. Let us show that in this case the internal forces have a potential; that is, $\delta A^{(i)}$ can be represented as $dU^{(i)}$.

Let us denote the distance between the points with the position vectors $\mathbf{r}_\mu$ and $\mathbf{r}_\nu$ by $\rho_{\mu\nu}$. It is obvious that $\rho_{\mu\nu}^2 = (\mathbf{r}_\nu - \mathbf{r}_\mu)^2$, and therefore

$$\rho_{\mu\nu} d\rho_{\mu\nu} = (\mathbf{r}_\nu - \mathbf{r}_\mu) \cdot d(\mathbf{r}_\nu - \mathbf{r}_\mu). \tag{4.6}$$

The projection of the force with which the point M_ν acts on the point M_μ onto the direction $(\mathbf{r}_\nu - \mathbf{r}_\mu)/\rho_{\mu\nu}$ is, by the assumption, the function of the distance $\rho_{\mu\nu}$ only. We denote this function by $f_{\mu\nu}(\rho_{\mu\nu})$. In this case, we represent the force itself as

$$f_{\mu\nu}(\rho_{\mu\nu})(\mathbf{r}_\nu - \mathbf{r}_\mu)/\rho_{\mu\nu}. \tag{4.7}$$

It is obvious that we get the force with which the point M_μ acts on the point M_ν by interchanging the indices μ and ν in expression (4.7). In so doing, on the basis of the Newton third law we have a vector that is equal to expression (4.7) with the opposite sign, and $\mathbf{r}_\mu - \mathbf{r}_\nu = -(\mathbf{r}_\nu - \mathbf{r}_\mu)$ and $\rho_{\mu\nu} = \rho_{\nu\mu}$, and hence

$$f_{\mu\nu}(\rho_{\mu\nu}) = f_{\nu\mu}(\rho_{\nu\mu}). \tag{4.8}$$

Taking into account relations (4.8) and (4.6) we can write the elementary work of the interaction forces of points M_ν and M_μ as

$$\delta A_{\mu\nu} = \frac{f_{\mu\nu}(\rho_{\mu\nu})[(\mathbf{r}_\nu - \mathbf{r}_\mu) \cdot d\mathbf{r}_\mu + (\mathbf{r}_\mu - \mathbf{r}_\nu) \cdot d\mathbf{r}_\nu]}{\rho_{\mu\nu}} =$$

$$= -\frac{f_{\mu\nu}(\rho_{\mu\nu})(\mathbf{r}_\nu - \mathbf{r}_\mu) \cdot d(\mathbf{r}_\nu - \mathbf{r}_\mu)}{\rho_{\mu\nu}} = -f_{\mu\nu}(\rho_{\mu\nu}) \, d\rho_{\mu\nu}.$$

The double sum $\sum_{\mu,\nu=1}^{n} \delta A_{\mu\nu}$ takes twice into account the work of each of the forces under consideration. Hence,

$$\delta A^{(i)} = \frac{1}{2} \sum_{\mu,\nu=1}^{n} \delta A_{\mu\nu} = -\frac{1}{2} \sum_{\mu,\nu=1}^{n} f_{\mu\nu}(\rho_{\mu\nu}) \, d\rho_{\mu\nu} = dU^{(i)}, \tag{4.9}$$

where $U^{(i)} = -\frac{1}{2}\sum_{\mu,\nu=1}^{n} \int f_{\mu\nu}(\rho_{\mu\nu})\,d\rho_{\mu\nu}$ is the *force function of internal forces*.
Relation (4.9) implies that

$$A^{(i)} = U^{(i)} - U_0^{(i)} = \Pi_0^{(i)} - \Pi^{(i)}, \qquad (4.10)$$

where $\Pi^{(i)} = -U^{(i)}$ and $\Pi_0^{(i)} = -U_0^{(i)}$ are the potential energies of the internal forces at instants t and t_0, respectively.

Assume that the external forces are also potential; that is, $\delta A^{(e)} = -d\Pi^{(e)}$, where $\Pi^{(e)}$ is the potential energy of external forces. Then

$$A^{(e)} = \Pi_0^{(e)} - \Pi^{(e)}. \qquad (4.11)$$

Hence, if the internal and external forces are potential, then their work does not depend on the shape of a path and is expressed by the difference of potential energies (see formulas (4.10) and (4.11)).

Substituting expressions (4.10) and (4.11) into relation (4.5) we obtain the *energy integral*

$$E \equiv T + \Pi^{(e)} + \Pi^{(i)} = T_0 + \Pi_0^{(e)} + \Pi_0^{(i)} = \text{const},$$

where E is *the total mechanical energy of the system*. A mechanical system for which there exists the energy integral is called *conservative*.

If the distances between the points of a system do not change, then such a system is called *invariable (rigid body)*. In this case $d\rho_{\mu\nu} = 0$, and according to relation (4.9) the work of internal forces is zero. The invariable systems include absolutely rigid bodies. By the above, an absolutely rigid body differs from a real one by the fact that the work of external forces applied to it is totally expended in variation of its kinetic energy. Thus, the model of an absolutely rigid body can be used only when we can neglect the work expended in the deformation of a body compared to the variation of its kinetic energy.

König's theorem. The value of the kinetic energy of a system depends essentially on the fact in which coordinate system the velocities of its points are measured. When formulating the work-energy principle both in differential and integral forms a certain frame of reference was fixed. We denote this system by $Ox_1x_2x_3$.

We introduce a new frame of reference $Cx_1''x_2''x_3''$, whose origin will be located at the center of mass of the mechanical system under discussion (Fig. 4). The position of the center of mass relative to the initial frame of reference $Ox_1x_2x_3$ is specified by the position vector

$$\mathbf{r}_c = \sum_\nu m_\nu \mathbf{r}_\nu / M, \quad M = \sum_\nu m_\nu.$$

We denote the position vector of the point M_ν by $\mathbf{r}_\nu'$ in the new frame of reference. The origin of this frame coincides with the center of mass, and hence

Fig. 4 König's theorem

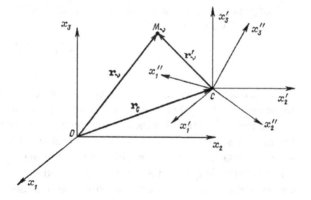

$$M\mathbf{r}'_c = \sum_\nu m_\nu \mathbf{r}'_\nu = 0.\tag{4.12}$$

If the instantaneous angular velocity of the frame of reference $Cx''_1x''_2x''_3$ is equal to $\boldsymbol{\omega}$ relative to the frame $Ox_1x_2x_3$, then the time derivatives of the position vector $\mathbf{r}'_\nu$ in the moving and fixed coordinate systems, as was shown in Sect. 1 of Chap. 3, Part I are related as

$$\frac{d\mathbf{r}'_\nu}{dt} = \frac{d^*\mathbf{r}'_\nu}{dt} + \boldsymbol{\omega} \times \mathbf{r}'_\nu.\tag{4.13}$$

We denote the velocity of the point M_ν relative to the moving frame of reference that is equal to $d^*\mathbf{r}'_\nu/dt$ by $\mathbf{v}''_\nu$.

Differentiating the equality $\mathbf{r}_\nu = \mathbf{r}_c + \mathbf{r}'_\nu$ and taking into account relation (4.13) we get

$$\mathbf{v}_\nu = \mathbf{v}_c + \mathbf{v}''_\nu + \boldsymbol{\omega} \times \mathbf{r}'_\nu,$$

or

$$\mathbf{v}_\nu = \mathbf{v}_c + \mathbf{v}'_\nu, \quad \mathbf{v}'_\nu = \mathbf{v}''_\nu + \boldsymbol{\omega} \times \mathbf{r}'_\nu,\tag{4.14}$$

where $\mathbf{v}'_\nu = d\mathbf{r}'_\nu/dt$ is the velocity of the point M_ν relative to the additional frame of reference $Cx'_1x'_2x'_3$ the axes of which are parallel to the axes of the initial frame. Differentiating equality (4.12) we find that

$$\sum_\nu m_\nu \mathbf{v}'_\nu = \sum_\nu m_\nu \mathbf{v}''_\nu + \boldsymbol{\omega} \times \left(\sum_\nu m_\nu \mathbf{r}'_\nu\right) = \sum_\nu m_\nu \mathbf{v}''_\nu = 0.\tag{4.15}$$

Substituting the velocities $\mathbf{v}_\nu$ in the form (4.14) into the expression for the kinetic energy of a system, we obtain

$$T = \frac{1}{2}\sum_\nu m_\nu \mathbf{v}_\nu^2 = \frac{1}{2}\sum_\nu m_\nu [\mathbf{v}_c^2 + 2\mathbf{v}_c \cdot \mathbf{v}'_\nu + (\mathbf{v}'_\nu)^2].$$

Taking into consideration relation (4.15) and that $M = \sum_\nu m_\nu$ is the mass of the whole system we have

$$T = \frac{M v_c^2}{2} + \frac{1}{2} \sum_\nu m_\nu (v'_\nu)^2 . \tag{4.16}$$

This equality represents *König's theorem*, which is formulated as follows: the kinetic energy of a system is equal to the sum of the kinetic energy of the center of mass on the assumption that the mass of the whole system is concentrated in it and the kinetic energy of a system in its motion relative the frame of reference that is moving translationally together with the center of mass.

If the system in question is a solid, whose motion is controlled by the velocity $\mathbf{v}_c$ of its center of mass and the angular velocity $\boldsymbol{\omega}$ of rotation about the center of mass, the velocity v'_ν of the elementary mass m_ν is equal to $\omega \rho_\nu$, where ρ_ν is the distance to the instantaneous axis of rotation passing through the center of mass. This implies that the second summand in formula (4.16) has the form

$$\frac{1}{2} \sum_\nu m_\nu (v'_\nu)^2 = \frac{J_\omega \omega^2}{2} .$$

Here $J_\omega = \sum_\nu m_\nu \rho_\nu^2$ is the inertia moment of a solid relative to the instantaneous axis of rotation passing through the center of mass. The mass in a solid is considered to be continuously distributed, that is why, in the strict sense, the summation should be replaced by the integration. It follows from the above that the kinetic energy should be calculated by the formula

$$T = \frac{M v_c^2}{2} + \frac{J_\omega \omega^2}{2} . \tag{4.17}$$

Substituting the kinetic energy of a system of the form (4.16) into equality (4.3) we get

$$d\left(\frac{M v_c^2}{2} \right) + d\left(\frac{1}{2} \sum_\nu m_\nu (v'_\nu)^2 \right) = \left(\sum_\nu \mathbf{F}_\nu^{(e)} \right) \cdot d\mathbf{r}_c +$$

$$+ \sum_\nu \mathbf{F}_\nu^{(e)} \cdot d\mathbf{r}'_\nu + \left(\sum_\nu \mathbf{F}_\nu^{(i)} \right) \cdot d\mathbf{r}_c + \sum_\nu \mathbf{F}_\nu^{(i)} \cdot d\mathbf{r}'_\nu . \tag{4.18}$$

The equation of motion of the center of mass (2.4) implies that

$$d\left(\frac{M v_c^2}{2} \right) = \left(\sum_\nu \mathbf{F}_\nu^{(e)} \right) \cdot d\mathbf{r}_c . \tag{4.19}$$

Using relation (4.19) and the fact that according to the Newton third law $\sum_\nu \mathbf{F}_\nu^{(i)} = 0$, we can write equality (4.18) as

$$d\left(\frac{1}{2}\sum_{\nu}m_{\nu}(v'_{\nu})^2\right) = \sum_{\nu}\mathbf{F}_{\nu}^{(e)}\cdot d\mathbf{r}'_{\nu} + \sum_{\nu}\mathbf{F}_{\nu}^{(i)}\cdot d\mathbf{r}'_{\nu}\,. \qquad (4.20)$$

Equation (4.20) is *the work-energy principle for a system in its motion relative to the frame of reference that is moving translationally together with the center of mass*. In other words, this theorem can be formulated as follows: the differential of the kinetic energy of a system in its motion relative to the center of mass is equal to the sum of elementary works of all the internal and external forces done in moving through displacements of their application points with respect to the center of mass.

5   Equilibrium Conditions of a Particle and a System

If in a given frame of reference the velocity of a particle is zero during some time interval, then the particle is said to be at a *state of rest* or in an *equilibrium*. This takes place only in the case when $\mathbf{F} = 0$ for the force acting on the point.

If the force $\mathbf{F}$ is the sum of k forces applied to the particle, we have in the case of equilibrium

$$\mathbf{F} = \sum_{i=1}^{k}\mathbf{F}_i = 0\,.$$

In this case the forces applied to the particle are said to counterbalance each other. It is obvious, that if the vector $\mathbf{F}$ is zero, then its components projected onto the axes of the chosen frame of reference also vanish. In particular, in the Cartesian coordinate system the equilibrium conditions read as

$$X_j = \sum_{i=1}^{k}X_{ij} = 0\,, \quad j = 1, 2, 3.$$

In a similar manner we can verify that in a curvilinear coordinate system the equilibrium conditions can be written as

$$Q_\sigma = \mathbf{F}\cdot\frac{\partial\mathbf{r}}{\partial q^\sigma} = X_j\frac{\partial x_j}{\partial q^\sigma} = 0\,, \quad \sigma = 1, 2, 3. \qquad (5.1)$$

If we move a particle that is at the state of rest to a new position and give an initial speed to it, then it starts moving. In so doing, two cases are possible. In one of them the particle stays within the limits of some vicinity of equilibrium state. In particular, it can oscillate about it or return to it after some period of time. In another case, for any infinitely small deviations from the state of equilibrium the particle always leaves the limits of the vicinity of the equilibrium state. In the first case the state of equilibrium is called *stable*, in the second case, *unstable*.

We can define mathematically the stable state of an equilibrium given by the coordinates $q^\sigma = 0$ as follows: an equilibrium of a particle is called stable if for any infinitely small $\varepsilon > 0$ we can take such a positive number $\delta > 0$ such that

$$|q^\sigma(t_0)| < \delta, \quad |\dot{q}^\sigma(t_0)| < \delta, \quad \sigma = 1, 2, 3,$$

for any $t > t_0$ the following inequalities hold

$$|q^\sigma(t_0)| < \varepsilon, \quad |\dot{q}^\sigma(t_0)| < \varepsilon, \quad \sigma = 1, 2, 3.$$

Here we assume that the time and the generalized coordinates are dimensionless variables.

If a particle is in a potential force field, then the generalized forces should be calculated by the formula

$$Q_\sigma = -\partial \Pi / \partial q^\sigma .$$

Equations (5.1) imply that in a state of equilibrium the potential energy Π has a stationary value. We shall prove that if in a state of equilibrium the potential energy of a stationary force field has an isolated minimum, then the equilibrium is stable. This is the statement of *Lagrange's theorem*.

For the sake of simplicity we suppose that in a state of equilibrium $q^\sigma = 0$ we have $\Pi(0) = 0$. In the case under discussion there exists an energy integral

$$\frac{mv^2}{2} + \Pi(q) = \frac{mv_0^2}{2} + \Pi(q_0) = \text{const}, \tag{5.2}$$

whence

$$\frac{mv^2}{2} = \frac{mv_0^2}{2} - \Delta\Pi, \tag{5.3}$$

where $\Delta\Pi = \Pi(q) - \Pi(q_0)$.

It follows from formula (5.3) that we get a maximum value that can be attained by $\Delta\Pi$ in the vicinity of the state of equilibrium when $\mathbf{v} = 0$.

Let this condition be satisfied at a point q^* for which

$$\Pi(q^*) = \frac{mv_0^2}{2} + \Pi(q_0) = C^* . \tag{5.4}$$

By the assumption the potential energy Π has an isolated minimum which is zero at the state of equilibrium, and so $\Pi(q_0) > 0$ and $\Pi(q) > 0$. Hence, using (5.2) and (5.4),

$$\Pi(q^*) > \Pi(q_0) > 0, \quad \Pi(q^*) > \Pi(q) > 0, \quad v \neq 0.$$

Equation $\Pi(q) = C^*$ corresponds to some equipotential surface S^*, whose limits a particle cannot leave with given q_0^σ and $\dot{q}_0^\sigma$. Other equipotential surfaces corresponding to $\Pi(q) = C < C^*$ lie inside S^*. This means that the state of equilibrium is stable.

The Lagrange theorem can be explained as follows. The vector $\nabla\Pi$ is oriented along the normal to an equipotential surface in the direction in which the function $\Pi(q)$ increases. As a particle is moving away from the state of equilibrium the function $\Pi(q)$ increases, the force $\mathbf{F} = -\nabla\Pi$ is directed inside the surface $\Pi(q) = C$; that is, it tends to return the particle to the state of equilibrium.

As an example we consider a load of mass m attached to a spring with stiffness c at an equilibrium. We direct the y-axis downwards and locate the origin at the state of static equilibrium. If the load deviates by the quantity y it is acted upon by the force $-cy$ which tends to return the load into the state of equilibrium. The potential energy of this force is $\Pi = cy^2/2$. Let us take the load out of equilibrium position by giving the initial velocity $\dot{y}_0 = v_0$. In this case formula (5.3) takes the form

$$\frac{mv^2}{2} = \frac{mv_0^2}{2} - \frac{cy^2}{2},$$

whence $\frac{dy}{dt} = \sqrt{v_0^2 - \frac{c}{m}y^2}$, and so

$$\frac{d\left(\frac{y}{v_0}\sqrt{\frac{c}{m}}\right)}{\sqrt{1 - \frac{c}{m}\left(\frac{y}{v_0}\right)^2}} = dt\sqrt{\frac{c}{m}},$$

and hence

$$\arcsin\left(\sqrt{\frac{c}{m}}\frac{y}{v_0}\right) = \sqrt{\frac{c}{m}}\,t + C_1.$$

As for $t = 0$ we have $y = 0$, then $C_1 = 0$ and

$$y = \frac{v_0}{\sqrt{\frac{c}{m}}}\sin\sqrt{\frac{c}{m}}\,t,$$

i.e., a particle oscillates about the state of stable equilibrium $y = 0$, in which the function $\Pi(y)$ has a minimum.

If we now come to consideration of equilibrium of a system of particles, then it is convenient to use a notion of representative point. It is obvious that in this case all the foregoing concerning a particle can be applied to a system of particles. We should remind, however, that indices j and σ take, then the values from 1 to $3n$.

A more detailed account of statics and equilibrium equations will be given in Chap. 10.

Chapter 6
Constrained Motion

N. N. Polyakhov, Sh. Kh. Soltakhanov, M. P. Yushkov⊙, and S. A. Zegzhda

In the present chapter, the motion equations of mechanical systems subject to either holonomic or nonholonomic constraints are derived not from variational principles, as is customary, but directly from the analysis of the restrictions imposed by the constraint equations on the acceleration of points in the system. We first consider in detail the constrained motion of one material point. Next, using the concept of a representative point, the above results are extended in a natural way to the problem of motion of a system of material points. They are further extended to mechanical systems consisting of material bodies. For this extension, we employ the concepts of a differentiable manifold and the tangent space to it.

Much attention is paid to the discussion of conditions for the ideality of nonholonomic constraints. We introduce the linear transformation of forces, which turns to be very constructive both in the derivation of the motion equations and in the definition of forces that secure the fulfillment of the equations constraints in case they are ideal.

The available forms of motion equations of nonholonomic systems are surveyed. The Maggi equations are shown to be the most general and convenient equations. We give the motion equations of systems subject to second-order linear nonholonomic constraints. The theory of constraint motion is extended to apply to the problem of controlled motion, in which the control is given by constraints depending on parameters. The theoretical part is supplemented by the solution of a number of problems from both holonomic and nonholonomic mechanics.

For convenience of presentation the material of the chapter is organized in two parts. In Part I we consider the constrained motion of a system of material points, and in Part II, we deal with general mechanical systems.

© Springer Nature Switzerland AG 2021   167
N. N. Polyakhov et al., *Rational and Applied Mechanics*,
Foundations of Engineering Mechanics,
https://doi.org/10.1007/978-3-030-64061-3_6

I) Constrained Motion of a System of Material Points

1   Constrained Motion of a point

The motion of a material point, which we have been considering so far, may be called the motion of a *free point* in the sense that, for any acting forces, from the arbitrariness in the choice of initial conditions, one may require that the point at a given time instant t at a given position with coordinates x_1, x_2, x_3 would have a given velocity with the projections $\dot{x}_1, \dot{x}_2, \dot{x}_3$. This is clear from the fact that the integral of the motion equations are of the form

$$\Phi_j(t, x, \dot{x}) = C_j, \quad x = (x_1, x_2, x_3), \quad \dot{x} = (\dot{x}_1, \dot{x}_2, \dot{x}_3), \quad j = \overline{1, 6},$$

whence it follows that for any given $t, x, \dot{x}$ one may find the corresponding C_j. In other words, no additional conditions are placed *a priori* on the coordinates and the velocity of a free point.

We shall assume that a point is *constrained* if its coordinates and velocity projections satisfy some a priori given conditions. For example, a point should remain during the entire time on the surface or on the line, which can be considered as the intersection of two surfaces. Such conditions are called *constraints* imposed on a point. Usually they are expressed in the form of some equalities or inequalities, that is, in the form of relations

$$\varphi(t, x, \dot{x}) \geqslant 0. \tag{1.1}$$

If the equality

$$\varphi(t, x, \dot{x}) = 0 \tag{1.2}$$

is satisfied during the motion, then the constraint is called *retaining or bilateral*, which means that the point during the entire motion must remain on the constraint. If at some time the equality is violated, then the point is said to leave the constraint. In this case, the constraint is called *releasing or unilateral*. The origin of these terms can be explained by the following example.

Let us assume that a point is on the surface given by the equation

$$f(t, x) = 0. \tag{1.3}$$

Then its coordinates should satisfy the surface equation, and so one may say that the surface 'retains' the point. If the coordinates of the point satisfy the condition $f(t, x) \geqslant 0$, then this means that it may leave the surface. As an example of such a constraint we may consider a convex surface with a heavy ball at rest near its vertex (Fig. 1). Under certain values of the initial point v_0 and depending on the form of the surface, the case is possible when the ball first rolls on the surface and then leaves it at a certain time instant. In the same way, expressions (1.1) and (1.2) can

Fig. 1 Motion on the
releasing constraint

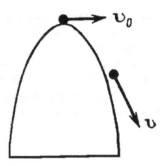

be considered in the 6-dimensional space of variables $(x_1, x_2, x_3, \dot{x}_1, \dot{x}_2, \dot{x}_3)$ with the above terminology.

If the time t enters explicitly formula (1.3), then the constraint is called *rheonomic* or *nonstationary*. To it there corresponds the image of a surface that deforms and travels in time. Otherwise, the constraint is called *scleronomic* or *stationary* .

The constraints expressed by formula (1.2), which contains the coordinates and first derivatives of the coordinates, are *differential constraints of the first order*. They can be either integrable or nonintegrable in the sense they can be reduced to an equation in the form (1.3). So, a constraint in the form $a_1\dot{x}_1 + a_2\dot{x}_2 + a_3\dot{x}_3 + a_0 = 0$ (or in the differential form $a_1 dx_1 + a_2 dx_2 + a_3 dx_3 + a_0 dt = 0$), where a_i are functions of the coordinates x_i, may or may fail to be reducible to the total differential depending on whether the functions $a_i(t, x)$ satisfy or not some conditions. For example, if

$$a_0 = \frac{\partial f(t, x)}{\partial t}, \quad a_1 = \frac{\partial f(t, x)}{\partial x_1}, \quad a_2 = \frac{\partial f(t, x)}{\partial x_2}, \quad a_3 = \frac{\partial f(t, x)}{\partial x_3},$$

then the constraint equation assumes the form

$$\frac{\partial f(t, x)}{\partial t} dt + \sum_{j=1}^{3} \frac{\partial f(t, x)}{\partial x_j} dx_j = 0.$$

This implies that $f(t, x) = $ const; that is, the constraint is integrable. In what follows, we shall adopt the summation convention: summation is implied over those suffixes which appear twice (called dummy suffixes).

Constraints of the form (1.3) are called *geometric, finite* or *holonomic* constraints. In the case when the differential constraints (1.2) are not integrable (that is, reducible to constraints of the form (1.3)), they are called *nonholonomic constraints of the first order*. A nonholonomic constraint is called *linear* if its equation depends linearly on the velocities. In the general case, the constraint equation (1.2) may be nonlinear both in coordinates and velocities.

If a constraint (1.2) is nonholonomic, then any two positive positions of a material point can be joined by a piecewise smooth curve of finite length along which the constraint equation is satisfied.[1] In the case when Eq. (1.2) is obtained by differentiation of the expression $f(t, x) = $ const in time, then it is not possible to come from any point to any point without violating the constraint (1.2). Indeed, by the choice of an arbitrary constant, the original point can be made to be any point. However, from it one may reach only a point which satisfies the equation $f(t, x) = $ const.

Clearly, any holonomic constraint expressible by a differentiable function can be represented in the differential form

$$\varphi(t, x, \dot{x}) \equiv \dot{f} = \frac{\partial f}{\partial t} + \frac{\partial f}{\partial x_1}\dot{x}_1 + \frac{\partial f}{\partial x_2}\dot{x}_2 + \frac{\partial f}{\partial x_3}\dot{x}_3 = 0, \qquad (1.4)$$

or

$$\varphi(t, x, \dot{x}) \equiv \frac{\partial f}{\partial t} + \nabla f \cdot \mathbf{v} = 0.$$

In the general case, the acceleration $\mathbf{w}^*$, which exerts the force $\mathbf{F}(t, x, \dot{x})$ on the original point, does not coincide with the acceleration $\mathbf{w}$, which obtains the point subject to the same force with a constraint. In other words, the constraint acts on a point as some force, which we shall call the *constraint reaction force* $\mathbf{R}$. It follows that the Newton equation for a constrained point should read as

$$m\,\mathbf{w} = \mathbf{F} + \mathbf{R}. \qquad (1.5)$$

If we assume that the force $\mathbf{R}$ is available, then the motion of the constrained point should agree with that of a free point subject to the given forces $\mathbf{F}$ and $\mathbf{R}$. This is the essence of the *principle of releasability from constraints*. We note that for such a free motion the constraint equation is an integral of the motion equations.

Let us consider the general constraint equation (1.2). Assuming that the function $\varphi(t, x, \dot{x})$ is differentiable in all its arguments, we may write

$$\dot{\varphi} = \frac{\partial \varphi}{\partial t} + \frac{\partial \varphi}{\partial x_j}\dot{x}_j + \frac{\partial \varphi}{\partial \dot{x}_j}\ddot{x}_j = 0, \quad j = \overline{1, 3},$$

or

$$\nabla'\varphi \cdot \mathbf{w} = -\frac{\partial \varphi}{\partial t} - \nabla\varphi \cdot \mathbf{v}, \qquad (1.6)$$

where

$$\nabla' = \frac{\partial}{\partial \dot{x}_j}\mathbf{i}_j, \quad j = \overline{1, 3}.$$

Multiplying Eq. (1.5) by $\nabla'\varphi$ and multiplying Eq. (1.6) by m, this gives

$$m\,\mathbf{w} \cdot \nabla'\varphi = \mathbf{F} \cdot \nabla'\varphi + \mathbf{R} \cdot \nabla'\varphi,$$

---

[1] For more detail, see Sect. 5 of the present chapter. In parallel with it, we note that the terms holonomic and nonholonomic constraints were first introduced by H. Herz.

$$m \, \mathbf{w} \cdot \nabla'\varphi = -m \left(\frac{\partial \varphi}{\partial t} + \nabla \varphi \cdot \mathbf{v} \right),$$

and therefore,

$$\mathbf{R} \cdot \nabla'\varphi = -\left(m \, \frac{\partial \varphi}{\partial t} + m \, \nabla \varphi \cdot \mathbf{v} + \mathbf{F} \cdot \nabla'\varphi \right). \tag{1.7}$$

Solving this equation in $\mathbf{R}$, we may write

$$\mathbf{R} = -\left(m \, \frac{\partial \varphi}{\partial t} + m \, \nabla \varphi \cdot \mathbf{v} + \mathbf{F} \cdot \nabla'\varphi \right) \frac{\nabla'\varphi}{|\nabla'\varphi|^2} + \mathbf{T}_0,$$

where $\mathbf{T}_0$ is an arbitrary vector orthogonal to the vector $\nabla'\varphi$. This formula can be conveniently represented in the form

$$\mathbf{R} = \Lambda \, \nabla'\varphi + \mathbf{T}_0 = \mathbf{N} + \mathbf{T}_0, \quad \mathbf{N} \cdot \mathbf{T}_0 = 0.$$

Substituting this expression into Eq. (1.7) for the vector $\mathbf{R}$, we see that the vector $\mathbf{T}_0$ can be excluded from it. Hence, constraint (1.2) can be satisfied for any vector $\mathbf{T}_0$. In particular, for the solution of the problem it suffices to assume that $\mathbf{T}_0 = 0$. Constraints with this property will be called *ideal constraints*. However, the physical method of realization of a constraint usually results in the appearance of the component $\mathbf{T}_0$, which in the majority of cases depends substantially on $\mathbf{N}$. In other words, a physically implemented constant turns out to be nonideal, and so, for the complete solution of the problem one needs to know, apart from the constraint equation, the physical law describing the setting of the vector $\mathbf{T}_0$.

The above analytic expression for the reaction $\mathbf{N}$ of the ideal constraint (1.2) was obtained here by analyzing the restrictions (1.6) imposed by a given constraint on the acceleration vector $\mathbf{w}$. These restrictions will be discussed from different positions in Sects. 5 and 7 of the present chapter.

Let us consider a particular case of the holonomic constraint (1.3). We write it in the form (1.2):

$$\varphi \equiv \dot{f} = \frac{\partial f}{\partial t} + \frac{\partial f}{\partial x_j} \dot{x}_j = 0.$$

Hence, in this case

$$\frac{\partial \varphi}{\partial \dot{x}_j} = \frac{\partial f}{\partial x_j},$$

and hence, for the holonomic constraint (1.3), the above vector $\nabla'\varphi$ agrees with the conventional vector ∇f.[2] Now the vector $\mathbf{N} = \Lambda \nabla f$ is directed along the normal

---

[2] The vector $\nabla'\varphi$ used in this subsection was introduced by N.N. Polyakhov (see *N.N. Polyakhov, Motion equations of mechanical systems under nonlinear and nonholonomic constraints in the general case // Vest. Leningr. Univ. no. 1 (1972), pp. 124–132 [in Russian]*). This vector may also be called the *generalized Hamilton operator*, because the classical 'nabla' Hamilton operator

Fig. 2 Motion on a scleronomic nonideal constraint

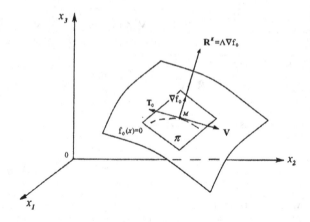

to the surface, as given by Eq. (1.3), while the vector $\mathbf{T}_0$ lies in the tangent plane π to this surface. This is demonstrated most clearly for scleronomic constraints (see Fig. 2, where the trajectory of a point lying on the surface $f(x) = 0$ is plotted by the dotted line). In particular, if the equation of scleronomic holonomic constraint specifies some material surface, on which a point M moves, then we have $\mathbf{T}_0 = 0$ if the surfaces is perfectly polished. Otherwise, one has to indicate the formation law of the vector $\mathbf{T}_0$, for example, specify Coulomb's friction law (for more details, see the next subsection).

2 Motion of a Material Point on a surface and on a line

Motion on a surface. Assume that point is forced to stay on the surface given by the Eq. (1.3). By the above, the reaction force $\mathbf{R}$ can be put in the form

$$\mathbf{R} = \Lambda\,\nabla f + \mathbf{T}_0 = \mathbf{N} + \mathbf{T}_0\,, \quad \nabla f \cdot \mathbf{T}_0 = 0\,, \tag{2.1}$$

and hence, the motion equation for the point is as follows:

$$m\,\mathbf{w} = \mathbf{F} + \Lambda\,\nabla f + \mathbf{T}_0\,, \tag{2.2}$$

the vectors $\mathbf{T}_0$ and ∇f being perpendicular.

If the surface (1.3) is fixed and material, then any motion on it is always accompanied by the friction force, which is directed opposite to the velocity vector. This is the force $\mathbf{T}_0$ (see Fig. 2). Experience shows that the sliding friction force is proportional to the magnitude of the normal component $\mathbf{N} = \Lambda\,\nabla f$ of the constraint reaction

---

follows from it as a particular case. The concept of the generalized Hamilton operator can be easily extended to nonholonomic constraints of any order.

vector, that is $|\mathbf{T}_0| = k\,|\mathbf{N}| = k\,|\Lambda\,\nabla f|$, where k is the proportionality factor, known as the *coefficient of dynamic friction*. This coefficient depends substantially on the material of the surfaces slipping with respect to each other, the method of surface treatment, and to a lesser extent, on the velocity. That the friction force is proportional to the normal pressure was established by Coulomb. This is the crux of the *Coulomb's law*.

By the above, Eq. (2.2) with a scleronomic constraint can be put in the form

$$m\,\mathbf{w} = \mathbf{F} + \Lambda\,\nabla f - k\,|\Lambda\,\nabla f|\,\mathbf{v}/|\mathbf{v}|\,, \tag{2.3}$$

or

$$m\,\mathbf{w} = \mathbf{F} + N_{n_1}\,\mathbf{n_1} - k\,|\mathbf{N}|\,\mathbf{v}/|\mathbf{v}|\,, \tag{2.4}$$

where $\mathbf{n}_1$ is the unit vector, which is normal to the surface. It is assumed that the vector $\mathbf{n}_1$ is orthogonal to the tangent plane in the direction where the nearest part of the surface is located.

Equation (2.3) in the projections to the Cartesian axes reads as

$$m\,\ddot{x}_i = X_i + \Lambda\,\frac{\partial f}{\partial x_i} - k\,|\Lambda\,\nabla f|\,\frac{\dot{x}_i}{|\mathbf{v}|}\,, \quad i = \overline{1,3}\,. \tag{2.5}$$

To obtain a system of four equations, from which the unknowns x_1, x_2, x_3 and Λ can be found, these equations should be augmented with the constraint equation $f(t, x_1, x_2, x_3) = 0$.

The inconvenience of the differential equations in the form (2.5) is that the unknown Λ enters all three equations. A more convenient form is obtained by writing Eq. (2.4) in the projections to the orthogonal vectors $\boldsymbol{\tau}, \mathbf{n_1}, \mathbf{p} = \boldsymbol{\tau} \times \mathbf{n}_1$, where $\boldsymbol{\tau}$ is the tangent vector to the trajectory of the point. Besides, we have

$$m\,\ddot{s} = F_\tau - k\,|N_{n_1}|\,\dot{s}/|\mathbf{v}|\,,$$
$$\frac{m\,v^2}{\rho}\,\cos\theta = F_{n_1} + N_{n_1}\,, \tag{2.6}$$
$$\frac{m\,v^2}{\rho}\,\sin\theta = F_p\,,$$

where s is the arc length, ρ is the curvature radius of the trajectory, θ is the angle between the unit normal vector $\mathbf{n}_1$ to the surface and the principal normal $\mathbf{n}$ of the trajectory.

If $\mathbf{F} = 0$, then Eq. (2.6) assume the form

$$m\,\ddot{s} = -k\,|N_{n_1}|\,\dot{s}/|\mathbf{v}|\,,$$

$$\frac{m\,v^2}{\rho}\cos\theta = N_{n_1}\,,$$

$$\frac{m\,v^2}{\rho}\sin\theta = 0\,.$$

By the above choice of the direction $\mathbf{n}_1$ the angle θ is smaller than $\pi/2$, which implies that $\theta = 0$ in these equations. Hence, they assume the form

$$m\,\ddot{s} = -k\,|N_{n_1}|\,\dot{s}/|\mathbf{v}|\,, \quad m\,v^2/\rho = N_{n_1}\,.$$

A curve lying on the surface and having the property that, at all its points, its principal normal vector $\mathbf{n}$ agrees with the normal vector to the surface $\mathbf{n}_1$ is called a *geodesic*. So, for $\mathbf{F} = 0$ a material point moves on a geodesic. In this case

$$\ddot{s} = -k\,v^2/\rho\,.$$

Setting $v = \dot{s}$ and taking into account that

$$\ddot{s} = \frac{d\dot{s}}{dt} = \frac{dv}{ds}\frac{ds}{dt} = v\,\frac{dv}{ds}\,,$$

we find that

$$v\,\frac{dv}{ds} = \ddot{s} = -k\,\frac{v^2}{\rho}\,.$$

As a result, we have $dv/v = -k\,ds/\rho$, and hence,

$$\ln\frac{v}{v_0} = -k\int_{s_0}^{s}\frac{ds}{\rho(s)} = -k\,\psi(s)\,,$$

that is $v = v_0\,e^{-k\,\psi(s)}$. From this formula it is seen that the velocity of a point decreases as it moves.

If the constraint (1.3) is effected without friction, then the above formulas with $\mathbf{F} = 0$ assume the form

$$m\,\dot{v} = 0\,, \quad v = \text{const}\,, \quad N_{n_1} = m\,v^2/\rho\,.$$

Let us now consider a motion on a fixed surface without friction in curvilinear coordinates. Assuming that the surface is smooth (that is, its tangent plane changes continuously), we choose on it a system of orthogonal Gaussian coordinates q^1, q^2. We assume that the third coordinate q^3 is given in the form $q^3 = f(x_1, x_2, x_3)$, where $f(x_1, x_2, x_3)$ is a function which, when it is equated to zero, gives the surface equation. In other words, we shall assume that a point moves on the coordinate surface. For simplicity, we shall be concerned with the particular case, when the principal basis $\{\mathbf{e}_i\}$, corresponding to the so-chosen system of curvilinear coordinates, is orthogonal at all points of the surface on which the motion is effected.

If a point is free, then in view of the orthogonality and stationarity of the basis, the equations of its motion (4.7) of Chap. 4 read as

$$m\,(g_{\rho\rho}\ddot{q}^{\,\rho} + \Gamma_{\rho,\sigma\tau}\,\dot{q}^{\sigma}\dot{q}^{\tau}) = Q_{\rho}\,, \quad \rho, \sigma, \tau = \overline{1, 3}\,. \tag{2.7}$$

The motion subject to the conditions $q^3 = 0$, $\dot{q}^3 = 0$, $\ddot{q}^{\,3} = 0$, can be found from the first two equations of the system (2.7), on assuming $q^3 \equiv 0$, that is, from the equations

$$m\,(g_{11}\ddot{q}^{\,1} + \Gamma_{1,\sigma\tau}\,\dot{q}^{\sigma}\dot{q}^{\tau}) = Q_1\,,$$
$$m\,(g_{22}\ddot{q}^{\,2} + \Gamma_{2,\sigma\tau}\,\dot{q}^{\sigma}\dot{q}^{\tau}) = Q_2\,, \quad \sigma, \tau = 1, 2\,. \tag{2.8}$$

Integrating these equations, we get the solution

$$q^{\sigma} = q^{\sigma}(t, C_1, C_2, C_3, C_4)\,, \quad \sigma = 1, 2\,,$$

from which one may also calculate the expressions $\Gamma_{3,\sigma\tau}\,\dot{q}^{\sigma}\dot{q}^{\tau}$, $\sigma, \tau = 1, 2$, on the left of the third equation of system (2.7). It may fail to be equal to $Q_3 = \mathbf{F} \cdot \mathbf{e}_3$. In this case the third equation should be written in the form

$$m\,\Gamma_{3,\sigma\tau}\,\dot{q}^{\sigma}\dot{q}^{\tau} - Q_3 = Q_{N_3}\,, \tag{2.9}$$

where Q_{N_3} is the generalized reaction.

Let us show that the generalized reaction Q_{N_3} is equal to the factor Λ from the original Eq. (2.2). Indeed, since $\mathbf{T}_0 = 0$ by the assumption, it follows from Eqs. (2.2) and (2.9) that

$$Q_{N_3} = \Lambda\,\nabla f \cdot \mathbf{e}_3 = \Lambda\,\nabla f \cdot \mathbf{e}_3^0 H_3\,.$$

The projection of the gradient to the direction $\mathbf{e}_3^0$ is the derivative in this direction, that is,

$$\nabla f \cdot \mathbf{e}_3^0 = \frac{\partial f}{\partial s_3} = \frac{1}{H_3}\frac{\partial f}{\partial q^3}\,.$$

Taking into account that $f(x_1, x_2, x_3) = q^3$, we find that $Q_{N_3} = \Lambda$, and so formula (2.9) can be put in the form

$$\Lambda = m\,\Gamma_{3,\sigma\tau}\,\dot{q}^{\sigma}\dot{q}^{\tau} - Q_3\,, \quad \sigma = 1, 2\,. \tag{2.10}$$

Formulas (2.8) are the differential equations of motion of a material point on a perfectly polished surface, while formula (2.10) determines the generalized reaction Λ that controls this motion.

So, the motion of a point on a retaining surface is determined by the two independent coordinates q^1, q^2. In this case, a point is said to have two degrees of freedom.

If one assumes that $\mathbf{T}_0 \neq 0$, then for the definiteness of the problem, one should specify the physical dependence $\mathbf{T}_0 = -k\,|\mathbf{N}|\,\mathbf{v}/|\mathbf{v}|$. The presence of this force is responsible for the appearance of the generalized forces

$$Q_{T_0 1} = \mathbf{T}_0 \cdot \mathbf{e}_1 = -k\,|\Lambda\,\nabla f|\,\mathbf{v}/|\mathbf{v}| \cdot \mathbf{e}_1\,,$$
$$Q_{T_0 2} = \mathbf{T}_0 \cdot \mathbf{e}_2 = -k\,|\Lambda\,\nabla f|\,\mathbf{v}/|\mathbf{v}| \cdot \mathbf{e}_2$$

in Eq. (2.8). However, formula (2.10) remains unchanged and it can be used to exclude Λ.

According to the above, if $\mathbf{F} = 0$ and $\mathbf{T}_0 = 0$ then the point moves on a geodesic with constant velocity. In fact, as it follows from Eq. (2.6), to obtain the motion on a geodesic it suffices to nullify not the entire force $\mathbf{F}$, but rather require that the component lying in the tangent plane to the surface be zero.

When using the curvilinear coordinates, if the force $\mathbf{F}$ acts only in the normal direction to the constraint surface, we have

$$Q_1 = 0\,, \quad Q_2 = 0\,, \quad \mathbf{F} = Q_3\,\mathbf{e}^3\,.$$

It follows that the differential equations of geodesics can be obtained from Eq. (2.8) by setting $Q_1 = 0$, $Q_2 = 0$ in (2.8):

$$g_{11}\ddot{q}^{\,1} + \Gamma_{1,\sigma\tau}\,\dot{q}^{\sigma}\dot{q}^{\tau} = 0\,,$$
$$g_{22}\ddot{q}^{\,2} + \Gamma_{2,\sigma\tau}\,\dot{q}^{\sigma}\dot{q}^{\tau} = 0\,, \quad \sigma, \tau = 1, 2\,.$$

These equations are derived in the differential geometry to find geodesics in the parametric form.

Motion on the line. If we are given surfaces defined by the equations

$$f^{\varkappa}(t, x_1, x_2, x_3) = 0\,, \quad \varkappa = 1, 2, \tag{2.11}$$

and if a point should be on the line of their intersection, then arguing as above we see that

$$\varphi^{\varkappa} = \dot{f}^{\varkappa}(t, x_1, x_2, x_3) = \frac{\partial f^{\varkappa}}{\partial x_{\alpha}}\,\dot{x}_{\alpha} = \frac{\partial f^{\varkappa}}{\partial t} + \nabla f \cdot \mathbf{v} = 0\,,$$
$$\dot{\varphi}^{\varkappa} = \frac{\partial \varphi^{\varkappa}}{\partial x_{\alpha}}\,\dot{x}_{\alpha} + \frac{\partial \varphi^{\varkappa}}{\partial \dot{x}_i}\,\ddot{x}_i = \frac{\partial \varphi^{\varkappa}}{\partial t} + \nabla \varphi^{\varkappa} \cdot \mathbf{v} + \nabla' \varphi^{\varkappa} \cdot \mathbf{w} = 0\,, \tag{2.12}$$

where $\alpha = \overline{0,3}$, $i = \overline{1,3}$, $x_0 = t$, $\nabla' \varphi^{\varkappa} = \nabla f^{\varkappa}$.

As before, the differential motion equation reads as

$$m\,\mathbf{w} = \mathbf{F} + \mathbf{R}\,. \tag{2.13}$$

This relation allows one to exclude from expressions (2.12) for $\dot{\varphi}^{\varkappa}$ the vector **w** and write them in the form

$$\mathbf{R} \cdot \nabla' \varphi^{\varkappa} \equiv R^{\varkappa} = -\left(m \frac{\partial \varphi^{\varkappa}}{\partial t} + m \nabla \varphi^{\varkappa} \cdot \mathbf{v} + \mathbf{F} \cdot \nabla' \varphi^{\varkappa} \right), \qquad \varkappa = 1, 2.$$

It follows if the vector **R** is written as the sum

$$\mathbf{R} = \Lambda_{\varkappa} \nabla' \varphi^{\varkappa} + \mathbf{T}_0,$$

where $\mathbf{T}_0$ is some unknown vector which is orthogonal to the vectors $\nabla' \varphi^{\varkappa} = \nabla f^{\varkappa}$, $\varkappa = 1, 2$, then the coefficients $\Lambda_{\varkappa}$ can be found from the system of equations

$$\Lambda_1 |\nabla' \varphi^1|^2 + \Lambda_2 \nabla' \varphi^1 \cdot \nabla' \varphi^2 = R^1,$$
$$\Lambda_1 \nabla' \varphi^2 \cdot \nabla' \varphi^1 + \Lambda_2 |\nabla' \varphi^2|^2 = R^2.$$

So, if the determinant of this system is nonzero, then the components $\Lambda_{\varkappa} \nabla' \varphi^{\varkappa}$ of the vector **R** can be uniquely determined from the constraint equations and the force **F**.

The tangent planes to the surfaces $f^1 = 0$ and $f^2 = 0$ intersect in a line, which at the point under consideration touches the curve, which is defined as the intersection of these surfaces. It follows that the unit tangent vector τ to this curve is perpendicular both to the vector ∇f^1 and to the vector ∇f^2. Hence, the component $\mathbf{T}_0$ of the reaction **R** can be put in the form $\mathbf{T}_0 = T_0 \tau$. Here, T_0 can be looked upon as the projection of the vector $\mathbf{T}_0$ to the selected direction of the vector τ.

If the physical realization of the constraints is such that $\mathbf{T}_0 = 0$, then the reactions are completely determined by their 'gradient' components:

$$\mathbf{N}_1 = \Lambda_1 \nabla f^1, \quad \mathbf{N}_2 = \Lambda_2 \nabla f^2,$$
$$\mathbf{N} = \mathbf{N}_1 + \mathbf{N}_2 = \Lambda_1 \nabla f^1 + \Lambda_2 \nabla f^2.$$

Such constraints are called *ideal*. If constraints are not ideal, then for the solution of the problem one needs to additionally specify the physical dependence T_0 on **N**.

The above definition of ideality of holonomic constraints can be more clearly formulated as follows. The holonomic constraint (1.3) is *ideal* if its reaction has no component lying in the tangent plane to the surface on which the point may be at a given time. Similarly, the system of holonomic constraints (2.11) is *ideal* if their reactions $\mathbf{R}_1$ and $\mathbf{R}_2$ do not have components directed along the tangent line to a curve on which the point may be at a given time. If constraints are scleronomic, then this curve agrees with the trajectory of the point; in the general case, this curve is distinct from it.

We note that these definitions were formulated in a form convenient for generalizations to the case of a representative point, that is, to the case of a mechanical system consisting of a finite number of material points.

In the Cartesian system of coordinates, Eq. (2.13) $\mathbf{T}_0 = 0$ reads as

$$m\,\ddot{x}_j = X_j + \Lambda_1 \frac{\partial f^1}{\partial x_j} + \Lambda_2 \frac{\partial f^2}{\partial x_j}, \quad j = \overline{1,3}.$$

These equations are known as the *Lagrange equations of the first kind* for a material point. Augmenting them with the constraint equations $f^1(t, x_1, x_2, x_3) = 0$, $f^2(t, x_1, x_2, x_3) = 0$, we arrive at five equations for the five unknowns x_1, x_2, x_3, Λ_1, Λ_2.

In the case of ideal scleronomic constraints, when the trajectory of a point is given by the constraint equations, it is appropriate to define the Newton equation (2.13) in its projections to the axes of the natural trihedron $\boldsymbol{\tau}$, $\mathbf{n}$, $\mathbf{b}$. Besides, we have

$$m\,\ddot{s} = F_\tau, \quad \frac{m\,v^2}{\rho} = F_n + N_n, \quad 0 = F_b + N_b. \tag{2.14}$$

We recall that the vector $\boldsymbol{\tau}$ is directed towards the positive direction of the angular coordinate s. Integrating the first equation, which does not contain the reaction $\mathbf{N}$, we shall find the function $v(t)$, which will allow us from the two successive formulas to find the components N_n and N_b of the vector $\mathbf{N}$. As was pointed out in Sect. 1 of Chap. 1, Eq. (2.14) are called the *Euler equations*.

Curvilinear motion of a point along a given line may also be studied with the use the curvilinear coordinates. Assume that the first coordinate is the parameter of the trajectory, while the two other ones (assuming that the constraints are scleronomic) are given in the form

$$q^2 = f^1(x_1, x_2, x_3), \quad q^3 = f^2(x_1, x_2, x_3).$$

For simplicity, let us assume that the principal basis of this system of coordinates is orthogonal at all points of the trajectory. Taking into account that the free motion in these coordinates is described by Eq. (2.7), we have, for the constrained motion ($q^2 \equiv 0$, $q^3 \equiv 0$),

$$m\,(g_{11}\,\ddot{q}^{\,1} + \Gamma_{1,11}\,(\dot{q}^1)^2) = Q_1,$$
$$m\,\Gamma_{2,11}\,(\dot{q}^1)^2 - Q_2 = \Lambda_1,$$
$$m\,\Gamma_{3,11}\,(\dot{q}^1)^2 - Q_3 = \Lambda_2.$$

If the coordinate q^1 is taken to be the arc length s, then $g_{11} = 1$, $\Gamma_{1,11} = 0$. Hence, the motion equation assumes the form

$$m\,\ddot{s} = F_\tau,$$

what might have been anticipated.

Mathematical pendulum. By a *circular mathematical pendulum* we shall mean a heavy point moving in the vertical plane on the circle. Such a motion will be executed, for example, by a weight of mass m suspended on a light inextensible rod of length l. The constraint equation is the equation of the circle $x_1^2 + x_2^2 - l^2 = 0$.

Fig. 3 Mathematical
pendulum

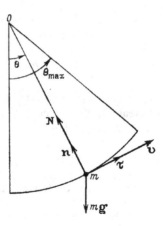

This constraint can be assumed to be ideal, because its reaction force $\mathbf{R} = \mathbf{N}$ is normal to this circle and equals the string tension. The motion equations in the vector form in the case of an ideal constraint and if there is no friction reads as

$$m\,\mathbf{w} = m\,\mathbf{g} + \mathbf{N}\,,$$

where $\mathbf{g}$ is the acceleration of gravity directed vertically downwards. We write this equation in the projections onto the tangent and normal planes to the trajectories; in other words, we shall use the Euler equations (2.14). Writing the arc length in the form $s = l\,\theta$, where θ is the angle of rotation of the pendulum (see Fig. 3), and taking into account that $\rho = l$, we find

$$\ddot{\theta} = -\frac{g}{l}\,\sin\theta\,, \quad m\,l\,\dot{\theta}^2 = -m\,g\,\cos\theta + N\,. \tag{2.15}$$

Denoting $\dot{\theta}$ by ω, we have

$$\dot{\omega} = -n^2\,\sin\theta\,, \quad n^2 = g/l\,,$$

or

$$\dot{\omega} = \omega\,\frac{d\omega}{d\theta} = -n^2\,\sin\theta\,,$$

and hence,

$$\frac{\omega^2}{2} - \frac{\omega_0^2}{2} = n^2\,(\cos\theta - \cos\theta_0)\,.$$

A mathematical pendulum may execute both circular and oscillatory motions. We shall be concerned with the oscillatory motion and assume that $\theta_0 = \theta_{\max}$, which implies that $\omega_0 = 0$. As a result, we have

$$\omega^2 = 2n^2 \left(\cos \theta - \cos \theta_{\max} \right),$$

and so,

$$\dot{\theta} = \omega = \pm n \sqrt{2 \left(\cos \theta - \cos \theta_{\max} \right)}, \tag{2.16}$$

the upper (lower) sign corresponding to the motion for which the angle θ increases (decreases). We assume that for $t = 0$ the angle $\theta = 0$ and $\dot{\theta} > 0$. Then, integrating Eq. (2.16), we find that

$$nt = \int_0^\theta \frac{d\theta}{\sqrt{2 \left(\cos \theta - \cos \theta_{\max} \right)}} = \int_0^\theta \frac{d\theta}{2 \sqrt{\sin^2 \frac{\theta_{\max}}{2} - \sin^2 \frac{\theta}{2}}} =$$

$$= \int_0^\theta \frac{d\theta}{2k \sqrt{1 - \frac{\sin^2 (\theta/2)}{k^2}}}, \tag{2.17}$$

where $k = \sin \left(\theta_{\max}/2 \right) > \sin \left(\theta/2 \right)$.

Changing from the variable θ to the variable α by the formula

$$\sin(\theta/2)/k = \sin \alpha,$$

we conclude that

$$\frac{1}{2k} \cos \frac{\theta}{2} d\theta = \cos \alpha \, d\alpha,$$

and so

$$\frac{1}{2k} \sqrt{1 - k^2 \sin^2 \alpha} \, d\theta = \cos \alpha \, d\alpha.$$

Now formula (2.17) assumes the form

$$\tau \equiv nt = \int_0^\alpha \frac{d\alpha}{\sqrt{1 - k^2 \sin^2 \alpha}}. \tag{2.18}$$

This is an elliptic integral of the first kind, which is not expressible in terms of elementary functions. With a fixed k and various values of α, one may find the function $\tau = F(\alpha, k)$ by evaluating integral (2.18) and then construct the inverse function $\alpha = \psi(\tau, k)$ and the function

$$\sin \alpha = \sin \psi(\tau, k) = \operatorname{sn}(\tau, k), \tag{2.19}$$

which is called the *Jacobi elliptic function*. Note that this function is a periodic function of τ.

We have $\sin \frac{\theta}{2} / \sin \frac{\theta_{\max}}{2} = \sin \alpha$, and hence, in view of formula (2.19) we get the law of motion of the pendulum in the form

$$\sin \frac{\theta(t)}{2} = \sin \frac{\theta_{\max}}{2} \sin \alpha = \sin \frac{\theta_{\max}}{2} \operatorname{sn}(nt, k) \,. \tag{2.20}$$

For small k it may be approximately assumed that $\sqrt{1 - k^2 \sin^2 \alpha} \approx 1$. Hence, $\tau \approx \alpha$ by (2.18), and therefore

$$\sin \alpha = \operatorname{sn}(\tau, k) \approx \sin \tau \,,$$

that is, for small k the elliptic function $\operatorname{sn}(\tau, k)$ is independent of k and agrees with $\sin \tau$. In this approximation, formula (2.20) assumes the form

$$\theta(t) = \theta_{\max} \sin nt \,.$$

The last term may also be derived directly by writing the motion equation for small θ in the form

$$\ddot{\theta} + n^2 \theta = 0 \,.$$

Coming back to the general case, we note that for $\theta = \theta_{\max}$

$$\sin \alpha = 1 \,, \quad \alpha = \pi/2 \,,$$

and hence,

$$nt_* = \int_0^{\pi/2} \frac{d\alpha}{\sqrt{1 - k^2 \sin^2 \alpha}} = \int_0^{\pi/2} \left[1 + \frac{k^2}{2} \sin^2 \alpha + \frac{1 \cdot 3}{2 \cdot 4} k^4 \sin^4 \alpha + \ldots \right] d\alpha =$$
$$= \frac{\pi}{2} \left[1 + \left(\frac{1}{2} \right)^2 k^2 + \left(\frac{1 \cdot 3}{2 \cdot 4} \right)^2 k^4 + \ldots \right] \,.$$

The quantity t_* is the time during which the pendulum moves from the initial position ($\theta = 0$) to the position in which $\theta = \theta_{\max}$ and $\omega = 0$. After this it begins to move in the opposite direction according to the law (2.16). It follows that $t_* = T_*/4$, where

$$T_* = 2\pi \sqrt{\frac{l}{g}} \left[1 + \left(\frac{1}{2} \right)^2 k^2 + \left(\frac{1 \cdot 3}{2 \cdot 4} \right)^2 k^4 + \ldots \right] \,,$$

is the period of oscillations of the pendulum. We have $k = \sin(\theta_{\max}/2)$, and hence for small $\theta_{\max}$,

$$T_* \approx 2\pi \sqrt{l/g} \,,$$

Fig. 4 Cycloidal pendulum

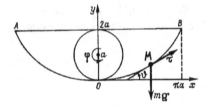

that is, the period is independent of $\theta_{\max}$. This is the property that small oscillations of the pendulum are *tautochrone (isochrone)*.

Cycloidal pendulum. Let us assume that a heavy point moves without friction in the vertical plane along a convex cycloid (for which at any point the tangent line is below the graph of the cycloid); see Fig. 4. In this case, the first Eq. (2.14) assumes the form

$$m\ddot{s} = -mg\sin\vartheta,$$

where s is the length of the arc OM, ϑ is the angle between the tangent line to the centroid and the x-axis. In the Cartesian coordinates, the cycloid has the equation

$$x = a\,(\varphi + \sin\varphi), \quad y = a\,(1 - \cos\varphi), \quad -\pi \leqslant \varphi \leqslant \pi, \qquad (2.21)$$

where φ is the angle through which the diameter of the disc rolling without slip along a fixed line AB rotates with respect to the initial vertical position, a is the disc radius. From Eq. (2.21) we have

$$dx = a\,(1 + \cos\varphi)\,d\varphi, \quad dy = a\,\sin\varphi\,d\varphi, \quad -\pi \leqslant \varphi \leqslant \pi. \qquad (2.22)$$

Further, we have

$$ds = \pm\sqrt{(dx)^2 + (dy)^2} = 2a\,\cos\frac{\varphi}{2}\,d\varphi,$$

$$s = 4a\,\sin\frac{\varphi}{2} = \pm 4a\,\sqrt{\frac{1 - \cos\varphi}{2}} = \pm\sqrt{8ay}, \quad -\pi \leqslant \varphi \leqslant \pi.$$

Besides, from Eq. (2.22) we also have

$$\tan\vartheta = dy/dx = \tan(\varphi/2), \quad \vartheta = \varphi/2,$$

and hence

$$ds = 4a\,\cos\vartheta\,d\vartheta = l_0\,\cos\vartheta\,d\vartheta, \quad l_0 = 4a,$$
$$\rho = ds/d\vartheta = 4a\,\cos\vartheta = l_0\,\cos\vartheta,$$
$$s = 4a\,\sin\vartheta = l_0\,\sin\vartheta.$$

Here, ρ is the curvature radius of the cycloid, l_0 is its maximal value. By the above, the motion equation assumes the form

$$\ddot{s} + \frac{g}{l_0} s = 0 \,.$$

Integrating, we obtain

$$s = C_1 \cos nt + C_2 \sin nt \,, \quad n = \sqrt{g/l_0} \,.$$

Suppose that for $t = 0$ the initial conditions are as follows: $s(0) = s_0 = 0$, $\dot{s}(0) = v_0 \neq 0$. Then

$$s = \frac{v_0}{n} \sin nt \,, \quad \dot{s} = v_0 \cos nt \,.$$

As the point M moves along the arc OB, the distance covered s and angle ϑ both increase, the velocity $v = \dot{s}$ is decreasing. At some point, to which there corresponds the angle $\vartheta_{max} < \pi/2$ and the moment t_*, the velocity vanishes. For such a point, we have

$$t = t_* = \pi/(2n) \,, \quad s = s_{max} = v_0/n \,, \quad v = 0 \,,$$
$$\sin \vartheta_{max} = s_{max}/l_0 = v_0/(nl_0) \,.$$

Hence, a necessary condition that ϑ_{max} be smaller than $\pi/2$ is that $v_0 < nl_0 = l_0 \sqrt{g/l_0} = \sqrt{gl_0}$.

So, for the cycloidal pendulum we finally obtain

$$s = s_{max} \sin nt \,, \quad \sin \vartheta = \sin \vartheta_{max} \sin nt \,.$$

The last formula shows that the period of oscillations of a cycloidal pendulum is independent of the angle ϑ_{max}, and hence, its motion is strictly tautochrone. Huygens, who found this property, proposed a method to execute oscillations on a cycloidal pendulum. This method is based on the fact that the evolute of a cycloid is the same cycloid, but translated along the x-axis with respect to the involute by the halved length of AB and lifted up by $2a$ (Fig. 5). Hence, if a string of length l_0 is fixed at a cusp point C of the cycloid-evolute and a material point is attached at the other

Fig. 5 Evolute and involute of a cycloidal pendulum

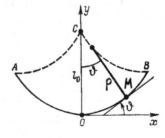

Fig. 6 Spherical pendulum

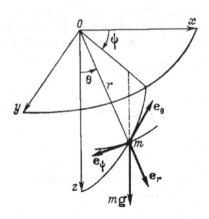

extremity, then as the pendulum swings, the string will go about the branches of the evolute, while the material point will move along the cycloid-involute with the same parameters. The string tension is as follows:

$$N = mg \cos \vartheta + \frac{mv^2}{\rho} = m \left(g \cos \vartheta + \frac{v_0^2 \cos^2 nt}{l_0 \cos \vartheta} \right).$$

Spherical pendulum. By a *spherical pendulum* we shall mean a heavy point which is forced to move along a sphere. In the rough, such a motion can be implemented by taking a rod, whose one end-point is supported with a spherical hinge, while the other (variable) end-point carries a weight of mass m far exceeding the mass of the rod, but whose sizes are much smaller than those of the rod length l.

For the analysis it is convenient to use the spherical coordinates r, θ, ψ (Fig. 6), which are related with the Cartesian ones by the relations

$$x = r \sin \theta \cos \psi, \quad y = r \sin \theta \sin \psi, \quad z = r \cos \theta. \qquad (2.23)$$

The angles $\theta = q^1$ and $\psi = q^2$ are free variable, the quantity $r = q^3$ is subject to the condition $q^3 = l$, $\dot{q}^3 = \ddot{q}^3 = 0$. The basis vectors $\mathbf{e}_\theta$, $\mathbf{e}_\psi$, $\mathbf{e}_r$ are orthogonal, the constraint is ideal, because its reaction (the string tension) is directed along the normal vector to the surface of the sphere of radius l. Hence, in this setting one may employ Eq. (2.8). The kinetic energy of the material point in the spherical coordinates is as follows:

$$T = \frac{m}{2} \left(r^2 \dot{\theta}^2 + (r^2 \sin^2 \theta) \dot{\psi}^2 + \dot{r}^2 \right).$$

Hence, Eq. (2.8) assume the form

$$\begin{aligned} m \left(l^2 \ddot{\theta} - l^2 \sin \theta \cos \theta \, \dot{\psi}^2 \right) &= Q_1, \\ m \left(l^2 \sin^2 \theta \, \ddot{\psi} + 2l^2 \sin \theta \cos \theta \, \dot{\theta} \, \dot{\psi} \right) &= Q_2. \end{aligned} \qquad (2.24)$$

The generalized forces Q_1 and Q_2 can be found from the expression for the potential energy $\Pi = -mgz = -mgl\cos\theta$:

$$Q_1 = -\frac{\partial\Pi}{\partial q^1} = -\frac{\partial\Pi}{\partial\theta} = -mgl\sin\theta\,, \quad Q_2 = -\frac{\partial\Pi}{\partial q^2} = -\frac{\partial\Pi}{\partial\psi} = 0\,.$$

Substituting these relations into Eq. (2.24), this gives

$$\begin{aligned} l\,(\ddot{\theta} - \dot{\psi}^2\sin\theta\cos\theta) &= -g\sin\theta\,, \\ \ddot{\psi}\sin\theta + 2\dot{\psi}\dot{\theta}\cos\theta &= 0\,. \end{aligned} \tag{2.25}$$

That the second equation can be integrated is seen by rewriting it in the form

$$\frac{d\dot{\psi}}{\dot{\psi}} = -2\,\frac{d(\sin\theta)}{\sin\theta}\,,$$

and hence,

$$\dot{\psi}\sin^2\theta = C = \text{const} = \omega_0\sin^2\theta_0\,, \quad \text{where} \quad \omega_0 = \dot{\psi}|_{t=0}\,. \tag{2.26}$$

This expression is the integral of angular momentum with respect to the z-axis. Using it to exclude $\dot{\psi}$ from the first equation of system (2.25), we find that

$$\ddot{\theta} - C^2\,\frac{\cos\theta}{\sin^3\theta} + \frac{g}{l}\sin\theta = 0\,, \tag{2.27}$$

and therefore,

$$\dot{\theta}\,d\dot{\theta} - C^2\,\frac{\cos\theta\,d\theta}{\sin^3\theta} + \frac{g}{l}\sin\theta\,d\theta = 0\,.$$

Integrating, this gives

$$\dot{\theta}^2 + \frac{C^2}{\sin^2\theta} - \frac{2g}{l}\cos\theta = C'\,, \tag{2.28}$$

which is the energy integral (which, however, could be derived directly).

We set $\cos\theta = u$. Now $\dot{u} = -\dot{\theta}\sin\theta$, and Eq. (2.28) can be put in the form

$$\dot{u}^2 = \left(\frac{2g}{l}u + C'\right)(1 - u^2) - C^2\,, \quad \text{or} \quad \frac{du}{dt} = \pm\sqrt{F(u)}\,,$$

where $F(u) = ((2g/l)\,u + C')(1 - u^2) - C^2$ is a polynomial of degree 3. Setting $t_0 = 0$, we have

$$t = \pm\int_{u_0}^{u} du/\sqrt{F(u)}\,. \tag{2.29}$$

From calculus we know that integrals of this type are not generally expressible in terms of elementary functions, but can be expressed in terms of elliptic integrals.

Let us now consider Eq. (2.26), which we rewrite in the form

$$(1 - u^2)\, d\psi = C\, dt\,.$$

Using formula (2.29), this gives

$$(1 - u^2)\, d\psi = \pm C\, \frac{du}{\sqrt{F(u)}}\,,$$

which implies that

$$\psi = \pm \int_{u_0}^{u} \frac{C\, du}{(1 - u^2)\, \sqrt{F(u)}} + \psi_0\,.$$

Having specified the initial conditions, we may find the constants C' and C, and then, inverting the elliptic integrals, find u and ψ qua functions of time.

For a more transparent visualization of the behaviour of the spherical pendulum, we set the initial conditions in the most simple, but fairly general form, which encompasses all the principally possible cases of motion of the pendulum.

The initial value of the angle ψ depends on the direction of the Ox-axis. Since this choice of the direction is immaterial in this problem, it may be assumed without loss of generality that the angle $\psi = \psi_0$ is zero with $t = t_0 = 0$.

The time variation of the angle ψ depends on the integral of angular momentum (2.26). If $\omega_0 = 0$, then we also have $C = 0$, which means that $\dot\psi = 0$ at all subsequent times, inasmuch as $\sin^2 \theta = 0$ for all $t > 0$ only if the pendulum is at rest. So, $\omega_0 = 0$ the spherical pendulum degenerates to the mathematical one.

From Eq. (2.26) it follows that the angle $\psi(t)$ increases monotonically if $\omega_0 > 0$ and decreases monotonically if $\omega_0 < 0$.

From the integral of angular momentum (2.26) one may also derive the following important conclusion regarding the behaviour of the angle θ: if $C \neq 0$, then $\sin^2 \theta$ is also nonzero during the entire motion. It follows that under any initial conditions the proper spherical pendulum cannot pass both through the lowermost position on the sphere (corresponding to $\theta = 0$) and through the uppermost position ($\theta = \pi$). Moreover, since by the physical considerations the angular velocity $\dot\psi(t) \neq 0$ may assume only finite values of the same sign, it follows from the integral (2.26) that the angle θ may vary only in some interval

$$0 \neq \theta_{\min} \leqslant \theta \leqslant \theta_{\max} \neq \pi\,.$$

Let us consider the coordinate $z = l \cos \theta$ of the spherical pendulum. In general, z oscillates. Assume that at the time from which we begin to examine the motion (that is, for $t = t_0 = 0$), the coordinate z reaches either its maximal or minimal value, which is z_0. Besides, $\dot z_0 = -l\dot\theta_0 \sin \theta_0 = 0$. Since $\sin \theta_0 \neq 0$, this means that $\dot\theta_0 = 0$.

Under the so-chosen initial conditions the energy integral (2.28) can be put in the form

$$\dot{\theta}^2 = \frac{2g}{l} (\cos \theta_0 - \cos \theta) \left[\frac{\omega_0^2 l \sin^2 \theta_0}{2g \sin^2 \theta} (\cos \theta_0 + \cos \theta) - 1 \right].$$

This formula shows that the angular velocity $\dot{\theta}$ can vanish either when $\theta = \theta_0$ or when the expression in the square brackets is zero. The latter observation leads to the following equation, which is quadratic in $\cos \theta$:

$$f(\cos \theta) \equiv \cos^2 \theta + \frac{\omega_0^2 l \sin^2 \theta_0}{2g} \cos \theta - \left(1 - \frac{\omega_0^2 l \sin^2 \theta_0 \cos \theta_0}{2g} \right) = 0.$$

Setting $\cos \theta = u$, $\cos \theta_0 = u_0$, $\omega_0^2 l \sin^2 \theta_0 / (4g) = \alpha$, we write it in the form

$$u^2 + 2\alpha u - 1 + 2\alpha u_0 = 0.$$

The roots of this equation $u_{1,2} = -\alpha \pm \sqrt{\alpha^2 - 2\alpha u_0 + 1}$ are real, since from the conditions $-1 < u_0 < 1$ we have

$$0 \leqslant (\alpha - 1)^2 < (\alpha - u_0)^2 < \alpha^2 - 2\alpha u_0 + 1 < (\alpha + 1)^2.$$

It also implies that

$$-u_0 = -\alpha + \alpha - u_0 < u_1 < -\alpha + \alpha + 1 = 1, \quad |u_2| > \alpha + |\alpha - 1|.$$

Let $\alpha > 1$. Then $|\alpha - 1| = \alpha - 1$ and $|u_2| > 2\alpha - 1 > 1$. If $\alpha \leqslant 1$, then $|\alpha - 1| = 1 - \alpha$ and

$$|u_2| > \alpha + 1 - \alpha = 1.$$

This being so,

$$-u_0 < u_1 < 1, \tag{2.30}$$

and since $|u_2| > 1$ for all α, we conclude that the root $u_2 = \cos \theta_2$ has no physical sense.

The above implies that the weight moves along a sphere remaining between two parallel horizontal discs of radii $l \sin \theta_0$ and $l \sin \theta_1$. As the weight reaches theses discs, the velocity $\dot{\theta}$ vanishes. We note that the mean horizontal plane $z = l (u_0 + u_1)/2$ is always below the sphere centre O, because $u_0 + u_1 > 0$ by inequality (2.30).

The plane $z = l u_0$ lies below the plane $z = l u_1$ if $u_0 > u_1$, that is, if $u_0 + \alpha > \sqrt{\alpha^2 - 2\alpha u_0 + 1}$. This inequality is satisfied under the condition that

$$\omega_0^2 l \cos \theta_0 > g. \tag{2.31}$$

This condition is possible only when $\cos \theta_0 > 0$, that is, when $0 < \theta_0 < \pi/2$.

If $\cos\theta_0 < 0$, then for any initial angular velocity ω_0 we have

$$\omega_0^2\, l \cos\theta_0 < g\,, \tag{2.32}$$

and hence, the plane $z = lu_0$ is above the sphere centre, the plane $z = lu_1$ lying below its centre.

For $\cos\theta_0 > 0$, if the angular velocity ω_0 is such that inequality (2.31) holds, the disc of radius $l \sin\theta_0$ is the lowest, while if inequality (2.32) holds, then the disc of radius $l \sin\theta_1$ becomes lowest. In the particular case, when

$$\omega_0^2\, l \cos\theta_0 = g\,, \tag{2.33}$$

the discs merge, the mass m runs along the circle with positive constant velocity $v_0 = \omega_0\, l \sin\theta_0$, the rod describing a cone. The pendulum thus obtained is called *conical*.

The transit time from one disc to another is expressed in terms of the integral (2.29), in which the upper limit is u_1. The integral is taken with the plus sign if $u_1 > u_0$ (with the minus sign if $u_1 < u_0$). The period T_* of oscillations of the pendulum in the vertical direction is obtained by doubling this time. Taking into account that the roots of the equation $f(u) = 0$ are also roots of the polynomial $F(u) = 0$, we find that

$$T_* = \pm 2 \sqrt{\frac{l}{2g}} \int_{u_0}^{u_1} \frac{du}{\sqrt{(u_0 - u)(u - u_1)(u - u_2)}}\,.$$

This integral can be expressed in terms of the elliptic integral of the first kind. Thus, in the general case the expression for the period is quite complicated.

Let us now consider the case when θ_0 and θ_1 are close to each other. We shall write the solution in the form

$$\theta = \theta_0 + \widetilde{\theta}$$

under the same initial conditions, where $\widetilde{\theta}$ is small in comparison with θ_0; it may be also assumed that

$$\sin\theta \approx \sin\theta_0 + \widetilde{\theta}\cos\theta_0\,, \quad \cos\theta \approx \cos\theta_0 - \widetilde{\theta}\sin\theta_0\,.$$

Substituting these quantities into the differential equation (2.27) and discarding the quantities in which $\widetilde{\theta}$ has power more than one, this gives

$$\ddot{\widetilde{\theta}} + k^2\widetilde{\theta} + H = 0\,, \tag{2.34}$$

where

$$k^2 = \omega_0^2\,(3\cos^2\theta_0 + \sin^2\theta_0) + g\cos\theta_0/l\,, \tag{2.35}$$

$$H = \sin\theta_0\,(g - \omega_0^2\, l \cos\theta_0)/l\,. \tag{2.36}$$

From Eq. (2.34) we have

$$\widetilde{\theta} = A \cos kt + B \sin kt - H/k^2 . \qquad (2.37)$$

Under the initial conditions $(\widetilde{\theta})_{t=0} = 0$, $(\dot{\widetilde{\theta}})_{t=0} = 0$, which correspond to $\dot{\theta} = 0$ with $\theta = \theta_0$, we find that $A = H/k^2$, $B = 0$, and hence,

$$\widetilde{\theta} = -\frac{H}{k^2}(1 - \cos kt) = -\frac{2H}{k^2} \sin^2 \frac{kt}{2} ,$$

$$\theta = \theta_0 - \frac{2H}{k^2} \sin^2 \frac{kt}{2} = \left(\theta_0 - \frac{H}{k^2}\right) + \frac{H}{k^2} \cos kt ,$$

that is, the pendulum executes small oscillations with period $2\pi/k$ and amplitude $\widetilde{A} = H/k^2$ near the position characterized by the $\theta_* = \theta_0 - H/k^2$. Besides, the plane of the pendulum rotates about the vertical axis with the angular velocity

$$\dot{\psi} = \frac{\omega_0 \sin^2 \theta_0}{\sin^2 \theta} = \omega_0 (1 - -2\widetilde{\theta} \cot \theta_0) = \omega_0 + \frac{4H\omega_0}{k^2} \cot \theta_0 \sin^2 \frac{kt}{2} .$$

The projection of the trajectories on the plane Oxy is represented by the equation

$$x^2 + y^2 = l^2 \sin^2 \theta = l^2 \sin^2 \theta_0 (1 + 2\widetilde{\theta} \cot \theta_0) = l^2 \sin^2 \theta_0 \left(1 - \frac{4H}{k^2} \cot \theta_0 \sin^2 \frac{kt}{2}\right),$$

which follows from formulas (2.23).

The solution (2.37) may also be used to investigate the motion of a spherical pendulum generated from that of the conical one, when the latter with $t = t_0 = 0$ is subject to small velocity $\dot{\theta} = \widetilde{\theta}_0$. Since here the relation (2.33) is satisfied, it follows from formula (2.36) that in this case $H = 0$ and the solution (2.37), which corresponds to the initial conditions $(\widetilde{\theta})_{t=0} = 0$, $(\dot{\widetilde{\theta}})_{t=0} = \widetilde{\theta}_0$, is as follows:

$$\widetilde{\theta} = \frac{\dot{\widetilde{\theta}}_0}{k} \sin kt.$$

Besides, in view of (2.33) formula (2.35) for the angular frequency can be put in the form

$$k^2 = \frac{g}{l \cos \theta_0}(1 + 3 \cos^2 \theta_0) = \omega_0^2 (1 + 3 \cos^2 \theta_0) .$$

If the angle θ_0 is small, then one may approximately put $\cos \theta_0 \approx 1$. Here, the period of revolution of a conical pendulum $2\pi/\omega_0$ becomes equal to the period T_1 of small oscillation of a mathematical pendulum of length l, that is, $T_1 = 2\pi \sqrt{l/g}$, while the period T_2 of small oscillations is equal to $T_2 = 2\pi/k = T_1/2$.

3 Constrained Motion of a system of Material Points. Constrained Motion of a representative Point. The Lagrange Equations

Constrained motion of a representative point. If the coordinates and velocities of points of a mechanical system satisfy, in addition to the differential motion equations, the following additional equalities or inequalities of the kind

$$\varphi^{\varkappa}(t, x, \dot{x}) \geqslant 0, \quad \varkappa = \overline{1, k},$$

then such a system is called *constrained*. Here, by $x = (x_1, \ldots, x_{3n})$ we mean the family of coordinates of n points of the system, which was introduced in Sect. 1 of Chap. 5.

Clearly, a single point from a constrained system is also constrained, and hence, by the results of Sect. 1 the differential equation of its motion should be written in the form

$$m_\nu \ddot{\mathbf{r}}_\nu = \mathbf{F}_\nu + \mathbf{R}'_\nu, \tag{3.1}$$

where ν is the number of a point, $\nu = \overline{1, n}$, $\mathbf{F}$· is the resultant of all external and internal forces applied to the point, $\mathbf{R}'$ is the resultant of the constraint reactions at this point. Setting

$$\mathbf{r}_\nu = x_{3\nu-2}\mathbf{i}_1 + x_{3\nu-1}\mathbf{i}_2 + x_{3\nu}\mathbf{i}_3,$$
$$\mathbf{F}_\nu = X_{3\nu-2}\mathbf{i}_1 + X_{3\nu-1}\mathbf{i}_2 + X_{3\nu}\mathbf{i}_3,$$
$$\mathbf{R}' = \mathbf{R}'_{3°-2}\mathbf{i}_1 + \mathbf{R}'_{3°-1}\mathbf{i}_2 + \mathbf{R}'_{3°}\mathbf{i}_3,$$
$$m_\mu = m_\nu \quad \text{for} \quad \mu = 3\nu - 2, \, 3\nu - 1, \, 3\nu,$$

we write system (3.1) in the form of $3n$ scalar equations:

$$m_\mu \ddot{x}_\mu = X_\mu + R'_\mu, \quad \mu = \overline{1, 3n}. \tag{3.2}$$

In order to employ in the analysis of the constrained motion the analytic machinery developed in the framework of the constrained motion of one point, we introduce in the $3n$-dimensional Euclidean space the Cartesian system of coordinates $Oy_1 \ldots y_{3n}$ with the orthogonal unit vectors $\mathbf{j}_1, \ldots, \mathbf{j}_{3n}$. In this Cartesian system of coordinates, one may introduce a representative point with the coordinate-s

$$y_\mu = \sqrt{\widetilde{m}_\mu}\, x_\mu, \quad \widetilde{m}_\mu = \frac{m_\mu}{M}, \quad M = \sum_{\nu=1}^{n} m_\nu \equiv \frac{1}{3}\sum_{\mu=1}^{3n} m_\mu, \quad \mu = \overline{1, 3n}.$$

Associating with the forces $\mathbf{F}_\nu$ and reactions $\mathbf{R}'$ the multidimensional vectors

$$\mathbf{Y} = Y_\mu \mathbf{j}_\mu, \quad Y_\mu = X_\mu / \sqrt{\tilde{m}_\mu}, \quad \mathbf{R} = R_\mu \mathbf{j}_\mu, \quad R_\mu = R'_\mu / \sqrt{\tilde{m}_\mu},$$
$$\mu = \overline{1, 3n},$$

paves the way for writing the system of scalar equations (3.2) in the form of one vector equation:

$$M\mathbf{W} = \mathbf{Y} + \mathbf{R}, \quad \mathbf{W} = \dot{\mathbf{V}} = \ddot{\mathbf{y}} = \ddot{y}_\mu \mathbf{j}_\mu. \tag{3.3}$$

If the constraints are expressed by the equations

$$\varphi^\varkappa(t, y, \dot{y}) = 0, \quad \varkappa = \overline{1, k}, \quad k < 3n, \tag{3.4}$$

then arguing as in the case of a single point, we obtain

$$\mathbf{R} = \Lambda_\varkappa \nabla' \varphi^\varkappa + \mathbf{T}_0 = \mathbf{N} + \mathbf{T}_0,$$

where $\nabla' = \frac{\partial}{\partial \dot{y}_\mu} \mathbf{j}_\mu$, $\mathbf{N} = \Lambda_\varkappa \nabla' \varphi^\varkappa$, $\mathbf{T}_0$ is an arbitrary vector orthogonal to the vector $\mathbf{N}$. By the above, Eq. (3.3) can be put in the form

$$M\mathbf{W} = \mathbf{Y} + \mathbf{N} + \mathbf{T}_0. \tag{3.5}$$

This equation, together with the constraint equations (3.4), forms the system of $3n + k$ scalar differential equations, from which one needs to determine $3n + k$ unknowns $y_1, y_2, \dots, y_{3n}, \Lambda_1, \Lambda_2, \dots, \Lambda_k$.

The vector $\mathbf{T}_0$ cannot be obtained from the form of the constraint equations. To obtain it, one should in addition specify the functional dependence of this vector bearing in mind the physical realization of the constraints. The most simple case is the one with $\mathbf{T}_0 = 0$, to which there correspond *ideal constraints*. However, not always such constraints can be physically realized. For example, integrable differential constraints realized with the help of a contact always feature friction. However, some constrains occurring in electro-mechanical systems are considered as ideal.

If the constraints are holonomic (that is, if they are expressed by the equations $f^\varkappa(t, y) = 0, \varkappa = \overline{1, k}$), then

$$\varphi^\varkappa \equiv \dot{f}^\varkappa = \frac{\partial f^\varkappa}{\partial t} + \frac{\partial f^\varkappa}{\partial y_\mu} \dot{y}_\mu,$$

and hence,

$$\frac{\partial \varphi^\varkappa}{\partial \dot{y}_\mu} = \frac{\partial f^\varkappa}{\partial y_\mu}, \quad \nabla' \varphi^\varkappa = \nabla f^\varkappa, \quad \mathbf{N} = \Lambda_\varkappa \nabla f^\varkappa.$$

Hence, the principal Eq. (3.5) assumes the form

$$M\mathbf{W} = \mathbf{Y} + \Lambda_\varkappa \nabla f^\varkappa + \mathbf{T}_0. \tag{3.6}$$

If the constraints are ideal, then $\mathbf{T}_0 = 0$. In this case, the three-dimensional reactions $\mathbf{R}'_\nu$, to which there corresponds the multi-dimensional vector $\mathbf{R} = \mathbf{N}$, have the components

$$R'_{\nu j} \equiv R'_\mu = \sqrt{\tilde{m}_\mu}\, R_\mu = \Lambda_\varkappa \sqrt{\tilde{m}_\mu}\, \frac{\partial f^\varkappa}{\partial(\sqrt{\tilde{m}_\mu}\, x_\mu)} = \Lambda_\varkappa \frac{\partial f^\varkappa}{\partial x_\mu},$$

$$\mu = 3(\nu - 1) + j, \quad j = 1, 2, 3, \quad \nu = \overline{1, n},$$

or

$$R'_{\nu j} = \Lambda_\varkappa \frac{\partial f^\varkappa}{\partial x_{\nu j}}, \quad x_{\nu j} = x_\mu. \tag{3.7}$$

Example 1 An important example of a holonomic constraint between two points of a system with numbers k and l is the constraint defined by the condition that the distance s_{kl} between these points varies according to a given law. Such a constraint is expressed by the equation

$$f(t, x) = (x_{k1} - x_{l1})^2 + (x_{k2} - x_{l2})^2 + (x_{k3} - x_{l3})^2 - s_{kl}^2(t) = 0. \tag{3.8}$$

We claim that this constraint can be assumed to be ideal, because the corresponding reactions $\mathbf{R}'_\nu$ can be put in the form (3.7). Indeed, the minimal additional forces $\mathbf{R}'_k$ and $\mathbf{R}'_l$, which should be applied to the masses m_k and m_l in order to keep them at a given distance, have equal magnitude, are directed opposite each other, and act along the line connecting these points, that is,

$$\mathbf{R}'_k = -\mathbf{R}'_l = \Lambda_* (\mathbf{r}_k - \mathbf{r}_l), \tag{3.9}$$

where Λ_* is some scalar quantity. At the same time, in accordance with formula (3.7), $R'_{\nu j} = \Lambda_\varkappa \frac{\partial f^\varkappa}{\partial x_{\nu j}}$, $j = 1, 2, 3$, and hence, for the constraint given by Eq. (3.8), we have

$$R'_{\nu j} = 0, \quad \nu \neq k, l,$$

$$R'_{kj} = \Lambda \frac{\partial f}{\partial x_{kj}} = 2\Lambda (x_{kj} - x_{lj}),$$

$$R'_{lj} = \Lambda \frac{\partial f}{\partial x_{lj}} = -2\Lambda (x_{kj} - x_{lj}),$$

$$j = 1, 2, 3,$$

or in the vector form

$$\mathbf{R}'_k = -\mathbf{R}'_l = 2\Lambda (\mathbf{r}_k - \mathbf{r}_l).$$

Comparing this expression, which follows from the assumption that the constraint be ideal, with expression (3.9), which was obtained from the condition of minimality of the additional forces that secure the fulfillment of the constraint equation, we see that in this case the assumption that the constraint be ideal is physically justified.

A mechanical system, in which the distances between its points are fixed, is called *rigid* or *invariable*. According to the results obtained for two arbitrary points of a system with a fixed distance between them, we conclude that the assumption that the system be rigid corresponds to imposing ideal constraints.

The Lagrange equations of the first kind. As was shown above, introduction of a representative point enables one to express the three-dimensional reactions $\mathbf{R}'_\nu$ in terms of the constraint equations. Now in view of (3.7) the system of Eq. (3.2) for ideal holonomic constraints is written in the form

$$m_\mu \ddot{x}_\mu = X_\mu + \Lambda_\varkappa \frac{\partial f^\varkappa}{\partial x_\mu}, \quad \mu = \overline{1, 3n}. \tag{3.10}$$

Augmenting this system of k equations with the constraints $f^\varkappa(t, x) = 0$, we obtain the closed system of equations with respect to the unknowns $x_1, x_2, ..., x_{3n}$, $\Lambda_1, \Lambda_2, ..., \Lambda_k$.

For ideal differential constraints, as given by equations (3.4), the components R'_μ of the reactions $\mathbf{R}'_\nu$ are as follows:

$$R'_\mu = \sqrt{\tilde{m}_\mu} \, R_\mu = \Lambda_\varkappa \sqrt{\tilde{m}_\mu} \frac{\partial \varphi^\varkappa}{\partial(\sqrt{\tilde{m}_\mu} \, \dot{x}_\mu)} = \Lambda_\varkappa \frac{\partial \varphi^\varkappa}{\partial \dot{x}_\mu}.$$

Hence, in our setting, we write Eq. (3.2) in the form

$$m_\mu \ddot{x}_\mu = X_\mu + \Lambda_\varkappa \frac{\partial \varphi^\varkappa}{\partial \dot{x}_\mu}, \quad \mu = \overline{1, 3n}. \tag{3.11}$$

Jointly with the constraint equations (3.4) these equations form a closed system.

The systems of Eqs. (3.10) and (3.11) are called the *Lagrange equations of the first kind*, respectively, for ideal *holonomic* and *nonholonomic systems*.

It is worth pointing out that if the functions $f^\varkappa$ depend nonlinearly on the coordinates x_μ, then it is convenient to consider the Lagrange equations of the first kind (3.10) jointly with the differentiated constraint equations

$$\varphi^\varkappa(t, x, \dot{x}) \equiv \dot{f}^\varkappa(t, x) = \frac{\partial f^\varkappa}{\partial t} + \frac{\partial f^\varkappa}{\partial x_\mu} \dot{x}_\mu = 0, \quad \varkappa = \overline{1, k}.$$

Similarly if the functions $\varphi^\varkappa$ depend nonlinearly on the velocities, then it is expedient to augment Eq. (3.11) with the relations

$$\psi^\varkappa(t, x, \dot{x}, \ddot{x}) \equiv \dot{\varphi}^\varkappa(t, x, \dot{x}) = \frac{\partial \varphi^\varkappa}{\partial t} + \frac{\partial \varphi^\varkappa}{\partial x_\mu} \dot{x}_\mu + \frac{\partial \varphi^\varkappa}{\partial \dot{x}_\mu} \ddot{x}_\mu = 0, \quad \varkappa = \overline{1, k}.$$

Let us now consider in more detail the situation when the functions $\varphi^\varkappa$ depend linearly on the velocities. This case frequently occurs in many practical problems and also applies to holonomic constraints which depend linearly or nonlinearly on the coordinates.

So, let us assume that the functions $\varphi^\varkappa$ are of the form

$$\varphi^\varkappa(t, x, \dot{x}) \equiv a_\mu^{l+\varkappa}(t, x)\dot{x}_\mu + b^{l+\varkappa}(t, x) = 0, \quad l = 3n - k, \quad \varkappa = \overline{1, k}.$$

Besides, $\partial\varphi^\varkappa/\partial\dot{x}_\mu = a_\mu^{l+\varkappa}(t, x)$, Eq. (3.11) assuming the form

$$m_\mu\ddot{x}_\mu = X_\mu + \Lambda_\varkappa a_\mu^{l+\varkappa}. \tag{3.12}$$

The use of the superscript $l + \varkappa$ will prove convenient in the sequel.

We shall assume that the rank of the matrix composed of the coefficients $a_\mu^{l+\varkappa}$ is k. Hence, one may employ k equations of (3.12) to exclude the unknowns $\Lambda_\varkappa$. The equations, from which $\Lambda_\varkappa$ are excluded, look like

$$(m_\mu\ddot{x}_\mu - X_\mu)b_{\lambda\mu} = 0, \quad \lambda = \overline{1, l}, \quad l = 3n - k. \tag{3.13}$$

Augmenting these equations with the constraint equations, we get a closed system of differential equations with respect to the unknowns functions $x_\mu(t)$.

Equation (3.13) show that in the case of ideal constraints there exist l such linear combinations of the Newton equations (3.2) that do not contain constraint reaction forces R'_μ.

The energy integral. For a constrained motion of one material point, under the condition that the holonomic constraints are ideal and scleronomic, any elementary displacement $d\mathbf{r} = \mathbf{v}dt$ is perpendicular to the constraint reaction. It follows that during the motion of the point, the constraint reaction does not perform any work, and hence, the total mechanical energy with constrained motion in a stationary potential force field is conserved. Motion of a spherical pendulum serves as an example of such a motion.

Changing from the consideration of the motion of one point to the consideration of mechanical systems consisting of finite numbers of material points, we introduce, as before, a representative point. In order to keep the discussion general, we shall assume that the constraints are differential and are given by Eq. (3.4). We shall show that if such constraints are ideal ($\mathbf{T}_0 = 0$) and if the functions $\varphi^\varkappa$ are homogeneous functions with respect to the velocities $\dot{y}_\mu$, then Eq. (3.5) have the energy integral, provided that the force $\mathbf{Y}$ has a potential U which is independent of time. Indeed, multiplying both sides of Eq. (3.5) by $d\mathbf{y} = \dot{y}dt$, we have

$$M\dot{\mathbf{y}} \cdot d\dot{\mathbf{y}} = \mathbf{Y} \cdot d\mathbf{y} + \mathbf{N} \cdot \dot{\mathbf{y}}dt + \mathbf{T}_0 \cdot \dot{\mathbf{y}}dt.$$

By the assumption $\mathbf{Y} = \nabla U$, and hence $\mathbf{Y} \cdot d\mathbf{y} = dU$, which implies that

$$d\left(\frac{M\mathbf{V}^2}{2}\right) = dU + \mathbf{N} \cdot \dot{\mathbf{y}}dt + \mathbf{T}_0 \cdot \dot{\mathbf{y}}dt.$$

However,

$$\mathbf{N} \cdot \dot{\mathbf{y}} = \Lambda_{\varkappa} \boldsymbol{\nabla}' \varphi^{\varkappa} \cdot \dot{\mathbf{y}} = \Lambda_{\varkappa} \frac{\partial \varphi^{\varkappa}}{\partial \dot{y}_{\mu}} \, \dot{y}_{\mu} \, . \tag{3.14}$$

Since by the assumption the functions $\varphi^{\varkappa}$ are homogeneous with respect to the velocities $\dot{y}_{\mu}$, by Euler's theorem on homogeneous functions we may write

$$\frac{\partial \varphi^{\varkappa}}{\partial \dot{y}_{\mu}} \, \dot{y}_{\mu} = p \, \varphi^{\varkappa} \, ,$$

where p is the homogeneity degree of the function $\varphi^{\varkappa}$. However, from the constraint equations we have $\varphi^{\varkappa} = 0$, showing that the expression (3.14) vanishes. Now we finally have

$$d \left(\frac{M \mathbf{V}^2}{2} - U \right) = \mathbf{T}_0 \cdot \dot{\mathbf{y}} dt \, ,$$

whence it follows that the energy integral $T - U = \text{const}$ for this kind of constraints will exist if they are ideal ($\mathbf{T}_0 = 0$).

The Lagrange equations of the second kind. Let us now consider the derivation of the motion equations of a reaction-free system under ideal restraining holonomic constraints. This problem for a representative point is solved in the same way as for a single isolated point.

Under ideal holonomic constraints ($\mathbf{T}_0 = 0$), Eq. (3.6) can be rewritten as follows:

$$M \mathbf{W} = \mathbf{Y} + \Lambda_{\varkappa} \boldsymbol{\nabla} f^{\varkappa} \, . \tag{3.15}$$

Let us introduce the curvilinear coordinates of a representative point as follows:

$$q^{\sigma} = q^{\sigma}(t, y) \, , \quad y = (y_1, \, \dots, \, y_{3n}) \, , \quad \sigma = \overline{1, 3n} \, .$$

Assuming that the solvability conditions are satisfied, we may write

$$y_{\rho} = y_{\rho}(t, q) \, , \quad q = (q^1, \, \dots, \, q^{3n}) \, , \quad \rho = \overline{1, 3n} \, .$$

Multiplying Eq. (3.15) by the vectors

$$\mathbf{e}_{\sigma} = \frac{\partial y_{\rho}}{\partial q^{\sigma}} \, \mathbf{i}_{\rho} \, , \quad \rho, \sigma = \overline{1, 3n} \, ,$$

of the principal basis of the above curvilinear system of coordinates, we find that

$$M W_{\sigma} = Q_{\sigma} + \Lambda_{\varkappa} \boldsymbol{\nabla} f^{\varkappa} \cdot \mathbf{e}_{\sigma} \, , \quad \sigma = \overline{1, 3n} \, , \tag{3.16}$$

where

$$MW_\sigma = M\mathbf{W} \cdot \mathbf{e}_\sigma = \frac{d}{dt}\frac{\partial T}{\partial \dot{q}^\sigma} - \frac{\partial T}{\partial q^\sigma} = M(g_{\sigma\tau}\ddot{q}^\tau + \Gamma_{\sigma,\alpha\beta}\dot{q}^\alpha\dot{q}^\beta),$$

$$\sigma, \tau = \overline{1,3n}, \quad \alpha, \beta = \overline{0,3n}, \quad q^0 = t, \quad \dot{q}^0 = 1, \tag{3.17}$$

$$Q_\sigma = \mathbf{Y} \cdot \mathbf{e}_\sigma = \sum_{\mu=1}^{3n} Y_\mu \frac{\partial y_\mu}{\partial q^\sigma} = \sum_{\mu=1}^{3n} X_\mu \frac{\partial x_\mu}{\partial q^\sigma} = \sum_{\nu=1}^{n} \mathbf{F}_\nu \cdot \frac{\partial \mathbf{r}_\nu}{\partial q^\sigma}.$$

Note that these formulas for the generalized forces Q_σ agree with the above expressions (1.10) of Chap. 5.

Since the functions q^σ can be chosen arbitrarily, we take k such functions in the form

$$q^{l+\varkappa} = f^\varkappa(t, y), \quad \varkappa = \overline{1,k},$$

where $f^\varkappa(t, y)$ are functions which, when are equated to zero, give the equations for k holonomic constraints:

$$f^\varkappa(t, y) = 0, \quad \varkappa = \overline{1,k}.$$

With such a choice of the curvilinear coordinates, we have

$$\nabla f^\varkappa \cdot \mathbf{e}_\sigma = \frac{\partial f^\varkappa}{\partial y_\mu}\frac{\partial y_\mu}{\partial q^\sigma} = \frac{\partial q^{l+\varkappa}}{\partial y_\mu}\frac{\partial y_\mu}{\partial q^\sigma} = \delta_\sigma^{l+\varkappa} = \begin{cases} 0, & \sigma \neq l+\varkappa, \\ 1, & \sigma = l+\varkappa. \end{cases} \tag{3.18}$$

Here, we write the system of Eq. (3.16) in the form of Lagrange equations of the second kind

$$\frac{d}{dt}\frac{\partial T}{\partial \dot{q}^\lambda} - \frac{\partial T}{\partial q^\lambda} = Q_\lambda, \quad \lambda = \overline{1,l}, \tag{3.19}$$

$$\frac{d}{dt}\frac{\partial T}{\partial \dot{q}^{l+\varkappa}} - \frac{\partial T}{\partial q^{l+\varkappa}} - Q_{l+\varkappa} = \Lambda_\varkappa, \quad \varkappa = \overline{1,k}, \tag{3.20}$$

or in the expanded form,

$$M(g_{\lambda\tau}\ddot{q}^\tau + \Gamma_{\lambda,\alpha\beta}\dot{q}^\alpha\dot{q}^\beta) = Q_\lambda, \quad \lambda = \overline{1,l}, \tag{3.21}$$

$$M(g_{l+\varkappa,\tau}\ddot{q}^\tau + \Gamma_{l+\varkappa,\alpha\beta}\dot{q}^\alpha\dot{q}^\beta) - Q_{l+\varkappa} = \Lambda_\varkappa, \quad \varkappa = \overline{1,k}. \tag{3.22}$$

Here, $\tau = \overline{1,l}$, $\alpha, \beta = \overline{0,l}$, because $q^{l+\varkappa} = \dot{q}^{l+\varkappa} = \ddot{q}^{l+\varkappa} = 0$, $\varkappa = \overline{1,k}$.

The system of Eq. (3.21) is a system of l differential equations for l unknowns functions $q^1, q^2, ..., q^l$. From formulas (3.22) one may find the generalized reactions $\Lambda_\varkappa$, $\varkappa = \overline{1,k}$.

The coordinates q^λ, as given by the formulas

$$q^\lambda = q^\lambda(t, y), \quad \lambda = \overline{1,l}, \tag{3.23}$$

will be chosen arbitrarily. However, the functions $q^\lambda(t, y)$ should be such that from relations (3.23), when considered jointly with the constraint equations

$$f^\varkappa(t, y) = 0, \quad \varkappa = \overline{1, k}, \tag{3.24}$$

one could uniquely express all the coordinates y_μ in terms of the independent parameters q^λ. These parameters, which uniquely determine the position of a holonomic mechanical system, are called the *generalized Lagrange coordinates*. The number of such coordinates is called the *number of the degrees of freedom* of a holonomic mechanical system. In practice, when choosing the coordinates q^λ, as a rule there is no need in writing down the explicit expressions for the functions $q^\lambda(t, y)$. For example, assume that a point lies on an expanding sphere of radius $r = r(t)$. As the generalized coordinates q^1, q^2 it is convenient to take the angles φ and ψ of the spherical system of coordinates.

We note that Eq. (3.19), which involve only independent Lagrange coordinates q^λ, have the form of *Lagrange equations of the second kind*, which were previously derived for the case of a free point. They can be also called the *equations of motion of holonomic mechanical systems*.

So, if we have a constraint system of n points subject to k ideal holonomic constraints, then one may always find a system of curvilinear coordinates $q^1, q^2, ..., q^l$, $q^{l+1}, q^{l+2}, ..., q^{3n}$, in which the motion equations (3.19) do not contain the unknown reactions. Integrating these equations (that is, finding the coordinates $q^1, q^2, ..., q^l$ as functions of time and initial data), one may from formulas (3.20) also find the generalized reactions $\Lambda_1, \Lambda_2, ..., \Lambda_k$ as functions of time.

This has the following geometrical interpretation. The family of constraint equations (3.24) is the equation of an l-dimensional surface in the $3n$-dimensional space. These constraint equations describe a surface on which should lie the representative point at a given time instant. From relations (3.23), (3.24), when considered jointly, it follows that the coordinates y_μ of points on this surface are functions of the parameters q^λ, $\lambda = \overline{1, l}$, and hence, the quantities q^λ can be looked upon as Gaussian coordinates. It also implies this surface has the equation in the parametric form

$$\mathbf{y} = \mathbf{y}(t, q^1, q^2, ..., q^l),$$

and hence, the vectors $\mathbf{e}_\lambda = \partial \mathbf{y}/\partial q^\lambda$, $\lambda = \overline{1, l}$, lie in the plane parallel to this surface, while in view of (3.18) the vectors $\nabla f^\varkappa$, $\varkappa = \overline{1, k}$ are orthogonal to this plane. It also implies that the reaction vector of the ideal holonomic constraints $\mathbf{R} = \Lambda_\varkappa \nabla f^\varkappa$ is orthogonal to the tangent plane to the surface on which the representative point must lie at a given time.

In the particular case, when the generalized forces Q_λ are zero 0, $\lambda = \overline{1, l}$, the representative point moves along the above l-dimensional surface by its inertia, and by (3.21) the differential equations of its trajectory read as

$$g_{\lambda\tau} \ddot{q}^\tau + \Gamma_{\lambda,\alpha\beta} \dot{q}^\alpha \dot{q}^\beta = 0, \quad \lambda, \tau = \overline{1, l}, \quad \alpha, \beta = \overline{0, l}.$$

If in addition the constraints are scleronomic, then the previous equations should be written as

$$g_{\lambda\tau}\ddot{q}^{\tau} + \Gamma_{\lambda,\sigma\tau}\dot{q}^{\sigma}\dot{q}^{\tau} = 0, \quad \lambda,\sigma,\tau = \overline{1,l}.$$

These equations agree with the differential equations of geodesic lines, as they are understood in the multi-dimensional differential geometry.

Derivations of the Lagrange equations of the first and second kinds based on the solution of a direct mechanical problem. We shall show that the Lagrange equations of the first and second kinds can be also obtained in a purely analytic fashion, that is, without using the representative point and the equation of its motion. This is achieved by the solution of a direct mechanical problem on finding the forces from the given motion considered in the curvilinear coordinates.

Let us assume that the position of a free mechanical system consisting of n material points is uniquely determined by specifying the generalized coordinates q^{σ}, which are related with the Cartesian coordinates x_{μ} of the system points by the relations

$$q^{\sigma} = f_{*}^{\sigma}(t,x), \quad \sigma = \overline{1,s}, \quad s = 3n. \tag{3.25}$$

Assume that the motion of the system is controlled by the family of functions q^{σ}. It is required to find which forces R'_{μ} in Eq. (3.2) need to be added to the forces X_{μ} to effect this motion.

The most simple way to determine the forces R'_{μ} is as follows. From relations (3.25) we find the coordinates x_{μ} *qua* composite functions of time:

$$x_{\mu} = x_{\mu}(t,q(t)), \quad \mu = \overline{1,3n}.$$

Calculating by these formulas the accelerations $\ddot{x}_{\mu}$, we obtain the family of all the required additional forces

$$R'_{\mu} = m_{\mu}\ddot{x}_{\mu} - X_{\mu}, \quad \mu = \overline{1,3n}. \tag{3.26}$$

The forces X_{μ} are assumed to be fixed functions of time, the coordinates and velocities of points of the system are reciprocal.

This solution of the direct mechanical problem does not explain, however, the effect of each of the functions $f_{*}^{\sigma}(t,x)$ on the forces R'_{μ}. To find the form of the solution, which rationally takes into account the transition to the new coordinates, we first consider the forces X_{μ}. We shall assume that these forces have a potential defined in the coordinates q^{σ}. Then

$$X_{\mu} = \frac{\partial U}{\partial x_{\mu}} = Q_{\sigma}\frac{\partial q^{\sigma}}{\partial x_{\mu}}, \tag{3.27}$$

where

$$Q_{\sigma} = \frac{\partial U}{\partial q^{\sigma}} = X_{\mu}\frac{\partial x_{\mu}}{\partial q^{\sigma}}. \tag{3.28}$$

The *linear transformation of forces* (3.27), (3.28), which relate the change to the new variables with the change to new forces, correspond to the case when the forces have a potential. It is necessary to extend these transformations to the case of any forces, because both the potential and nonpotential forces are included into the second Newton law as terms of equal right. Applying these formulas to the forces R'_μ, this gives

$$R'_\mu = \Lambda^*_\sigma \frac{\partial q^\sigma}{\partial x_\mu}, \tag{3.29}$$

$$\Lambda^*_\sigma = R'_\mu \frac{\partial x_\mu}{\partial q^\sigma}. \tag{3.30}$$

We consider the quantities Λ^*_σ as new unknowns, using which it is appropriate to express the sought-for forces R'_μ under the condition that the motion is defined in the coordinates q^σ. Taking into account (3.26), we write the system of Eq. (3.30) in the form

$$\Lambda^*_\sigma = (m_\mu \ddot{x}_\mu - X_\mu) \frac{\partial x_\mu}{\partial q^\sigma}, \quad \sigma = \overline{1, 3n}. \tag{3.31}$$

To exclude from considerations the functions $x_\mu(t, q)$ and expressions for Λ^*_σ in terms of the given functions $q^\sigma(t)$ and their derivatives, we shall use the transformation

$$m_\mu \ddot{x}_\mu \frac{\partial x_\mu}{\partial q^\sigma} = \frac{d}{dt} \left(m_\mu \dot{x}_\mu \frac{\partial x_\mu}{\partial q^\sigma} \right) - m_\mu \dot{x}_\mu \frac{d}{dt} \frac{\partial x_\mu}{\partial q^\sigma}. \tag{3.32}$$

Since

$$\dot{x}_\mu = \frac{\partial x_\mu}{\partial t} + \frac{\partial x_\mu}{\partial q^\tau} \dot{q}^\tau, \quad \mu, \tau = \overline{1, 3n},$$

$$\frac{d}{dt} \frac{\partial x_\mu}{\partial q^\sigma} = \frac{\partial^2 x_\mu}{\partial q^\sigma \partial t} + \frac{\partial^2 x_\mu}{\partial q^\sigma \partial q^\tau} \dot{q}^\tau = \frac{\partial}{\partial q^\sigma} \left(\frac{\partial x_\mu}{\partial t} + \frac{\partial x_\mu}{\partial q^\tau} \dot{q}^\tau \right),$$

we have

$$\frac{\partial x_\mu}{\partial q^\sigma} = \frac{\partial \dot{x}_\mu}{\partial \dot{q}^\sigma}, \quad \frac{d}{dt} \frac{\partial x_\mu}{\partial q^\sigma} = \frac{\partial \dot{x}_\mu}{\partial q^\sigma}.$$

These formulas are called the *Lagrange relations*. Now the transformations (3.32) can be put in the form

$$m_\mu \ddot{x}_\mu \frac{\partial x_\mu}{\partial q^\sigma} = \frac{d}{dt}\left(m_\mu \dot{x}_\mu \frac{\partial \dot{x}_\mu}{\partial \dot{q}^\sigma}\right) - m_\mu \dot{x}_\mu \frac{\partial \dot{x}_\mu}{\partial q^\sigma} =$$

$$= \frac{d}{dt}\frac{\partial}{\partial \dot{q}^\sigma}\left(\sum_{\mu=1}^{3n}\frac{m_\mu \dot{x}_\mu^2}{2}\right) - \frac{\partial}{\partial q^\sigma}\left(\sum_{\mu=1}^{3n}\frac{m_\mu \dot{x}_\mu^2}{2}\right) = \tag{3.33}$$

$$= \frac{d}{dt}\frac{\partial T}{\partial \dot{q}^\sigma} - \frac{\partial T}{\partial q^\sigma} = M(g_{\sigma\tau}\ddot{q}^\tau + \Gamma_{\sigma,\alpha\beta}\dot{q}^\alpha \dot{q}^\beta),$$

$$T = \frac{M}{2}g_{\alpha\beta}\dot{q}^\alpha \dot{q}^\beta, \quad \mu, \sigma, \tau, = \overline{1,3n}, \quad \alpha, \beta = \overline{0,3n}.$$

In view of expressions (3.17), (3.33) Eq. (3.31) assume the form

$$\frac{d}{dt}\frac{\partial T}{\partial \dot{q}^\sigma} - \frac{\partial T}{\partial q^\sigma} = Q_\sigma + \Lambda_\sigma^*, \tag{3.34}$$

or, equivalently,

$$M(g_{\sigma\tau}\ddot{q}^\tau + \Gamma_{\sigma,\alpha\beta}\dot{q}^\alpha \dot{q}^\beta) = Q_\sigma + \Lambda_\sigma^*. \tag{3.35}$$

From Eq. (3.34) it follows that the unknowns Λ_σ^* are the additional generalized forces, which in combination with the generalized forces Q_σ govern the motion according to the given law $q^\sigma = q^\sigma(t)$.

For a more detailed analysis of the concept of a generalized force, we compare the motion equations of a free mechanical system in different systems of coordinates: x_μ, y_ρ, q^σ. We have

$$m_\mu \ddot{x}_\mu \equiv \frac{d}{dt}\frac{\partial T}{\partial \dot{x}_\mu} - \frac{\partial T}{\partial x_\mu} = X_\mu,$$

$$M\ddot{y}_\rho \equiv \frac{d}{dt}\frac{\partial T}{\partial \dot{y}_\rho} - \frac{\partial T}{\partial y_\rho} = Y_\rho, \tag{3.36}$$

$$MW_\sigma \equiv \frac{d}{dt}\frac{\partial T}{\partial \dot{q}^\sigma} - \frac{\partial T}{\partial q^\sigma} = Q_\sigma,$$

$$\mu, \rho, \sigma = \overline{1,3n}.$$

The force action on the system in the whole, which is controlled by the right-hand sides of these equations (that is, by the gamily of the quantities X_μ, Y_ρ, Q_σ), shall be considered as one object. We denote it by $\mathbf{Y}$.

Intrinsically, the quantity $\mathbf{Y}$ is of vector structure, because the joint action of the two systems of forces X_μ^{I} and X_μ^{II} is equivalent to the action of one system of forces for which $X_\mu = X_\mu^{\mathrm{I}} + X_\mu^{\mathrm{II}}$. Accordingly, one may write

$$\mathbf{Y} = \mathbf{Y}^{\mathrm{I}} + \mathbf{Y}^{\mathrm{II}}.$$

The components X_μ, Y_ρ, Q_σ of the same force action $\mathbf{Y}$ are related by the relations

$$Y_\rho = X_\mu \frac{\partial x_\mu}{\partial y_\rho} \equiv \frac{X_\rho}{\sqrt{\tilde{m}_\rho}}, \qquad Q_\sigma = X_\mu \frac{\partial x_\mu}{\partial q^\sigma}. \tag{3.37}$$

An object of vector nature, whose components are transformed by the law (3.37) when changing to a new system of coordinates, is called a *covariant vector* or briefly a *covector*.

Having proved that the family of the right-hand sides of the motion equations (3.36) is a covector, one may assert that the left-hand sides of these equations also define a covector. If we denote it by $M\mathbf{W}$, then the law of motion can be written in the form

$$M\mathbf{W} = \mathbf{Y}. \tag{3.38}$$

Here, M is the mass of the entire system, $\mathbf{W}$ is the covector characterizing its acceleration. For a free mechanical system, consisting of a finite number of points, it was shown that the quantities $\mathbf{W}$ and $\mathbf{Y}$ can be looked upon as vectors of the $3n$-dimensional Euclidean space, in which the representative point moves according to the law (3.38).

In the sequel, the motion law (3.38) of a mechanical system of the form of the second Newton law for a single material point will be called for convenience the *Newton law* or the *Newton equation*.

Let us go back to Eq. (3.35), which enable one to determine the additional generalized forces Λ_σ^* that control the given motion. We point out the following important fact, which relates the interrelationship between the generalized coordinates and the generalized forces.

Theorem. *A motion for which one of the generalized coordinates is a given function of time can be secured by introduction of one additional generalized force corresponding to this coordinate.*

Let us prove this theorem. For definiteness, we assume that the coordinate q^{3n} is the given time function. Assuming that all Λ_σ^* (except Λ_{3n}^*) are zero, we have

$$M(g_{\sigma\tau}\ddot{q}^\tau + \Gamma_{\sigma,\alpha\beta}\dot{q}^\alpha\dot{q}^\beta) = Q_\sigma,$$
$$\sigma = \overline{1, 3n-1}, \quad \tau = \overline{1, 3n}, \quad \alpha, \beta = \overline{0, 3n}, \tag{3.39}$$

$$M(g_{3n,\tau}\ddot{q}^\tau + \Gamma_{3n,\alpha\beta}\dot{q}^\alpha\dot{q}^\beta) = Q_{3n} + \Lambda_{3n}^*, \quad \tau = \overline{1, 3n}, \quad \alpha, \beta = \overline{0, 3n}. \tag{3.40}$$

Integrating the system of Eq. (3.39) involving the unknown functions $q^1(t), q^2(t), ...,$ $q^{3n-1}(t)$, we get al.l the generalized coordinates q^τ, $\tau = \overline{1, 3n}$, as functions of time. The sought-for additional force Λ_{3n}^*, which secures the given motion in the coordinate q^{3n}, can be found from expression (3.40). However, it should be pointed out that by applying to a system an additional generalized force Λ_{3n}^*, we change the motion in all other coordinates, because Eq. (3.39) involve the given function $q^{3n}(t)$.

This assertion does not mean that specifying the motion in some coordinate has no effect on the motion in the other coordinates. It only means that there is no need to apply other additional forces, except for the force corresponding to a chosen coordinate.

As a direct consequence of this theorem we mention the following more general fact that the motion given in several coordinates can be realized by the same number of the corresponding additional generalized forces.

Let us now use the linear systems (3.29), (3.30) for the study of a restricted motion with holonomic constraints, which are given by the equations $f^{\varkappa}(t, x) = 0$, $\varkappa = \overline{1, k}$. We include the functions $f^{\varkappa}(t, x)$ into the system of functions $f_*^{\varkappa}(t, x)$ by setting

$$q^{l+\varkappa} = f_*^{l+\varkappa}(t, x) = f^{\varkappa}(t, x) = 0, \quad \varkappa = \overline{1, k}, \quad l = 3n - k. \tag{3.41}$$

It turns out that the motion in the coordinates $q^{l+\varkappa}$ is given and in the most simple form. It was shown that such a motion can be secured by the additional generalized forces $\Lambda_{l+\varkappa}^* = \Lambda_{\varkappa}$. In view of Eqs. (3.2) and (3.41) in this case system (3.29) assumes the form

$$m_{\mu}\ddot{x}_{\mu} = X_{\mu} + \Lambda_{\varkappa} \frac{\partial f^{\varkappa}}{\partial x_{\mu}}, \quad \mu = \overline{1, 3n}$$

(the Lagrange equations of the first kind).

The system of Eq. (3.34) splits into two subsystems

$$\frac{d}{dt} \frac{\partial T}{\partial \dot{q}^{\lambda}} - \frac{\partial T}{\partial q^{\lambda}} = Q_{\lambda}, \quad \lambda = \overline{1, l}, \tag{3.42}$$

$$\frac{d}{dt} \frac{\partial T}{\partial \dot{q}^{l+\varkappa}} - \frac{\partial T}{\partial q^{l+\varkappa}} - Q_{l+\varkappa} = \Lambda_{\varkappa}, \quad \varkappa = \overline{1, k}. \tag{3.43}$$

These are the Lagrange equations of the second kind. The first subsystem, which has no reactions, can serve to determine the unknown Lagrange coordinates q^{λ}. The second one can be used to find the generalized reactions $\Lambda_{\varkappa}$. Here, it is to be taken into account, that in order to calculate $\Lambda_{\varkappa}$ one should express the kinetic energy T in terms of all the coordinates q^{σ} and all the velocities $\dot{q}^{\sigma}$, assuming formally that there are no constraints. Once the left-hand sides of Eq. (3.43) are calculated, we put in them $q^{l+\varkappa} = \dot{q}^{l+\varkappa} = \ddot{q}^{l+\varkappa} = 0$, $\varkappa = \overline{1, k}$.

Equation (3.30) can be looked upon as a solution to the system of linear equations (3.29) with respect to the unknowns Λ_{σ}^*, while Eq. (3.29) can be regarded as the solution to the system of Eq. (3.30) with respect to the unknowns R_{μ}'. Taking into account this relation between these systems, we shall call them *reciprocal*. This being so, the Lagrange equations of the first and second kinds, which follows, respectively, from Eqs. (3.29) and (3.30) with $\Lambda_{\lambda}^*=0$, $\lambda = \overline{1, l}$, are *reciprocal systems*.

A motion for which the constraint equations are satisfied is possible not only with $\Lambda_{\lambda}^*=0$, $\lambda = \overline{1, l}$, but also with $\Lambda_{\lambda}^* \neq 0$. However, if the generalized forces Λ_{λ}^* are not specified, then the problem becomes indeterminate. If they are given in some or other way, they should be subsumed as the forces Q_{λ}.

Comparing this derivation of the Lagrange equations of the first and second kind with that depending on the introduction of a representative point, we see that the constraints are *ideal* if they do not cause additional generalized forces corresponding to the free coordinates. This unfurls from the new relevant side the physical content of the concept of holonomic constraints, which are in general not rheonomic.

Motion equations of holonomic systems subject to forces with potential. The canonical equations. Assume that forces X_μ have a potential U, that is,

$$X_\mu = \frac{\partial U}{\partial x_\mu} = -\frac{\partial \Pi}{\partial x_\mu}.$$

Besides,

$$Q_\lambda = X_\mu \frac{\partial x_\mu}{\partial q^\lambda} = -\frac{\partial \Pi}{\partial x_\mu}\frac{\partial x_\mu}{\partial q^\lambda} = -\frac{\partial \Pi}{\partial q^\lambda}, \quad \lambda = \overline{1, l}. \tag{3.44}$$

Let us introduce the function $L = T - \Pi$, which is known as the *Lagrange function* or the *kinetic potential of a system*. The potential energy Π depends only on the coordinates and time, and hence,

$$\frac{\partial L}{\partial \dot{q}^\lambda} = \frac{\partial T}{\partial \dot{q}^\lambda}. \tag{3.45}$$

Taking into account relations (3.44), (3.45), the Lagrange equations of the second kind (3.42) can be put in the form

$$\frac{d}{dt}\frac{\partial L}{\partial \dot{q}^\lambda} - \frac{\partial L}{\partial q^\lambda} = 0, \quad \lambda = \overline{1, l}, \quad l = 3n - k. \tag{3.46}$$

In the practical application of such equations there is generally no need to write down explicit expressions for the constraint equations. Instead, one should only determine the number of degrees of freedom and then find the parameters q^λ in order that the Lagrange function L in these coordinates would be of the most simple form. First of all, one needs to check whether one may find the parameters q^λ so that some of them would not enter the function L, because in this way the Lagrange equation in the coordinate q^τ (which is not explicitly involved in the function L) has the *cyclic integral* $\partial L/\partial \dot{q}^\tau = $ const. The corresponding curvilinear coordinate q^τ is called the *cyclic coordinate*.

According to the results of Chap. 4, system (6.4) of Chap. 4 can be reduced to system (6.10) of Chap. 4. Similarly, system (3.46) is equivalent to the following system of *canonical equations* or the system of *Hamilton equations*

$$\dot{p}_\lambda = -\frac{\partial H}{\partial q^\lambda}, \quad \dot{q}^\lambda = \frac{\partial H}{\partial p_\lambda}, \quad \lambda = \overline{1, l},$$

where

$$H = p_\lambda \dot{q}^\lambda - L$$

the *Hamilton function*.

To conclude this subsection, we make the following **important observation**. The Lagrange equations of the second kind (3.46) were obtained for a mechanical system consisting of a finite number of material points. Taking into account that the Lagrange function L can be introduced for any mechanical system consisting both of absolutely rigid and deformable bodies for which the positions of all points are uniquely determined by defining the unknown Lagrange coordinates q^λ, $\lambda = \overline{1, l}$, it is natural to assume that the dynamics of the mechanical system is also in this case described by Eq. (3.46). This extension of these equations should be looked upon as an additional postulate analogous to other postulates in physics.

4   Examples of Applications of the Lagrange Equations of the Second Kind

Example 2 Two equal weights m joined by a spring of rigidity c move without friction along a fixed ring of radius r lying in the horizontal plane. The spring length in the undeformed state is l. Find the motion of the system if at the initial time, when the spring is undeformed, the velocity of the first weight is $v_{1,0}$, the velocity of the second weight is $v_{2,0}$. The numbering of the weights corresponds to the positive direction of the angular coordinate (Fig. 7).

Let us assume that the weights are small in comparison with the ring radius. Here, the velocities of all points of each of the weights can be assumed to be approximately equal (of values, respectively, v_1 and v_2). We assume also that the spring mass can be neglected. Under this statement of the problem, the kinetic energy of the system admits the simple expression

$$T = \frac{mv_1^2}{2} + \frac{mv_2^2}{2}.$$

The potential energy of the system, which equals the potential energy of the spring deformation, is calculated by the formula $\Pi = c\Delta^2/2$, where Δ is the extension of the spring.

We assume that the losses of energy by friction can be neglected, and hence, the constraints can be assumed to be ideal. The motion equations in this case read as (3.46).

The system has two degrees of freedom. Its position is uniquely determined by specifying the angles φ_1 and φ_2, which correspond to the position on the circle of the first and second weights. The potential energy depends on one variable parameter Δ, which is expressed in terms of the difference $\varphi_2 - \varphi_1$. Hence, in order to find the cyclic integral, we change to the new variables

Fig. 7 Two weights with a
spring on a ring

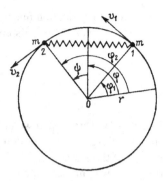

$$\varphi = \frac{\varphi_1 + \varphi_2}{2}, \quad \psi = \frac{\varphi_2 - \varphi_1}{2},$$

or

$$\varphi_1 = \varphi - \psi, \quad \varphi_2 = \varphi + \psi.$$

The angle φ describes the position of the centre of mass, the angle ψ is related with the distance between the weights $\rho = 2r \sin \psi$. Hence,

$$\Delta = \rho - l = 2r (\sin \psi - \sin \psi_0),$$

where $\psi_0 = \arcsin(l/(2r))$.

We write the Lagrange function L in the new variables as follows:

$$L = \frac{mr^2 (\dot{\varphi} - \dot{\psi})^2}{2} + \frac{mr^2 (\dot{\varphi} + \dot{\psi})^2}{2} - 2cr^2 (\sin \psi - \sin \psi_0)^2.$$

The Lagrange equations of the second kind corresponding to the given function L read as

$$\ddot{\varphi} = 0, \quad \ddot{\psi} + \frac{2c}{m} (\sin \psi - \sin \psi_0) \cos \psi = 0.$$

Integrating the first equation, we find that

$$\varphi = \varphi_0 + \dot{\varphi}_0 t, \quad \text{where} \quad \dot{\varphi}_0 = \frac{v_{1,0} + v_{2,0}}{2r}.$$

In the consideration of the equation with respect to the angle ψ, we shall be concerned only with the case when the difference $\alpha = \psi - \psi_0$ can be assumed to be small (that is, here one may put $\sin \psi - \sin \psi_0 = \alpha \cos \psi_0$). The equations for α now reads as

$$\ddot{\alpha} + \omega^2 \alpha = 0, \quad \omega^2 = \frac{2c}{m} \cos^2 \psi_0 = \frac{c(4r^2 - l^2)}{2mr^2}.$$

Integrating this equation under the initial conditions $\alpha(0) = \alpha_0 = 0$, $\dot{\alpha}(0) = \dot{\alpha}_0 = \dot{\psi}_0 = (v_{2,0} - v_{1,0})/(2r)$, this gives

$$\psi = \psi_0 + \frac{v_{2,0} - v_{1,0}}{2r\omega} \sin \omega t .$$

Example 3 Three equal weights m, which are connected in sequence by two equal springs of rigidity c, slip along the horizontal line Ox without friction. Find the motion of the system if at the initial moment, when the springs are undeformed, the rightmost weight has velocity v_0, while the velocities of the two remaining weights are zero.

The system has three degrees of freedom. It position is controlled by the coordinates x_1, x_2, x_3 of the centres of mass of, respectively, the first, second, and third weights.

$$x_{2,0} = x_{1,0} + l, \quad x_{3,0} = x_{2,0} + l, \quad \dot{x}_{1,0} = \dot{x}_{2,0} = 0, \quad \dot{x}_{3,0} = v_0. \tag{4.1}$$

Here, l are the distances between the centres of mass when the springs are undeformed (Fig. 8).

The kinetic and potential energies of the system in these coordinates are as follows:

$$T = \frac{m\dot{x}_1^2}{2} + \frac{m\dot{x}_2^2}{2} + \frac{m\dot{x}_3^2}{2}, \quad \Pi = \frac{c(x_2 - x_1 - l)^2}{2} + \frac{c(x_3 - x_2 - l)^2}{2}.$$

Considering the reduced expression for the potential energy, we see that in this case it is appropriate to use the formulas

$$x = x_2, \quad \delta_1 = x_2 - x_1 - l, \quad \delta_2 = x_3 - x_2 - l, \tag{4.2}$$

to change to the new coordinates. As a result, we have $x_1 = x - \delta_1 - l$, $x_2 = x$, $x_3 = x + \delta_2 + l$. In coordinates (4.2) the Lagrange function L reads as

$$L = \frac{m\dot{x}^2}{2} + \frac{m(\dot{x} - \dot{\delta}_1)^2}{2} + \frac{m(\dot{x} + \dot{\delta}_2)^2}{2} - \frac{c\delta_1^2}{2} - \frac{c\delta_2^2}{2}.$$

Let us fist consider the Lagrange equation in the x-coordinate, on which the function L does not depend. We have

$$3\ddot{x} - \ddot{\delta}_1 + \ddot{\delta}_2 = 0. \tag{4.3}$$

Fig. 8 Three weights with strings

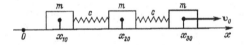

Integrating this equation under the initial conditions (4.1), this establishes

$$3x - \delta_1 + \delta_2 = 3x_{2,0} + v_0 t.$$

Let us now write down the equations with respect to the coordinates δ_1 and δ_2:

$$m(\ddot{\delta}_1 - \ddot{x}) + c\,\delta_1 = 0, \quad m(\ddot{\delta}_2 + \ddot{x}) + c\,\delta_2 = 0.$$

Summing these equations and then subtracting the second one from the first one, it follows by (4.3) that

$$\ddot{\xi} + \omega_1^2 \xi = 0, \quad \ddot{\eta} + \omega_2^2 \eta = 0,$$

where $\xi = \delta_1 + \delta_2$, $\eta = \delta_1 - \delta_2$, $\omega_1^2 = c/m$, $\omega_2^2 = 3c/m$.

Integrating the above equations and taking into account that in view of formulas (4.1), (4.2) the initial conditions for the coordinates ξ and η read as $\xi_0 = 0$, $\dot{\xi}_0 = v_0$, $\eta_0 = 0$, $\dot{\eta}_0 = -v_0$, we find that

$$\xi = \frac{v_0}{\omega_1} \sin \omega_1 t, \quad \eta = -\frac{v_0}{\omega_2} \sin \omega_2 t.$$

In the initial coordinates x_1, x_2 and x_3, assuming for simplicity that $x_{2,0} = 0$, we finally obtain

$$x_1 = \frac{v_0 t}{3} - \frac{v_0}{2\omega_1} \sin \omega_1 t + \frac{v_0}{6\omega_2} \sin \omega_2 t - l,$$

$$x_2 = \frac{v_0 t}{3} - \frac{v_0}{3\omega_2} \sin \omega_2 t,$$

$$x_3 = \frac{v_0 t}{3} + \frac{v_0}{2\omega_1} \sin \omega_1 t + \frac{v_0}{6\omega_2} \sin \omega_2 t + l.$$

Example 4 Three weights are attached at an inextensible string of length $4a$. The first and third weights have the mass m, the mass of the second weight is M (Fig. 9). The string is supported at its extremities so that its initial and finial segments are at the equilibrium position form an angle α_0 with the vertical line, while the middle segment forms an angle β_0. Write the Lagrange equation in the case when the weight M moves along the line. Integrate the resulting equation assuming that the departures of the system from the equilibrium position are small.

This system, which consists of three material points, has only one degree of freedom under the above conditions. In our case we may introduce a representative point, write down all the constraint equations in the form $f^{\varkappa}(t, y)$, $\varkappa = \overline{1, 5}$, and use Eq. (3.6) for the study. In solving this problem in practice it is more appropriate to use the Lagrange equation of the second kind. In this case, there is not need to write all the constraint equations and consider their reactions. In order to compose the Lagrange equation, it suffices to find the kinetic and potential energies due to the gravity force.

Fig. 9 Three weights on a string

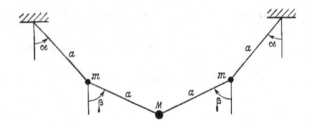

As the generalized coordinate we take the displacement q of the weight M, as references from the equilibrium position in the vertical direction downwards. Assume that in the displaced positions the string has the angles α and β, respectively, with the vertical position. From Fig. 9 it is seen that these angles *qua* functions of the parameter q are given by the relations

$$a(\cos \alpha + \cos \beta) - a(\cos \alpha_0 + \cos \beta_0) = q\,,$$
$$\sin \alpha + \sin \beta - \sin \alpha_0 - \sin \beta_0 = 0\,. \tag{4.4}$$

The kinetic energy of the system is as follows:

$$T = M\dot{q}^2/2 + ma^2\,\dot{\alpha}^2\,.$$

In order to express $\dot{\alpha}$ in terms of $\dot{q}$, we differentiate relations (4.4) in time. As a result, we have

$$-a(\dot{\alpha}\sin\alpha + \dot{\beta}\sin\beta) = \dot{q}\,,\quad \dot{\alpha}\cos\alpha + \dot{\beta}\cos\beta = 0\,. \tag{4.5}$$

Considering expressions (4.5) as a system of linear algebraic equations with respect to the unknowns $\dot{\alpha}$ and $\dot{q}$ and solving it, we find that

$$\dot{\alpha} = \frac{\dot{q}\cos\beta}{a\sin(\beta - \alpha)}\,,\quad \dot{\beta} = -\frac{\dot{q}\cos\alpha}{a\sin(\beta - \alpha)}\,. \tag{4.6}$$

It follows that the kinetic energy of the system can be represented in the form

$$T_* = \frac{M_*\dot{q}^2}{2}\,,\quad M_* = M + \frac{2m\cos^2\beta}{\sin^2(\beta - \alpha)}\,.$$

Here, M_* is the reduced mass.

Now let us find the generalized force Q corresponding to the coordinate q. In this case, the gravity forces of weights are the active forces, the strings tensions are the constraint reactions. By the assumption, the strings are inextensible, and hence the work of the constraint reactions is zero. Hence, the constraints are ideal. The potential energy of the gravity force is equal to the weight multiplied by the displacement of the weight in the vertical direction. Hence, the potential energy of the system is as follows:

$$\Pi = -Mgq - 2mga(\cos\alpha - \cos\alpha_0).$$

Differentiating this expression in q, we find that

$$Q = -\frac{d\Pi}{dq} = Mg - 2mga\sin\alpha\,\frac{d\alpha}{dq}.$$

From (4.6) it follows that

$$\frac{d\alpha}{dq} = \frac{\cos\beta}{a\sin(\beta - \alpha)},$$

and hence,

$$Q = Mg - \frac{2mg\sin\alpha\cos\beta}{\sin(\beta - \alpha)}. \tag{4.7}$$

At the equilibrium position $Q = 0$, and hence,

$$2m = \frac{M\sin(\beta_0 - \alpha_0)}{\sin\alpha_0\cos\beta_0}. \tag{4.8}$$

Expressions (4.4) show that the trigonometric functions of the angles α and β, which enter the expressions for T and Q, depends in a fairly involved fashion on the parameter q. Hence, in the general case the Lagrange equation with respect to the coordinate q is bulky and nonlinear. For sufficiently small q this problem can be linearized. For small q, the reduced mass $M_*(q)$ can be assumed to be constant and equal to $M_*(0)$. Taking into account relation (4.8), we find that

$$M_*(q) \approx M_*(0) = 2m\left(\frac{M}{2m} + \frac{\cos^2\beta_0}{\sin^2(\beta_0 - \alpha_0)}\right) =$$

$$= \frac{2m\cos\beta_0}{\sin^2(\beta_0 - \alpha_0)}\left(\sin\alpha_0\sin(\beta_0 - \alpha_0) + \cos\beta_0\right) =$$

$$= \frac{2m\cos\beta_0}{\sin^2(\beta_0 - \alpha_0)}\left(-\frac{1}{2}\cos\beta_0 + \frac{1}{2}\cos(2\alpha_0 - \beta_0) + \cos\beta_0\right) =$$

$$= \frac{2m\cos\beta_0\cos\alpha_0\cos(\beta_0 - \alpha_0)}{\sin^2(\beta_0 - \alpha_0)}.$$

So,

$$M_*(0) = \frac{2m\cos\beta_0\cos\alpha_0\cos(\beta_0 - \alpha_0)}{\sin^2(\beta_0 - \alpha_0)}.$$

Expanding the generalized force Q in a series in powers of q, we have, to the first order of small quantities,

$$Q(q) \approx Q'_q(0)\,q.$$

The generalized force Q, as given in the form (4.7), is a composite function of q. Its derivative in q can be found by the formula

$$Q'_q = \frac{\partial Q}{\partial \alpha}\frac{d\alpha}{dq} + \frac{\partial Q}{\partial \beta}\frac{d\beta}{dq}.$$ (4.9)

Using (4.6), this gives

$$\frac{d\alpha}{dq} = \frac{\cos\beta}{a\sin(\beta-\alpha)}, \quad \frac{d\beta}{dq} = -\frac{\cos\alpha}{a\sin(\beta-\alpha)}.$$

Differentiating (4.7) in α, we find that

$$\frac{\partial Q}{\partial \alpha} = -2\,mg\left(\frac{\cos\alpha\cos\beta}{\sin(\beta-\alpha)} + \frac{\sin\alpha\cos\beta\cos(\beta-\alpha)}{\sin^2(\beta-\alpha)}\right) =$$

$$= -\frac{2\,mg\cos\beta(\cos\alpha\sin(\beta-\alpha) + \sin\alpha\cos(\beta-\alpha))}{\sin^2(\beta-\alpha)} = -\frac{2\,mg\cos\beta\sin\beta}{\sin^2(\beta-\alpha)}.$$

So, we have

$$\frac{\partial Q}{\partial \alpha} = -\frac{2\,mg\cos\beta\sin\beta}{\sin^2(\beta-\alpha)}.$$

A similar analysis shows that

$$\frac{\partial Q}{\partial \beta} = \frac{2\,mg\sin\alpha\cos\alpha}{\sin^2(\beta-\alpha)}.$$

Substituting these expressions into (4.9), we find that

$$-Q'_q(0) = C_*(0) = \frac{2\,mg(\cos^2\beta_0\sin\beta_0 + \cos^2\alpha_0\sin\alpha_0)}{a\sin^3(\beta_0-\alpha_0)}.$$

The Lagrange equation of the second kind, which corresponds to these approximate expressions for T and Q, reads as

$$M_*(0)\,\ddot{q} = -C_*(0)\,q,$$ (4.10)

or

$$\ddot{q} + \omega^2 q = 0, \quad \omega^2 = \frac{C_*(0)}{M_*(0)} = \frac{g(\cos^2\beta_0\sin\beta_0 + \cos^2\alpha_0\sin\alpha_0)}{a\cos\beta_0\cos\alpha_0\sin(\beta_0-\alpha_0)\cos(\beta_0-\alpha_0)}.$$ (4.11)

Under the initial conditions

$$q(0) = q_0, \quad \dot{q}(0) = \dot{q}_0$$ (4.12)

the solution to Eq. (4.11) reads as

$$q(t) = q_0 \cos \omega t + \frac{\dot{q}_0}{\omega} \sin \omega t .$$

In the actual fact, the oscillations for this system are damping. The energy dissipates due to the air resistance and due to the internal friction in the strings during their deformation. Hence it is difficult to consider in practice the second factor pointing out that the constraints can be regarded ideal only approximately, we shall consider only the first factor.

Let us assume that the force of air resistance during the motion of the weights M and m is proportional to the first power of the velocity, respectively, with the coefficients k_1 and k_2. We first consider the weight M. As the weights M deflect by dq, the resistance force $-k_1\dot{q}$ attached to the weight performs the work[3]

$$\delta A_1 = -k_1 \dot{q} \, dq .$$

In a similar way, the elementary work of the resistance forces attached to the weight m is written as

$$\delta A_2 = -2k_2 a^2 \dot{\alpha} \, d\alpha . \tag{4.13}$$

The quantities $d\alpha$ and dq are related between each other in the same way as $\dot{\alpha}$ is related with $\dot{q}$. Hence, in view of (4.6) expression (4.13) can be represented in the form

$$\delta A_2 = -\frac{2k_2 \cos^2 \beta}{\sin^2(\beta - \alpha)} \dot{q} \, dq .$$

Hence, the generalized force corresponding to these air resistance forces reads as $Q_* = -k(q)\dot{q}$, where

$$k(q) = k_1 + \frac{2k_2 \cos^2 \beta}{\sin^2(\beta - \alpha)} .$$

For small oscillations this coefficient can be assumed to be constant

$$k(q) \approx k(0) = k_1 + \frac{2k_2 \cos^2 \beta_0}{\sin^2(\beta_0 - \alpha_0)} .$$

Augmenting the restoring force $(-C_*(0) q)$ in the Lagrange equation (4.10) the resistance force $Q_* = -k(0)\dot{q}$ thus obtained, we find that

$$\ddot{q} + 2n\dot{q} + \omega^2 q = 0 , \quad 2n = k(0)/M_*(0) .$$

The solution to this equation under the initial conditions (4.12) reads as

---

[3] The method of finding the generalized force by using the expression for the elementary work on a virtual displacement is given in Sect. 1 of Chap. 9.

Fig. 10 Two weights
connected by a string

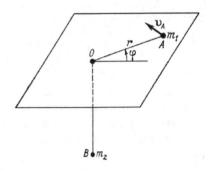

$$q(t) = e^{-nt} \left(q_0 \cot \omega_1 t + \frac{\dot{q}_0 + nq_0}{\omega_1} \sin \omega_1 t \right), \quad \omega_1 = \sqrt{\omega^2 - n^2}.$$

Example 5 Two point weights m_1 and m_2 are attached at the extremities A and B of a string which is passed through a hole O in a smooth horizontal table surface (Fig. 10). The first weight remains on the table surface, while the second one moves in a vertical line passing through the point O. At initial time $OA = r_0$, the velocity of the weight m_2 is zero, the velocity v_0 of the weight m_1 is directed orthogonally to the initial position of the segment OA of the string. Find the motion of the system if the string is considered weightless, inextensible and absolutely flexible.

The system has two degrees of freedom. As the generalized Lagrange coordinates, one may take the polar coordinates r and φ of the point A. Now the kinetic and potential energies of the system read as

$$T = \frac{(m_1 + m_2)\dot{r}^2}{2} + \frac{m_1 r^2 \dot{\varphi}^2}{2}, \quad \Pi = m_2 g (r - r_0).$$

The system of Lagrange equations (3.46) in this case assumes the form

$$\frac{d}{dr} (m_1 r^2 \dot{\varphi}) = 0, \quad (m_1 + m_2)\ddot{r} - m_1 r \dot{\varphi}^2 = -m_2 g.$$

From the first equation it follows that there exists the integral

$$r^2 \dot{\varphi} = C = \text{const}. \tag{4.14}$$

At the initial time $r = r_0$ and $r_0 \dot{\varphi}_0 = v_0$, and hence $C = r_0 v_0$. We note that the so-obtained integral of the motion equations (4.14) is cylindrical, because the Lagrange function

$$L = T - \Pi = \frac{(m_1 + m_2)\dot{r}^2}{2} + \frac{m_1 r^2 \dot{\varphi}^2}{2} - m_2 g (r - r_0)$$

does not contain the generalized coordinate φ.

In view of the integral (4.14) and the relation

$$\ddot{r} = \frac{d\dot{r}}{dt} = \frac{d\dot{r}}{dr}\frac{dr}{dt} = \frac{d(\dot{r}^2/2)}{dr}, \tag{4.15}$$

the second Lagrange equation can be reduced to the equation with separated variables

$$(m_1 + m_2)\, d\left(\frac{\dot{r}^2}{2}\right) = \left(m_1 \frac{r_0^2 v_0^2}{r^3} - m_2 g\right) dr. \tag{4.16}$$

Integrating Eq. (4.16) and taking into account that $\dot{r}(0) = \dot{r}_0 = 0$ *ab initio*, we have

$$\frac{(m_1 + m_2)\,\dot{r}^2}{2} = m_2 g(r_0 - r) + \frac{m_1 v_0^2}{2r^2}\,(r^2 - r_0^2).$$

Factoring the right-hand side of this expression, this gives

$$\dot{r}^2 = \frac{2m_2 g}{r^2(m_1 + m_2)}\,(r_0 - r)(r - r_1)(r - r_2), \tag{4.17}$$

where

$$r_1 = \frac{\lambda + \sqrt{\lambda^2 + 8\lambda}}{4}\, r_0, \quad r_2 = \frac{\lambda - \sqrt{\lambda^2 + 8\lambda}}{4}\, r_0 < 0, \quad \lambda = \frac{m_1 v_0^2}{r_0 m_2 g}.$$

From (4.17) it follows that the product $(r_0 - r)(r - r_1)(r - r_2)$ with $r \neq r_k$, $k = 0, 1, 2$, must be positive. However $r - r_2 > 0$, and hence $r_0 - r$ and $r - r_1$ must be of the same sign. It follows that the radius r lies between r_0 and r_1. Besides, using the inequalities

$$\lambda + \sqrt{\lambda^2 + 8\lambda} > 4, \quad \lambda > 1,$$
$$\lambda + \sqrt{\lambda^2 + 8\lambda} < 4, \quad \lambda < 1,$$

we have

$$\begin{aligned} r_0 \leqslant r \leqslant r_1, \quad \lambda > 1, \\ r_1 \leqslant r \leqslant r_0, \quad \lambda < 1. \end{aligned} \tag{4.18}$$

For $\lambda = 1$, when $r_1 = r = r_0$, the point A moves in a circle, while the point B does not change its position. In this case, the centrifugal force $m_1 v_0^2/r_0$, when applied to the mass m_1, is equal to the gravity force $m_2 g$ of the second mass. If these forces are not equal, then the acceleration of the point B at the initial time is different from zero. Using (4.16), (4.15), we find that

$$\ddot{r}_0 = \frac{m_2 g}{m_1 + m_2}\,(\lambda - 1).$$

Hence, from inequalities (4.18) it follows that the point B for $t > 0$ moves up ($\lambda > 1$) or down ($\lambda < 1$) until it stops at a position where $r = r_1$. Next the backward motion occurs with a stop at $r = r_0$ and the motion pattern repeats. Hence, the function $r(t)$ is periodic. Let T_* be its period. Since with the same r the values $\dot{r}$ have the same magnitude for the down and up motion, but are oppositely directed, it follows that in the interval $0 \leqslant t \leqslant T_*$ the function $r(t)$ is symmetric with respect to the line $t = T_*/2$. Thus, the periodic function $r(t)$ is determined from its values from the interval $0 \leqslant t \leqslant T_*/2$. From Eq. (4.17) it follows that the relation between r and t in this interval is given by the definite integral

$$t = \pm \sqrt{\frac{m_1 + m_2}{2m_2 g}} \int_{r_0}^{r} \frac{r\,dr}{\sqrt{(r_0 - r)(r - r_1)(r - r_2)}}.$$

The plus sign corresponds to the upward motion ($r > r_0$), and the minus sign, to the downward motion ($r < r_0$). From integral tables[4] we see that the function $t(r)$ for $r < r_0$ can be written in the form

$$t(r) = \frac{2r_2}{\sqrt{r_0 - r_2}} F(x, p) + \frac{2r_0}{r_1} \sqrt{r_0 - r_2}\, E(x, p),$$

where $F(x, p)$ and $E(x, p)$ are the elliptic integrals of, respectively, first and second kind

$$F(x, p) = \int_0^x \frac{d\alpha}{\sqrt{1 - p^2 \sin^2 \alpha}}, \quad E(x, p) = \int_0^x \sqrt{1 - p^2 \sin^2 \alpha}\, d\alpha.$$

Here,

$$x = \arcsin \sqrt{\frac{r_0 - r}{r_0 - r_1}}, \quad p = \sqrt{\frac{r_0 - r_1}{r_0 - r_2}} = \sqrt{\frac{4 - \lambda - \sqrt{\lambda^2 + 8\lambda}}{4 - \lambda + \sqrt{\lambda^2 + 8\lambda}}}.$$

In this problem it can be assumed without loss of generality that the time is referenced to the instant from which the point B begins to move downwards, and hence one may assert that the function $t(r)$, as expressed through elliptic integrals, completely describe the process of oscillations of the point B. In particular, the period of its oscillations is as follows

$$T_* = 2t(r_1) = \frac{2r_2}{\sqrt{r_0 - r_2}} F\left(\frac{\pi}{2}, p\right) + \frac{2r_0}{r_1} E\left(\frac{\pi}{2}, p\right) = \frac{2r_2}{\sqrt{r_0 - r_2}} K(p) + \frac{2r_0}{r_1} E(p),$$

---

[4] See *I.S. Gradshteyn, I.M. Ryzhik*, Tables of integrals, series, and products, Academic Press, Boston, MA, 1994; formula (3.1325).

where $K(p)$ and $E(p)$ are the complete elliptic integrals.

Example 6 A homogeneous cylinder of elliptic cross-section with semi-axes a and b rolls on a horizontal plane support without slip. Neglecting the force of rolling friction, find the motion of the cylinder, if the velocity of its centre of mass of the cylinder at the lowest position is v_0. It is assumed that the velocity is such that the cylinder moves without separation from the plane.

The motion of a cylinder is plane parallel, and hence, kinematically this problem can be reduced to the roll on the line MN (Fig. 11). We consider a moving system of coordinates Cxy in which the ellipse equation has the canonical form

$$\frac{x^2}{a^2} + \frac{y^2}{b^2} = 1.$$ (4.19)

The line MN, on which the ellipse rolls, will be considered as a tangent line to it, which passes through the point P with coordinates (x, y). The point P is the instantaneous centre of velocities, and hence its velocity is zero. We let φ denote the angle between the line MN and the Cx-axis. If the roll is without slips, then the angle φ completely defines the cylinder position, and hence it can be taken to be a generalized Lagrange coordinate.

By König's theorem, the kinetic energy of the cylinder is as follows

$$T = \frac{Mv_c^2}{2} + \frac{J_c\dot\varphi^2}{2},$$

where M is the cylinder mass, v_c is the velocity of its centre of mass C, $J_c = M(a^2 + b^2)/4$ is the cylinder moment of inertia with respect to its axis.

As was already pointed out, the point P of contact with coordinates (x, y) is an instantaneous centre of velocities, and hence,

$$v_c^2 = r^2\dot\varphi^2, \quad r^2 = x^2 + y^2.$$

Fig. 11 Cylinder rolling on a plane support

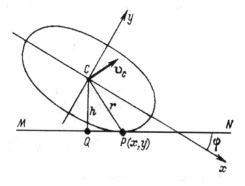

Let us express r^2 in terms of φ. We have

$$\tan \varphi = \frac{dy}{dx}, \quad \text{and besides,} \quad \frac{xdx}{a^2} + \frac{ydy}{b^2} = 0.$$

It follows that

$$x^2 b^4 - y^2 a^4 \tan^2 \varphi = 0. \tag{4.20}$$

Considering equality (4.20) and the ellipse Eq. (4.19) as a system of linear algebraic equations with respect to the unknowns x^2 and y^2 and solving it, we find that

$$r^2 = r^2(\sin^2 \varphi) = \frac{(a^4 - b^4)\sin^2 \varphi + b^4}{(a^2 - b^2)\sin^2 \varphi + b^2}.$$

Thus, the kinetic energy of the system can be written in the form

$$T = Mg_{11}(\sin^2 \varphi)\dot\varphi/2, \quad \text{where} \quad g_{11}(\sin^2 \varphi) = r^2(\sin^2 \varphi) + (a^2 + b^2)/4.$$

The potential energy reads as $\Pi = Mgh$, where h is the distance of the centre of mass from the plane on which the elliptic cylinder rolls. It is known, that the distance of the origin from the line given by the equation

$$Ax' + By' + C = 0,$$

where $C < 0$, is given by the formula

$$h = -\frac{C}{\sqrt{A^2 + B^2}}. \tag{4.21}$$

Since the equation of the tangent line to the ellipse at a point $P(x, y)$ reads as

$$\frac{xx'}{a^2} + \frac{yy'}{b^2} - 1 = 0,$$

it follows from formula (4.21) that the rise of the cylinder centre of mass is

$$h = 1 \Big/ \sqrt{\frac{x^2}{a^4} + \frac{y^2}{b^4}}.$$

Substituting into this formula the values x^2 and y^2, as obtained by solving the system of Eqs. (4.19), (4.20), we find that

$$h = h(\sin^2 \varphi) = \sqrt{(a^2 - b^2)\sin^2 \varphi + b^2}.$$

In this case, the Lagrange equation has the form of the energy integral

$$T + \Pi = T_0 + \Pi_0 \,, \tag{4.22}$$

where T_0 and Π_0 are, respectively, the kinetic and potential energy of the cylinder at $t_0 = 0$, when the cylinder centre of mass is in the lowest position, its velocity is v_0. Besides, $\varphi(0) = 0$, $r(0) = h(0) = b$, $\dot{\varphi}(0) = v_0/b > 0$, because one may always assume that the velocity v_0 has the same direction as the Cx-axis. From the above initial conditions we find that

$$T_0 = M(5\,b^2 + a^2)\,v_0^2/(8b^2)\,, \quad \Pi_0 = Mgb\,.$$

Taking into account these expressions, we write the energy integral (4.22) in the form

$$\dot{\varphi}^2 = \frac{1}{f^2(\sin^2 \varphi)} = \frac{(5b^2 + a^2)\,v_0^2 - 8gb^2(h(\sin^2 \varphi) - b)}{4b^2 g_{11}(\sin^2 \varphi)}\,. \tag{4.23}$$

Considering this equality as a differential equation with respect to the function $\varphi(t)$ and integrating it under the condition $\dot{\varphi} > 0$, this gives

$$t = \int_0^\varphi f(\sin^2 \alpha)\,d\alpha\,, \quad \frac{dt}{d\varphi} = f(\sin^2 \varphi) > 0\,. \tag{4.24}$$

From the energy integral (4.23), it follows that the angular velocity $\dot{\varphi}$ for $T_0 > Mg(a - b)$ does not vanish during the motion, and hence, the sign of $\dot{\varphi}$ for $t > 0$ agrees with that of $\dot{\varphi}$ for $t = 0$. From the statement of the problem $\dot{\varphi}(0) > 0$, and hence, equality (4.24) in the case under consideration holds for any $\varphi > 0$.

Taking into account that

$$f(\sin^2 \alpha) = f(\sin^2(n\pi - \alpha)) = f(\sin^2(n\pi + \alpha))$$

for any integer n, we may put the integral (4.24) in the form

$$t = \int_0^\varphi f(\sin^2 \alpha)\,d\alpha = \int_0^{n\pi} f(\sin^2 \alpha)\,d\alpha + \int_{n\pi}^{n\pi\pm\beta} f(\sin^2 \alpha)\,d\alpha = 2nt_* \pm J(\beta)\,, \tag{4.25}$$

where $0 \leqslant \beta \leqslant \pi/2$, $n\pi = \varphi \mp \beta$, $J(\beta) = \int_0^\beta f(\sin^2 \alpha)\,d\alpha$, $t_* = J(\pi/2)$ is the time during which the cylinder centre of mass changes from the lowest to the uppermost position. Relation (4.25) implies that in order to find the motion law (that is, to define the function $\varphi(t)$ for any $t > 0$), it suffices to define it for $0 \leqslant t \leqslant t_*$.

Let us consider the case $T_0 < Mg(a - b)$. Here, as follows from energy integral (4.23), the angular velocity $\dot{\varphi}$ vanishes if φ satisfies the equation

$$(5\,b^2 + a^2)\,v_0^2 - -8gb^2\left(\sqrt{(a^2 - b^2)\sin^2 \varphi + b^2} - b\right) = 0\,. \tag{4.26}$$

The motion in this case is periodic, the period T_* can be found from the formula

$$T_* = 4 \int\limits_0^{\varphi_{\max}} f(\sin^2 \alpha)\, d\alpha,$$

where $\varphi_{\max}$ is the smallest positive root of Eq. (4.26). The relation between t and φ in the first quarter of the period is given by integral (4.24). The function $\varphi(t)$ thus constructed for $0 \leqslant t \leqslant T_*/4$ can be extended by taking into account that in this case, as for a mathematical pendulum, the function φ in the interval $0 \leqslant t \leqslant T_*/2$ is symmetric with respect to the line $t = T_*/4$, and besides,

$$\varphi(t + T_*/2) = -\varphi(t).$$

For small oscillations of this system, the relation between t and φ, as given by integral (4.24), can be simplified, but this is a fairly involved task. It is much better in this case to employ the Lagrange equation composed from approximate expressions for the kinetic energy and the generalized force. Assuming that in the formula for kinetic energy the metric coefficient $g_{11}(\sin^2 \varphi)$ is the constant $g_{11}(0)$, we have

$$T = M(5b^2 + a^2)\dot{\varphi}^2/8.$$

The generalized force

$$Q_\varphi = -\frac{d\Pi}{d\varphi} = -\frac{Mg(a^2 - b^2)\sin 2\varphi}{2\sqrt{(a^2 - b^2)\sin^2 \varphi + b^2}}$$

for small angles φ can be put in the form $Q_\varphi = -Mg(a^2 - b^2)\varphi/b$. The Lagrange equation reads as

$$\ddot{\varphi} + \omega^2 \varphi = 0, \quad \omega^2 = \frac{4g(a^2 - b^2)}{(5b^2 + a^2)b}.$$

Taking into that for $t = 0$ the angle $\varphi = 0$ and $\dot{\varphi}(0) = v_0/b$, we have

$$\varphi(t) = \frac{v_0}{b\omega} \sin \omega t.$$

Example 7 Two cylindrical shafts of mass P_1 and P_2 are rolled down along two planes inclined at angles α and β, respectively, to the horizon (Fig. 12). The shafts are connected by an inextensible string whose end-points are fixed on the shafts. Find the string tension and its acceleration during the motion on the inclined planes. The shafts are assumed to be homogeneous round cylinders. The weight of the string can be neglected.

Fig. 12 Two cylinders on inclined slopes

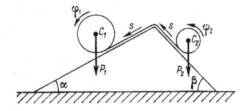

The system has three degrees of freedom. As the generalized Lagrange coordinates one may take the string displacement s and the roll angles φ_1 and φ_2 of the cylinders of weights P_1 and P_2, respectively. The positive direction of the coordinates are indicated by arrows.

By König's theorem, the kinetic energy of this system can be written in the form

$$T = \frac{P_1 v_1^2}{2g} + \frac{J_1 \dot{\varphi}_1^2}{2} + \frac{P_2 v_2^2}{2g} + \frac{J_2 \dot{\varphi}_2^2}{2}.$$

Here, v_1, v_2 are the velocities of the centres of mass of the cylinders, J_1, J_2 are their moments of inertia

$$J_k = P_k r_k^2 / (2 g), \quad k = 1, 2,$$

r_k is the cylinder radius. The velocity v_1 is composed from the string velocity $\dot{s}$ and that of the cylinder centre $r_1 \dot{\varphi}_1$ with respect to it. Similarly, for the second cylinder we have $v_2 = \dot{s} + r_2 \dot{\varphi}_2$. Hence,

$$T = \frac{P_1(\dot{s} + r_1 \dot{\varphi}_1)^2}{2g} + \frac{P_1 r_1^2 \dot{\varphi}_1^2}{4g} + \frac{P_2(\dot{s} + r_2 \dot{\varphi}_2)^2}{2g} + \frac{P_2 r_2^2 \dot{\varphi}_2^2}{4g}.$$

Let us now find the potential energy due to gravity of the first and second cylinders.[5] For the first cylinder, as the string moves by s and it rotates by φ_1, the centre of mass moves down by the quantity $(s + r_1 \varphi_1) \sin \alpha$, while for the second one it rises by the quantity $(s + r_2 \varphi_2) \sin \beta$. Hence, the potential energy in this problem is as follows:

$$\Pi = -P_1 (s + r_1 \varphi_1) \sin \alpha + P_2 (s + r_2 \varphi_2) \sin \beta.$$

From the above expressions for the kinetic and potential energy it follows that the Lagrange equations of the second kind for the coordinates s, φ_1 and φ_2 assume the form

$$\frac{P_1}{g} (\ddot{s} + r_1 \ddot{\varphi}_1) + \frac{P_2}{g} (\ddot{s} + r_2 \ddot{\varphi}_2) = P_1 \sin \alpha - P_2 \sin \beta, \qquad (4.27)$$

---

[5] A different more general method of finding the generalized forces in this problem is considered in Example 1 of Chap. 9.

$$\frac{P_1}{g}(\ddot{s} + r_1\ddot{\varphi}_1) + \frac{P_1 r_1}{2g}\ddot{\varphi}_1 = P_1 \sin\alpha, \tag{4.28}$$

$$\frac{P_2}{g}(\ddot{s} + r_2\ddot{\varphi}_2) + \frac{P_2 r_2}{2g}\ddot{\varphi}_2 = -P_2 \sin\beta. \tag{4.29}$$

Subtracting Eqs. (4.28), (4.29) from Eq. (4.27), we find that

$$P_1 r_1 \ddot{\varphi}_1 + P_2 r_2 \ddot{\varphi}_2 = 0.$$

Hence, using Eq. (4.27), we see that sought-for acceleration of the string is as follows:

$$\ddot{s} = \frac{(P_1 \sin\alpha - P_2 \sin\beta)g}{P_1 + P_2}. \tag{4.30}$$

Substituting the value of $\ddot{s}$ into Eq. (4.28), this gives

$$\ddot{\varphi}_1 = \frac{2g}{3r_1}\left(\sin\alpha - \frac{P_1 \sin\alpha - P_2 \sin\beta}{P_1 + P_2}\right). \tag{4.31}$$

In the same way one finds the angular acceleration $\ddot{\varphi}_2$ of the second cylinder.

In order to determine the string tension F we shall use the equation of motion of the centre of mass of the first cylinder. In the projection to the motion direction it reads as

$$\frac{P_1}{g}(\ddot{s} + r_1\ddot{\varphi}_1) = P_1 \sin\alpha - F,$$

hence, using (4.30), (4.31), we find that

$$F = \frac{P_1 P_2(\sin\alpha + \sin\beta)}{3(P_1 + P_2)}.$$

We conclude this chapter by pointing out that the machinery of the Lagrange equations of the second kind is efficient in many different problems in mechanics and is capable of giving answers to principal problems related with the character of motion of the mechanical system under consideration.

5 The Motion Equations of a Nonholonomic System of Material Points in the Generalized Coordinates. The Maggi Equations

Quasi-velocities. Nonholonomic bases. Ideal nonholonomic constraints. The generalized Lagrange coordinates were shown above as being handy both from the practical and theoretical positions. Using them one may describe the force action on a system of material points from the general position and write, following this approach, the second Newton's law in the most perfect analytic form as the Lagrange equations of the second kind, which were first derived for mechanical systems with holonomic constraints. The generalized coordinates will also prove useful in the subsequent analysis of systems of material points with nonholonomic constraints.

Assume that the position of a mechanical system of n material points can be uniquely determined by introducing the generalized coordinates $q^1, q^2, ..., q^{3n}$, which can be looked upon as curvilinear coordinates of the representative point. The number of the generalized coordinates, which is $3n$ in this setting, will be denoted by s.

The motion of the system is supposed to be subject to k constraints, which in the generalized coordinates are expressed by the equations

$$\varphi^{\varkappa}(t, q, \dot{q}) = 0, \quad \varkappa = \overline{1, k}. \tag{5.1}$$

This system of constraints will be assumed to be nonholonomic; that is, we shall suppose that Eq. (5.1) cannot be integrated nether separately nor jointly. In this case, any two positions of a mechanical system can be connected by a piecewise-smooth curve of finite length, the constraint equations being satisfied along this curve This theorem[6] is capable to distinguish the differential second-order constraints, as defined in the general case in the form (5.1), judging from whether decreases or not (if present) the dimension of the set of those positions, at which the system may arrive from a given position. If all the constraints in (5.1) are holonomic, then the dimension of this set decreases by k units in comparison with the case when there are no such constraints. Contrariwise, the constraints (5.1) are nonholonomic if this dimension does not change. We have already mentioned that term "holonomic", which was introduced by H. Herz, comes from the Greek words "$o' \lambda o \varsigma$" ("whole", which means here "integrable") and "$\nu o' \mu o \varsigma$" ("a law"). Use of this term in relations (5.1) in the holonomic setting means that in this case the differential laws (5.1) reduce by k the number of independent parameters, which uniquely specify the position of the mechanical system.

We shall assume that the nonholonomic constraints (5.1) are ideal. Then, according to Sect. 3, the motion equation of the representative point in the vector form reads as

$$M\mathbf{W} = \mathbf{Y} + \Lambda_{\varkappa} \nabla' \varphi^{\varkappa}, \quad \nabla' \varphi^{\varkappa} = (\partial \varphi^{\varkappa} / \partial \dot{q}^{\sigma}) \mathbf{e}^{\sigma}. \tag{5.2}$$

Let us introduce the functions

$$v_*^{\rho} = v_*^{\rho}(t, q, \dot{q}), \quad \rho = \overline{1, s}, \tag{5.3}$$

---

[6] See L.A. Pars, Treatise on Analytical Dynamics. London, Heinemann, 1965, § 1.8; P.K. Rashevskii, "Connection of any two points of totally nonholonomic space by admissible line," Uch. Zap. Ped. Inst. im. K. Libknecht. Ser. Fiz.-Mat. Nauk, 1938. issue 2. P. 83–94 [in Russian]; W.L. Chow, "Systeme von linearen partiellen differentialen Gleichungen erster Ordnung," Math. Ann. 1939. Bd. 117. S. 98–105.

and shall consider the values of v_*^ρ as new variables with respect to the variables $\dot{q}^\sigma$, assuming that t and q are parameters. Supposing that the solvability conditions are satisfied, we get

$$\dot{q}^\sigma = \dot{q}^\sigma(t, q, v_*), \quad v_* = (v_*^1, v_*^2, \dots, v_*^s). \tag{5.4}$$

Thus, together with the generalized velocities $\dot{q} = (\dot{q}^1, \dots, \dot{q}^s)$, we have introduced the variables $v_* = (v_*^1, \dots, v_*^s)$, which can be looked upon as new generalized velocities. However, whereas the generalized velocities $\dot{q}^\sigma$, $\sigma = \overline{1, s}$, were in correspondence with the generalized coordinates q^σ, $\sigma = \overline{1, s}$, now the new variables v_*^ρ, $\rho = \overline{1, s}$, may fail to be in correspondence with some new generalized coordinates. This is why v_*^ρ, $\rho = \overline{1, s}$, are called *quasi-velocities* (*pseudo-velocities*). For example, in the study of the rotation of a rigid body about a fixed point, it is convenient to introduce the Euler angles ψ, θ, φ, which are the generalized coordinates specifying the position of the rigid body. To these generalized coordinates there correspond the generalized velocities $\dot{\psi}$, $\dot{\theta}$, $\dot{\varphi}$. However, on the other hand, to specify the value of the angular velocity of a rigid body ω one may introduce the projections of this vector ω_x, ω_y, ω_z to the moving axes. However, with these quantities one cannot assign the derivatives of some angles of rotation. Hence, the above characteristics of the rotation rate ω_x, ω_y, ω_z are to be called the quasi-velocities or pseudo-velocities, rather than the generalized velocities. The difference here is that in the former case we use the Latin words, and in the latter, the Greek word.

Calculating the partial differential of these functions with fixed t and q^σ, we find that

$$\delta' v_*^\rho = \frac{\partial v_*^\rho(t, q, \dot{q})}{\partial \dot{q}^\sigma} \delta' \dot{q}^\sigma, \quad \delta' \dot{q}^\sigma = \frac{\partial \dot{q}^\sigma(t, q, v_*)}{\partial v_*^\tau} \delta' v_*^\tau,$$

$$\rho, \sigma, \tau = \overline{1, s},$$

and therefore,

$$\delta' v_*^\rho = \frac{\partial v_*^\rho}{\partial \dot{q}^\sigma} \frac{\partial \dot{q}^\sigma}{\partial v_*^\tau} \delta' v_*^\tau = \delta_\tau^\rho \delta' v_*^\tau,$$

where

$$\delta_\tau^\rho = \frac{\partial v_*^\rho}{\partial \dot{q}^\sigma} \frac{\partial \dot{q}^\sigma}{\partial v_*^\tau} = \begin{cases} 1, & \rho = \tau, \\ 0, & \rho \neq \tau. \end{cases} \tag{5.5}$$

The assumption that functions (5.3), (5.4) have continuous partial derivatives enables one to introduce two systems of linearly independent vectors

$$\varepsilon_\tau = \frac{\partial \dot{q}^\sigma}{\partial v_*^\tau} \mathbf{e}_\sigma, \quad \varepsilon^\rho = \frac{\partial v_*^\rho}{\partial \dot{q}^\tau} \mathbf{e}^\tau, \quad \rho, \sigma, \tau = \overline{1, s}. \tag{5.6}$$

The first system is introduced using the principal basis, and the second one, using the reciprocal basis of the original curvilinear system of coordinates q^σ.

We recall that the vectors $\mathbf{e}_\sigma$ and $\mathbf{e}^\tau$ for the representative point are given by the formulas

$$\mathbf{e}_\sigma = \frac{\partial \mathbf{y}}{\partial q^\sigma} = \frac{\partial y_\mu}{\partial q^\sigma} \mathbf{i}_\mu \,, \quad g_{\sigma\tau} = \mathbf{e}_\sigma \cdot \mathbf{e}_\tau \,, \quad \mathbf{e}^\tau = g^{\tau\sigma} \mathbf{e}_\sigma \,,$$

which are completely similar to those considered in the analysis of kinematics of one point. The scalar product of the vectors $\mathbf{a}$ and $\mathbf{b}$, which are defined, respectively, in terms of the covariant and contravariant components $\mathbf{a} = a_\tau \mathbf{e}^\tau$, $\mathbf{b} = b^\sigma \mathbf{e}_\sigma$, reads as

$$\mathbf{a} \cdot \mathbf{b} = a_\sigma b^\sigma \,. \tag{5.7}$$

Hence, using (5.5), (5.6), this gives

$$\varepsilon^\rho \cdot \varepsilon_\tau = \delta_\tau^\rho = \begin{cases} 1 \,, & \rho = \tau \,, \\ 0 \,, & \rho \neq \tau \,. \end{cases}$$

It follows that the system of vectors ε^ρ can be looked upon as a reciprocal basis with respect to the basis defined in terms of the vectors ε_τ. The vectors (5.6) are called the *vectors of nonholonomic bases*.

From the arbitrariness in the choice of the functions $v_*^\rho(t, q, \dot{q})$ we may take as the variables $v_*^{l+\varkappa}$ the constraint functions $\varphi^\varkappa(t, q, \dot{q})$; that is, we set

$$v_*^{l+\varkappa} = \varphi^\varkappa(t, q, \dot{q}) \,, \quad l = s - k \,, \quad \varkappa = \overline{1, k} \,.$$

In this case, we have by formulas (5.6)

$$\varepsilon^{l+\varkappa} = \frac{\partial \varphi^\varkappa}{\partial \dot{q}^\tau} \mathbf{e}^\tau = \nabla' \varphi^\varkappa \,, \quad \varkappa = \overline{1, k} \,. \tag{5.8}$$

Differentiating the constraint equations (5.1), we find that

$$\frac{\partial \varphi^\varkappa}{\partial t} + \frac{\partial \varphi^\varkappa}{\partial q^\sigma} \dot{q}^\sigma + \frac{\partial \varphi}{\partial \dot{q}^\sigma} \ddot{q}^\sigma = 0 \,, \quad \varkappa = \overline{1, k} \,. \tag{5.9}$$

In accordance with expression (1.12) of Chap. 5, the acceleration vector $\mathbf{W}$ has the contravariant representation

$$\mathbf{W} = (\ddot{q}^\sigma + \Gamma_{\alpha\beta}^\sigma \dot{q}^\alpha \dot{q}^\beta) \mathbf{e}_\sigma \,, \quad \sigma = \overline{1, s} \,, \quad \alpha, \beta = \overline{0, s} \,, \tag{5.10}$$

where $\Gamma_{\alpha\beta}^\sigma = g^{\sigma\tau} \Gamma_{\tau,\alpha\beta}$, $q^0 = t$. Using (5.7), (5.8), (5.10), we find that

$$\varepsilon^{l+\varkappa} \cdot \mathbf{W} = \frac{\partial \varphi^\varkappa}{\partial \dot{q}^\sigma} (\ddot{q}^\sigma + \Gamma_{\alpha\beta}^\sigma \dot{q}^\alpha \dot{q}^\beta) \,, \quad \varkappa = \overline{1, k} \,.$$

Hence, in view of (5.9),

Fig. 13 Decomposition of
the acceleration in vectors of
nonholonomic bases

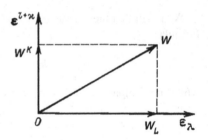

$$\mathbf{W} \cdot \varepsilon^{l+\varkappa} = \chi^{l+\varkappa}(t, q, \dot{q}), \quad \varkappa = \overline{1, k},$$

$$\chi^{l+\varkappa}(t, q, \dot{q}) = -\frac{\partial \varphi^{\varkappa}}{\partial t} - \frac{\partial \varphi^{\varkappa}}{\partial q^{\sigma}} \dot{q}^{\sigma} + \frac{\partial \varphi^{\varkappa}}{\partial \dot{q}^{\sigma}} \Gamma^{\sigma}_{\alpha\beta} \dot{q}^{\alpha} \dot{q}^{\beta}. \tag{5.11}$$

The tuple of k scalar quantities $\chi^{l+\varkappa}$ defines some vector $\mathbf{W}^K$ in the space with the basis $\{\varepsilon^{l+1}, \varepsilon^{l+2}, \dots, \varepsilon^{l+k}\}$. Decomposing the total acceleration $\mathbf{W}$ into the vector $\mathbf{W}^K$ and the orthogonal vector $\mathbf{W}_L$, we get

$$\mathbf{W} = \mathbf{W}_L + \mathbf{W}^K, \quad \mathbf{W}_L \cdot \mathbf{W}^K = 0, \tag{5.12}$$

where $\mathbf{W}_L = \widetilde{W}_L^{\lambda} \varepsilon_{\lambda}$, $\lambda = \overline{1, l}$, $\mathbf{W}^K = \widetilde{W}_{l+\varkappa}^K \varepsilon^{l+\varkappa}$, $\varkappa = \overline{1, k}$. Here, the tilde symbol points out that the components of the acceleration vector are calculated in nonholonomic bases. This decomposition of the vector $\mathbf{W}$ is illustrated in Fig. 13. The subscripts L and K of the vectors $\mathbf{W}_L$ and $\mathbf{W}^K$ indicate, respectively, the dimension l and k of the subspaces that contain them.

To calculate $\widetilde{W}_L^{\lambda}$, we multiply the left- and right-hand sides of equality (5.12) by ε_{μ}, $\mu = \overline{1, l}$. This gives us the following system of linear algebraic equations

$$\widetilde{W}_L^{\lambda} \varepsilon_{\lambda} \cdot \varepsilon_{\mu} = \mathbf{W} \cdot \varepsilon_{\mu}, \quad \lambda, \mu = \overline{1, l},$$

with respect to the unknowns $\widetilde{W}_L^{\lambda}$. Applying Cramer's rule to the solution of this system, we express $\widetilde{W}_L^{\lambda}$ in terms of the scalar products $\mathbf{W} \cdot \varepsilon_{\mu}$, $\mu = \overline{1, l}$. Thus, we reach the important conclusion that the vector $\mathbf{W}_L$ is completely determined by the tuple of quantities $\mathbf{W} \cdot \varepsilon_{\mu}$, $\mu = \overline{1, l}$. A similar analysis shows that the vector $\mathbf{W}^K$ is determined by the quantities $\mathbf{W} \cdot \varepsilon^{l+\varkappa}$, $\varkappa = \overline{1, k}$. From (5.11) it follows that with given $t, q^{\sigma}, \dot{q}^{\sigma}$ the values of $\mathbf{W} \cdot \varepsilon^{l+\varkappa}$ can be found using the coefficients $\Gamma^{\sigma}_{\alpha\beta}$, which in turn can be found from the elements $g_{\alpha\beta}$ and the constraint equations. Bases on this, we may say that the vector $\mathbf{W}^K$ lies in that space in which the vector of acceleration (*qua* a function of time, position of the system, its velocities) is completely determined by the coefficients $g_{\alpha\beta}$ and the constraint equations. At the same time, nothing definite can be said about the effect of the constraints on the vector $\mathbf{W}_L$, because it can be excluded from (5.11) and can be written in the form

$$\mathbf{W}^K \cdot \varepsilon^{l+\varkappa} = \chi^{l+\varkappa}(t, q, \dot{q}), \quad \varkappa = \overline{1, k}.$$

Let us consider Newton's law as applied to the vector $\mathbf{W}^K$:

$$M\mathbf{W}^K = \mathbf{Y}^K + \mathbf{R}^K . \tag{5.13}$$

Here,

$$\mathbf{Y}^K = \widetilde{Y}^K_{l+\varkappa} \mathbf{e}^{l+\varkappa} , \quad \mathbf{R}^K = \widetilde{R}^K_{l+\varkappa} \mathbf{e}^{l+\varkappa} .$$

The vector $\mathbf{R}^K$ is added to the active force $\mathbf{Y}^K$ in order to get in sum the vector $M\mathbf{W}^K$, which is completely determined by the constraint equations.

To the acceleration $\mathbf{W}_L$ there corresponds, in general, the Newton equation

$$M\mathbf{W}_L = \mathbf{Y}_L + \mathbf{R}_L , \tag{5.14}$$

where

$$\mathbf{Y}_L = \widetilde{Y}^\lambda_L \boldsymbol{\varepsilon}_\lambda , \quad \mathbf{R}_L = \widetilde{R}^\lambda_L \boldsymbol{\varepsilon}_\lambda , \quad \lambda = \overline{1, l} .$$

According to the above, the vector $\mathbf{W}_L$ can be excluded from Eq. (5.11), and hence one cannot find the effect of the constraints on the vector $\mathbf{W}_L$ by merely using their mathematical equations. As a result, Eq. (5.14) holds for any $\mathbf{R}_L$ and, in particular, for $\mathbf{R}_L = 0$. Hence, the vector $\mathbf{W}_L$, which is perpendicular to the vector $\mathbf{W}^K$, is independent of the constraints, and thus the Newton's law in the corresponding subspace reads as

$$M\mathbf{W}_L = \mathbf{Y}_L . \tag{5.15}$$

The constraints that have no effect on the vector $\mathbf{W}_L$ are called *ideal*. These constraints, which are completely determined by their analytic representations, allow one to write down the second Newton's law in the same form (5.15) as for an unconstrained system. Here it is important that this law is written in the context of the subspace orthogonal to the subspace of reactions, in which the acceleration is completely determined by the constraint equations.

Adding Eqs. (5.13) and (5.15), this establishes

$$M\mathbf{W} = \mathbf{Y} + \mathbf{R}^K .$$

A comparison of the equation in this form with Eq. (5.2) shows that, for ideal nonholonomic constraints,

$$\mathbf{R} = \mathbf{R}^K = \Lambda_\varkappa \nabla' \varphi^\varkappa .$$

The motion equations of nonholonomic systems. The Maggi equations. Let us go back to the vector equality (5.15). The component $\mathbf{a}_L = a^\lambda_L \boldsymbol{\varepsilon}_\lambda$ of an arbitrary vector $\mathbf{a}$ is completely determined by the set of quantities $\mathbf{a} \cdot \boldsymbol{\varepsilon}_\lambda$, $\lambda = \overline{1, l}$, and hence one Eq. (5.15) is equivalent to the system of l equations

$$(M\mathbf{W} - \mathbf{Y}) \cdot \boldsymbol{\varepsilon}_\lambda = 0 , \quad \lambda = \overline{1, l} . \tag{5.16}$$

These equations, which express in the scalar form the second Newton's law (5.15), will be called the *dynamic equations of constrained motion*. Using (5.7) and substituting into Eq. (5.16) the vectors ε_λ, as expressed in terms of the vectors $\mathbf{e}_\sigma$ of the principal basis, we have

$$(MW_\sigma - Q_\sigma)\frac{\partial \dot{q}^\sigma}{\partial v_*^\lambda} = 0, \quad \lambda = \overline{1, l}. \tag{5.17}$$

The resulting motion equations (5.17) are called the *Maggi equations*.[7]

According to the above, under ideal constraints we have

$$MW - Y = R = R^K = \Lambda_\varkappa \nabla'\varphi^\varkappa = \Lambda_\varkappa \varepsilon^{l+\varkappa}.$$

It follows that

$$(MW - Y) \cdot \varepsilon_{l+\varkappa} = \Lambda_\varkappa, \quad \varkappa = \overline{1, k}.$$

Taking into account the relations (5.6), (5.7), these equations can be put in the form

$$(MW_\sigma - Q_\sigma)\frac{\partial \dot{q}^\sigma}{\partial v_*^{l+\varkappa}} = \Lambda_\varkappa, \quad \varkappa = \overline{1, k}. \tag{5.18}$$

Equations (5.17), (5.18), when considered jointly, form a system of linear algebraic equations with respect to the unknowns generalized forces

$$R_\sigma = MW_\sigma - Q_\sigma, \quad \sigma = \overline{1, s}.$$

When adding these equations to the forces Q_σ the motion satisfies the constraint equations. We write this system in the form

$$\beta_\rho^\sigma R_\sigma = \Lambda_\rho^*, \quad \rho, \sigma = \overline{1, s},$$

where

$$\beta_\rho^\sigma = \frac{\partial \dot{q}^\sigma}{\partial v_*^\rho}, \quad \Lambda_\rho^* = \begin{cases} 0, & \rho = \overline{1, l}, \\ \Lambda_{\rho-l}, & \rho = \overline{l+1, s}. \end{cases}$$

Taking into account the complete analogy of this system with system (3.30), we have

$$R_\sigma = \Lambda_\rho^* \frac{\partial v_*^\rho}{\partial \dot{q}^\sigma} = \Lambda_\varkappa \frac{\partial \varphi^\varkappa}{\partial \dot{q}^\sigma}, \quad \sigma = \overline{1, s}.$$

---

[7] Equations in this form were derived by G.A. Maggi in 1896 for the case of linear nonholonomic first-order constraints using the generalized d'Alembert–Lagrange principle. In 1931 these equations were obtained by A. Przeborski in the case of nonlinear first-order constraints and linear second-order constraints using the same generalized principle.

This is a system of the *Lagrange equations of the first kind in curvilinear coordinates for nonholonomic systems*

$$MW_\sigma = Q_\sigma + \Lambda_\varkappa \frac{\partial \varphi^\varkappa}{\partial \dot{q}^\sigma}, \quad \sigma = \overline{1, s};$$

this system is also frequently called the system of *Lagrange equations of the second kind with multipliers for nonholonomic systems*.

Expressions (5.16), (5.17) define the vector MW_L. Using (5.9), (5.10) and (5.11), the vector W^K can be found by differentiating the constraint equations. Hence, augmenting Eq. (5.17) with the constraint equations, we arrive at a closed system of equations with respect to the vector MW. The vector W can be put in the form (5.10), and hence the problem of motion satisfying the constraint equations can be reduced to obtaining a system of differential equations that is solved with respect to the generalized accelerations:

$$\ddot{q}^\sigma = F^\sigma(t, q, \dot{q}), \quad \sigma = \overline{1, s}.$$

To reduce the problem to such a system one first needs to find the partial derivatives $\partial \dot{q}^\sigma / \partial v_*^\rho = \beta_\rho^\sigma$ qua functions of t, q and $\dot{q}$. To this aim, the system of functions $v_*^{l+\varkappa} = \varphi^\varkappa(t, q, \dot{q})$ should be augmented with l functions $v_*^\lambda(t, q, \dot{q})$, which are taken to have the physical sense of velocities which are characteristic for the problem under study. Introducing the function $v_*^\rho(t, q, \dot{q})$, $\rho = \overline{1, s}$, we calculate their derivatives

$$\frac{\partial v_*^\rho}{\partial \dot{q}^\sigma} = \alpha_\sigma^\rho(t, q, \dot{q}), \quad \rho, \sigma = \overline{1, s}.$$

The coefficients of the matrix $[\beta_\rho^\sigma]$, which is inverse of the matrix $[\alpha_\sigma^\rho]$, are found by Cramer's rule. Thus, to find the functions $\beta_\rho^\sigma (t, q, \dot{q})$ it suffices to know only the original functions $v_*^\rho(t, q, \dot{q})$, $\rho = \overline{1, s}$. This is very important, because with nonlinear dependences of the constraint equations on the generalized velocities, the calculation of the functions $\dot{q}^\sigma(t, q, v_*)$ may prove to be fairly involved. Having determined the coefficients β_λ^σ from equations (5.17) and taking into account that the expressions MW_σ are calculated by the formulas

$$MW_\sigma = \frac{d}{dt} \frac{\partial T}{\partial \dot{q}^\sigma} - \frac{\partial T}{\partial q^\sigma} = M(g_{\sigma\tau}\ddot{q}^\tau + \Gamma_{\sigma,\alpha\beta} \dot{q}^\alpha \dot{q}^\beta),$$

$$\sigma, \tau = \overline{1, s}, \quad \alpha, \beta = \overline{0, s},$$

we have

$$(M(g_{\sigma\tau}\ddot{q}^\tau + \Gamma_{\sigma,\alpha\beta} \dot{q}^\alpha \dot{q}^\beta) - Q_\sigma) \beta_\lambda^\sigma (t, q, \dot{q}) = 0, \quad \lambda = \overline{1, l}. \tag{5.19}$$

Augmenting this system with k constraint equations (5.1), we get the closed system of s differential equations for the functions $q^\sigma(t)$, $\sigma = \overline{1, s}$.

Special form of quasi-velocities. In many cases it is expedient to introduce the new variables v_*^ρ:

$$v_*^\lambda = \dot{q}^\lambda, \quad v_*^{l+\varkappa} = \varphi^\varkappa(t, q, \dot{q}), \quad \lambda = \overline{1, l}, \quad \varkappa = \overline{1, k}. \tag{5.20}$$

Here, it is assumed that the conditions, with which Eq. (5.20) can be solved in $\dot{q}^\sigma$, are satisfied. From (5.20) it readily follows that

$$\beta_\rho^\lambda = \begin{cases} 1, & \lambda = \rho, \\ 0, & \lambda \neq \rho, \end{cases} \quad \lambda = \overline{1, l}, \quad \rho = \overline{1, s}. \tag{5.21}$$

To determine the remaining coefficients β_ρ^{l+i}, we shall calculate the partial differential of the function φ^i with fixed t and q^σ:

$$\delta' v_*^{l+i} = \frac{\partial \varphi^i}{\partial \dot{q}^\lambda} \delta' \dot{q}^\lambda + \frac{\partial \varphi^i}{\partial \dot{q}^{l+j}} \delta' \dot{q}^{l+j}, \quad \lambda = \overline{1, l}, \quad i, j = \overline{1, k}.$$

Considering these relations as a system of k linear algebraic equations with respect to the unknowns $\delta' \dot{q}^{l+j}$ and solving it by Carmer's law, we find that

$$\delta' \dot{q}^{l+j} = \frac{A_i^j}{\det[a_j^i]} \left(\delta' v_*^{l+i} - \frac{\partial \varphi^i}{\partial \dot{q}^\lambda} \delta' v_*^\lambda \right),$$

where $a_j^i = \partial \varphi^i / \partial \dot{q}^{l+j}$ ($i, j = \overline{1, k}$), A_i^j is the algebraic complement of a_j^i in the determinant $\det[a_j^i]$. Besides, by the adopted notation we have $\delta' \dot{q}^{l+j} = \beta_\rho^{l+j} \delta' v_*^\rho$, and hence

$$\beta_\lambda^{l+j} = -\frac{A_i^j \frac{\partial \varphi^i}{\partial \dot{q}^\lambda}}{\det[a_j^i]}, \quad \beta_{l+i}^{l+j} = \frac{A_i^j}{\det[a_j^i]}. \tag{5.22}$$

Substituting into the system of Eq. (5.19) the calculated values of the coefficients β_λ^σ and taking into account that among them there are 1's or 0's, we see that

$$M(g_{\lambda\sigma} + g_{l+j,\sigma} \beta_\lambda^{l+j}) \ddot{q}^\sigma + M(\Gamma_{\lambda,\alpha\beta} + \Gamma_{l+j,\alpha\beta} \beta_\lambda^{l+j}) \dot{q}^\alpha \dot{q}^\beta = Q_\lambda + Q_{l+j} \beta_\lambda^{l+j},$$

or

$$M(\tilde{g}_{\lambda\sigma} \ddot{q}^\sigma + \tilde{\Gamma}_{\lambda,\alpha\beta} \dot{q}^\alpha \dot{q}^\beta) = \tilde{Q}_\lambda, \quad \lambda = \overline{1, l}. \tag{5.23}$$

In the same way, we reduce formulas (5.18) to the form

$$M(\tilde{g}_{l+i,\sigma} \ddot{q}^\sigma + \tilde{\Gamma}_{l+i,\alpha\beta} \dot{q}^\alpha \dot{q}^\beta) - \tilde{Q}_{l+i} = \Lambda_i, \quad i = \overline{1, k}, \tag{5.24}$$

where

$$\widetilde{g}_{\lambda\sigma} = g_{\lambda\sigma} + g_{l+j,\sigma}\,\beta_\lambda^{l+j}\,, \quad \widetilde{\Gamma}_{\lambda,\alpha\beta} = \Gamma_{\lambda,\alpha\beta} + \Gamma_{l+j,\alpha\beta}\,\beta_\lambda^{l+j}\,,$$

$$\widetilde{g}_{l+i,\sigma} = g_{l+j,\sigma}\,\beta_{l+i}^{l+j}\,, \quad \widetilde{\Gamma}_{l+i,\alpha\beta} = \Gamma_{l+j,\alpha\beta}\,\beta_{l+i}^{l+j}\,,$$

$$\widetilde{Q}_\lambda = Q_\lambda + Q_{l+j}\,\beta_\lambda^{l+j}\,, \quad \widetilde{Q}_{l+i} = Q_{l+j}\,\beta_{l+i}^{l+j}\,,$$

$$\lambda = \overline{1,l}\,, \quad \sigma = \overline{1,s}\,, \quad \alpha,\beta = \overline{0,s}\,, \quad i,j = \overline{1,k}\,.$$

Thus, in the case of nonholonomic first-order constraints, the form of motion equations is the same as for holonomic constraints, the only difference is that the functions $\widetilde{g}_{\lambda\sigma}$ and $\widetilde{\Gamma}_{\lambda,\alpha\beta}$ are distinct from $g_{\lambda\sigma}$ and $\Gamma_{\lambda,\alpha\beta}$. If the constraints are linear with respect to the generalized velocities, then $\widetilde{g}_{\lambda\sigma}$ and $\widetilde{\Gamma}_{\lambda,\alpha\beta}$ are independent of $\dot{q}^\sigma$.

To completely solve the mechanical problem it is required to augment system (5.23) with k constraint equations (5.1). As a result, we get the system of s differential equations for the unknowns functions q^σ, $\sigma = \overline{1,s}$.

If constraints are nonlinear in $\dot{q}^\sigma$ it is convenient to write the equations of these constraints as follows:

$$\dot{\varphi}^\varkappa = \frac{\partial\varphi^\varkappa}{\partial t} + \frac{\partial\varphi^\varkappa}{\partial q^\sigma}\dot{q}^\sigma + \frac{\partial\varphi^\varkappa}{\partial\dot{q}^\sigma}\ddot{q}^\sigma = 0\,, \quad \varkappa = \overline{1,k}\,.$$

Adding these conditions to system (5.23) and solving the resulting set of equations with respect to $\ddot{q}^\sigma$, we have the system of differential equations of the form

$$\ddot{q}^\sigma = F^\sigma(t,q,\dot{q})\,, \quad \sigma = \overline{1,s}\,. \tag{5.25}$$

Setting $\dot{q}^\sigma = \widehat{v}^\sigma$, we write the system of equations in the normal form

$$\frac{d\widehat{v}^\sigma}{dt} = F^\sigma(t,q,\widehat{v})\,, \quad \frac{dq^\sigma}{dt} = \widehat{v}^\sigma\,, \quad \sigma = \overline{1,s}\,, \tag{5.26}$$

which is equivalent to system (5.25). Here, q^σ, $\widehat{v}^\sigma$, $\sigma = \overline{1,s}$, are unknowns functions. Such a form of expression is quite convenient both in the qualitative examination of the problem and in its numerical solution.

We note that in solving (5.26) the initial conditions should satisfy the constraint equations $\varphi^\varkappa(t_0,q_0,\widehat{v}_0) = 0$.

Having determined the functions $q^\sigma(t)$ from system (5.26) and substituting them into Eq. (5.24), we find the generalized reactions $\Lambda_i(t)$.

Example 8 *Motion of a two-mass system with holonomic and nonholonomic constraints* (application of the Maggi equations). Let us consider the motion in the horizontal plane Oxy of two points $M_1(x_1,y_1)$ and $M_2(x_2,y_2)$ of mass m, which are connected by a light inextensible rod of length $2l$ (Fig. 14a). A short horizontal runner with curved end-points (a skate) is fixed perpendicularly to the rod at the point C, which is the midpoint of the rod. The runner has sharp edges and hence it may move

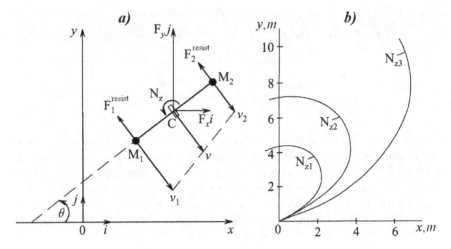

Fig. 14 Two-mass system with a nonholonomic constraint

without friction along the edge; however, it hinders the motion in the perpendicular direction. It is assumed that because of a sufficiently small runner length and since its end-points are rounded, the system may freely rotate about its center.

The motion of points is subject to the holonomic constraint

$$(x_2 - x_1)^2 + (y_2 - y_1)^2 = (2l)^2,$$

and hence, the position of the system is uniquely determined by three parameters. For the generalized coordinates, we take the Cartesian coordinates x, y of the rod midpoint and the angle of rotation θ of the rod about the Oz-axis:

$$q^1 = x, \qquad q^2 = \theta, \qquad q^3 = y. \tag{5.27}$$

We have

$$\dot{x}_1 = \dot{x} + \dot{\theta} l \sin \theta, \qquad \dot{y}_1 = \dot{y} - \dot{\theta} l \cos \theta,$$
$$\dot{x}_2 = \dot{x} - \dot{\theta} l \sin \theta, \qquad \dot{y}_2 = \dot{y} + \dot{\theta} l \cos \theta. \tag{5.28}$$

Let us consider the derivation of the nonholonomic constraint equations. The point C of the rod midpoint has a runner, and hence it may have only the velocity perpendicular to the rod axis. In Sect. 5 of Chap. 2 it was shown that the projections of the velocities of any two points of a rigid body to the line passing through them are equal. Because of the presence of the skate, the velocity $\mathbf{v}$ of the rod midpoint has no projection to the rod axis, and hence the velocities $\mathbf{v}_1$ and $\mathbf{v}_2$ of the points M_1 and M_2 also do not have such a projection. This can be written as

$$\frac{\dot{x}_1}{\dot{x}_2} = \frac{\dot{y}_1}{\dot{y}_2}.$$

Hence, using formulas (5.28),

$$\dot{\theta} \left(\dot{x} \cos \theta + \dot{y} \sin \theta \right) = 0.$$

This equation is satisfied with $\dot{\theta} = 0$ or when

$$\dot{x} \cos \theta + \dot{y} \sin \theta = 0. \qquad (5.29)$$

In the case $\dot{\theta} = 0$, the angle θ is constant, and hence we have a plane-parallel motion with rectilinear motion of the point C. Such a motion is effected with a long runner, which hinders any rotation of the system about the point C. Since in this example we consider the case of a short runner, we specify the nonholonomic constraint in the form (5.29). (We note that due to the presence of the nonholonomic constraint (5.29) the system of Fig. 14a, is capable in particular of interpreting the motion of an ice skater standing vertically on one skate, an in the case $\dot{\theta} = 0$, the motion of a speed skater on racing skates.)

It is worth pointing out that the constraint (5.29) holds both with $\dot{\theta} = 0$ and with $\dot{\theta} \neq 0$. Hence, one cannot say that the equation $\dot{\theta}(\dot{x} \cos \theta + \dot{y} \sin \theta) = 0$ is more general than Eq. (5.29), and hence, it cannot be looked upon as an example of a nonlinear nonholonomic constraint. This is why in the classical nonholonomic mechanics it is assumed that the constraints reflecting the no-slip condition of a body when it rolls on a fixed surface are expressed linearly with respect to the generalized velocities.

To compose the motion equations, we first write down the expression for the kinetic energy T. In our case, it follows by (5.28) that

$$T = \frac{m}{2} \left(\dot{x}_1^2 + \dot{y}_1^2 + \dot{x}_2^2 + \dot{y}_2^2 \right) = m \left(\dot{x}^2 + \dot{y}^2 + \dot{\theta}^2 l^2 \right).$$

Using this expression, we find that

$$MW_1 = 2m\ddot{x}, \qquad MW_2 = 2ml^2\ddot{\theta}, \qquad MW_3 = 2m\ddot{y}, \qquad (5.30)$$

where $M = 2m$ is the mass of the representative point.

According to general theory, the new velocities v_*^1, v_*^2, v_*^3 are derived by the formilas $v_*^1 = \dot{q}^1 \equiv \dot{x}$, $v_*^2 = \dot{q}^2 \equiv \dot{\theta}$, $v_*^3 = \dot{x} \cos \theta + \dot{y} \sin \theta$, and hence

$$\dot{x} \equiv \dot{q}^1 = v_*^1, \qquad \dot{\theta} \equiv \dot{q}^2 = v_*^2, \qquad \dot{y} \equiv \dot{q}^3 = \frac{v_*^3 - v_*^1 \cos \theta}{\sin \theta}. \qquad (5.31)$$

From expressions (5.30), (5.31) it follows that in this setting the Maggi equations (5.17) reads as

$$2m\ddot{x} - Q_1 + (2m\ddot{y} - Q_3)(-\cot\theta) = 0,$$
$$2ml^2\ddot{\theta} - Q_2 = 0. \tag{5.32}$$

We note that here the second equation agrees with the standard Lagrange second-kind equation corresponding to the generalized coordinate θ, because the nonholonomic constraint equation (5.29) does not contain the velocity $\dot{\theta}$.

The system of Eq. (5.32) must be augmented with the constraint equation (5.29). Differentiating it in time, we find that

$$\ddot{x}\cos\theta - \dot{x}\dot{\theta}\sin\theta + \ddot{y}\sin\theta + \dot{y}\dot{\theta}\cos\theta = 0. \tag{5.33}$$

Solving the system of Eq. (5.32) and (5.33) as a system of linear algebraic inhomogeneous equations in $\ddot{x}$, $\ddot{y}$, $\ddot{\theta}$ and writing the results as a system of six differential first-kind equations, we see that

$$\dot{x} = v_x, \quad \dot{y} = v_y, \quad \dot{\theta} = \omega_z,$$
$$\dot{v}_x = \omega_z(v_x \sin\theta - v_y \cos\theta)\cos\theta + (Q_1 \sin\theta - Q_3 \cos\theta)\sin\theta/(2m),$$
$$\dot{v}_y = \omega_z(v_x \sin\theta - v_y \cos\theta)\sin\theta - (Q_1 \sin\theta - Q_3 \cos\theta)\cos\theta/(2m),$$
$$\dot{\omega}_z = Q_2/(2ml^2).$$

This normal form of a system of differential equations is convenient for numerical integration.

To calculate the generalized reaction of the nonholonomic constraint we have by (5.18)

$$\Lambda = (2m\ddot{y} - Q_3)/\sin\theta.$$

Let us consider the motion of the system under the force $\mathbf{F} = F_x\mathbf{i} + F_y\mathbf{j}$ applied to the point C, assuming that the moment $\mathbf{N} = N_z\mathbf{k}$ is present. In addition, let us consider the resistance forces $\mathbf{F}_1^{\text{resist}} = -\mu\mathbf{v}_1$, $\mathbf{F}_2^{\text{resist}} = -\mu\mathbf{v}_2$ ($\mu = \text{const}$), as applied to the points M_1, M_2 (Fig. 14,a). Here, to the generalized coordinates (5.27) there correspond the following generalized forces:

$$Q_1 \equiv Q_x = F_x - 2\mu\dot{x}, \quad Q_2 \equiv Q_\theta = N_z - 2\mu l^2\dot{\theta}, \quad Q_3 \equiv Q_y = F_y - 2\mu\dot{y}.$$

In calculations it was assumed that $m = 7\,\text{kg}$, $l = 1\,\text{m}$, $\mu = 0.6\,\text{N·s/m}$, $F_x = F_y = 2\,\text{N}$. Figure 14b depicts three trajectories of the point C in 15 s with $N_{z1} = 1\,\text{N·m}$, $N_{z2} = 0.65\,\text{N·m}$, $N_{z3} = 0.3\,\text{N·m}$. The initial data are zero.

6 The Appell Equations for A system of Material Points

The Appell equations. Let us consider another possible form of Eq. (5.17). The derivation of these equations will depend on transformations (5.20).

The coordinate vectors $\mathbf{e}_\sigma$ can be put in the form

$$\mathbf{e}_\sigma = \frac{\partial \mathbf{y}}{\partial q^\sigma} = \frac{\partial \mathbf{V}}{\partial \dot{q}^\sigma} = \frac{\partial \mathbf{W}}{\partial \ddot{q}^\sigma}, \qquad (6.1)$$

inasmuch as

$$\mathbf{V} = \dot{\mathbf{y}} = \frac{\partial \mathbf{y}}{\partial t} + \frac{\partial \mathbf{y}}{\partial q^\sigma} \dot{q}^\sigma, \quad \mathbf{W} = \dot{\mathbf{V}} = \frac{\partial \mathbf{V}}{\partial t} + \frac{\partial \mathbf{V}}{\partial q^\sigma} \dot{q}^\sigma + \frac{\partial \mathbf{V}}{\partial \dot{q}^\sigma} \ddot{q}^\sigma.$$

Based on (6.1), expression for $M W_\sigma$ can be written as

$$M W_\sigma = \frac{d}{dt} \frac{\partial T}{\partial \dot{q}^\sigma} - \frac{\partial T}{\partial q^\sigma} = M\mathbf{W} \cdot \mathbf{e}_\sigma = M\mathbf{W} \cdot \frac{\partial \mathbf{W}}{\partial \ddot{q}^\sigma} = \frac{\partial S(t, q, \dot{q}, \ddot{q})}{\partial \ddot{q}^\sigma}. \qquad (6.2)$$

Here, $S = M\mathbf{W}^2/2$ is the *Appell function*. From the last formula it follows that the second-kind Lagrange equation (3.19), (3.20), which hold only for a holonomic system, is equivalent to the equations

$$\frac{\partial S}{\partial \ddot{q}^\lambda} = Q_\lambda, \quad \lambda = \overline{1, l}, \qquad \frac{\partial S}{\partial \ddot{q}^{l+\varkappa}} = Q_{l+\varkappa} + \Lambda_\varkappa, \quad \varkappa = \overline{1, k},$$

which are called the *Appell equations*. The following convenient formula for the function $S(t, q, \dot{q}, \ddot{q})$ is worth noting:

$$S(t, q, \dot{q}, \ddot{q}) = \frac{M\mathbf{W} \cdot \mathbf{W}}{2} = \frac{M}{2} W_\sigma W^\sigma =$$

$$= \frac{M}{2} (g_{\sigma\tau} \ddot{q}^\tau + \Gamma_{\sigma,\alpha\beta} \dot{q}^\alpha \dot{q}^\beta)(\ddot{q}^\sigma + \Gamma^\sigma_{\alpha\beta} \dot{q}^\alpha \dot{q}^\beta),$$

$$\sigma, \tau = \overline{1, s}, \quad \alpha, \beta = \overline{0, s}, \quad q^0 = t, \quad \dot{q}^0 = 1.$$

If the constraints are nonholonomic and are of the first order, then by (6.2) one may write Eq. (5.17) in the form

$$\frac{\partial S}{\partial \ddot{q}^\sigma} \frac{\partial \dot{q}^\sigma}{\partial v_*^\lambda} = Q_\sigma \frac{\partial \dot{q}^\sigma}{\partial v_*^\lambda}, \quad \lambda = \overline{1, l}, \quad \sigma = \overline{1, s}. \qquad (6.3)$$

We recall that the functions $\partial \dot{q}^\sigma / \partial v_*^\lambda = \beta_\lambda^\sigma(t, q, \dot{q})$ in transformations (5.20) are calculated by formulas (5.21), (5.22), which were obtained on differentiating relations (5.20).

Equation (6.3) should be augmented with the constraint equations

$$\varphi^{\varkappa}(t, q, \dot{q}) = 0, \quad \varkappa = \overline{1, k}. \tag{6.4}$$

Considering the system of Eq. (6.3) jointly with the constraint equations, we get a closed system of ordinary differential equations with respect to the unknowns functions $q^{\sigma}(t)$, the number $s = l + k$ of which is the number of equations.

Having found the functions $q^{\sigma}(t)$ from the solution of system (6.3), (6.4), we shall use the formulas (5.18) to write down the generalized reactions

$$\Lambda_{\varkappa} = \frac{\partial S}{\partial \ddot{q}^{\sigma}} \frac{\partial \dot{q}^{\sigma}}{\partial v_*^{l+\varkappa}} - Q_{\sigma} \frac{\partial \dot{q}^{\sigma}}{\partial v_*^{l+\varkappa}}.$$

The Appell equations (6.3) with nonholonomic first-order constraints can be written in a more general form, which, as we shall show later, also holds for linear nonholonomic second-order constraints.

Differentiating (5.20) in t, we find that

$$\dot{v}_*^{\lambda} = \ddot{q}^{\lambda}, \quad \lambda = \overline{1, l},$$

$$\dot{v}_*^{l+\varkappa} = \psi^{\varkappa}(t, q, \dot{q}, \ddot{q}) = \dot{\varphi}^{\varkappa}(t, q, \dot{q}) =$$

$$= \frac{\partial \varphi^{\varkappa}}{\partial t} + \frac{\partial \varphi^{\varkappa}}{\partial q^{\sigma}} \dot{q}^{\sigma} + \frac{\partial \varphi^{\varkappa}}{\partial \dot{q}^{\sigma}} \ddot{q}^{\sigma}, \quad \varkappa = \overline{1, k}.$$

Let us introduce the functions w_*^{ρ} by the formulas

$$w_*^{\lambda} = \ddot{q}^{\lambda}, \quad \lambda = \overline{1, l},$$

$$w_*^{l+\varkappa}(t, q, \dot{q}, \ddot{q}) = \frac{\partial \varphi^{\varkappa}}{\partial \dot{q}^{\sigma}} \ddot{q}^{\sigma} = \psi^{\varkappa}(t, q, \dot{q}, \ddot{q}) - \psi^{\varkappa}(t, q, \dot{q}, 0), \quad \varkappa = \overline{1, k}. \tag{6.5}$$

A comparison of relations (5.20) and (6.5) shows that

$$\frac{\partial w_*^{\rho}}{\partial \ddot{q}^{\sigma}} = \frac{\partial v_*^{\rho}}{\partial \dot{q}^{\sigma}}, \quad \rho, \sigma = \overline{1, s},$$

and hence

$$\frac{\partial \ddot{q}^{\sigma}}{\partial w_*^{\rho}} = \frac{\partial \dot{q}^{\sigma}}{\partial v_*^{\rho}} = \beta_{\rho}^{\sigma}(t, q, \dot{q}), \quad \rho, \sigma = \overline{1, s}.$$

Thus, Eq. (6.3) can be put in the form

$$\frac{\partial S}{\partial \ddot{q}^{\sigma}} \frac{\partial \ddot{q}^{\sigma}}{\partial w_*^{\lambda}} = Q_{\sigma} \frac{\partial \ddot{q}^{\sigma}}{\partial w_*^{\lambda}}, \quad \lambda = \overline{1, l}, \quad \sigma = \overline{1, s}.$$

The left-hand sides of these formulas can be looked upon as the partial derivatives $\partial S / \partial w_*^\lambda$ of the Appell function

$$S = S(t, q, \dot{q}(t, q, v_*), \ddot{q}(t, q, v_*, w_*)),$$

which is a composite function of w_*^λ.

We have thus arrived at a more general form of the *Appell equations*

$$\frac{\partial S}{\partial w_*^\lambda} = Q_\lambda^*, \quad Q_\lambda^* = Q_\sigma \frac{\partial \ddot{q}^\sigma}{\partial w_*^\lambda}, \quad \lambda = \overline{1, l}. \tag{6.6}$$

Relation between the Appell equations and the Gauss principle.[8] We recall that the vector $\mathbf{W}$ can be written as

$$\mathbf{W} = (\ddot{q}^\sigma + \Gamma_{\alpha\beta}^\sigma \dot{q}^\alpha \dot{q}^\beta)\, \mathbf{e}_\sigma.$$

Hence, we have

$$\frac{\partial \mathbf{W}}{\partial w_*^\lambda} = \frac{\partial \ddot{q}^\sigma}{\partial w_*^\lambda}\, \mathbf{e}_\sigma, \quad \mathbf{Y} = Q_\sigma\, \mathbf{e}^\sigma,$$

and so Q_λ^* can be rewritten in the form

$$Q_\lambda^* = \mathbf{Y} \cdot \frac{\partial \mathbf{W}}{\partial w_*^\lambda} = \frac{\partial (\mathbf{Y} \cdot \mathbf{W})}{\partial w_*^\lambda}. \tag{6.7}$$

Substituting expressions (6.7) into the Appell equations (6.6), this establishes

$$\frac{\partial \left(\frac{M\mathbf{W}^2}{2} - \mathbf{Y} \cdot \mathbf{W} \right)}{\partial w_*^\lambda} = 0, \quad \lambda = \overline{1, l}.$$

Instead of the function $R = M\mathbf{W}^2/2 - \mathbf{Y} \cdot \mathbf{W}$, let us consider the function

$$Z = \frac{M\mathbf{W}^2}{2} - \mathbf{Y} \cdot \mathbf{W} + \frac{1}{2M} \mathbf{Y}^2 = \frac{M}{2} \left(\mathbf{W} - \frac{\mathbf{Y}}{M} \right)^2 > 0,$$

for which

$$\frac{\partial Z}{\partial w_*^\lambda} = \frac{\partial R}{\partial w_*^\lambda},$$

inasmuch as $\partial \mathbf{Y}^2 / \partial w_*^\lambda = 0$. In these notation, the motion equations (6.6) assume the form

---

[8] The Gauss principle will be discussed from different positions in Chap. 9.

$$\frac{\partial Z}{\partial w_*^\lambda} = \frac{\partial Z}{\partial \ddot{q}^\sigma} \frac{\partial \ddot{q}^\sigma}{\partial w_*^\lambda} = 0, \quad \lambda = \overline{1,l}.$$

These equations can be written as the scalar products

$$\nabla''Z \cdot \varepsilon_\lambda = 0, \quad \lambda = \overline{1,l}, \tag{6.8}$$

where

$$\nabla''Z = \frac{\partial Z}{\partial \ddot{q}^\sigma} \mathbf{e}^\sigma, \quad \varepsilon_\lambda = \frac{\partial \ddot{q}^\sigma}{\partial w_*^\lambda} \mathbf{e}_\sigma = \beta_\lambda^\sigma \mathbf{e}_\sigma.$$

Comparing Eq. (6.8) with Eq. (5.16), we see that

$$M\mathbf{W} - \mathbf{Y} = \nabla''Z.$$

We now claim that from Eq. (6.8) it follows that the function $Z(\mathbf{W})$ with $\mathbf{W}$ corresponding to the actual motion has the smallest value in comparison with those of Z_1, as obtained for any other accelerations $\mathbf{W}_1$, which are also kinematically possible with the same $t, q, \dot{q}$.

In Sect. 5 it was shown that the vector of acceleration can be written as the sum

$$\mathbf{W} = \mathbf{W}_L + \mathbf{W}^K, \quad \mathbf{W}_L = \widetilde{W}_L^\lambda \varepsilon_\lambda, \quad \mathbf{W}^K = \widetilde{W}_{l+\varkappa}^K \varepsilon^{l+\varkappa};$$

moreover, with fixed $t, q, \dot{q}$ the vector $\mathbf{W}^K$ is completely determined by the constraint equations, the vector $\mathbf{W}_L$ remaining arbitrary. In other words, any acceleration $\mathbf{W}_L$ is kinematically possible. Taking this into account, the difference between the kinematically possible $\mathbf{W}_1$ and the actual $\mathbf{W}$ accelerations can be written in the form

$$\mathbf{W}_1 - \mathbf{W} = \mathbf{W}_{1L} + \mathbf{W}^K - \mathbf{W}_L - \mathbf{W}^K = \widetilde{W}_\Delta^\lambda \varepsilon_\lambda, \quad \lambda = \overline{1,l},$$

where $\widetilde{W}_\Delta^\lambda$ may have arbitrary values. Substituting this expression for $\mathbf{W}_1$ into the function Z and taking into account (6.8), this gives

$$\begin{aligned}
Z_1 &= \frac{M}{2}\left(\mathbf{W} - \frac{\mathbf{Y}}{M} + \widetilde{W}_\Delta^\lambda \varepsilon_\lambda\right)^2 = \\
&= Z + \nabla''Z \cdot \widetilde{W}_\Delta^\lambda \varepsilon_\lambda + \frac{M}{2}(\widetilde{W}_\Delta^\lambda \varepsilon_\lambda)^2 = \\
&= Z + \frac{M}{2}(\widetilde{W}_\Delta^\lambda \varepsilon_\lambda)^2 > Z, \quad \mathbf{W}_1 \neq \mathbf{W}.
\end{aligned} \tag{6.9}$$

The condition $Z_1 > Z$ is obtained here from the equations of constrained motion (6.8). We note that from expressions (6.9) it also follows that the function $Z(\mathbf{W}_1)$, as defined on the set of kinematically possible accelerations, has a minimum with acceleration $\mathbf{W}$, which satisfies Eq. (6.8) or (5.16).

The function $Z(\mathbf{W})$ has the following interesting interpretation. Let us introduce some time interval τ and multiply $Z(\mathbf{W})$ by $\tau^4/4$. We have

$$Z_g = \frac{\tau^4}{4} Z = \frac{\tau^4}{4} \frac{M}{2} \left(\mathbf{W} - \frac{\mathbf{Y}}{M} \right)^2.$$

Expressing the $3n$-dimensional vectors $\mathbf{W}$ and $\mathbf{Y}$ in terms of their Cartesian coordinates we have, by formulas (1.4), (1.5) of Chap. 5,

$$Z_g = \frac{\tau^4}{4} \sum_{\mu=1}^{3n} \frac{M}{2} \left(\ddot{y}_\mu - \frac{Y_\mu}{M} \right)^2 = \frac{\tau^4}{4} \sum_{\mu=1}^{3n} \frac{M}{2} \left(\sqrt{\tilde{m}_\mu} \ddot{x}_\mu - \frac{X_\mu}{M\sqrt{\tilde{m}_\mu}} \right)^2 =$$

$$= \frac{\tau^4}{4} \sum_{\mu=1}^{3n} \frac{M\tilde{m}_\mu}{2} \left(\ddot{x}_\mu - \frac{X_\mu}{M\tilde{m}_\mu} \right)^2 = \frac{\tau^4}{4} \sum_{\nu=1}^{n} \frac{m_\nu}{2} \left(\mathbf{w}_\nu - \frac{\mathbf{F}_\nu}{m_\nu} \right)^2,$$

and so

$$Z_g = \sum_{\nu=1}^{n} \frac{m_\nu}{2} \Delta \mathbf{g}_\nu^2, \quad \Delta \mathbf{g}_\nu = \left(\mathbf{w}_\nu - \frac{\mathbf{F}_\nu}{m_\nu} \right) \frac{\tau^2}{2}.$$

If a material point of mass m_ν would be unconstrained, then under the force $\mathbf{F}_\nu$ it would get the acceleration $\mathbf{w}_\nu^* = \mathbf{F}_\nu/m_\nu$. Having at a given time t the velocity $\mathbf{v}_\nu$ and acceleration $\mathbf{w}_\nu^*$, this point executes a motion, which for sufficiently small τ, can be approximately written as

$$\mathbf{g}_\nu^* = \mathbf{v}_\nu \tau + \frac{\mathbf{F}_\nu}{m_\nu} \frac{\tau^2}{2}.$$

The acceleration of this point with the presence of constraints is $\mathbf{w}_\nu$, and hence the actual motion is given by

$$\mathbf{g}_\nu = \mathbf{v}_\nu \tau + \mathbf{w}_\nu \frac{\tau^2}{2}.$$

Calculating the difference $\mathbf{g}_\nu - \mathbf{g}_\nu^*$, we find the vector $\Delta \mathbf{g}_\nu$, which characterizes the departure of the constrained motion from the unconstrained one at a given time.

The above shows that the function Z_g is equal to the sum of quantities, which are proportional to the squared deflections of the points of the system. This function looks like the famous expression, which by Gauss's proposal laid the basis for the theory of data observations by the method of least squares. In mechanics, the function Z_g was also introduced by Gauss, who called it a *constraint*.[9] Gauss also put forward the principle, which says that "The motion of a system of material points, which are arbitrary related with each other and subject to any influences, takes place in every

---

[9] In the notation of constraints, the letter "Z" comes from the German word "*der Zwang*", meaning "*a constraint*".

moment in maximum accordance with the free movement or under least constraint...".
He continues "the measure of constraint, ... is considered as the sum of products of
mass and the square of the deviation to the free motion".[10]

Gauss' principle can be interpreted geometrically. According to Sect. 1 of
Chap. 5, the motion of a mechanical system consisting of n material points, can
be looked upon as the motion of one representative point in the $3n$-dimensional
Euclidean space. The vector of acceleration of the representative point, as well as
the vector of acceleration of the material point, can be written as the sum of the
tangential and normal accelerations:

$$\mathbf{W} = \ddot{s}\boldsymbol{\tau} + \dot{s}^2 K \mathbf{n}.$$

Here, s is the angular coordinate, K is the path curvature. It follows that the Gauss
function Z with $\mathbf{Y} = 0$ can be written as

$$Z = \frac{M}{2}\mathbf{W}^2 = \frac{M}{2}(\ddot{s}^2 + K^2\dot{s}^4).$$

Assume now that the motion of the system is subject to k stationary holonomic
ideal constraints. Now the representative point executes a motion along a $(3n - k)$-
dimensional surface. The reactions of constraints $\mathbf{R} = \mathbf{N}$ is directed along the normal
vector to the surface, and hence, the vector $\mathbf{R}$ is orthogonal to the unit tangent vector $\boldsymbol{\tau}$.
As a result, the constrained motion equation $M\mathbf{W} = \mathbf{Y} + \mathbf{R}$ in the projection to the
tangent line to the path has the form

$$M\ddot{s} = Y_\tau, \quad Y_\tau = \mathbf{Y} \cdot \boldsymbol{\tau}.$$

In the absence of active forces ($Y_\tau = 0$), we have $\dot{s}^2 = |\mathbf{V}|^2 = \text{const}$, and the Gauss
function reads as

$$Z = \frac{M}{2}K^2|\mathbf{V}|^4, \quad |\mathbf{V}| = \text{const}.$$

It follows that to the minimal value of Z there corresponds the minimal curvature K.
Thus, in the absence of active forces, the representative point moves with constant
velocity along a curve of smallest curvature.

---

[10] See: *C. F. Gauss*, "Über ein neues allgemeines Grundgesetz der Mechanik," Journal für die reine
und angewandte Mathematik, 1829. Vol. IV. S. 233.

II) Constrained Motion of General Mechanical Systems

7 Use of the Tangent Space in the Study of the Constrained Motion of General Mechanical Systems

The vector motion equation of a general unconstrained mechanical system.
Assume that the motion of an unconstrained mechanical system is written in the generalized coordinates q^σ, $\sigma = \overline{1, s}$, by the Lagrange equations of second kind

$$
\frac{d}{dt}\frac{\partial T}{\partial \dot{q}^\sigma} - \frac{\partial T}{\partial q^\sigma} = Q_\sigma, \qquad T = \frac{M}{2}\, g_{\alpha\beta}\, \dot{q}^\alpha \dot{q}^\beta,
$$
$$
\sigma = \overline{1, s}, \qquad \alpha, \beta = \overline{0, s}, \qquad q^0 = t, \qquad \dot{q}^0 = 1,
$$

(7.1)

where Q_σ is the generalized force corresponding to the coordinate q^σ, M is the mass of the entire system. Let us find the vector representation of these differential equations.

Consider the manifold of all positions of the mechanical system under study, which it may occupy at a given time t. On this manifold we fix some point with coordinates q^σ, $\sigma = \overline{1, s}$. Assume that the old and new q_*^ρ, $\rho = \overline{1, s}$, coordinates of this point are expressed in terms of each other by the formulas

$$
q^\sigma = q^\sigma(t, q_*), \quad q_*^\rho = q_*^\rho(t, q), \qquad \rho, \sigma = \overline{1, s},
$$

or in the differential form

$$
\delta q^\sigma = \frac{\partial q^\sigma}{\partial q_*^\rho}\,\delta q_*^\rho, \quad \delta q_*^\rho = \frac{\partial q_*^\rho}{\partial q^\sigma}\,\delta q^\sigma, \qquad \rho, \sigma = \overline{1, s}.
$$

The quantities δq^σ and δq_*^ρ in the above relations are called the *contravariant components of the tangent vector* $\delta \mathbf{y}$, the entire set of vectors $\delta \mathbf{y}$ is known as the *tangent space* to the above manifold at a given point.[11] It is expedient to write the vector $\delta \mathbf{y}$ in the form

$$
\delta \mathbf{y} = \delta q^\sigma\, \mathbf{e}_\sigma = \delta q_*^\rho\, \mathbf{e}_\rho^*, \qquad \rho, \sigma = \overline{1, s},
$$

and consider the sets of vectors $\mathbf{e}_\sigma$ and $\mathbf{e}_\rho^*$ as the principal bases of the tangent space in the coordinate systems q^σ and q_*^ρ.

The Euclidean structure in the tangent space will be introduced using the invariance with respect to the positive definite quadratic form

$$
(\delta \mathbf{y})^2 = g_{\sigma\tau}\,\delta q^\sigma\,\delta q^\tau = g_{\sigma_*\tau_*}^*\,\delta q_*^{\sigma_*}\,\delta q_*^{\tau_*}, \qquad \sigma, \tau, \sigma_*, \tau_* = \overline{1, s}.
$$

---

[11] See the book: *B.A. Dubrovin, A.T. Fomenko, S.P. Novikov*, Modern geometry. Methods and applications. New York: Springer. 1984.

Here, $g_{\sigma\tau}$ and $g^*_{\sigma_*\tau_*}$ are the coefficients involved in the expression for the kinetic energy, respectively, in the coordinates q^σ and q^ρ_* ($\rho, \sigma = \overline{1, s}$). In terms of these coefficients we have the metric tensor, which enables one to write the scalar product of the vectors $\mathbf{a} = a^\sigma \mathbf{e}_\sigma = a^{\sigma_*}_* \mathbf{e}^*_{\sigma_*}$ and $\mathbf{b} = b^\tau \mathbf{e}_\tau = b^{\tau_*}_* \mathbf{e}^*_{\tau_*}$ in the form

$$\mathbf{a} \cdot \mathbf{b} = g_{\sigma\tau} a^\sigma b^\tau = g^*_{\sigma_*\tau_*} a^{\sigma_*}_* b^{\tau_*}_*, \quad g_{\sigma\tau} = \mathbf{e}_\sigma \cdot \mathbf{e}_\tau, \quad g^*_{\sigma_*\tau_*} = \mathbf{e}^*_{\sigma_*} \cdot \mathbf{e}^*_{\tau_*},$$
$$\sigma, \tau, \sigma_*, \tau_* = \overline{1, s}.$$

The components δq^σ, $\sigma = \overline{1, s}$, of the tangent vector $\delta \mathbf{y}$ are also called *variations of the coordinates* q^σ or *virtual (possible)* displacements. By definition, the generalized forces Q_σ from the system of Eq. (7.1) are coefficients under the variations of coordinates δq^σ in the expression for possible elementary work δA. Using the continuous numbering $\mu = 1, 2, 3, \ldots$ to denote both the Cartesian coordinates of the action points of forces and the projection of these forces, we have

$$\delta A = X_\mu \, \delta x_\mu.$$

Next, since

$$\delta x_\mu = \frac{\partial x_\mu}{\partial q^\sigma} \delta q^\sigma = \frac{\partial x_\mu}{\partial q^\rho_*} \delta q^\rho_*,$$

we have

$$\delta A = Q_\sigma \, \delta q^\sigma = Q^*_\rho \, \delta q^\rho_*, \tag{7.2}$$

where

$$Q_\sigma = X_\mu \frac{\partial x_\mu}{\partial q^\sigma}, \quad Q^*_\rho = X_\mu \frac{\partial x_\mu}{\partial q^\rho_*} = Q_\sigma \frac{\partial q^\sigma}{\partial q^\rho_*}.$$

Expression (7.2) is a linear invariant differential form of the vector $\delta \mathbf{y}$. Its coefficients Q_σ and Q^*_ρ in the coordinates q^σ and q^ρ_*, respectively, are the *components of covariant vector* $\mathbf{Y}$ (see the previous footnote). Using the Euclidean structure of the tangent space, we write δA as the scalar product

$$\delta A = \mathbf{Y} \cdot \delta \mathbf{y}, \quad \mathbf{Y} = Q_\sigma \mathbf{e}^\sigma, \quad \sigma = \overline{1, s},$$

where $\mathbf{e}^\sigma$, $\sigma = \overline{1, s}$, are vectors of the reciprocal basis,

$$\mathbf{e}^\sigma \cdot \mathbf{e}_\tau = \delta^\sigma_\tau = \begin{cases} 0, & \sigma \neq \tau, \\ 1, & \sigma = \tau. \end{cases}$$

Hence and since $g_{\sigma\tau} = \mathbf{e}_\sigma \cdot \mathbf{e}_\tau$, we find that

$$\mathbf{e}_\tau = g_{\sigma\tau} \mathbf{e}^\sigma, \quad \mathbf{e}^\sigma = g^{\sigma\tau} \mathbf{e}_\tau.$$

The coefficient $g^{\sigma\tau}$ are entries of the inverse to the matrix with entries $g_{\sigma\tau}$.

Introduction of the covariant vector **Y** from the expression for the possible elementary work δA enables one to consider the system of Eq. (7.1) as a single vector equality

$$M\mathbf{W} = \mathbf{Y}. \tag{7.3}$$

Here,

$$\mathbf{W} = W_\sigma \mathbf{e}^\sigma = \frac{1}{M}\left(\frac{d}{dt}\frac{\partial T}{\partial \dot{q}^\sigma} - \frac{\partial T}{\partial q^\sigma}\right)\mathbf{e}^\sigma =$$

$$= \left(g_{\sigma\tau}\ddot{q}^\tau + \Gamma_{\sigma,\alpha\beta}\dot{q}^\alpha\dot{q}^\beta\right)\mathbf{e}^\sigma = W^\sigma \mathbf{e}_\sigma = \left(\ddot{q}^\sigma + \Gamma^\sigma_{\alpha\beta}\dot{q}^\alpha\dot{q}^\beta\right)\mathbf{e}_\sigma,$$

$$\Gamma^\sigma_{\alpha\beta} = g^{\sigma\tau}\Gamma_{\tau,\alpha\beta} = \frac{1}{2}g^{\sigma\tau}\left(\frac{\partial g_{\tau\beta}}{\partial q^\alpha} + \frac{\partial g_{\tau\alpha}}{\partial q^\beta} - \frac{\partial g_{\alpha\beta}}{\partial q^\tau}\right), \tag{7.4}$$

$$\tau, \sigma = \overline{1, s}, \qquad \alpha, \beta = \overline{0, s}.$$

Thus, using formulas (7.3) and (7.4) one may introduce in the tangent space the vector of acceleration **W** for an arbitrary mechanical system with s degrees of freedom.

Motion of a general constrained mechanical system. Ideal constraints. Let us now proceed with the study of the constrained motion. According to the principle of releasability from constraints, application of constraints results in the appearance of the reaction force **R**, and hence the second Newton's law assumes the form

$$M\mathbf{W} = \mathbf{Y} + \mathbf{R}.$$

The reaction force is related to the presence of the acceleration form the constraints. Here, one first has to find the effect of the constraints on the formation of the vector **W**. We first consider nonlinear nonholonomic first-order constraints in the form

$$\varphi^\varkappa(t, q, \dot{q}) = 0, \qquad \varkappa = \overline{1, k}.$$

Differentiating these constraints in time, this establishes

$$\psi^\varkappa(t, q, \dot{q}, \ddot{q}) \equiv a^{l+\varkappa}_\sigma(t, q, \dot{q})\ddot{q}^\sigma + a^{l+\varkappa}_0(t, q, \dot{q}) = 0,$$

$$\varkappa = \overline{1, k}, \qquad l = s - k. \tag{7.5}$$

We note that in this form one may also define nonholonomic linear second-order constraints. Double differentiation in time of the holonomic constraints leads to relations (7.5)

The introduction of the tangent space and the vector **W** in it (defined by formulas (7.4)) enables one to rewrite the system of Eq. (7.5) in the vector form

$$\varepsilon^{l+\varkappa} \cdot \mathbf{W} = \chi^\varkappa(t, q, \dot{q}),$$

$$\varepsilon^{l+\varkappa} = a^{l+\varkappa}_\sigma \mathbf{e}^\sigma, \qquad \chi^\varkappa = -a^{l+\varkappa}_0 + a^{l+\varkappa}_\sigma \Gamma^\sigma_{\alpha\beta}\dot{q}^\alpha\dot{q}^\beta, \tag{7.6}$$

$$\varkappa = \overline{1, k}, \qquad \alpha, \beta = \overline{0, s}.$$

The vectors $\varepsilon^{l+\varkappa}$, $\varkappa = \overline{1,k}$, which correspond to constraints (7.5), are assumed to be linearly independent. This makes it possible to introduce in the s-dimensional tangent space the subspace $\mathbb{R}^K$ with the basis of these vectors (the K-space). Now the entire space decomposes as the direct sum of this subspace and its orthogonal complement $\mathbb{R}_L$ with the basis ε_λ, $\lambda = \overline{1,l}$ (the L-space). Moreover,

$$\varepsilon_\lambda \cdot \varepsilon^{l+\varkappa} = 0, \qquad \lambda = \overline{1,l}, \qquad \varkappa = \overline{1,k}. \tag{7.7}$$

Here, the formation of the vectors ε_λ, $\lambda = \overline{1,l}$, can be explained as follows.

For an arbitrary mechanical system with constraints (7.5), the choice of the quantities w_*^λ, $\lambda = \overline{1,l}$, which are expressible in terms of the generalized accelerations $\ddot{q}^\sigma$, $\sigma = \overline{1,s}$, by the relations

$$w_*^\lambda = a_\sigma^\lambda(t, q, \dot{q}) \ddot{q}^\sigma, \qquad \lambda = \overline{1,l},$$

is free. Augmenting them with the equations

$$w_*^{l+\varkappa} = a_\sigma^{l+\varkappa}(t, q, \dot{q}) \ddot{q}^\sigma, \qquad \varkappa = \overline{1,k},$$

we get a closed system of equations with respect to the accelerations $\ddot{q}^\sigma$, $\sigma = \overline{1,s}$. Assuming that the solvability conditions for this system of equations are satisfied, we have

$$\ddot{q}^\sigma = \beta_\rho^\sigma(t, q, \dot{q}) w_*^\rho, \qquad \rho, \sigma = \overline{1,s}.$$

Now the vectors ε_λ, as given in the form

$$\varepsilon_\lambda = \frac{\partial \ddot{q}^\sigma}{\partial w_*^\lambda} \mathbf{e}_\sigma = \beta_\lambda^\sigma \mathbf{e}_\sigma, \qquad \lambda = \overline{1,l}, \tag{7.8}$$

satisfy conditions (7.7). Indeed, since $w_*^{l+\varkappa}$ and w_*^λ are independent, we may write

$$\frac{\partial w_*^{l+\varkappa}}{\partial w_*^\lambda} = \frac{\partial w_*^{l+\varkappa}}{\partial \ddot{q}^\sigma} \frac{\partial \ddot{q}^\sigma}{\partial w_*^\lambda} = a_\sigma^{l+\varkappa} \beta_\lambda^\sigma = \varepsilon^{l+\varkappa} \cdot \varepsilon_\lambda = 0, \qquad \varkappa = \overline{1,k}, \qquad \lambda = \overline{1,l}.$$

We note that this decomposition of the tangent space by the constraint equations corresponds to fixed values of the variables $t, q^\sigma, \dot{q}^\sigma$ ($\sigma = \overline{1,s}$). The bases just introduced will be called *nonholonomic*.

Substituting the acceleration $\mathbf{W}$, as given in the form (in what follows, the tilde over components of vectors indicates that they are written for the above nonholonomic bases),

$$\mathbf{W} = \mathbf{W}_L + \mathbf{W}^K,$$
$$\mathbf{W}_L = \widetilde{W}^\lambda \varepsilon_\lambda, \qquad \mathbf{W}^K = \widetilde{W}_{l+\varkappa} \varepsilon^{l+\varkappa}, \qquad \mathbf{W}_L \cdot \mathbf{W}^K = 0, \tag{7.9}$$

into Eq. (7.6), this gives

$$\widetilde{W}_{l+\varkappa^*} = h_{\varkappa^*\varkappa}\,\chi^\varkappa(t,q,\dot{q})\,, \qquad \varkappa, \varkappa^* = \overline{1,k}\,,$$

where $h_{\varkappa^*\varkappa}$ are entries of the matrix inverse to the matrix with entries $h^{\varkappa\varkappa^*}$,

$$h^{\varkappa\varkappa^*} = \varepsilon^{l+\varkappa} \cdot \varepsilon^{l+\varkappa^*}\,, \qquad \varkappa, \varkappa^* = \overline{1,k}\,.$$

The vectors $\varepsilon^{l+\varkappa}$, $\varkappa = \overline{1,k}$, are linearly independent, and hence

$$|h^{\varkappa\varkappa^*}| \neq 0\,. \tag{7.10}$$

Using expressions (7.9), we write the second Newton's law by the two equations

$$MW_L = Y_L + R_L\,,$$
$$\mathbf{Y}_L = \widetilde{Q}^\lambda \varepsilon_\lambda\,, \qquad \mathbf{R}_L = \mathcal{R}^\lambda \varepsilon_\lambda\,, \qquad \lambda = \overline{1,l}\,,$$
$$MW^K = Y^K + R^K\,, \tag{7.11}$$
$$\mathbf{Y}^K = \widetilde{Q}_{l+\varkappa}\varepsilon^{l+\varkappa}\,, \qquad \mathbf{R}^K = \Lambda_\varkappa \varepsilon^{l+\varkappa}\,, \qquad \varkappa = \overline{1,k}\,.$$

Here, $\mathbf{R} = \mathbf{R}_L + \mathbf{R}^K$ is the reaction of constraints, the components $\mathcal{R}_{l+\varkappa}$ of the vector $\mathbf{R}^K$ are deliberately denoted by $\Lambda_\varkappa$, because they will turn out to be Lagrange multipliers. Under condition (7.10) it follows from (7.9)–(7.10) that the vector $\mathbf{W}^K$ is uniquely determined by the constraint equations *qua* a function of the variables $t, q^\sigma, \dot{q}^\sigma$. Thus, in the K-space the law of motion is governed by the constraint equations and assumes the form (7.6). The component of the reaction $\mathbf{R}^K$, which appears here, is calculated using the second equation of the system (7.11).

The mathematical form of the constraint equations cannot have a direct effect on the vector $\mathbf{W}_L$, because this vector can be excluded from Eq. (7.6). Hence, the constraints may have only indirect effect on the component of the acceleration $\mathbf{W}_L$ in terms of the vector $\mathbf{R}_L$. In particular, the constraint equations may hold also with $\mathbf{R}_L = 0$. Such constraints are called *ideal*. This being so, *the effect of ideal constraints on the acceleration $\mathbf{W}$ is completely determined by their analytic representations.*

It is worth mentioning that in order to find out the effect of constraints on the formation of the reaction force it was necessary to write all types of constraints in a unique differential form (7.5). Namely, because of this form of the constraint equations it proved possible to show that the entire space of constraint equations decomposes into two orthogonal subspaces. Moreover, in passing we have found the analytic expressions for the reactions of constraints.[12]

---

[12] Such results for nonholonomic constraints were first published in the paper by *N.N. Polyakhov, S.A. Segzhda, and M.P. Yushkov*, "Dynamic equations as necessary conditions for minimality of the Gauss constraint," in *Vibrations and stability of mechanical systems. Appl. Mechanics.* Issue 5. Leningrad, Izd-vo Leningr. Univ., 1981, pp. 9–16 [in Russian]; later in 1985 they appeared as a part of the first edition of this monograph. In 1992 the same results were obtained with the help of the matrix

To conclude the subsection, we point out that in the case when the nonholonomic constraints $\varphi^{\varkappa}(t, q, \dot{q}) = 0$, $\varkappa = \overline{1, k}$, the coefficients $a_{\sigma}^{l+\varkappa}$ in formulas (7.5) are as follows:

$$a_{\sigma}^{l+\varkappa} = \frac{\partial \varphi^{\varkappa}}{\partial \dot{q}^{\sigma}}.$$

Hence, the above vectors $\varepsilon^{l+\varkappa}$ turn out to be equal to the generalized Hamilton operators

$$\varepsilon^{l+\varkappa} = \frac{\partial \varphi^{\varkappa}}{\partial \dot{q}^{\sigma}} \, \mathbf{e}^{\sigma} = \nabla' \varphi^{\varkappa}, \quad \varkappa = \overline{1, k}. \tag{7.12}$$

In turn, if holonomic constraints are given by the formulas $f^{\varkappa}(t, q) = 0$, $\varkappa = \overline{1, k}$, then, on differentiating them two times in time, we get

$$a_{\sigma}^{l+\varkappa} = \frac{\partial f^{\varkappa}}{\partial q^{\sigma}},$$

and hence,

$$\varepsilon^{l+\varkappa} \equiv \mathbf{e}_{*}^{l+\varkappa} = \frac{\partial f^{\varkappa}}{\partial q^{\sigma}} \, \mathbf{e}^{\sigma} = \nabla f^{\varkappa}, \quad \varkappa = \overline{1, k}. \tag{7.13}$$

Let us now examine one the most important questions of the nonholonomic mechanics—this being the question about the formation of the constraint reaction.

8   Reaction of Ideal Constraints

Finding the minimal additional force to secure the given motion conditions. Two statements of the problem are possible in the study of the constrained motion. In the first one, the mechanical system of general form, as subject to an active force $\mathbf{Y}$, moves to satisfy the ideal constraint equations. It is required to ascertain the character of motion of the system subject to given initial conditions. In this setting of the problem, the ultimate task is to write down the system of differential motion equations not involving the unknowns reactions constraints.

However, a different statement of the problem is also possible. Assume, for example, that some considerations require that a material point under a force $\mathbf{Y}$ would move along the surface of a prescribed form. This problem is not necessarily produced by a direct contact. This statement requires finding the force $\mathbf{R}$ (in addition to $\mathbf{Y}$) to produce a motion satisfying the effective requirement. Under this approach, the force $\mathbf{R}$ is called the *control* force; the motion is the *control*.

---

calculus by *F.E. Udwadia and R.E. Kalaba*, "A new perspective on constrained motion," Proceedings of the Royal Society. London. 1992. Vol. A439. Issue 1906. pp. 407–410. Analytic representations of the Lagrange multipliers as functions of time and the generalized coordinates and velocities for holonomic systems were derived in the early 1900s by G.K. Suslov and A.M. Lyapunov.

This problem for a single point can be extended in a natural way to general mechanical systems. It is worth pointing out that in such approach to motion, the equations that were earlier called the constraint equations now play as the mathematical expression of the conditions to be satisfied by the motion. Such equations are generally called the *equations of motion program.*

So, let us assume that there is a mechanical system of general form, that is, a system involving material bodies, rather than just material points. Assume that the equations of motion program are given in the form

$$\psi^{\varkappa}(t, q, \dot{q}, \ddot{q}) = a_{\sigma}^{l+\varkappa}(t, q, \dot{q})\ddot{q}^{\,\sigma} + a_0^{l+\varkappa}(t, q, \dot{q}) = 0,$$
$$\varkappa = \overline{1, k}, \quad l = s - k.$$
(8.1)

We recall that in this form one may write down time-differentiated nonholonomic constraints and two times time-differentiated holonomic constraints. It is required to find out which control forces R_{σ} should be added to the given forces Q_{σ} in order that the motion would satisfy the program given by Eq. (8.1). The forces R_{σ} need to be determined as functions depending on time, the position of the system, and its velocities.

Consider an additional requirement on the minimality of the magnitude of the vector **R** *qua* a vector from the tangent space, whose metric is defined from the expression for the kinetic energy of the system. We claim that in this case the required functions $R_{\sigma}(t, q, \dot{q})$ can be found, and in a unique fashion.

Indeed, if the program motion (8.1) is considered as a nonholonomic constraint, then the sought-for control force can be interpreted as the reaction force **R** of the constraint equations (8.1). However,

$$\mathbf{R} = \mathbf{R}^K + \mathbf{R}_L.$$

Moreover, $\mathbf{R}_L = 0$ if the constraints are ideal, and hence, $\mathbf{R} = \mathbf{R}^K = \Lambda_{\varkappa}\varepsilon^{l+\varkappa}$, that is the reaction (the control force) is of the smallest magnitude. In this case, the vector motion equation reads as

$$M\mathbf{W} = \mathbf{Y} + \Lambda_{\varkappa}\varepsilon^{l+\varkappa}.$$

Since $M\mathbf{W}$ is equal to the vector $\mathbf{Y} + \mathbf{R}$, it follows that

$$\mathbf{R} \cdot \varepsilon_{\lambda} = 0, \quad \lambda = \overline{1, l},$$
$$\mathbf{R} \cdot \varepsilon^{l+\varkappa} = M\chi^{\varkappa}(t, q, \dot{q}) - \mathbf{Y} \cdot \varepsilon^{l+\varkappa}, \quad \varkappa = \overline{1, k}.$$

Substituting into these equations the vectors ε_{λ} and $\varepsilon^{l+\varkappa}$ of the form (7.8) and (7.6), respectively, we have

$$R_{\sigma}\beta_{\lambda}^{\sigma}(t, q, \dot{q}) = 0, \quad \lambda = \overline{1, l},$$
$$R^{\tau}a_{\tau}^{l+\varkappa}(t, q, \dot{q}) = M\chi^{\varkappa}(t, q, \dot{q}) - Q^{\tau}a_{\tau}^{l+\varkappa}(t, q, \dot{q}),$$
$$\varkappa = \overline{1, k}, \quad \tau = \overline{1, s},$$
(8.2)

where $R^\tau = \mathbf{R} \cdot \mathbf{e}^\tau$, $Q^\tau = \mathbf{Y} \cdot \mathbf{e}^\tau$. Taking into account that $R^\tau = g^{\tau\sigma} R_\sigma$, $Q^\tau = g^{\tau\sigma} Q_\sigma$, system (8.2) assumes the form

$$
\begin{aligned}
\beta_\lambda^\sigma R_\sigma &= 0, \qquad \lambda = \overline{1, l}, \quad l = s - k, \\
a_\tau^{l+\varkappa} g^{\tau\sigma} R_\sigma &= b^\varkappa, \qquad \varkappa = \overline{1, k}, \quad \sigma, \tau = \overline{1, s}.
\end{aligned}
\tag{8.3}
$$

Here, $b^\varkappa = M \chi^\varkappa - a_\tau^{l+\varkappa} g^{\tau\sigma} Q_\sigma$.

System (8.3) is a system of linear algebraic equations with respect to the unknowns R_σ. This system can be solved as follows.

Let is consider the case when the pseudo-accelerations are given as follows:

$$
\begin{aligned}
w_*^\lambda &= \ddot{q}^{\,\lambda}, \qquad \lambda = \overline{1, l}, \\
w_*^{l+\varkappa} &= a_\sigma^{l+\varkappa} \ddot{q}^{\,\sigma}, \qquad \varkappa = \overline{1, k}.
\end{aligned}
$$

In this case, we have

$$
\beta_\lambda^\mu = \begin{cases} 0, & \mu \neq \lambda, \\ 1, & \mu = \lambda, \end{cases}
$$

$$
\lambda, \mu = \overline{1, l},
$$

and hence, the first l equations of (8.3) can be rewritten as

$$
R_\lambda = -\beta_\lambda^{l+\varkappa^*} R_{l+\varkappa^*}, \qquad \lambda = \overline{1, l}, \quad \varkappa^* = \overline{1, k}.
\tag{8.4}
$$

Substituting (8.4) into the last k equations of system (8.3), we find that

$$
r^{\varkappa\varkappa^*} R_{l+\varkappa^*} = b^\varkappa, \qquad \varkappa, \varkappa^* = \overline{1, k},
\tag{8.5}
$$

where $r^{\varkappa\varkappa^*} = a_\tau^{l+\varkappa} (g^{\tau, l+\varkappa^*} - \beta_\lambda^{l+\varkappa^*} g^{\tau\lambda})$. In turn, the solution to system (8.5) can be found by Cramer's rule

$$
R_{l+\varkappa^*} = \frac{A_{\varkappa\varkappa^*} b^\varkappa}{\det[r^{\varkappa\varkappa^*}]}, \qquad \varkappa, \varkappa^* = \overline{1, k}.
\tag{8.6}
$$

Here, $A_{\varkappa\varkappa^*}$ is the algebraic complement of the element $r^{\varkappa\varkappa^*}$ in the determinant $\det[r^{\varkappa\varkappa^*}]$.

This being so, we have found R_σ *qua* functions of t, q and $\dot{q}$, because by formulas (8.4) and (8.6) they can be expressed in terms of the coefficients $g_{\alpha\beta}(t, q)$, $a_\alpha^{l+\varkappa}(t, q, \dot{q})$, $\alpha, \beta = \overline{0, s}$, $\varkappa = \overline{1, k}$, and the generalized forces $Q_\sigma(t, q, \dot{q})$.

9 Equations of the Constrained Motion of General Mechanical Systems

In this subsection, we write, respectively, $f_0^\varkappa(t, q) = 0$, $f_1^\varkappa(t, q, \dot{q}) = 0$, $f_2^\varkappa(t, q, \dot{q}, \ddot{q}) = 0$, for holonomic constraints, nonholonomic constraints, and for second-order nonholonomic constraints.

The motion equations of holonomic system. The motion of a general mechanical system with s degrees of freedom will be described in curvilinear coordinates $q = (q^1, ..., q^s)$, in which the principal basis and the reciprocal basis are as follows:

$$\{\mathbf{e}_1, ..., \mathbf{e}_s\}, \quad \{\mathbf{e}^1, ..., \mathbf{e}^s\}. \tag{9.1}$$

The vectors of these bases clearly satisfy the relations

$$\mathbf{e}^\rho \cdot \mathbf{e}_\sigma = \delta_\sigma^\rho = \begin{cases} 1, & \rho = \sigma, \\ 0, & \rho \neq \sigma. \end{cases} \tag{9.2}$$

Assume that the motion of a system is subject to ideal holonomic constraints

$$f_0^\varkappa(t, q) = 0, \quad \varkappa = \overline{1, k}. \tag{9.3}$$

Hence, by formulas (7.13) the vector motion equation in the tangent space reads as

$$M\mathbf{W} = \mathbf{Y} + \Lambda_\varkappa \nabla f_0^\varkappa. \tag{9.4}$$

Let us introduce the new system of curvilinear coordinates $q_* = (q_*^1, ..., q_*^s)$ and define the transition formulas between the two systems under consideration as follows:

$$q_*^\rho = q_*^\rho(t, q), \quad q^\sigma = q^\sigma(t, q_*), \quad \rho, \sigma = \overline{1, s}. \tag{9.5}$$

Based on formulas (9.5), we write two systems of vectors

$$\mathbf{e}_*^\rho = \frac{\partial q_*^\rho}{\partial q^{\sigma_*}} \mathbf{e}^{\sigma_*}, \quad \mathbf{e}_\sigma^* = \frac{\partial q^\tau}{\partial q_*^\sigma} \mathbf{e}_\tau, \quad \rho, \sigma, \sigma_*, \tau = \overline{1, s}. \tag{9.6}$$

These vectors form the principal and the reciprocal bases of the new system of coordinates $q_* = (q_*^1, ..., q_*^s)$, because by formulas (9.2)

$$\mathbf{e}_*^\rho \cdot \mathbf{e}_\sigma^* = \frac{\partial q_*^\rho}{\partial q^{\sigma_*}} \mathbf{e}^{\sigma_*} \cdot \frac{\partial q^\tau}{\partial q_*^\sigma} \mathbf{e}_\tau = \frac{\partial q_*^\rho}{\partial q^\tau} \frac{\partial q^\tau}{\partial q_*^\sigma} = \delta_\sigma^\rho = \begin{cases} 1, & \rho = \sigma, \\ 0, & \rho \neq \sigma. \end{cases}$$

We shall write the transition formulas (9.5) in a more detailed form on setting

$$q_*^\lambda = q_*^\lambda(t, q), \quad \lambda = \overline{1, l}, \quad l = s - k,$$
$$q_*^{l+\varkappa} = q_*^{l+\varkappa}(t, q) \equiv f_0^\varkappa(t, q), \quad \varkappa = \overline{1, k}. \tag{9.7}$$

It is worth pointing out that here the functions of (t, q) in the first formulas (with $\lambda = \overline{1, l}$) are chosen by the researcher, while the following ones (with $l + \varkappa$, $\varkappa = \overline{1, k}$) take into account the constraint equations (9.3), and hence the coordinates $q_*^{l+\varkappa}$, $\varkappa = \overline{1, k}$, vanish in the course of motion.

By formulas (9.6) and (9.7) the constraint equations (9.3) define the k vectors

$$\mathbf{e}_*^{l+\varkappa} = \frac{\partial q_*^{l+\varkappa}}{\partial q^\sigma} \mathbf{e}^\sigma = \frac{\partial f_0^\varkappa}{\partial q^\sigma} \mathbf{e}^\sigma = \nabla f_0^\varkappa, \quad \varkappa = \overline{1, k}, \tag{9.8}$$

which form the reciprocal basis for the K-space. By (9.6) the orthogonal L-space is formed by the vectors from the principal basis

$$\mathbf{e}_\lambda^* = \frac{\partial q^\tau}{\partial q_*^\lambda} \mathbf{e}_\tau, \quad \lambda = \overline{1, l}.$$

Taking into account formulas (9.8), we rewrite the motion equation (9.4) in the form

$$M\mathbf{W} = \mathbf{Y} + \Lambda_\varkappa \mathbf{e}_*^{l+\varkappa}. \tag{9.9}$$

Multiplying it by the vectors $\mathbf{e}_\sigma^*$, $\sigma = \overline{1, s}$, we get two groups of the *Lagrange equations of the second kind*:

$$MW_\lambda^* = Q_\lambda^*, \quad \lambda = \overline{1, l}, \quad l = s - k, \tag{9.10}$$

$$MW_{l+\varkappa}^* = Q_{l+\varkappa}^* + \Lambda_\varkappa, \quad \varkappa = \overline{1, k}.$$

We recall that

$$MW_\sigma = \frac{d}{dt} \frac{\partial T}{\partial \dot{q}^\sigma} - \frac{\partial T}{\partial q^\sigma}, \quad \sigma = \overline{1, s}. \tag{9.11}$$

Using formulas (9.8), Eq. (9.9) can be rewritten as follows:

$$M\mathbf{W} = \mathbf{Y} + \Lambda_\varkappa \frac{\partial f_0^\varkappa}{\partial q^\tau} \mathbf{e}^\tau.$$

Multiplying it by the vectors $\mathbf{e}_\sigma$, $\sigma = \overline{1, s}$, we find

$$MW_\sigma = Q_\sigma + \Lambda_\varkappa \frac{\partial f_0^\varkappa}{\partial q^\sigma}, \quad \sigma = \overline{1, s}. \tag{9.12}$$

The system of s scalar equations (9.12) contains $s + k$ unknowns $q^1, \ldots, q^s, \Lambda_1, \ldots, \Lambda_k$. This system should be augmented with the constraint equations (9.3). This is analogous to the case of first-kind Lagrange equations for a motion of a system of material points. Hence, the system of equations (9.12) may be called the *Lagrange equations of the first kind in curvilinear coordinates*. In the literature, this system is usually called the *Lagrange equations of the second kind with multipliers*.

The motion equations for nonholonomic systems. Assume now that the motion of a general mechanical system described by the curvilinear coordinates $q = (q^1, \ldots, q^s)$ with the bases (9.1) is subject to the ideal nonholonomic constraints

$$f_1^{\varkappa}(t, q, \dot{q}) = 0, \quad \varkappa = \overline{1, k}. \tag{9.13}$$

By formulas (7.12), the vector equation of the constrained motion reads as

$$M\mathbf{W} = \mathbf{Y} + \Lambda_{\varkappa} \nabla' f_1^{\varkappa}. \tag{9.14}$$

Let us explain the appearance of the vectors $\nabla' f_1^{\varkappa}$ with somewhat different positions. In the case of the constrained motion of a mechanical system under nonholonomic constraints (9.13), it is not sufficient to introduce a new system of curvilinear coordinates, and so sometimes together with the vector of generalized velocities $\dot{q} = (\dot{q}^1, \ldots, \dot{q}^s)$ one has to introduce a new vector of pseudo-velocities (quasi-velocities) $v_* = (v_*^1, \ldots, v_*^s)$. We define the transition formulas between them, assuming that now t and q are parameters:

$$v_*^{\rho} = v_*^{\rho}(t, q, \dot{q}), \quad \dot{q}^{\sigma} = \dot{q}^{\sigma}(t, q, v_*), \quad \rho, \sigma = \overline{1, s}. \tag{9.15}$$

From transformations (9.15) one may construct two systems of vectors

$$\varepsilon^{\rho} = \frac{\partial v_*^{\rho}}{\partial \dot{q}^{\sigma_*}} \mathbf{e}^{\sigma_*}, \quad \varepsilon_{\sigma} = \frac{\partial \dot{q}^{\tau}}{\partial v_*^{\sigma}} \mathbf{e}_{\tau}, \quad \rho, \sigma, \sigma_*, \tau = \overline{1, s}. \tag{9.16}$$

It is worth pointing out that these vectors are calculated for a particular position of the system $q = (q^1, \ldots, q^s)$, which it occupies at a given time t and having the generalized velocities $\dot{q} = (\dot{q}^1, \ldots, \dot{q}^s)$. In other words, vectors (9.16) are calculated with the phase state of the mechanical system at time t.

The above vectors (9.16) can be called the vectors of *nonholonomic bases*, since in view of formulas (9.2) they have the property

$$\varepsilon^{\rho} \cdot \varepsilon_{\sigma} = \frac{\partial v_*^{\rho}}{\partial \dot{q}^{\sigma_*}} \mathbf{e}^{\sigma_*} \cdot \frac{\partial \dot{q}^{\tau}}{\partial v_*^{\sigma}} \mathbf{e}_{\tau} = \frac{\partial v_*^{\rho}}{\partial \dot{q}^{\tau}} \frac{\partial \dot{q}^{\tau}}{\partial v_*^{\sigma}} = \delta_{\sigma}^{\rho} = \begin{cases} 1, & \rho = \sigma, \\ 0, & \rho \neq \sigma. \end{cases}$$

If in transformations (9.15) the constraint equations (9.13) are taken into account by the formulas

$$v_*^\lambda = v_*^\lambda(t, q, \dot{q}), \quad \lambda = \overline{1, l}, \quad l = s - k,$$

$$v_*^{l+\varkappa} = v_*^{l+\varkappa}(t, q, \dot{q}) \equiv f_1^\varkappa(t, q, \dot{q}), \quad \varkappa = \overline{1, k},$$

then the last vectors of the reciprocal nonholonomic basis assume the form

$$\varepsilon^{l+\varkappa} = \frac{\partial v_*^{l+\varkappa}}{\partial \dot{q}^\sigma} \mathbf{e}^\sigma = \frac{\partial f_1^\varkappa}{\partial \dot{q}^\sigma} \mathbf{e}^\sigma = \nabla' f_1^\varkappa(t, q, \dot{q}), \quad \varkappa = \overline{1, k}. \tag{9.17}$$

The vectors (9.17) form in the s-dimensional tangent space the k-dimensional K-space; by formulas (9.16) the orthogonal l-dimensional space L is given by the vectors

$$\varepsilon_\lambda = \frac{\partial \dot{q}^\tau}{\partial v_*^\lambda} \mathbf{e}_\tau, \quad \lambda = \overline{1, l}.$$

It is worth noting that this decomposition of the original s-dimensional tangent space into the direct sum of the subspaces K and L is defined with fixed t, q and $\dot{q}$.

Now in view of formulas (9.17) the vector equation (9.14) assumes the form

$$M\mathbf{W} = \mathbf{Y} + \Lambda_\varkappa \varepsilon^{l+\varkappa}.$$

Multiplying this equation by the vectors $\varepsilon_\rho = \frac{\partial \dot{q}^\tau}{\partial v_*^\rho} \mathbf{e}_\tau$, $\rho = \overline{1, s}$, we get two systems of *Maggi equations*:

$$(MW_\sigma - Q_\sigma)\frac{\partial \dot{q}^\sigma}{\partial v_*^\lambda} = 0, \quad \lambda = \overline{1, l}, \quad l = s - k, \tag{9.18}$$

$$(MW_\sigma - Q_\sigma)\frac{\partial \dot{q}^\sigma}{\partial v_*^{l+\varkappa}} = \Lambda_\varkappa, \quad \varkappa = \overline{1, k}. \tag{9.19}$$

We recall that here MW_σ can be put in the form (9.11).

The l equations in (9.18) involve, in particular, all the generalized coordinates $q^1, \ldots, q^s$, and hence they are to be integrated jointly with the constraint equations (9.13). For purposes of numerical integration, it is more convenient to differentiate in time. After specifying the initial conditions one may obtain the solution of this closed system of differential equations

$$q^\sigma = q^\sigma(t), \quad \sigma = \overline{1, s}. \tag{9.20}$$

Substituting solution (9.20) in formulas (9.19), we find the variation of the generalized reactions

$$\Lambda_\varkappa = \Lambda_\varkappa(t), \quad \varkappa = \overline{1, k}.$$

From these formulas one may find, in particular, the conditions for lifting constraints (9.13).

We note that from the above it follows that, in a motion of a nonholonomic systems, the number $l = s - k$ does not mean the number of degrees of freedom of the mechanical system. Depending on the specification of the initial conditions, by integrating the Maggi equations (9.18) jointly with the constraint equations (9.13) the mechanical system at a given time t may occupy any position $q^1, \dots, q^s$. Hence, for nonholonomic systems the number of degrees of freedom is s, while l defines the number of independent quasi-velocities $v_*^1, \dots, v_*^l$.

Let us now use formulas (9.17) to rewrite the vector motion equation in the form

$$M\mathbf{W} = \mathbf{Y} + \Lambda_\varkappa \frac{\partial f_1^\varkappa}{\partial \dot{q}^\tau} \, \mathbf{e}^\tau.$$

Multiplying it by the vectors $\mathbf{e}_\sigma$, $\sigma = \overline{1, s}$, we get the *Lagrange equations of the first kind in curvilinear coordinates for nonholonomic systems*:

$$MW_\sigma = Q_\sigma + \Lambda_\varkappa \frac{\partial f_1^\varkappa}{\partial \dot{q}^\sigma}, \quad \sigma = \overline{1, s}. \tag{9.21}$$

These equations are also called the *Lagrange equations of the second kind with multipliers for nonholonomic systems*.

The motion equations of general nonholonomic systems with linear nonholonomic second-order constraints. Assume that the motion of a general mechanical system, whose motion is described by the curvilinear coordinates $q = (q^1, \dots, q^s)$ with bases (9.1), is subject to linear nonholonomic second-order constraints[13]

$$f_2^\varkappa(t, q, \dot{q}, \ddot{q}) \equiv a_{2\sigma}^{l+\varkappa}(t, q, \dot{q}) \, \ddot{q}^\sigma + a_{2,0}^{l+\varkappa}(t, q, \dot{q}) = 0, \quad \varkappa = \overline{1, k}. \tag{9.22}$$

Assuming that $t, q, \dot{q}$ are parameters, we introduce the transformation formulas

$$w_*^\rho = w_*^\rho(t, q, \dot{q}, \ddot{q}), \quad \ddot{q}^\sigma = \ddot{q}^\sigma(t, q, \dot{q}, w_*), \quad \rho, \sigma = \overline{1, s} \tag{9.23}$$

between the generalized accelerations $\ddot{q} = (\ddot{q}^1, \dots, \ddot{q}^s)$ and the *quasi-accelerations* (*pseudo-accelerations*) $w_* = (w_*^1, \dots, w_*^s)$. Using the transition formulas (9.23), we construct two systems of vectors

$$\varepsilon^\rho = \frac{\partial w_*^\rho}{\partial \ddot{q}^{\sigma_*}} \, \mathbf{e}^{\sigma_*} = a_{2\sigma_*}^{l+\varkappa} \mathbf{e}^{\sigma_*}, \quad \varepsilon_\sigma = \frac{\partial \ddot{q}^\tau}{\partial w_*^\sigma} \, \mathbf{e}_\tau, \quad \rho, \sigma, \sigma_*, \tau = \overline{1, s}. \tag{9.24}$$

---

[13] At present only one mechanical example of second-order constraint is known: notion of a heavy point at the end of a string spiraling about a vertical circular cylinder (see the paper by F. Kitzka, "An example for the application of a nonholonomic constraint of 2nd order in particle mechanics," ZAMM. 1986. Vol. 66. Issue 7. S. 312–314). However, the class of problems under consideration becomes much wider if the form (9.22) specifies the program of motion.

The vectors (9.24) form the reciprocal and principal *nonholonomic bases*, because by formulas (9.2)

$$\varepsilon^\rho \cdot \varepsilon_\sigma = \frac{\partial w_*^\rho}{\partial \ddot{q}^{\,\sigma_*}} \mathbf{e}^{\sigma_*} \cdot \frac{\partial \ddot{q}^{\,\tau}}{\partial w_*^\sigma} \mathbf{e}_\tau = \frac{\partial w_*^\rho}{\partial \ddot{q}^{\,\tau}} \frac{\partial \ddot{q}^{\,\tau}}{\partial w_*^\rho} = \delta_\sigma^\rho = \begin{cases} 1, & \rho = \sigma, \\ 0, & \rho \neq \sigma. \end{cases}$$

Let us give the transition formulas (9.23) in the form

$$w_*^\lambda = w_*^\lambda(t, q, \dot{q}, \ddot{q}), \quad \lambda = \overline{1, l}, \quad l = s - k,$$
$$w_*^{l+\varkappa} = f_2^\varkappa(t, q, \dot{q}, \ddot{q}), \quad \varkappa = \overline{1, k}. \tag{9.25}$$

Constraints (9.22) holding, the last quasi-velocities $w_*^{l+\varkappa}$, $\varkappa = \overline{1, k}$, become zero in the process of motion. Formulas (9.25) give the vectors

$$\varepsilon^{l+\varkappa} = \frac{\partial w_*^{l+\varkappa}}{\partial \ddot{q}^{\,\sigma}} \mathbf{e}^\sigma = \frac{\partial f_2^\varkappa}{\partial \ddot{q}^{\,\sigma}} \mathbf{e}^\sigma = \nabla'' f_2^\varkappa, \quad \varkappa = \overline{1, k}, \tag{9.26}$$

which form the reciprocal basis for the K-space singled out in the s-dimensional tangent space by the constraint equations (9.22). The orthogonal complement in the form of the L-space is defined, in view of formulas (9.24), by vectors of its principal basis

$$\varepsilon_\lambda = \frac{\partial \ddot{q}^{\,\tau}}{\partial w_*^\lambda} \mathbf{e}_\tau, \quad \lambda = \overline{1, l}.$$

By formulas (9.26), the vector motion equation with ideal nonholonomic second-order constraints (9.22) is of the form

$$M\mathbf{W} = \mathbf{Y} + \Lambda_\varkappa \varepsilon^{l+\varkappa}. \tag{9.27}$$

Multiplying it by the vectors

$$\varepsilon_\rho = \left(\partial \ddot{q}^{\,\tau} / \partial w_*^\rho \right) \mathbf{e}_\tau, \quad \rho = \overline{1, s},$$

we get two groups of *generalized Maggi equations*[14]:

$$(MW_\sigma - Q_\sigma) \frac{\partial \ddot{q}^{\,\sigma}}{\partial w_*^\lambda} = 0, \quad \lambda = \overline{1, l}, \quad l = s - k, \tag{9.28}$$

$$(MW_\sigma - Q_\sigma) \frac{\partial \ddot{q}^{\,\sigma}}{\partial w_*^{l+\varkappa}} = \Lambda_\varkappa, \quad \varkappa = \overline{1, k}. \tag{9.29}$$

---

[14] These equations were first published in the papers: *A. Przeborski. Die allgemeinsten Gleichungen der klassischen Dynamik //* Math. Zeitschrift. 1931–1932. Bd. 36. H. 2. S. 184–194; *G. Hamel. Nichtholonome Systeme höherer Art //* Sitzungsberichte der Berliner Math. Gesellschaft. 1938. Bd. 37. S. 41–52.

We recall that the left-hand sides of these equations can be written in an expanded form with the help of formulas (9.11).

Integrating, with given initial conditions, Eq. (9.28) jointly with the constraint equations (9.22), we get the motion equations of the general mechanical system under consideration,

$$q^\sigma = q^\sigma(t), \quad \sigma = \overline{1, s}. \tag{9.30}$$

Substituting the functions (9.30) thus obtained into formulas (9.29), we obtain the law of variation of the generalized reactions

$$\Lambda_\varkappa = \Lambda_\varkappa(t), \quad \varkappa = \overline{1, k}. \tag{9.31}$$

Using functions (9.31) one may, in particular, find conditions for lifting a mechanical system from constraints (9.22).

As in the previous paragraph, in constraints (9.22) the number $l = s - k$ is equal to the number of independent quasi-accelerations w_*^λ, $\lambda = \overline{1, l}$ (rather than the number of degrees of freedom of the general nonholonomic mechanical system under consideration).

If now we rewrite Eq. (9.27) in the form

$$M\mathbf{W} = \mathbf{Y} + \Lambda_\varkappa \frac{\partial f_2^\varkappa}{\partial \ddot{q}^\tau} \mathbf{e}^\tau \tag{9.32}$$

and multiply Eq. (9.32) by the vectors $\mathbf{e}_\sigma$, $\sigma = \overline{1, s}$, then by (9.2) we get the *Lagrange equations of the first kind in curvilinear coordinates under second-order constraints* (the *Lagrange equations of the second kind with multipliers under second-order constraints*)[15]:

$$MW_\sigma = Q_\sigma + \Lambda_\varkappa \frac{\partial f_2^\varkappa}{\partial \ddot{q}^\sigma}, \quad \sigma = \overline{1, s}. \tag{9.33}$$

Integrating, with given initial conditions, Eq. (9.33) jointly with the constraint equations (9.22), we get the motion equations of the mechanical system.

To conclude this subsection, we point out once again that in the present setting the system of vectors of the principal and the reciprocal nonholonomic bases is defined for a concrete phase state of the system $(q, \dot{q})$ at a time t under consideration.

Example 9 *Motion of an ice skater* (composition of the Maggi equations). Let us consider the motion by an inclined ice skater, which stands on a short skate A (Fig. 15). Historically, S.A. Chaplygin was the first to set a similar problem of determining the plane-parallel motion of a sledge with thin bent runners; the sledge can rotate about the lower point of the runner (*the Chaplygin slegde*).

---

[15] See the paper by *H. Hamel* in the previous footnote.

Fig. 15 Motion of an
inclined ice skater (the
Chaplygin sledge)

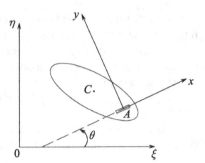

Let us introduce the movable $Axyz$ and fixed $O\xi\eta\zeta$ systems of coordinates. We assume that there is the resistance force $\mathbf{F}_{\text{resist}} = -\kappa_1 \mathbf{v}_C$ and the moment of resistance $\mathbf{N}_{\text{resist}} = -\kappa_2 \boldsymbol{\omega}$; C is the center of mass of the ice skater.

Since the ice skater may move only along the skate (possibly simultaneously rotating on it) the constraint imposed on this system is that the velocity of the point A is always directed along the moving axis Ax, that is, its projection v_{Ay} to the axis Ay is zero at each time. We let $\mathbf{i}_1$, $\mathbf{j}_1$, $\mathbf{k}_1$ denote the unit vectors of this fixed system of coordinates $O\xi\eta\zeta$, and denote by ξ_C, η_C the coordinates of the centroid in the fixed system coordinates. The coordinates of the centroid in the moving system coordinates $Axyz$ are assumed to be equal: $x_C = \alpha$, $y_C = \beta$.

For the generalized coordinates of the system, we take the coordinates of A and the angle between the axes Ax and $O\xi$:

$$q^1 = \xi, \qquad q^3 = \eta, \qquad q^2 = \theta.$$

Let us write down the constraint equation. To this aim, we shall express the constraint in the projections of the vector $\mathbf{v}_A$ to the fixed axes $O\xi\eta$, taking into account that

$$\mathbf{v}_A = v_{A\xi}\mathbf{i}_1 + v_{A\eta}\mathbf{j}_1 = \dot{\xi}\mathbf{i}_1 + \dot{\eta}\mathbf{j}_1 \,.$$

The projection of the vector $\mathbf{v}_A$ to the axis Ay is of the form

$$v_{Ay} = -\dot{\xi}\sin\theta + \dot{\eta}\cos\theta \,,$$

and hence, the constraint equation $v_{Ay} = 0$ reads as

$$f_1(t, q^1, q^2, q^3, \dot{q}^1, \dot{q}^2, \dot{q}^3) \equiv -\dot{\xi}\sin\theta + \dot{\eta}\cos\theta = 0 \,. \tag{9.34}$$

The kinetic energy, which can be found by König's theorem, is given by

$$T = \frac{1}{2}M\big[(\dot{\xi} - \dot{\theta}(\alpha\sin\theta + \beta\cos\theta))^2 + (\dot{\eta} + \dot{\theta}(\alpha\cos\theta - \beta\sin\theta))^2 + k_C^2\dot{\theta}^2\big], \tag{9.35}$$

where k_C is the radius of gyration of the body with respect to the axis passing through its center of gravity and perpendicular to the motion plane; M is the mass of the system.

The generalized forces acting on the system are as follows:

$$Q_\xi = -\kappa_1 \dot{\xi}, \qquad Q_\eta = -\kappa_1 \dot{\eta}, \qquad Q_\theta = -\kappa_2 \dot{\theta}. \tag{9.36}$$

The constraint equation (9.34) can be put in the form

$$\dot{\xi} \tan \theta - \dot{\eta} = 0. \tag{9.37}$$

Let us introduce the quasi-velocities as follows:

$$v_*^1 = \dot{\xi}, \qquad v_*^2 = \dot{\theta}, \qquad v_*^3 = \dot{\xi} \tan \theta - \dot{\eta}.$$

Expressing the generalized velocities in terms of the quasi-velocities, we obtain the inverse transformation

$$\dot{\xi} = v_*^1, \qquad \dot{\theta} = v_*^2, \qquad \dot{\eta} = v_*^1 \tan \theta - v_*^3.$$

Using these formulas one finds the derivatives

$$\frac{\partial \dot{q}^1}{\partial v_*^1} = 1, \qquad \frac{\partial \dot{q}^2}{\partial v_*^1} = 0, \qquad \frac{\partial \dot{q}^3}{\partial v_*^1} = \tan \theta,$$

$$\frac{\partial \dot{q}^1}{\partial v_*^2} = 0, \qquad \frac{\partial \dot{q}^2}{\partial v_*^2} = 1, \qquad \frac{\partial \dot{q}^3}{\partial v_*^2} = 0,$$

$$\frac{\partial \dot{q}^1}{\partial v_*^3} = 0, \qquad \frac{\partial \dot{q}^2}{\partial v_*^3} = 0, \qquad \frac{\partial \dot{q}^3}{\partial v_*^3} = 1.$$

Next, employing the already calculated coefficients in the Maggi equations (9.18) we may, after some simplifications, write the differential motion equations for our system:

$$\ddot{\xi} + \ddot{\eta} \tan \theta - \ddot{\theta} \frac{\beta}{\cos \theta} - \dot{\theta}^2 \frac{\alpha}{\cos \theta} = -\frac{\kappa_1}{M}(\dot{\xi} + \dot{\eta} \tan \theta),$$

$$\gamma^2 \ddot{\theta} + \ddot{\eta}(\alpha \tan \theta - \beta \sin \theta) - \ddot{\xi}(\alpha \sin \theta + \beta \cos \theta) = -\frac{\kappa_2}{M} \dot{\theta}. \tag{9.38}$$

These equations must be integrated jointly with the constraint equation (9.37).

Figure 16 gives the results of numerical integration of the system of differential equations within 10 s. In calculations, it was assumed that

Fig. 16 Time-distance
graphs of an inclined ice
skater

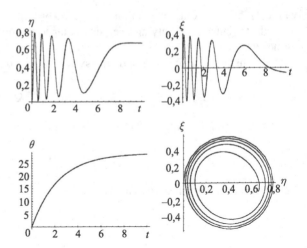

$$\gamma^2 = 0.07 \text{ m}^2, \quad \kappa_1/M = 1 \text{ s}^{-1}, \quad \kappa_2/M = 0.02 \text{ m}^2 \cdot \text{s}^{-1},$$

$$\xi(0) = 0, \quad \dot{\xi}(0) = 5 \text{ m} \cdot \text{s}^{-1}, \quad \eta(0) = 0, \quad \dot{\eta}(0) = 0,$$

$$\theta(0) = 0, \quad \dot{\theta}(0) = 12.5 \text{ s}^{-1}, \quad \alpha = 0, \quad \beta = 0.$$

Moreover, using equations (9.19) one may also easily write down the expression for the generalized reaction of the nonholonomic constraint

$$\frac{\Lambda}{M} = \ddot{\xi} \tan \theta + \dot{\xi} \dot{\theta} \frac{1}{\cos^2 \theta} + \ddot{\theta}(\alpha \cos \theta - \beta \sin \theta) - \dot{\theta}^2(\alpha \sin \theta + \beta \cos \theta) + \frac{\kappa_1}{M} \dot{\xi} \tan \theta.$$

Substituting the above motion equations on the right of this formula, we get an expression for the generalized reactions *qua* a funtion of time. From this function one may determine, in particular, the moments of possible liberation of a mechanical from a the imposed nonholonomic constraint.

Example 10 *Motion of a wheeled robot* (composition of the Maggi equations and the Lagrange equations of the first kind). Let us consider the motion of a wheeled robot (Fig. 17), consisting of a body of mass M_1 and the forward axle of mass M_2. Let J_1 and J_2 be, correspondingly, their moments of inertia with respect to the vertical axes passing through their centers of mass. The forward axle may rotate about its vertical axis, which passes through its center. We shall neglect the individual masses of the wheels and the rear axle. The motion of the robot is triggered by the force $F_1(t)$, which acts along its longitudinal axis Cx, and by the twisting moment $L_1(t)$, which rotates the forward axle. Here, $F_1(t), L_1(t)$ are given time functions. In addition, we take into account the resistance force $F_2(v_C)$, which points in the opposite direction to the velocity $\mathbf{v}_C$ of the body center of mass C, and consider the moment of resistance $L_2(\dot{\theta})$ (which is applied to the forward axle and is directed opposite to the angular velocity of its rotation) and the restoring torque $L_3(\theta)$.

Fig. 17 Schematics of a wheeled robot

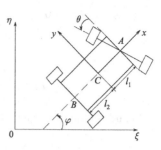

Let us compose the Maggi equations for the study of the motion of this system.

The motion of the robot in the horizontal plane will be studied with respect to the fixed coordinate system $O\xi\eta\zeta$. The position of the robot will be given in the generalized coordinates: $q^1 = \varphi$ is the angle between the longitudinal axis Cx of the body and the $O\xi$-axis, $q^2 = \theta$ is the angle between the forward axle and the perpendicular to the axis Cx, $q^3 = \xi_C$, $q^4 = \eta_C$ are the coordinates of the point C.

The motion is subject to two nonholonomic constraints, which express the absence of sideslip in the forward and backward axes. Their equations can be written as formula (9.34) from Example 9:

$$-\dot{\xi}_B \sin\varphi + \dot{\eta}_B \cos\varphi = 0,$$
$$-\dot{\xi}_A \sin(\varphi + \theta) + \dot{\eta}_A \cos(\varphi + \theta) = 0. \tag{9.39}$$

Here, $\xi_A, \eta_A, \xi_B, \eta_B$ are the coordinates of the centers of mass of the forward and backward axles of the robot. Assume that the distances of the centroids of these axles of the body centroid are, respectively, l_1 and l_2. Now the equation of the nonholonomic constraints (9.39) can be put in the form

$$f_1^1 \equiv -\dot{\xi}_C \sin\varphi + \dot{\eta}_C \cos\varphi - l_2\dot{\varphi} = 0,$$
$$f_1^2 \equiv -\dot{\xi}_C \sin(\varphi + \theta) + \dot{\eta}_C \cos(\varphi + \theta) + l_1\dot{\varphi}\cos\theta = 0. \tag{9.40}$$

The kinetic energy of the system, which is composed of the kinetic energies of the body and forward axle, is calculated by the formula

$$2T = M^*(\dot{\xi}_C^2 + \dot{\eta}_C^2) + J^*\dot{\varphi}^2 + J_2\dot{\theta}^2 + 2J_2\dot{\varphi}\dot{\theta} + 2M_2l_1\dot{\varphi}(-\dot{\xi}_C \sin\varphi + \dot{\eta}_C \cos\varphi),$$
$$M^* = M_1 + M_2, \qquad J^* = J_1 + J_2 + M_2l_1^2. \tag{9.41}$$

The generalized forces, which act on the robot, can be written as

$$Q_1 \equiv Q_\varphi = 0 \,,$$

$$Q_2 \equiv Q_\theta = L_1(t) - L_2(\dot\theta) - L_3(\theta) \,,$$

$$Q_3 \equiv Q_{\xi_C} = F_1(t) \cos\varphi - F_2(v_C)\dot\xi_C/v_C \,,$$
(9.42)

$$Q_4 \equiv Q_{\eta_C} = F_1(t) \sin\varphi - F_2(v_C)\dot\eta_C/v_C \,, \quad v_C = \sqrt{\dot\xi_C^2 + \dot\eta_C^2} \,.$$

Let us introduce the quasi-velocities by the formulas

$$v_*^1 = \dot\varphi \,, \qquad v_*^2 = \dot\theta \,,$$

$$v_*^3 = -l_2\dot\varphi - \dot\xi_C \sin\varphi + \dot\eta_C \cos\varphi \,,$$

$$v_*^4 = l_1\dot\varphi \cos\theta - \dot\xi_C \sin(\varphi + \theta) + \dot\eta_C \cos(\varphi + \theta) \,,$$

and write the inverse transformation

$$\dot q^1 \equiv \dot\varphi = v_*^1 \,, \qquad \dot q^2 \equiv \dot\theta = v_*^2 \,,$$

$$\dot q^3 \equiv \dot\xi_C = \beta_1^3 v_*^1 + \beta_3^3 v_*^3 + \beta_4^3 v_*^4 \,,$$
(9.43)

$$\dot q^4 \equiv \dot\eta_C = \beta_1^4 v_*^1 + \beta_3^4 v_*^3 + \beta_4^4 v_*^4 \,,$$

where

$$\beta_1^3 = (l_1 \cos\varphi \cos\theta + l_2 \cos(\varphi + \theta)/\sin\theta \,,$$

$$\beta_3^3 = \cos(\varphi + \theta)/\sin\theta \,, \qquad \beta_4^3 = -\cos\varphi/\sin\theta \,,$$
(9.44)

$$\beta_1^4 = (l_1 \sin\varphi \cos\theta + l_2 \sin(\varphi + \theta))/\sin\theta \,,$$

$$\beta_3^4 = \sin(\varphi + \theta)/\sin\theta \,, \qquad \beta_4^4 = -\sin\varphi/\sin\theta \,.$$

The first Maggi equation in our case assumes the form

$$(MW_1 - Q_1)\frac{\partial \dot q^1}{\partial v_*^1} + (MW_3 - Q_3)\frac{\partial \dot q^3}{\partial v_*^1} + (MW_4 - Q_4)\frac{\partial \dot q^4}{\partial v_*^1} = 0 \,. \qquad (9.45)$$

Since the constraint equations does not involve the velocity $\dot\theta$, the second Maggi equation becomes the second-kind Lagrange equation

$$MW_2 - Q_2 = 0 \,. \tag{9.46}$$

The term MW_σ can be found in terms of the kinetic energy (9.41) by the formulas

$$MW_\sigma = \frac{d}{dt}\frac{\partial T}{\partial \dot q^\sigma} - \frac{\partial T}{\partial q^\sigma} \,, \qquad \sigma = \overline{1,4} \,.$$

As a result, using formulas (9.41)–(9.44), the motion equations of robot (9.45), (9.46) can be put in the expanded form

$$\left[J^* + M_2 l_1 (l_1 - \beta_1^3 \sin \varphi + \beta_1^4 \cos \varphi)\right] \ddot{\varphi} + J_2 \ddot{\theta} + (M^* \beta_1^3 - M_2 l_1 \sin \varphi) \ddot{\xi}_C +$$
$$+ (M^* \beta_1^4 + M_2 l_1 \cos \varphi) \ddot{\eta}_C = M_2 l_1 \dot{\varphi}^2 (\beta_1^3 \cos \varphi + \beta_1^4 \sin \varphi) +$$
$$+ \left[F_1(t) \cos \varphi - F_2(v_C) \dot{\xi}_C / v_C\right] \beta_1^3 + \left[F_1(t) \sin \varphi - F_2(v_C) \dot{\eta}_C / v_C\right] \beta_1^4 , \tag{9.47}$$
$$J_2 (\ddot{\theta} + \ddot{\varphi}) = L_1(t) - L_2(\dot{\theta}) - L_3(\theta) .$$

If one is given the initial conditions and the analytic expressions for the functions $F_1(t)$, $F_2(v_C)$, $L_1(t)$, $L_2(\dot{\theta})$, $L_3(\theta)$, then after numerical integration of the nonlinear system the differential equations (9.40), (9.47) one may find the law of motion of the robot,

$$\varphi = \varphi(t), \qquad \theta = \theta(t), \qquad \xi_C = \xi_C(t), \qquad \eta_C = \eta_C(t) . \tag{9.48}$$

It is worth pointing out that the above motion of the robot may also be considered as a simplified mathematical model of motion of a car in a turn.[16] Thus, as a numerical example we consider the motion of a hypothetical small passenger car with

$$M_1 = 1000 \text{ kg}, \quad M_2 = 110 \text{ kg}, \quad J_1 = 1500 \text{ kg} \cdot \text{m}^2 ,$$
$$J_2 = 30 \text{ kg} \cdot \text{m}^2, \quad l_1 = 0.75 \text{ m}, \quad l_2 = 1.65 \text{ m} ,$$

under the following force characteristics:

$$F_1(t) = 2500 \text{ N}, \quad F_2(v_C) = \kappa_2 v_C , \quad \kappa_2 = 100 \text{ N} \cdot \text{s} \cdot \text{m}^{-1} ,$$
$$L_1(t) = 15 \text{ N} \cdot \text{m}, \quad L_2(\dot{\theta}) = \kappa_1 \dot{\theta}, \quad \kappa_1 = 0.5 \text{ N} \cdot \text{m} \cdot \text{s} ,$$
$$L_3(\theta) = \kappa_3 \theta, \quad \kappa_3 = 100 \text{ N} \cdot \text{m} .$$

Figure 18 gives the results of numerical solution of the nonlinear system of differential equations (9.40), (9.47). The following initial data were used:

$$\varphi(0) = 0, \ \dot{\varphi}(0) = 0, \ \theta(0) = \pi/180 \text{ rad}, \ \dot{\theta}(0) = 0, \ \xi_C(0) = 0,$$
$$\dot{\xi}_C(0) = 0.00176856 \text{ m} \cdot \text{s}^{-1}, \ \eta_C(0) = 0, \ \dot{\eta}_C(0) = 0.000018008 \text{ m} \cdot \text{s}^{-1} .$$

Let us now proceed with finding the generalized reactions. The second group of the Maggi equations (9.19) reads as

$$\Lambda_1 = (M W_3 - Q_3) \frac{\partial \dot{q}^3}{\partial v_*^3} + (M W_4 - Q_4) \frac{\partial \dot{q}^4}{\partial v_*^3} ,$$
$$\Lambda_2 = (M W_3 - Q_3) \frac{\partial \dot{q}^3}{\partial v_*^4} + (M W_4 - Q_4) \frac{\partial \dot{q}^4}{\partial v_*^4} ,$$

---

[16] This very approach was used by P.S. Lineikin in the paper "On the roll of a car", Tr. Saratovsk. Avt.-Dor. Inst., 1939. no. 5, pp. 3–22 [in Russian].

Fig. 18 Time-distance
graphs of a wheeled robot

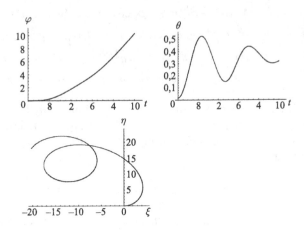

Fig. 19 Possible types of
motion of a wheeled robot

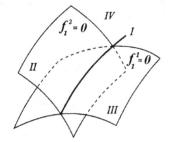

or in the expanded form

$$\Lambda_1 = [M^*\ddot{\xi}_C - M_2 l_1(\ddot{\varphi}\sin\varphi + \dot{\varphi}^2\cos\varphi) - F_1(t)\cos\varphi + F_2(v_C)\dot{\xi}_C/v_C]\beta_3^3 +$$
$$+ [M^*\ddot{\eta}_C + M_2 l_1(\ddot{\varphi}\cos\varphi - \dot{\varphi}^2\sin\varphi) - F_1(t)\sin\varphi + F_2(v_C)\dot{\eta}_C/v_C]\beta_3^4,$$
$$\Lambda_2 = [M^*\ddot{\xi}_C - M_2 l_1(\ddot{\varphi}\sin\varphi + \dot{\varphi}^2\cos\varphi) - F_1(t)\cos\varphi + F_2(v_C)\dot{\xi}_C/v_C]\beta_4^3 +$$
$$+ [M^*\ddot{\eta}_C + M_2 l_1(\ddot{\varphi}\cos\varphi - \dot{\varphi}^2\sin\varphi) - F_1(t)\sin\varphi + F_2(v_C)\dot{\eta}_C/v_C]\beta_4^4.$$

Substituting (9.48) into these formulas, we find the law of variation of the generalized reactions $\Lambda_i = \Lambda_i(t)$, $i = 1, 2$. Using these functions one may test when the nonholonomic constraints (9.40) are satisfied. If the reaction forces are equal to the Coulomb friction forces, then such constraints may fail to hold and the robot (car) may start sliding along the axes of its wheels.

Examining the possibilities of wheels sideslip, which is usually accompanied by an emergency situation, is important when considering the motion of a car in a turn. Let us examine the possible types of motion of the above mechanical model of a car. Figure 19 depicts two surfaces in the phase space of the variables q^σ, $\dot{q}^\sigma$, $\sigma = \overline{1, 4}$. The first one corresponds to the constraint given by the first equation of (9.40), and the second one, to the second equation.

If such nonholonomic constraints are simultaneously satisfied, then the point in the phase space must lie on the intersection line of these surfaces. This corresponds to the first type of motion of a car (the solid line I in Fig. 19). If the first constraint is violated and the second constraint holds, then the representative point is on the surface $f_1^2 = 0$ (the second type of motion). If the second constraint is relaxed, but the first constraint $f_1^1 = 0$ is sill satisfied, then the representative point lies on the surface $\varphi^1 = 0$ (the third type of motion). If both constraints are relaxed, then the representative point is away from the surfaces, the car is moving if there are lateral friction forces acting on the front and back axles (the fourth type of motion). From each type of motion, the representative point may go into any other type of motion.[17]

It is also possible to compose the first-kind Lagrange equations in the curvilinear coordinates for nonholonomic systems (see Eq. (9.21)). In our problem they read as

$$J^* \ddot{\varphi} + J_2 \ddot{\theta} - M_2 l_1 \ddot{\xi}_C \sin \varphi + M_2 l_1 \ddot{\eta}_C \cos \varphi = -\Lambda_1 l_2 + \Lambda_2 l_1 \cos \theta \,,$$

$$J_2 (\ddot{\theta} + \ddot{\varphi}) = L_1(t) - L_2(\dot{\theta}) - L_3(\theta) \,,$$

$$M^* \ddot{\xi}_C - M_2 l_1 \ddot{\varphi} \sin \varphi - M_2 l_1 \dot{\varphi}^2 \cos \varphi =$$

$$= F_1 \cos \varphi - L_2(\dot{\theta}) - \Lambda_1 \sin \varphi - \Lambda_2 \sin(\varphi + \theta) \,,$$

$$M^* \ddot{\eta}_C + M_2 l_1 \ddot{\varphi} \cos \varphi - M_2 l_1 \dot{\varphi}^2 \sin \varphi =$$

$$= F_1 \sin \varphi - k_2 \dot{\eta}_C + \Lambda_1 \cos \varphi + \Lambda_2 \cos(\varphi + \theta) \,.$$

The above four equations contain four unknown generalized coordinates and two unknown Lagrange multipliers, and hence they ought to be solved jointly with the constraint equations (9.40). This fact is characteristic of the first-kind Lagrange equations. Differentiating in time the constraint equations and using them to exclude the generalized reactions from the reduced Lagrange equations, we arrive at the Maggi motion equations (9.47) and the formulas for Λ_1 and Λ_2.

Example 11 *Rolling of an ellipsoid along a rough plane* (composition of the Maggi equations). We note that the specific form of the Maggi equations depends essentially on the choice of the variables v_*^ρ. A successful choice thereof may considerably simplify the calculations related with the reduction of the problem to the system of differential equations in the normal form.

As an example, consider the rolling of a homogeneous rigid body of ellipsoidal form on the fixed plane. The center of ellipsoid coincident with the center of mass will be taken as the origin of the moving coordinates system $Cxyz$, whose axes are rigidly fixed within its axes (Fig. 20). Assume that the plane π, on which the ellipsoid rolls, coincides with the plane $O\xi\eta$ of the fixed system of coordinates $O\xi\eta\zeta$. We let

---

[17] A detailed analysis of these types of motion for front- and rear-wheel drive vehicles were considered in Appendix E to the book *Sh.Kh. Soltakhanov, M.P. Yushkov, and S.A. Zegzhda*, Mechanics of non-holonomic systems. A new class of control systems. Berlin-Heidelberg: Springer-Verlag. 2009. 329 p.

Fig. 20 Ellipsoid on the
plane

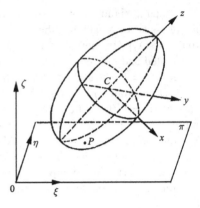

ξ, η, ζ denote the coordinates of the center of the ellipsoid with respect to the fixed frame. The velocity of the contact point P can be found from the formula

$$\mathbf{v}_P = \mathbf{v}_C + \boldsymbol{\omega} \times \overrightarrow{CP}.$$

For a rolling without sliding the velocity of the point P is zero, and hence, the constraint equation can be put in the form

$$\mathbf{v}_C + \boldsymbol{\omega} \times \overrightarrow{CP} = \dot{\xi}\mathbf{i}_\xi + \dot{\eta}\mathbf{i}_\eta + \dot{\zeta}\mathbf{i}_\zeta + \begin{vmatrix} \mathbf{i}_\xi & \mathbf{i}_\eta & \mathbf{i}_\zeta \\ \omega_\xi & \omega_\eta & \omega_\zeta \\ \xi_0 & \eta_0 & \zeta_0 \end{vmatrix} = 0. \tag{9.49}$$

Here, ξ_0, η_0, ζ_0 are the coordinates of the point P is the frame $C\xi_1\eta_1\zeta_1$, whose axes ξ_1, η_1, ζ_1 are parallel, respectively, to the axes ξ, η, ζ of the fixed system of coordinates. It can be shown that the quantities ξ_0, η_0, ζ_0 can be found by the formulas

$$-\xi_0\zeta = (a^2 - b^2)\sin\theta\cos\psi\sin\varphi\cos\varphi + (c^2 - a^2\sin^2\varphi - b^2\cos^2\varphi)\sin\psi\cos\theta\sin\theta,$$

$$-\eta_0\zeta = (a^2 - b^2)\sin\psi\sin\theta\sin\varphi\cos\varphi + (a^2\sin^2\varphi + b^2\cos^2\varphi - c^2)\cos\psi\cos\theta\sin\theta,$$

$$\zeta_0 = -\zeta = -\sqrt{a^2\sin^2\theta\sin^2\varphi + b^2\sin^2\theta\cos^2\varphi + c^2\cos^2\theta},$$

where a, b, c are the semi-axes of the ellipsoid, ψ, θ, φ are the Euler angles giving the orientation of the frame $Cxyz$ with respect to the frame $C\xi_1\eta_1\zeta_1$.

The vector equation (9.49) is equivalent to three scalar equation for the nonholonomic constraints in our problem:

$$f_1^1 \equiv \dot{\xi} + \omega_\eta \zeta_0 - \omega_\zeta \eta_0 = 0 \, ,$$
$$f_1^2 \equiv \dot{\eta} + \omega_\zeta \xi_0 - \omega_\xi \zeta_0 = 0 \, , \tag{9.50}$$
$$f_1^3 \equiv \dot{\zeta} + \omega_\xi \eta_0 - \omega_\eta \xi_0 = 0 \, .$$

In this problem, as generalized Lagrange coordinates one may take the coordinates ξ, η, ζ of the center of mass and the Euler angles ψ, θ, φ. To calculate the kinetic energy of the ellipsoid in these coordinates, we shall employ König's theorem. Now

$$T = \frac{M}{2}(\dot{\xi}^2 + \dot{\eta}^2 + \dot{\zeta}^2) + \frac{J_\omega \omega^2}{2} \, .$$

The quantity $J_\omega \omega^2$ can be given in the form

$$J_\omega \omega^2 = A\omega_x^2 + B\omega_y^2 + C\omega_z^2 \, ,$$

where A, B, C are moments of inertia of the ellipsoid with respect to the axes x, y, z, respectively. By the assumption, the ellipsoid is a homogeneous rigid body, and hence

$$A = \frac{M(b^2 + c^2)}{5} \, , \qquad B = \frac{M(c^2 + a^2)}{5} \, , \qquad C = \frac{M(a^2 + b^2)}{5} \, .$$

The projections ω_x, ω_y, ω_z of the vector ω to the axis of the moving coordinate system $Cxyz$ are as follows:

$$\omega_x = \dot{\psi} \sin\theta \sin\varphi + \dot{\theta} \cos\varphi \, ,$$
$$\omega_y = \dot{\psi} \sin\theta \cos\varphi - \dot{\theta} \sin\varphi \, ,$$
$$\omega_z = \dot{\psi} \cos\theta + \dot{\varphi} \, .$$

From the above formulas one may calculate the covariant components of the vector $M\mathbf{W}$,

$$MW_\xi = M\ddot{\xi} \, , \qquad MW_\eta = M\ddot{\eta} \, , \qquad MW_\zeta = M\ddot{\zeta} \, ,$$
$$MW_\varphi = \frac{d}{dt}\frac{\partial T}{\partial \dot{\varphi}} - \frac{\partial T}{\partial \varphi} \, , \qquad MW_\psi = \frac{d}{dt}\frac{\partial T}{\partial \dot{\psi}} - \frac{\partial T}{\partial \psi} \, , \qquad MW_\theta = \frac{d}{dt}\frac{\partial T}{\partial \dot{\theta}} - \frac{\partial T}{\partial \theta} \, .$$

The expressions for W_φ, W_ψ and W_θ cumbersome, and hence will not be given here.

The quantities ω_ξ, ω_η, ω_ζ from the constraint equations (9.50) are expressed in terms of the formulas

$$\omega_\xi = \dot{\varphi} \sin\psi \sin\theta + \dot{\theta} \cos\psi \, ,$$
$$\omega_\eta = \dot{\varphi} \cos\psi \sin\theta + \dot{\theta} \sin\psi \, ,$$
$$\omega_\zeta = \dot{\varphi} \cos\theta + \dot{\psi} \, .$$

Hence, if one takes $v_*^1 = \dot{\xi}$, $v_*^2 = \dot{\eta}$, $v_*^3 = \dot{\zeta}$, $v_*^{3+\varkappa} = f_1^\varkappa$, $\varkappa = \overline{1,3}$, then, due to the involved dependence of the functions $f_1^\varkappa$ on the velocities $\dot{q}^\sigma$, the expressions $\partial \dot{q}^\sigma / \partial v_*^\lambda$ in this problem turn out to be fairly clumsy. Hence, the resulting Maggi equations will also be bulky. The problem becomes substantially more simple if the angular velocities ω_ξ, ω_η, ω_ζ are taken as free variables v_*^λ. It may be shown that if the quasi-velocities v_*^ρ are given by the formulas

$$v_*^1 = \omega_\xi \,, \qquad\qquad v_*^2 = \omega_\eta \,, \qquad\qquad v_*^3 = \omega_\zeta \,,$$
$$v_*^4 = \dot{\xi} + \omega_\eta \zeta_0 - \omega_\zeta \eta_0 \,, \quad v_*^5 = \dot{\eta} + \omega_\zeta \xi_0 - \omega_\xi \zeta_0 \,, \quad v_*^6 = \dot{\zeta} + \omega_\xi \eta_0 - \omega_\eta \xi_0 \,,$$

then

$$\frac{\partial \dot{\xi}}{\partial \omega_\xi} = 0 \,, \quad \frac{\partial \dot{\eta}}{\partial \omega_\xi} = \zeta_0 \,, \quad \frac{\partial \dot{\zeta}}{\partial \omega_\xi} = -\eta_0 \,,$$

$$\frac{\partial \dot{\varphi}}{\partial \omega_\xi} = \frac{\sin\psi}{\sin\theta} \,, \quad \frac{\partial \dot{\psi}}{\partial \omega_\xi} = -\frac{\sin\psi\cos\theta}{\sin\theta} \,, \quad \frac{\partial \dot{\theta}}{\partial \omega_\xi} = \cos\psi \,,$$

$$\frac{\partial \dot{\xi}}{\partial \omega_\eta} = -\zeta_0 \,, \quad \frac{\partial \dot{\eta}}{\partial \omega_\eta} = 0 \,, \quad \frac{\partial \dot{\zeta}}{\partial \omega_\eta} = \xi_0 \,,$$

$$\frac{\partial \dot{\varphi}}{\partial \omega_\eta} = -\frac{\cos\psi}{\sin\theta} \,, \quad \frac{\partial \dot{\psi}}{\partial \omega_\eta} = \frac{\cos\psi\cos\theta}{\sin\theta} \,, \quad \frac{\partial \dot{\theta}}{\partial \omega_\eta} = \sin\psi \,,$$

$$\frac{\partial \dot{\xi}}{\partial \omega_\zeta} = \eta_0 \,, \quad \frac{\partial \dot{\eta}}{\partial \omega_\zeta} = -\xi_0 \,, \quad \frac{\partial \dot{\zeta}}{\partial \omega_\zeta} = 0 \,,$$

$$\frac{\partial \dot{\varphi}}{\partial \omega_\zeta} = 0 \,, \quad \frac{\partial \dot{\psi}}{\partial \omega_\zeta} = 1 \,, \quad \frac{\partial \dot{\theta}}{\partial \omega_\zeta} = 0 \,.$$

Hence, substituting these expressions into the Maggi equations, they can be written down explicitly.

This example shows how complicated are the problems pertaining to a roll of one body on the surface of another one even in the assumption that the constraint (9.49) is ideal.[18]

---

[18] The dynamics of bodies in contact with a hard surfaces is treated in the book: *A.P. Markeev, Dynamics of body contacted with a rigid surface.* Moscow: Izdat "Nauka", 1992 [in Russian]. New theory of interaction of a rolling rigid body with a deformable surface was proposed in the paper by *V.F. Zhuravlev,* "The model of dry friction in the problem of the rolling of rigid bodies," J. Appl. Math. Mech., vol. 62, issue 5, 1998, pp. 705–710 [in Russian].

10   Derivation of the Most Useful Forms of Motion Equations of Nonholonomic Systems from the Maggi Equations

There is a good deal of various forms of different motion equations for nonholonomic systems. Let us consider the principal ones and show that they can be obtained from the Maggi equations.

The Chaplygin and Voronets equations. Assume that a system is subject to stationary linear nonholonomic constraints, which can be put in the form

$$\dot{q}^{l+\varkappa} = \beta_\lambda^{l+\varkappa}(q)\,\dot{q}^\lambda, \qquad \lambda = \overline{1, l}, \quad \varkappa = \overline{1, k}. \tag{10.1}$$

Hence, setting

$$v_*^\lambda = \dot{q}^\lambda, \quad \lambda = \overline{1, l},$$
$$v_*^{l+\varkappa} = \dot{q}^{l+\varkappa} - \beta_\lambda^{l+\varkappa}(q)\,\dot{q}^\lambda, \quad \varkappa = \overline{1, k},$$

we have

$$\frac{\partial \dot{q}^\mu}{\partial v_*^\lambda} = \delta_\lambda^\mu = \begin{cases} 1, & \mu = \lambda, \\ 0, & \mu \neq \lambda, \end{cases} \qquad \lambda, \mu = \overline{1, l},$$

$$\frac{\partial \dot{q}^{l+\varkappa}}{\partial v_*^\lambda} = \beta_\lambda^{l+\varkappa}, \qquad \lambda = \overline{1, l}, \quad \varkappa = \overline{1, k}.$$

From these expressions it follows that under nonholonomic constraints of form (10.1) the Maggi equations (9.18) can be put in the form

$$M w_\lambda + M w_{l+\varkappa}\beta_\lambda^{l+\varkappa} = Q_\lambda + Q_{l+\varkappa}\beta_\lambda^{l+\varkappa},$$
$$\lambda = \overline{1, l}, \quad \varkappa = \overline{1, k}. \tag{10.2}$$

We assume that kinetic energy T does not depend on the generalized coordinates $q^{l+\varkappa}$ and $Q_{l+\varkappa} = 0$ ($\varkappa = \overline{1, k}$). Hence, Eq. (10.2) can be written as

$$\frac{d}{dt}\frac{\partial T}{\partial \dot{q}^\lambda} - \frac{\partial T}{\partial q^\lambda} + \beta_\lambda^{l+\varkappa}\frac{d}{dt}\frac{\partial T}{\partial \dot{q}^{l+\varkappa}} = Q_\lambda, \qquad \lambda = \overline{1, l}. \tag{10.3}$$

Let us transform Eq. (10.3). Using the constraint equations (10.1), we exclude all the velocities $\dot{q}^{l+\varkappa}$, from the expressions for the kinetic energy T and define the resulting expression for the kinetic energy by T_*. In this case, we have

$$\frac{\partial T_*}{\partial \dot{q}^\lambda} = \frac{\partial T}{\partial \dot{q}^\lambda} + \frac{\partial T}{\partial \dot{q}^{l+\varkappa}}\frac{\partial \dot{q}^{l+\varkappa}}{\partial \dot{q}^\lambda} = \frac{\partial T}{\partial \dot{q}^\lambda} + \frac{\partial T}{\partial \dot{q}^{l+\varkappa}}\beta_\lambda^{l+\varkappa}, \tag{10.4}$$

$$\frac{\partial T_*}{\partial q^\lambda} = \frac{\partial T}{\partial q^\lambda} + \frac{\partial T}{\partial \dot{q}^{l+\varkappa}} \frac{\partial \dot{q}^{l+\varkappa}}{\partial q^\lambda} = \frac{\partial T}{\partial q^\lambda} + \frac{\partial T}{\partial \dot{q}^{l+\varkappa}} \frac{\partial \beta_\mu^{l+\varkappa}}{\partial q^\lambda} \dot{q}^\mu,$$

$$\lambda, \mu = \overline{1,l}.$$

(10.5)

We assume that the coefficients $\beta_\lambda^{l+\varkappa}$ are independent of $q^{l+\varkappa}$, $\varkappa = \overline{1,k}$. Hence, differentiating (10.4) in time,

$$\frac{d}{dt}\frac{\partial T_*}{\partial \dot{q}^\lambda} = \frac{d}{dt}\frac{\partial T}{\partial \dot{q}^\lambda} + \beta_\lambda^{l+\varkappa}\frac{d}{dt}\frac{\partial T}{\partial \dot{q}^{l+\varkappa}} + \frac{\partial T}{\partial \dot{q}^{l+\varkappa}}\frac{d\beta_\lambda^{l+\varkappa}}{dt} =$$

$$= \frac{d}{dt}\frac{\partial T}{\partial \dot{q}^\lambda} + \beta_\lambda^{l+\varkappa}\frac{d}{dt}\frac{\partial T}{\partial \dot{q}^{l+\varkappa}} + \frac{\partial T}{\partial \dot{q}^{l+\varkappa}}\frac{\partial \beta_\lambda^{l+\varkappa}}{\partial q^\mu}\dot{q}^\mu,$$

$$\lambda, \mu = \overline{1,l}.$$

(10.6)

Using formulas (10.6) and (10.5) to find $d(\partial T/\partial \dot{q}^\lambda)/dt$ and $\partial T/\partial q^\lambda$ and substituting it into Eq. (10.3), this gives

$$\frac{d}{dt}\frac{\partial T_*}{\partial \dot{q}^\lambda} - \frac{\partial T_*}{\partial q^\lambda} - \frac{\partial T}{\partial \dot{q}^{l+\varkappa}}\left(\frac{\partial \beta_\lambda^{l+\varkappa}}{\partial q^\mu} - \frac{\partial \beta_\mu^{l+\varkappa}}{\partial q^\lambda}\right)\dot{q}^\mu = Q_\lambda,$$

$$\varkappa = \overline{1,k}, \qquad \lambda, \mu = \overline{1,l}.$$

(10.7)

These equations were obtained by *S.A. Chaplygin*.[19]

If in (10.7), using the constraint equations (10.1), exclude the dependent velocities $\dot{q}^{l+1}, \dot{q}^{l+2}, \ldots, \dot{q}^{l+k}$ in the expressions $\partial T/\partial \dot{q}^{l+\varkappa}$, then we obtain a complete system of l equations with respect to the unknowns functions $q^1, q^2, , \ldots, q^l$. Thus, the Chaplygin equations are capable of finding $q^1(t), q^2(t), \ldots, q^l(t)$ (independently constraints (10.1)) and after finding the remaining $q^{l+1}(t), q^{l+2}(t), \ldots, q^{l+k}(t)$ from Eq. (10.1).

Assume that the coefficients $\beta_\lambda^{l+\varkappa}$ satisfy the conditions

$$\frac{\partial \beta_\mu^{l+\varkappa}}{\partial q^\lambda} - \frac{\partial \beta_\lambda^{l+\varkappa}}{\partial q^\mu} = 0, \qquad \varkappa = \overline{1,k}, \qquad \lambda, \mu = \overline{1,l}.$$

(10.8)

Hence, from the assumption that the coefficients $\beta_\lambda^{l+\varkappa}$ are independent of $q^{l+\varkappa}$, ($\varkappa = \overline{1,k}$), it follows that they cannot be put in the form

$$\beta_\lambda^{l+\varkappa} = \frac{\partial V^{l+\varkappa}}{\partial q^\lambda}, \qquad \lambda = \overline{1,l}, \qquad \varkappa = \overline{1,k}.$$

(10.9)

Here, $V^{l+\varkappa}$ are functions of the coordinates $q^1, q^2, \ldots, q^l$. Substituting (10.9) into Eq. (10.1), this gives

---

[19] See: *S.A. Chaplygin*. "Motion of a heavy body of revolution on a horizontal plane", Trudy Otd. Fiz. Nauk. Ob. Lyub. Estestv. 1897. Vol. 9, issue 1. (See Memoirs, vol. 1, 1948) [in Russian].

$$q^{l+\varkappa} = V^{l+\varkappa}(q^1, q^2, \ldots, q^l), \qquad \varkappa = \overline{1, k}.$$

Thus, the coordinates $q^{l+\varkappa}$ can be expressed from the other ones. Hence, under conditions (10.8) the motion is described by the standard Lagrange equations.

Let us now derive the motion equations in the form obtained by P.V. Voronets.[20] Let us consider a mechanical system with constraints in the form (10.1), but without additional assumptions leading to the Chaplygin equations. The Maggi equations (10.2) in the case when the kinetic energy T depends on all coordinates read as

$$\frac{d}{dt}\frac{\partial T}{\partial \dot{q}^\lambda} - \frac{\partial T}{\partial q^\lambda} + \left(\frac{d}{dt}\frac{\partial T}{\partial \dot{q}^{l+\varkappa}} - \frac{\partial T}{\partial q^{l+\varkappa}}\right)\beta_\lambda^{l+\varkappa} = Q_\lambda + Q_{l+\varkappa}\beta_\lambda^{l+\varkappa},$$
$$\varkappa = \overline{1, k}, \qquad \lambda = \overline{1, l}. \tag{10.10}$$

In order to reduce these equations to the Vornets equations, we shall argue as above. Relations (10.5) retain their form. Using the fact that now coefficients $\beta_\lambda^{l+\varkappa}$ depend on all q^σ, expression (10.6) assumes the form

$$\frac{d}{dt}\frac{\partial T_*}{\partial \dot{q}^\lambda} = \frac{d}{dt}\frac{\partial T}{\partial \dot{q}^\lambda} + \beta_\lambda^{l+\varkappa}\frac{d}{dt}\frac{\partial T}{\partial \dot{q}^{l+\varkappa}} + \frac{\partial T}{\partial \dot{q}^{l+\varkappa}}\frac{\partial \beta_\lambda^{l+\varkappa}}{\partial q^\mu}\dot{q}^\mu +$$
$$+ \frac{\partial T}{\partial \dot{q}^{l+\varkappa}}\frac{\partial \beta_\lambda^{l+\varkappa}}{\partial q^{l+\nu}}\beta_\mu^{l+\nu}\dot{q}^\mu, \qquad \varkappa, \nu = \overline{1, k}, \quad \lambda, \mu = \overline{1, l}. \tag{10.11}$$

In this case, in parallel with (10.5) and (10.11), one should take into account the equalities

$$\beta_\lambda^{l+\varkappa}\frac{\partial T_*}{\partial q^{l+\varkappa}} = \beta_\lambda^{l+\varkappa}\left(\frac{\partial T}{\partial q^{l+\varkappa}} + \frac{\partial T}{\partial \dot{q}^{l+\nu}}\frac{\partial \beta_\mu^{l+\nu}}{\partial q^{l+\varkappa}}\dot{q}^\mu\right).$$

Using this expression and relations (10.5) and (10.11) one may write Eq. (10.10) in the form

$$\frac{d}{dt}\frac{\partial T_*}{\partial \dot{q}^\lambda} - \frac{\partial T_*}{\partial q^\lambda} - \beta_\lambda^{l+\varkappa}\frac{\partial T_*}{\partial q^{l+\varkappa}} - \frac{\partial T}{\partial \dot{q}^{l+\varkappa}}\beta_{\lambda\mu}^{l+\varkappa}\dot{q}^\mu =$$
$$= Q_\lambda + Q_{l+\varkappa}\beta_\lambda^{l+\varkappa}, \quad \lambda, \mu = \overline{1, l}, \quad \varkappa = \overline{1, k}, \tag{10.12}$$

where

$$\beta_{\lambda\mu}^{l+\varkappa} = \frac{\partial \beta_\lambda^{l+\varkappa}}{\partial q^\mu} - \frac{\partial \beta_\mu^{l+\varkappa}}{\partial q^\lambda} + \frac{\partial \beta_\lambda^{l+\varkappa}}{\partial q^{l+\nu}}\beta_\mu^{l+\nu} - \frac{\partial \beta_\mu^{l+\varkappa}}{\partial q^{l+\nu}}\beta_\lambda^{l+\nu}.$$

Equation (10.12) are called the *Voronets equations*. Augmenting the motion equations (10.12) with the constraint equations (10.1), we get the system of differential equations for the functions $q^\sigma(t)$, $\sigma = \overline{1, s}$.

---

[20] See: *P.V. Voronets*, "Sur les équations du mouvement pour les systémés non holonômes," Mat. Sb., 1901, vol. 22, no. 4, pp. 659–686.

In the case of motion of a constrained system under forces with potential U, Eq. (10.12) assume the form

$$\frac{d}{dt}\frac{\partial T_*}{\partial \dot{q}^\lambda} - \frac{\partial (T_* + U)}{\partial q^\lambda} - \beta_\lambda^{l+\varkappa}\frac{\partial (T_* + U)}{\partial q^{l+\varkappa}} - \frac{\partial T}{\partial \dot{q}^{l+\varkappa}}\beta_{\lambda\mu}^{l+\varkappa}\dot{q}^\mu = 0,$$

$$\lambda, \mu = \overline{1, l}, \quad \varkappa = \overline{1, k}.$$

In the particular case, when the coordinates $q^{l+1}, q^{l+2}, \dots, q^{l+k}$ corresponding to the excluded velocities do not explicitly enter expressions for the kinetic and potential energy, as well as in the constraint equations, the Voronets equations (10.12) agree with the Chaplygin equations (10.7).

Equations in quasi-coordinates (the Hamel–Novoselov, Voronets–Hamel, Poincaré–Chetaev equations). It is known that the projections of the vector of instantaneous angular velocity $\boldsymbol{\omega}$ to fixed axes cannot be considered as derivatives of some new angles, which uniquely specify the position of the rigid body. Likewise, it may happen that the quantities v_*^ρ, which are in a one-to-one correspondence with the generalized velocities $\dot{q}^\sigma$, cannot be considered as derivatives of some new coordinates q_*^ρ. Hence, the quantities v_*^ρ are called the *quasi-velocities*, and the variables π^ρ, as introduced via the formulas

$$\pi^\rho = \int\limits_t^{t_0} v_*^\rho dt,$$

are called the *quasi-coordinates*.

In the expression for the kinetic energy T, we shall replace the generalized velocities $\dot{q}^\sigma$ by the quasi-velocities v_*^ρ. Let T^* be the function thus obtained. Let us find how using the function T^* one may write the Maggi equations

$$\left(\frac{d}{dt}\frac{\partial T}{\partial \dot{q}^\sigma} - \frac{\partial T}{\partial q^\sigma} - Q_\sigma\right)\frac{\partial \dot{q}^\sigma}{\partial v_*^\lambda} = 0, \quad \sigma = \overline{1, s}, \quad \lambda = \overline{1, l}. \tag{10.13}$$

Taking into account the relations

$$\frac{\partial T^*}{\partial v_*^\lambda} = \frac{\partial T}{\partial \dot{q}^\sigma}\frac{\partial \dot{q}^\sigma}{\partial v_*^\lambda}, \quad \frac{\partial T^*}{\partial q^\sigma} = \frac{\partial T}{\partial q^\sigma} + \frac{\partial T}{\partial \dot{q}^\rho}\frac{\partial \dot{q}^\rho}{\partial q^\sigma},$$

$$\rho, \sigma = \overline{1, s}, \quad \lambda = \overline{1, l},$$

we see that

$$\left(\frac{d}{dt}\frac{\partial T}{\partial \dot{q}^\sigma}\right)\frac{\partial \dot{q}^\sigma}{\partial v_*^\lambda} = \frac{d}{dt}\left(\frac{\partial T}{\partial \dot{q}^\sigma}\frac{\partial \dot{q}^\sigma}{\partial v_*^\lambda}\right) -$$

$$-\frac{\partial T}{\partial \dot{q}^\sigma}\frac{d}{dt}\frac{\partial \dot{q}^\sigma}{\partial v_*^\lambda} = \frac{d}{dt}\frac{\partial T^*}{\partial v_*^\lambda} - \frac{\partial T}{\partial \dot{q}^\sigma}\frac{d}{dt}\frac{\partial \dot{q}^\sigma}{\partial v_*^\lambda}, \tag{10.14}$$

$$\frac{\partial T}{\partial q^\sigma}\frac{\partial \dot{q}^\sigma}{\partial v_*^\lambda} = \frac{\partial \dot{q}^\sigma}{\partial v_*^\lambda}\left(\frac{\partial T^*}{\partial q^\sigma} - \frac{\partial T}{\partial \dot{q}^\rho}\frac{\partial \dot{q}^\rho}{\partial q^\sigma}\right) =$$
$$= \frac{\partial \dot{q}^\sigma}{\partial v_*^\lambda}\frac{\partial T^*}{\partial q^\sigma} - \frac{\partial T}{\partial \dot{q}^\rho}\frac{\partial \dot{q}^\rho}{\partial q^\sigma}\frac{\partial \dot{q}^\sigma}{\partial v_*^\lambda}.$$
(10.15)

Swapping the summation indexes ρ and σ in the double sum on the right of (10.15), this gives

$$\frac{\partial T}{\partial q^\sigma}\frac{\partial \dot{q}^\sigma}{\partial v_*^\lambda} = \frac{\partial \dot{q}^\sigma}{\partial v_*^\lambda}\frac{\partial T^*}{\partial q^\sigma} - \frac{\partial T}{\partial \dot{q}^\sigma}\frac{\partial \dot{q}^\sigma}{\partial q^\rho}\frac{\partial \dot{q}^\rho}{\partial v_*^\lambda}.$$
(10.16)

Let us consider the operator

$$\frac{\partial}{\partial \pi^\rho} = \frac{\partial \dot{q}^\sigma}{\partial v_*^\rho}\frac{\partial}{\partial q^\sigma}, \qquad \rho, \sigma = \overline{1, s},$$
(10.17)

which in the case when one may set $v_*^\rho = \dot{\pi}^\rho = \dot{q}_*^\rho$ is transformed into the partial differentiation operator in the new coordinate q_*^ρ, because now we have

$$\frac{\partial \dot{q}^\sigma}{\partial v_*^\rho}\frac{\partial}{\partial q^\sigma} = \frac{\partial \dot{q}^\sigma}{\partial \dot{q}_*^\rho}\frac{\partial}{\partial q^\sigma} = \frac{\partial q^\sigma}{\partial q_*^\rho}\frac{\partial}{\partial q^\sigma} = \frac{\partial}{\partial q_*^\rho}.$$

In view of (10.17) expression (10.16) reads as

$$\frac{\partial T}{\partial q^\sigma}\frac{\partial \dot{q}^\sigma}{\partial v_*^\lambda} = \frac{\partial T^*}{\partial \pi^\lambda} - \frac{\partial T}{\partial \dot{q}^\sigma}\frac{\partial \dot{q}^\sigma}{\partial \pi^\lambda}.$$

Hence, from (10.14) it follows that the Maggi equations (10.13) can be put in the form

$$\frac{d}{dt}\frac{\partial T^*}{\partial v_*^\lambda} - \frac{\partial T^*}{\partial \pi^\lambda} - \frac{\partial T}{\partial \dot{q}^\sigma}\left(\frac{d}{dt}\frac{\partial \dot{q}^\sigma}{\partial v_*^\lambda} - \frac{\partial \dot{q}^\sigma}{\partial \pi^\lambda}\right) = Q_\lambda^*,$$
(10.18)
$$\sigma = \overline{1, s}, \qquad \lambda = \overline{1, l}.$$

Here,

$$Q_\lambda^* = Q_\sigma \frac{\partial \dot{q}^\sigma}{\partial v_*^\lambda}.$$
(10.19)

Equation (10.18) are sometimes called the *Chaplygin-type equations*.[21]

Let us consider a particular case, when the generalized velocities $\dot{q}^\sigma$ are related to the quasi-velocities v_*^ρ by stationary homogeneous linear relations

$$v_*^\rho = \alpha_\sigma^\rho(q)\,\dot{q}^\sigma, \qquad \dot{q}^\sigma = \beta_\rho^\sigma(q)\,v_*^\rho,$$
$$\rho, \sigma = \overline{1, s};$$
(10.20)

[21] See: *V.S. Novoselov*, "Application of nonlinear nonholonomic coordinates in analytic mechanics," Uch. Zap. LGU, 1957, no. 217, vyp. 31, pp. 50–83 [in Russian].

the constraint equations read as

$$v_*^{l+\varkappa} \equiv \alpha_\sigma^{l+\varkappa}(q)\,\dot{q}^\sigma = 0\,, \qquad \varkappa = \overline{1,k}\,. \tag{10.21}$$

In this case, using expressions (10.20) and operator (10.17), and taking into account that upon differentiating one may put $v_*^{l+\varkappa} = 0$ $(\varkappa = \overline{1,k})$, this gives

$$\frac{d}{dt}\frac{\partial \dot{q}^\sigma}{\partial v_*^\lambda} = \frac{d}{dt}\beta_\lambda^\sigma(q) = \frac{\partial \beta_\lambda^\sigma}{\partial q^\rho}\,\dot{q}^\rho = \frac{\partial \beta_\lambda^\sigma}{\partial q^\rho}\,\beta_\mu^\rho\,v_*^\mu =$$

$$= v_*^\mu \frac{\partial \dot{q}^\rho}{\partial v_*^\mu}\frac{\partial \beta_\lambda^\sigma}{\partial q^\rho} = v_*^\mu \frac{\partial \beta_\lambda^\sigma}{\partial \pi^\mu}\,, \qquad \rho, \sigma = \overline{1,s}\,, \quad \lambda, \mu = \overline{1,l}\,;$$

$$\frac{\partial \dot{q}^\sigma}{\partial \pi^\lambda} = \frac{\partial \dot{q}^\rho}{\partial v_*^\lambda}\frac{\partial \dot{q}^\sigma}{\partial q^\rho} = \frac{\partial \dot{q}^\rho}{\partial v_*^\lambda}\frac{\partial \beta_\mu^\sigma}{\partial q^\rho}\,v_*^\mu =$$

$$= v_*^\mu \frac{\partial \beta_\mu^\sigma}{\partial \pi^\lambda}\,, \qquad \rho, \sigma = \overline{1,s}\,, \quad \lambda, \mu = \overline{1,l}\,.$$

Hence, Eq. (10.18) assume the form

$$\frac{d}{dt}\frac{\partial T^*}{\partial v_*^\lambda} - \frac{\partial T^*}{\partial \pi^\lambda} - \frac{\partial T}{\partial \dot{q}^\sigma}\left(\frac{\partial \beta_\lambda^\sigma}{\partial \pi^\mu} - \frac{\partial \beta_\mu^\sigma}{\partial \pi^\lambda}\right)v_*^\mu = Q_\lambda^*\,, \tag{10.22}$$

$$\sigma = \overline{1,s}\,, \qquad \lambda, \mu = \overline{1,l}\,.$$

These equations are usually called the *Chaplygin equations in quasi-coordinates*.[22] We note that Eqs. (10.18) and (10.22) should be considered jointly with the equations of nonholonomic constraints.

Equations (10.18) and (10.22) involve both the function T^* and the function T. Let us now reduce the Maggi equations (10.13) to the form involving only the function T. From the relations

$$\frac{\partial T}{\partial \dot{q}^\sigma} = \frac{\partial T^*}{\partial v_*^\rho}\frac{\partial v_*^\rho}{\partial \dot{q}^\sigma}\,, \qquad \rho, \sigma = \overline{1,s}\,,$$

it follows that

$$\left(\frac{d}{dt}\frac{\partial T}{\partial \dot{q}^\sigma}\right)\frac{\partial \dot{q}^\sigma}{\partial v_*^\lambda} = \frac{\partial \dot{q}^\sigma}{\partial v_*^\lambda}\frac{d}{dt}\left(\frac{\partial T^*}{\partial v_*^\rho}\frac{\partial v_*^\rho}{\partial \dot{q}^\sigma}\right) =$$

$$= \left(\frac{d}{dt}\frac{\partial T^*}{\partial v_*^\rho}\right)\frac{\partial v_*^\rho}{\partial \dot{q}^\sigma}\frac{\partial \dot{q}^\sigma}{\partial v_*^\lambda} + \frac{\partial T^*}{\partial v_*^\rho}\frac{\partial \dot{q}^\sigma}{\partial v_*^\lambda}\frac{d}{dt}\frac{\partial v_*^\rho}{\partial \dot{q}^\sigma}\,.$$

Next, we have

$$\frac{\partial v_*^\rho}{\partial \dot{q}^\sigma}\frac{\partial \dot{q}^\sigma}{\partial v_*^\lambda} = \delta_\lambda^\rho = \begin{cases} 1\,, & \rho = \lambda\,, \\ 0\,, & \rho \neq \lambda\,, \end{cases}$$

---

[22] See the paper by V.S. Novoselov in the previous footnote and the book by *Ju.I. Neimark and N.A. Fufaev*, Dynamics of Nonholonomic Systems (Translations of Mathematical Monographs, Vol. 33). Published by American Mathematical Society. 2004. 518 p.

and hence,

$$\left(\frac{d}{dt}\frac{\partial T}{\partial \dot{q}^\sigma}\right)\frac{\partial \dot{q}^\sigma}{\partial v_*^\lambda} = \frac{d}{dt}\frac{\partial T^*}{\partial v_*^\lambda} + \frac{\partial T^*}{\partial v_*^\rho}\frac{\partial \dot{q}^\sigma}{\partial v_*^\lambda}\frac{d}{dt}\frac{\partial v_*^\rho}{\partial \dot{q}^\sigma}. \tag{10.23}$$

Further, taking into account the relations

$$\frac{\partial T}{\partial q^\sigma} = \frac{\partial T^*}{\partial q^\sigma} + \frac{\partial T^*}{\partial v_*^\rho}\frac{\partial v_*^\rho}{\partial q^\sigma}$$

and operator (10.17), we see that

$$\frac{\partial T}{\partial q^\sigma}\frac{\partial \dot{q}^\sigma}{\partial v_*^\lambda} = \frac{\partial T^*}{\partial \pi^\lambda} + \frac{\partial T^*}{\partial v_*^\rho}\frac{\partial \dot{q}^\sigma}{\partial v_*^\lambda}\frac{\partial v_*^\rho}{\partial q^\sigma}.$$

Hence, from formulas (10.19) and (10.23) it follows that the Maggi equations (10.13) can be put in the form

$$\frac{d}{dt}\frac{\partial T^*}{\partial v_*^\lambda} - \frac{\partial T^*}{\partial \pi^\lambda} + \frac{\partial T^*}{\partial v_*^\rho}\frac{\partial \dot{q}^\sigma}{\partial v_*^\lambda}\left(\frac{d}{dt}\frac{\partial v_*^\rho}{\partial \dot{q}^\sigma} - \frac{\partial v_*^\rho}{\partial q^\sigma}\right) = Q_\lambda^*, \tag{10.24}$$

$$\rho, \sigma = \overline{1, s}, \qquad \lambda = \overline{1, l}.$$

Equations (10.18) and (10.24) apply both to holonomic and to holonomic systems, both with velocity-linear and velocity-nonlinear ideal constraints. For the case when the time is implicit in both the kinetic energy and in the constraint equations, equations (10.18) and (10.24) were obtained by G. Hamel, and in the general setting, by V.S. Novoselov.[23] Hence, these equations will be called the *Hamel–Novoselov equations.*

In the case when the quasi-velocities are introduced by the formulas (10.20) and the constraints are given by Eq. (10.21), we have

$$\frac{\partial \dot{q}^\sigma}{\partial v_*^\lambda}\frac{d}{dt}\frac{\partial v_*^\rho}{\partial \dot{q}^\sigma} = \beta_\lambda^\sigma \frac{d\alpha_\sigma^\rho}{dt} = \beta_\lambda^\sigma \frac{\partial \alpha_\sigma^\rho}{\partial q^\tau}\dot{q}^\tau = \beta_\lambda^\sigma \beta_\mu^\tau \frac{\partial \alpha_\sigma^\rho}{\partial q^\tau}v_*^\mu,$$

$$\frac{\partial \dot{q}^\sigma}{\partial v_*^\lambda}\frac{\partial v_*^\rho}{\partial q^\sigma} = \beta_\lambda^\sigma \frac{\partial \alpha_\tau^\rho}{\partial q^\sigma}\dot{q}^\tau = \beta_\lambda^\sigma \beta_\mu^\tau \frac{\partial \alpha_\tau^\rho}{\partial q^\sigma}v_*^\mu,$$

$$\rho, \sigma, \tau = \overline{1, s}, \qquad \lambda, \mu = \overline{1, l}.$$

Hence, in this case Eq. (10.24) assume the form

---

[23] See: *G. Hamel.* Nichtholonome Systeme höherer Art // Sitzungberichte der Berliner Mathemat. Gesellschaft. 1938. Bd. 37. S. 41–52; *V.S. Novoselov,* "Application of nonlinear nonholonomic coordinates in analytic mechanics," Uch. Zap. LGU, 1957, no. 217, vyp. 31, pp. 50–83 [in Russian]; *V.S. Novoselov,* "Extended motion equations of nonlinear nonholonomic systems," ibid., pp. 84–89 [in Russian].

$$\frac{d}{dt}\frac{\partial T^*}{\partial v_*^\lambda} - \frac{\partial T^*}{\partial \pi^\lambda} + c_{\lambda\mu}^\rho v_*^\mu \frac{\partial T^*}{\partial v_*^\rho} = Q_\lambda^*,$$

$$c_{\lambda\mu}^\rho = \left(\frac{\partial \alpha_\sigma^\rho}{\partial q^\tau} - \frac{\partial \alpha_\tau^\rho}{\partial q^\sigma}\right)\beta_\lambda^\sigma \beta_\mu^\tau, \tag{10.25}$$

$$\rho, \sigma, \tau = \overline{1, s}, \qquad \lambda, \mu = \overline{1, l}.$$

In the setting $l = s$ these equations, as well as the expressions for coefficients $c_{\sigma\tau}^\rho$, were first obtained by P. V. Voronets in 1901 (see the above footnote). In 1904 Hamel[24] obtains the above results with $l < s$. This is why such equations are generally called the *Voronets–Hamel*, even though Hamel called them the Euler–Lagrange equations. Sometimes such equations are also called the *Hamel–Boltzmann equations*.

At the same time as the papers by P. V. Voronets, A. Poincaré[25] put forward the equations very similar to equations (10.25). The *Poincaré equations* correspond to the case when in Eq. (10.25) with $l = s$ the coefficients $c_{\sigma\tau}^\rho$ are constant and the forces are expressed in terms of the force function U:

$$Q_\tau^* = \beta_\tau^\sigma \frac{\partial U}{\partial q^\sigma}, \qquad \sigma, \tau = \overline{1, s}.$$

In this case, Eq. (10.25) can be put in the form proposed by A. Poincaré:

$$\frac{d}{dt}\frac{\partial L^*}{\partial v_*^\tau} = c_{\sigma\tau}^\rho v_*^\sigma \frac{\partial L^*}{\partial v_*^\rho} + \beta_\tau^\sigma \frac{\partial L^*}{\partial q^\sigma}, \qquad L^*(q, v_*) = T^* + U, \tag{10.26}$$

$$\rho, \sigma, \tau = \overline{1, s}.$$

Poincaré's derivation of the motion equations (10.26) depend on the results from group theory. Poincaré's approach was further developed by N. G. Chetaev, L. M. Markhashov, V. V. Rumyantsev, and Fam Guen, who extended the Poincaré equations to the setting when the coefficients $c_{\sigma\tau}^\rho$ are nonconstant and the motion is subject to both potential and nonpotential forces. In addition, Rumyantsev also examined the case of nonlinear nonholonomic first-order constraints. He derived Eq. (10.26), which describe the motion of nonholonomic systems—such equations are called the *Poincaré–Chetaev-Rumyantsev equations*.[26]

The Udwadia–Kalaba equations. Using the generalized Moore–Penrose inversion of matrices, F. E. Udwadia and R. E. Kalaba derived the dynamic equations with respect to all generalized coordinates, the resulting equations did not involve

---

[24] See: *G. Hamel*, "Die Lagrange-Eulerischen Gleichungen der Mechanik," Zeitschrift für Mathematik und Physik. 1904. Bd. 50. S. 1–50.

[25] See: *H. Poincaré*, "Sur une forme nouvelle des équations de la mécanique," Comptes Rendus. 1901. Vol. 132, pp. 369–371.

[26] See: *N. G. Chetaev*, "On Poincaré equations," Appl. Math. Mech., 1941, vol. V, issue 2, pp. 253–262 [in Russian]; *V. V. Rumyantsev*, "General equations of analytic mechanics," J. Appl. Math. Mech., 1994, vol. 58, issue 3, pp. 3–16 [in Russian].

Lagrange multipliers.[27] We note that the generalized Moore–Penrose inversion plays the same important role in the matrix derivation of the motion equations as the decomposition by the constraint equations

$$f_1^{\varkappa}(t, q, \dot{q}) = 0, \quad \varkappa = \overline{1, k}, \tag{10.27}$$

of the entire s-dimensional space into the direct sum of the K- and L- spaces. Above it was shown that using such a partition (9.19) we arrived at the generalized reactions. Substituting them into the second-kind Lagrange equations (9.21) with multipliers, this gives

$$A_{\sigma\tau}(t, q, \dot{q}) \, \ddot{q}^{\tau} = B_{\sigma}(t, q, \dot{q}),$$

$$A_{\sigma\tau} = M \left(g_{\sigma\tau} - g_{\sigma^*\tau} \frac{\partial \dot{q}^{\sigma^*}}{\partial v_*^{l+\varkappa}} \frac{\partial \varphi^{\varkappa}}{\partial \dot{q}^{\sigma}} \right),$$

$$B_{\sigma} = Q_{\sigma} - Q_{\sigma^*} \frac{\partial \dot{q}^{\sigma^*}}{\partial v_*^{l+\varkappa}} \frac{\partial \varphi^{\varkappa}}{\partial \dot{q}^{\sigma}} + M \, \Gamma_{\sigma^*, \alpha\beta} \dot{q}^{\alpha} \dot{q}^{\beta} \frac{\partial \dot{q}^{\sigma^*}}{\partial v_*^{l+\varkappa}} \frac{\partial \varphi^{\varkappa}}{\partial \dot{q}^{\sigma}} -$$

$$- M \, \Gamma_{\sigma, \alpha\beta} \dot{q}^{\alpha} \dot{q}^{\beta}, \quad \sigma, \sigma^*, \tau = \overline{1, s}, \quad \varkappa = \overline{1, k}.$$

These formulas also imply the *Udwadia–Kalaba equations*

$$\ddot{q}^{\tau} = A^{\tau\sigma}(t, q, \dot{q}) \, B_{\sigma}(t, q, \dot{q}), \quad \sigma, \tau = \overline{1, s},$$

where $A^{\tau\sigma}$ are entries of the matrix inverse to the matrix $(A_{\sigma\tau})$.

The above Udwadia–Kalaba equations were obtained here for the case of classical nonlinear nonholonomic constraints. Using the expressions for the generalized reactions (9.29), which were derived in the generalized Maggi equations, it would be possible to write down analogous equations also for the case of linear nonholonomic second-order constraints. These very equations were obtained in the matrix form in the paper from the first footnote of this paragraph. We also note that such equations can be obtained by excluding the generalized reactions $\Lambda_{\varkappa}$, $\varkappa = \overline{1, k}$, from the generalized Lagrange equations of the second kind with multipliers (9.33).[28]

Example 12 *Motion of an inclined ice skater (the Chaplygin sledge)* (application of the Chaplygin equations). Let us go back to the study of the motion from Example 9 in Sect. 9. We shall retain the above notion and use Fig. 15, but we shall compose the Chaplygin equations (10.7), rather than the Maggi equations (9.18).

---

[27] See the papers: *F.E. Udwadia, R.E. Kalaba*, "A new perspective on constrained motion," Proceedings of the Royal Society. London. 1992. Vol. A439. Issue 1906. P. 407–410; *E.H. Moore*, "On the reciprocal of the general algebraic matrix," Bidl. Am. math. Soc. 1920. Vol. 26. P. 394–395; *R. Penrose*, "A generalized inverse of matrices," Proc. Camb. phil. Soc. 1955. Vol. 51. P. 406–413.

[28] See the paper: *S.A. Zegzhda, N.V. Naumova, Sh.Kh. Soltakhanov, and M.P. Yushkov*, "Relationship between the Udwadia–Kalaba equations and the generalized Lagrange and Maggi equations," ISSN 1063-4541, Vestnik St. Petersburg University: Mathematics, 2016, volume 49, issue 1, pp. 81–84. ©Allerton Press, Inc., 2016. DOI: 10.3103/S1063454116010143.

As before, for the generalized coordinates of the mechanical system we take the coordinates ξ, η of the point A (the skate) and the rotation angle θ of the ice skater (the sledge) with respect to the system $O\xi\eta$:

$$q^1 = \xi, \quad q^2 = \theta, \quad q^3 = \eta. \tag{10.28}$$

The chosen numbering scheme of the generalized coordinates will be discussed below.

The motion of the system is subject to the nonholonomic constraint (9.34):

$$\dot{\eta} = \dot{\xi} \tan\theta. \tag{10.29}$$

The convenience in the choice of the generalized coordinates for this constraint is manifested in notation (10.28), because in this case the rule of using the indexes and their ranges from formulas (10.1)–(10.7) is retained. Indeed, we have $s = 3$, $k = 1$, $l = 2$, the constraint (10.29) assuming the form (10.1) if one takes $\beta_1^3(q) = \tan\theta$, $\beta_2^3(q) = 0$.

Clearly, to prevent the motion of the ice skater (the sledge) in the direction of the Ay-axis, the corresponding force $\mathbf{R}_A$ should be applied to the skate from the direction of the fixed plane $O\xi\eta$ (this force is perpendicular to the skate sharp edge). This force is the reaction of the constraint (10.29). If the plane $O\xi\eta$ acts on the skate only by the force $\mathbf{R}_A$, then the constraint is ideal.

By König's theorem, the kinetic energy of the system is of the form (see (9.35)):

$$T = \frac{M}{2}\{[\dot{\xi} - \dot{\varphi}(\alpha\sin\varphi + \beta\cos\varphi)]^2 + \\
+[\dot{\eta} + \dot{\varphi}(\alpha\cos\varphi - \beta\sin\varphi)]^2 + k_C^2\dot{\varphi}^2\}. \tag{10.30}$$

We recall that here M is the mass of the sledge, $k_C^2 = J_C/M$ is the radius of gyration of the sledge with respect to the axis parallel to the $Q\zeta$-axis and passing through the centroid C; α and β are the coordinates of the center of mass in the system Axy.

As is required in the general theory, the kinetic energy in this problem is dependent of $q^3 = \eta$, and it is assumed that $Q_3 = 0$. Using the constraint equation (10.29) to exclude the generalized velocity $\dot{q}^3 = \dot{\eta}$ from the expression for the kinetic energy T, this establishes

$$T_* = \frac{M}{2}\{[\dot{\xi} - \dot{\theta}(\alpha\sin\theta + \beta\cos\theta)]^2 + \\
+[\dot{\xi}\tan\theta + \dot{\theta}(\alpha\cos\theta - \beta\sin\theta)]^2 + k_C^2\dot{\theta}^2\}. \tag{10.31}$$

Now employing expressions (10.30) and (10.31) one may write down the Chaplygin equations (10.7) (here, we recall that $\gamma^2 = \alpha^2 + \beta^2 + k_C^2$):

$$\frac{\ddot{\xi}}{\cos^2\theta} + \frac{2\dot{\xi}\dot{\theta}\tan\theta}{\cos^2\theta} - \frac{\ddot{\theta}\beta}{\cos\theta} - \frac{\beta\dot{\theta}^2\sin\theta}{\cos^2\theta} -$$

$$-[\dot{\xi}\tan\theta + \dot{\theta}(\alpha\cos\theta - \beta\sin\theta]\frac{\dot{\theta}}{\cos^2\theta} = \frac{Q_1}{M}, \qquad (10.32)$$

$$-\frac{\beta\ddot{\xi}}{\cos\theta} + \gamma^2\ddot{\theta} + \frac{\alpha\dot{\xi}\dot{\theta}}{\cos\theta}\sin^2\theta - \frac{\beta\dot{\xi}\sin\theta}{\cos^2\theta} = \frac{Q_2}{M}.$$

Here, in the derivative $\partial T/\partial\dot{\eta}$, the generalized velocity $\dot{\eta}$ was expressed in terms of $\dot{\xi}$ with the help of the constraint equation (10.29). The generalized forces are given by formulas (9.36). Simplifying, one may write system (10.32) in the form

$$\ddot{\xi} - \ddot{\theta}\beta\cos\theta + \dot{\xi}\dot{\theta}\tan\theta - \dot{\theta}^2\alpha\cos\theta = (Q_1\cos^2\theta)/M,$$
$$-\ddot{\xi}\beta\cos\theta + \ddot{\theta}\gamma^2\cos^2\theta + \dot{\xi}\dot{\theta}(\alpha\cos\theta - \beta\sin\theta) = (Q_2\cos^2\theta)/M. \qquad (10.33)$$

From the Chaplygin equations (10.33) one may find ξ and θ *qua* functions of time; afterwards, from the constraint equation (10.29) one may find the law of variation of the coordinate η. We note that the derivation of the Chaplygin equations involved more computations than that of the Maggi equations for the same problem in Sect. 9.

For purposes of numerical integration, it is expedient to write the time-differentiated constraint equation as

$$\ddot{\xi}\tan\theta - \ddot{\eta} + \dot{\xi}\dot{\theta}/\cos^2\theta = 0. \qquad (10.34)$$

Considering this equation and system (10.33) as an algebraic system with respect to the unknowns $\ddot{\xi}$, $\ddot{\eta}$, $\ddot{\theta}$, one may write Eqs. (10.33), (10.34) in the normal form, which is amenable to well-proven numerical methods of solution.

To conclude this example, we compare the equations obtained in the Chaplygin form (10.33) and in the Maggi form (9.38). Using Chaplygin's method, we replace in system (9.38) the quantities $\dot{\eta}$ and $\ddot{\eta}$ by their expressions from the nonholonomic constraint equation (10.29). Now

$$\ddot{\xi} + \tan\theta\left(\ddot{\xi}\tan\theta + \dot{\xi}\dot{\theta}\frac{1}{\cos^2\theta}\right) - \ddot{\theta}\frac{\beta}{\cos\theta} - \dot{\theta}^2\frac{\alpha}{\cos\theta} = -\frac{\kappa_1}{M}(\dot{\xi} + \dot{\xi}\tan^2\theta),$$

$$\gamma^2\ddot{\theta} + \left(\ddot{\xi}\tan\theta + \dot{\xi}\dot{\theta}\frac{1}{\cos^2\theta}\right)(\alpha\cos\theta - \beta\sin\theta) -$$

$$-\ddot{\xi}(\alpha\sin\theta + \beta\cos\theta) = -\frac{\kappa_2}{M}\dot{\theta}.$$

Transforming, we arrive at the system

$$\ddot{\xi}\frac{1}{\cos^2\theta} + \dot{\xi}\dot{\theta}\frac{\tan\theta}{\cos^2\theta} - \ddot{\theta}\frac{\beta}{\cos\theta} - \dot{\theta}^2\frac{\alpha}{\cos\theta} = -\frac{\kappa_1}{M}\dot{\xi},$$

$$\gamma^2\ddot{\theta} - \ddot{\xi}\frac{\beta}{\cos\theta} + \dot{\xi}\dot{\theta}\frac{(\alpha\cos\theta - \beta\sin\theta)}{\cos^2\theta} = -\frac{\kappa_2}{M}\dot{\theta}.$$

Multiplying these equations by $\cos^2\theta$, we readily obtain the Chaplygin equations (10.33).

Example 13 *Motion of a wheeled robot* (application of the Hamel–Boltzmann equations). Let us compose the Hamel–Boltzmann equations (10.25) describing the motion of a wheeled robot from Sect. 9, Example 10 (see Fig. 17). The motion of the system is subject to constraint (9.40), the kinetic energy and the generalized forces are given by formulas (9.41) and (9.42).

Let us introduce the quasi-velocities by the formulas

$$v_*^1 = \dot\varphi, \qquad v_*^2 = \dot\theta,$$
$$v_*^3 = -\dot\xi_C \sin\varphi + \dot\eta_C \cos\varphi - l_2\dot\varphi,$$
$$v_*^4 = -\dot\xi_C \sin(\varphi + \theta) + \dot\eta_C \cos(\varphi + \theta) + l_1\dot\varphi\cos\theta,$$
(10.35)

that is, in formulas (10.20) the coefficients $\alpha_\sigma^\rho(q)$, $\rho, \sigma = \overline{1,4}$, are of the form

$$\alpha_1^1 = 1, \quad \alpha_2^2 = 1, \quad \alpha_1^3 = -l_2, \quad \alpha_3^3 = -\sin\varphi, \quad \alpha_4^3 = \cos\varphi,$$
$$\alpha_2^4 = l_1\cos\theta, \quad \alpha_3^4 = -\sin(\varphi + \theta), \quad \alpha_4^4 = \cos(\varphi + \theta).$$

To formulas (10.35) there corresponds the inverse transformation

$$\dot q^1 \equiv \dot\varphi = v_*^1, \qquad \dot q^2 \equiv \dot\theta = v_*^2,$$
$$\dot q^3 \equiv \dot\xi_C = \beta_1^3 v_*^1 + \beta_3^3 v_*^3 + \beta_4^3 v_*^4,$$
$$\dot q^4 \equiv \dot\eta_C = \beta_1^4 v_*^1 + \beta_3^4 v_*^3 + \beta_4^4 v_*^4,$$
(10.36)

where

$$\beta_1^3 = (l_1\cos\varphi\cos\theta + l_2\cos(\varphi + \theta)/\sin\theta,$$
$$\beta_3^3 = \cos(\varphi + \theta)/\sin\theta, \qquad \beta_4^3 = -\cos\varphi/\sin\theta,$$
$$\beta_1^4 = (l_1\sin\varphi\cos\theta + l_2\sin(\varphi + \theta))/\sin\theta,$$
$$\beta_3^4 = \sin(\varphi + \theta)/\sin\theta, \qquad \beta_4^4 = -\sin\varphi/\sin\theta.$$

The remaining coefficients α_σ^ρ and β_ρ^σ are zeros.

We have thus obtained the matrices (α_σ^ρ) and (β_ρ^σ) in transformations (10.20). Now we may calculate the coefficients of nonholonomicity by the formulas (10.25):

$$c_{13}^3 = -c_{31}^3 = \beta_3^3\cos\varphi + \beta_3^4\sin\varphi,$$
$$c_{14}^3 = -c_{41}^3 = \beta_4^3\cos\varphi + \beta_4^4\sin\varphi,$$
$$c_{21}^3 = -c_{12}^3 = l_1\sin\theta + \beta_1^3\cos(\varphi + \theta) + \beta_1^4\sin(\varphi + \theta),$$
(10.37)
$$c_{13}^4 = -c_{31}^4 = c_{23}^4 = -c_{32}^4 = \beta_3^3\cos(\varphi + \theta) + \beta_3^4\sin(\varphi + \theta),$$
$$c_{14}^4 = -c_{41}^4 = c_{24}^4 = -c_{42}^4 = \beta_4^3\cos(\varphi + \theta) + \beta_4^4\sin(\varphi + \theta).$$

The remaining $c^{l+\varkappa}_{\lambda\lambda^*}$ are zeros.

By formulas (10.19) we have

$$Q_1^* = (F_1(t)\cos\varphi - F_2(v_C)\dot{\xi}_C/v_C)\beta_1^3 + (F_1(t)\sin\varphi - F_2(v_C)\dot{\eta}_C/v_C)\beta_1^4,$$
$$Q_2^* = L_1 - L_2 - L_3,$$
$$Q_3^* = (F_1(t)\cos\varphi - F_2(v_C)\dot{\xi}_C/v_C)\beta_3^3 + (F_1(t)\sin\varphi - F_2(v_C)\dot{\eta}_C/v_C)\beta_3^4,$$
$$Q_4^* = (F_1(t)\cos\varphi - F_2(v_C)\dot{\xi}_C/v_C)\beta_4^3 + (F_1(t)\sin\varphi - F_2(v_C)\dot{\eta}_C/v_C)\beta_4^4.$$
(10.38)

Using formulas (10.36) and (9.41), we find the expression for T^*:

$$2T^* = (v_*^1)^2\Big(M^*((\beta_1^3)^2 + (\beta_1^4)^2) + J^* + M_2l_1^2 + 2M_2l_1(\beta_1^4\cos\varphi - \beta_1^3\sin\varphi)\Big)+$$
$$+(v_*^2)^2 J_2 + (v_*^3)^2 M^*((\beta_3^3)^2 + (\beta_3^4)^2) + (v_*^4)^2 M^*((\beta_4^3)^2 + (\beta_4^4)^2)+$$
$$+v_*^1 v_*^2 2 J_2 + v_*^1 v_*^3\Big(2M^*(\beta_1^3\beta_3^3 + \beta_1^4\beta_3^4) + 2M_2l_1(\beta_3^4\cos\varphi - \beta_3^3\sin\varphi)\Big)+$$
$$+ v_*^1 v_*^4\Big(2M^*(\beta_1^3\beta_4^3 + \beta_1^4\beta_4^4) + 2M_2l_1(\beta_4^4\cos\varphi - \beta_4^3\sin\varphi)\Big)+$$
$$+v_*^3 v_*^4 2M^*(\beta_3^3\beta_4^3 - \beta_3^4\beta_4^4).$$

Suppressing cumbersome calculations, we shall give the Hamel–Boltzmann equations (10.25) for our problem

$$\left[J^* + M_2l_1^2 + 2M_2l_1l_2 + M^*\left(\frac{l_2^2 + l_1^2\cos^2\theta + 2l_1l_2\cos^2\theta}{\sin^2\theta}\right)\right]\ddot{\varphi} + J_2\ddot{\theta}-$$
$$-\frac{(l_1 + l_2)^2 M^*\cos\theta}{\sin^3\theta}\dot{\varphi}\dot{\theta} = \left(F_1(t)\cos\varphi - \frac{F_2(v_C)\dot{\xi}_C}{v_C}\right)\beta_1^3+$$
$$+\left(F_1(t)\sin\varphi - \frac{F_2(v_C)\dot{\eta}_C}{v_C}\right)\beta_1^4,$$
$$J_2(\ddot{\varphi} + \ddot{\theta}) = L_1(t) - L_2(\dot{\theta}) - L_3(\theta).$$
(10.39)

which were obtained using formulas (10.17), (10.37), (10.38). We note that this system of equations has to be solved jointly with the constraint equations (9.40).

It is now interesting to compare the Hamel–Boltzmann equations (10.39) with the above Maggi equations (9.47). The second equations of these systems are seen to agree. If one determines the expressions $\dot{\xi}_C$ and $\ddot{\eta}_C$ from the constraint equations and substitute them into the first equation of system (9.47), then one obtains the first equation of system (10.39).

It should be noted that composing the Hamel–Boltzmann equations requires a considerable amount of computation in comparison with the application of the Maggi equations.

Comparing Examples 9, 10 and 12, 13 we conclude that in solving practical problems the use of the Maggi equations is superior to other possible representations of the motion equations of nonholonomic systems. The Maggi equations can be composed almost as cheaply as the second-kind Lagrange equations. With ideal nonholonomic constraints, the Maggi equations split into two groups. Upon specifying the initial conditions, from the first group jointly with the constraint equations one may find the law of motion of a nonholonomic system. Afterwards, the generalized reactions can be obtained from the second group.

In the Chaplygin, Hamel–Boltzmann and other similar equations, the authors were trying to single out the Lagrange operator, the remaining ones on the left were characterizing the nonholonomicity of the system. Hence, in the case when the constraints are integrable, the differential equations change to the conventional second-kind Lagrange equations of holonomic mechanics. Equations (10.7), (10.25) and similar equations were developed for specific types of (usually linear) nonholonomic constraints of type (10.1), (10.21); they proved useful in solving the corresponding problems. As a rule, from such equations one was capable of finding the smallest number of motion equations. For example, the left-hand side of the Chaplygin equations (10.7) contains only the unknowns q^1, ..., q^l, and hence, on integrating these equations the remaining coordinates q^{l+1}, ..., q^s could be obtained from the constraint equations (10.1).

In contrast, the Maggi equations were pointed out to hold for any nonholonomic constraints, and in particular, for those which are nonlinear with respect to the generalized velocities. It is important to note here that the composition of the differential motion equations (9.18) calls for a unified approach which is standard for all problems: following the choice of generalized coordinates q^1, ..., q^s, one writes down the left-hand sides of the conventional Lagrange equations of the second kind; next one introduces the transition formulas from the generalized velocities to the quasi-velocities (the latter taking into account the expressions for the nonholonomic constraints); further, one writes down the inverse transformation, and on differentiating it in the quasi-velocities composes the motion equations (9.18), which are linear combination of the Lagrange equations of the second kind. Here, two computationally handy remarks are worth noting.

First, for purposes of the numerical integration of system (9.18) jointly with constraints (9.13) it is advantageous to first differentiate the constraints in time and obtain equations which are linear with respect to the generalized accelerations. These equations and Maggi equations form a system of linear inhomogeneous algebraic equations with respect to $\ddot{q}^1$, ..., $\ddot{q}^s$. Solving this system, we shall obtain a system of differential equations ready for numerical integration.

Second, in the case of nonlinear nonholonomic constraints (9.13), derivation of the inverse transformation (that is, the second group of formulas in (9.15)) from the first group of formulas (9.15) may involve certain difficulties. Hence, to avoid such challenges, it is recommended to use the first group of formulas (9.15) to compose the matrix of the derivatives $(\partial v_*^\rho / \partial \dot{q}^\sigma)$, $\rho, \sigma = \overline{1, s}$, and then find the inverse matrix $(\partial \dot{q}^\sigma / \partial v_*^\rho)$, $\rho, \sigma = \overline{1, s}$, whose elements will be used to cook the Maggi equations.

11 Motion Control Using Constraints Depending on the Parameters

According to the above, applying nonholonomic constraints to a system, which was early considered as free, is equivalent to applying the reaction force $\mathbf{R} = \Lambda_\varkappa \nabla'\varphi^\varkappa + \mathbf{T}_0$; the solution of the problem could be satisfied using the "minimal" reaction

$$\mathbf{R} = \Lambda_\varkappa \nabla'\varphi^\varkappa, \quad \mathbf{T}_0 = 0.$$

There is a more general approach to the study of constraint motion. Let us assume that we have equations of the form

$$\varphi^\varkappa(t, q, \dot{q}, u, \dot{u}) = 0, \quad \varkappa = \overline{1, k},$$

where $u = \{u^1, \ldots, u^m\}$ and $u^1, \ldots, u^m$ are some parameters. With these conditions the following statement of the problem is possible: match parameters $u^1, u^2, \ldots, u^m$ *qua* functions of time in order that the motion of the system would obey the equations of the form

$$\pi^\mu(t, q, \dot{q}) = 0, \quad \mu = \overline{1, m}. \tag{11.1}$$

Here, it is assumed that in the equations there appear no additional forces (save the force $\mathbf{R} = \Lambda_\varkappa \nabla'\varphi^\varkappa$).

The above problem is not always solvable. Indeed, if Eq. (11.1) are looked upon as conventional constraint equations, then they would be satisfied by introducing the force of the form

$$\Pi = \Lambda_\mu^* \nabla'\pi^\mu \quad \mu = \overline{1, m}$$

into the Newton equation. If here one puts $\Lambda_\mu^* = 0$, $\mu = \overline{1, m}$, then this would mean that the motion control is effected only by the action of the generalized forces $\Lambda_\varkappa$, $\varkappa = \overline{1, k}$. Each of them has a direct effect on the value of the function $\varphi^\varkappa$ and may have an implicit effect on the value of the function $\pi^\varkappa$ if $\nabla'\pi^\mu \cdot \nabla'\varphi^\varkappa \neq 0$. Hence, if the conditions

$$\nabla'\pi^\mu \cdot \nabla'\varphi^\varkappa \neq 0, \quad \mu = \overline{1, m}, \quad \varkappa = \overline{1, k},$$

are not satisfied, then one cannot guarantee the solvability of this problem.

In this case, the concept of a constraint becomes more meaningful. For a better distinction of the functions $\varphi^\varkappa$ and π^μ from the functions, which were usually used to define the equations of conventional constraints, we shall adopt the following terminology.

Equation (11.1) will be called the *equations of motion program*, the functions $\varphi^\varkappa$ will be called the *control functions*, the corresponding forces $\mathbf{R} = \Lambda_\varkappa \nabla'\varphi^\varkappa$ will be called the *control forces* effected via constraints, and the parameters $u^1, u^2, \ldots, u^m$

will be called the *control parameters*. It is natural to call the motion subject to the above condition the *controlled motion*.

With a more general setting of the motion control problem, the parameters u^μ and their derivatives $\dot{u}^\mu$ can be also introduced into the expressions for the program motion. This program can be put in the form

$$\pi^\mu(t, q, \dot{q}, u, \dot{u}) = 0, \quad \mu = \overline{1, m}.$$

It is worth pointing out that the number of equations expressing the program of motion should be such that the number of the equations in the problem would agree with the number of the unknowns. To satisfy this requirement, it suffices to require that the number of parameters $u^1, u^2, \ldots, u^m$ would be equal to the number of the equations expressing the program. If the number of control parameters is larger than the number of equations for the program of motion, then the additional equations for solving the above problem could be obtained by introducing some conditions, which in particular may contain requirements about the extremality of separate characteristics of motion.

The control of motion can be effected also via the force $\mathbf{Y}$ on assuming that it depends on u^μ, $\dot{u}^\mu$, $\ddot{u}^\mu$; in other words, $\mathbf{Y}$ is a given function of the form

$$\mathbf{Y} = \mathbf{Y}(t, q, \dot{q}, u, \dot{u}, \ddot{u}).$$

The second derivatives of the control parameters are introduced into the force $\mathbf{Y}$ in order to take into account the case when the parameters u^μ define the motion of a movable system; that is, when the inertia forces are the part of the control forces.

Thus, in the general case a similar problem on the controlled motion is reduced to solving the system of equations

$$M\mathbf{W} = \mathbf{Y}(t, q, \dot{q}, u, \dot{u}, \ddot{u}) + \Lambda_\varkappa \nabla' \varphi^\varkappa,$$
$$\varphi^\varkappa(t, q, \dot{q}, u, \dot{u}) = 0, \quad \varkappa = \overline{1, k},$$
$$\pi^\mu(t, q, \dot{q}, u, \dot{u}) = 0, \quad \mu = \overline{1, m}.$$

Here, $q^1, q^2, \ldots, q^s$, $\Lambda_1, \Lambda_2, \ldots, \Lambda_k$, $u^1, u^2, \ldots, u^m$ are the unknowns (altogether, there are $s + k + m$ unknowns). The number of the equations is also $s + k + m$.

If it is required to find the character of motion of the system and ascertain the control parameters *qua* functions of time, then the unknowns $\Lambda_\varkappa$ can be excluded, writing the Newton equations in projections to the directions

$$\varepsilon_\lambda = \beta_\lambda^\sigma \mathbf{e}_\sigma, \quad \lambda = \overline{1, l}, \quad l = s - k, \quad \sigma = \overline{1, s},$$

where the coefficients β_λ^σ are calculated by the formulas (5.21), (5.22), in which u^μ and $\dot{u}^\mu$ enter as parameters.

In this case, the motion equations (5.19) free from the unknowns $\Lambda_\varkappa$ read as

$$(M \, (g_{\sigma\tau} \, \ddot{q}^{\,\tau} + \Gamma_{\sigma,\alpha\beta} \, \dot{q}^{\alpha}\dot{q}^{\beta}) - Q_{\sigma}(t,q,\dot{q},u,\dot{u},\ddot{u})) \, \beta^{\sigma}_{\lambda}(t,q,\dot{q},u,\dot{u}) = 0,$$
$$\lambda = \overline{1,l}, \quad \sigma,\tau = \overline{1,s}, \quad \alpha,\beta = \overline{0,s}.$$

These equations may also be reduced to equations of the form (5.23); that is, they can be written in the form

$$M \, (\widetilde{g}_{\lambda\sigma} \, \ddot{q}^{\,\sigma} + \widetilde{\Gamma}_{\lambda,\alpha\beta} \, \dot{q}^{\alpha}\dot{q}^{\beta}) = \widetilde{Q}_{\lambda}, \quad \lambda = \overline{1,l}, \quad l = s - k. \tag{11.2}$$

To obtain a closed system with respect to the unknowns $q^{\sigma}(t)$ and $u^{\mu}(t)$, it is required to augment the system (11.2) with the equations of control constraints

$$\varphi^{\varkappa}(t,q,\dot{q},u,\dot{u}) = 0, \quad \varkappa = \overline{1,k}, \tag{11.3}$$

and the equations of program motion

$$\pi^{\mu}(t,q,\dot{q},u,\dot{u}) = 0, \quad \mu = \overline{1,m}. \tag{11.4}$$

If the functions $\varphi^{\varkappa}$ and π^{μ} are nonlinear in $\dot{q}^{\sigma}$ and $\dot{u}^{\mu}$, then Eqs. (11.3), (11.4) should be differentiated in time and then added to the system (11.2). Differentiating, this gives

$$\dot{\varphi}^{\varkappa} = \frac{\partial\varphi^{\varkappa}}{\partial t} + \frac{\partial\varphi^{\varkappa}}{\partial q^{\sigma}} \, \dot{q}^{\sigma} + \frac{\partial\varphi^{\varkappa}}{\partial\dot{q}^{\sigma}} \, \ddot{q}^{\,\sigma} + \frac{\partial\varphi^{\varkappa}}{\partial u^{\gamma}} \, \dot{u}^{\gamma} + \frac{\partial\varphi^{\varkappa}}{\partial\dot{u}^{\gamma}} \, \ddot{u}^{\gamma} = 0,$$
$$\varkappa = \overline{1,k}, \quad \sigma = \overline{1,s}, \quad \gamma = \overline{1,m}, \tag{11.5}$$

$$\dot{\pi}^{\mu} = \frac{\partial\pi^{\mu}}{\partial t} + \frac{\partial\pi^{\mu}}{\partial q^{\sigma}} \, \dot{q}^{\sigma} + \frac{\partial\pi^{\mu}}{\partial\dot{q}^{\sigma}} \, \ddot{q}^{\,\sigma} + \frac{\partial\pi^{\mu}}{\partial u^{\gamma}} \, \dot{u}^{\gamma} + \frac{\partial\pi^{\mu}}{\partial\dot{u}^{\gamma}} \, \ddot{u}^{\gamma} = 0,$$
$$\gamma,\mu = \overline{1,m}, \quad \sigma = \overline{1,s}. \tag{11.6}$$

We shall assume that the quantities $\widetilde{Q}_{\lambda}$ depend linearly on $\ddot{u}^{\gamma}$. Now the set of Eqs. (11.2), (11.5) and (11.6) can be looked upon as a system of linear algebraic equations with respect to the unknowns $\ddot{q}^{\,\sigma}$ and $\ddot{u}^{\gamma}$. Solving this system, consisting of $s + m$ equations, by the Cramer formulas or excluding the unknowns in succession (which is sometimes more convenient in practice), we find that

$$\ddot{q}^{\,\sigma} = F^{\sigma}(t,q,\dot{q},u,\dot{u}), \quad \sigma = \overline{1,s},$$
$$\ddot{u}^{\gamma} = \Phi^{\gamma}(t,q,\dot{q},u,\dot{u}), \quad \gamma = \overline{1,m}. \tag{11.7}$$

Note that if the above problem would be unsolvable, then the above system of linear algebraic equations would not be uniquely solvable, and hence, the system of equations (11.2), (11.5) and (11.6) would not be reducible to a system of the form (11.7).

Denoting $\dot{q}^{\sigma} = \widehat{v}^{\sigma}$, $\sigma = \overline{1,s}$, $\dot{u}^{\gamma} = \widehat{v}^{s+\gamma}$, $\gamma = \overline{1,m}$, we write the of system Eq. (11.7) in the normal form:

Fig. 21 Control of a disk
with plate

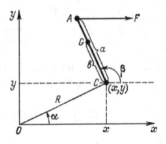

$$\frac{d\widehat{v}^{\sigma}}{dt} = F^{\sigma}(t, q, u, \widehat{v}), \quad \frac{dq^{\sigma}}{dt} = \widehat{v}^{\sigma}, \quad \widehat{v} = (\widehat{v}^{1}, \widehat{v}^{2}, \dots, \widehat{v}^{s+m}),$$

$$\frac{d\widehat{v}^{s+\gamma}}{dt} = \Phi^{\gamma}(t, q, u, \widehat{v}), \quad \frac{du^{\gamma}}{dt} = \widehat{v}^{s+\gamma}, \quad \sigma = \overline{1, s}, \quad \gamma = \overline{1, m}. \tag{11.8}$$

We recall that in solving system (11.8) the initial conditions should satisfy the conditions

$$\varphi^{\varkappa}(t_0, q_0, u_0, \widehat{v}_0) = 0, \quad \varkappa = \overline{1, k},$$

$$\pi^{\mu}(t_0, q_0, u_0, \widehat{v}_0) = 0, \quad \mu = \overline{1, m}.$$

Equation (11.4) are now called the *program of motion*. According to H. Béghin[29] these are the *servoconstraint equations*. V.I. Kirgetov[30] proposed to call Eq. (11.4) the *conditional constraints*. H. Béghin, P. Appell[31] and others in solving such problems used the d'Alembert–Lagrange principle. Above we have shown that any controlled motion can be treated using the Newton's law and the equations for the control constraints and the program of motion.

Let us consider two examples. These problems were first examined by H. Béghin (see the above footnote).

Example 14 *Control of a disk with plate.* The plate $\sum$ lying in a fixed horizontal plane is pin-connected at a point C with a round disk $\sum_1$, which rotates in the same plane about the fixed center O (Fig. 21). At a point A lying on the line connecting the point C with the center of gravity G of the plane $\sum$, the plate is subject to a constant force F, which is parallel to the fixed line Ox. The motion of the disk $\sum_1$ is automatically regulated so that the lines OC and CA remain perpendicular to each other.

In this problem, we shall consider the controlled motion of a plate $\sum$. As generalized coordinates, we take the coordinate (x, y) of the point C and the angle β between

[29] See *H. Béghin*, Étude théorique des compas gyrostatiques Anschütz et Sperry, Paris, 1922.

[30] See *V.I. Kirgetov*, "The motion of controlled mechanical systems with prescribed constraints (servoconstraints)," J. Appl. Math. Mech. USSR, 31, 431–466 (1967) [in Russian].

[31] *P. Appell*, Traité de Mécanique Rationnelle, T. 2, Gauthier-Villars, 1953.

the Ox-axis and the line CA. The control parameter is the angle of rotation α of the disk $\sum_1$.

So, we shall use the following generalized notation:

$$q^1 = \beta\,, \quad q^2 = x\,, \quad q^3 = y\,, \quad u^1 = \alpha\,, \quad s = 3\,, \quad m = 1\,.$$

The control constraint is the pin-connection of the disk $\sum_1$ and the plate $\sum$ at the point C. This constraint is given by the equations

$$x = R\cos\alpha\,, \quad y = R\sin\alpha\,, \tag{11.9}$$

from which it follows that the point C of the plate moves only along a circle of radius R. We assume that there is no friction at the joint and that the constraint is ideal.

Differentiating equations (11.9), this establishes

$$\varphi^1 = \dot{x} + \dot{\alpha}R\sin\alpha = 0\,,$$
$$\varphi^2 = \dot{y} - \dot{\alpha}R\cos\alpha = 0\,.$$

By the condition of the problem it is required to find under which law $\alpha = \alpha(t)$ the motion of the plate $\sum$ would correspond to the program $\beta = \alpha - \pi/2$.

The kinetic energy of the plate is as follows:

$$T = \frac{Mv_G^2}{2} + \frac{Mk^2\dot{\beta}^2}{2}\,.$$

Here, M is the mass of the plate, v_G is the velocity of the plate center of gravity, Mk^2 is the moment of inertia of the plate with respect to the point G. From Fig. 21 it follows that the squared velocity of the point G can be written in the form

$$v_G^2 = (\dot{x} - b\dot{\beta}\sin\beta)^2 + (\dot{y} + b\dot{\beta}\cos\beta)^2\,.$$

Substituting this relation into the expression for the kinetic energy, this gives

$$T = \frac{M}{2}(\dot{x}^2 + \dot{y}^2 + (k^2 + b^2)\dot{\beta}^2 + 2\dot{y}b\dot{\beta}\cos\beta - 2\dot{x}b\dot{\beta}\sin\beta)\,.$$

The motion equation of the plate in the tangent space reads as

$$M\mathbf{W} = \mathbf{Y} + \Lambda_1\mathbf{\nabla}'\varphi^1 + \Lambda_2\mathbf{\nabla}'\varphi^2\,,$$

where $\mathbf{Y}$ is the vector corresponding to the force F;

$$\nabla'\varphi^1 = \frac{\partial\varphi^1}{\partial\widehat{v}^2}\,\mathbf{e}^2 = \frac{\partial\varphi^1}{\partial\dot{x}}\,\mathbf{e}^2\,,$$

$$\nabla'\varphi^2 = \frac{\partial\varphi^2}{\partial\widehat{v}^3}\,\mathbf{e}^3 = \frac{\partial\varphi^2}{\partial\dot{y}}\,\mathbf{e}^3\,.$$

The force F has the potential $\Pi = -Fx_A = -F(x + a\cos\beta)$, hence, the vector $\mathbf{Y}$ can be written as follows:

$$\mathbf{Y} = -\nabla\Pi = -\frac{\partial\Pi}{\partial\beta}\,\mathbf{e}^1 - \frac{\partial\Pi}{\partial x}\,\mathbf{e}^2 = -Fa\sin\beta\,\mathbf{e}^1 + F\mathbf{e}^2 = Y_1\mathbf{e}^1 + Y_2\mathbf{e}^2\,.$$

Following the general approach, we introduce the nonholonomic basis using the functions

$$v_*^1 = \dot{q}^1 = \dot{\beta}\,,$$
$$v_*^2 = \varphi^1 = \dot{q}^2 + \dot{\alpha}R\sin\alpha\,,$$
$$v_*^3 = \varphi^2 = \dot{q}^3 - \dot{\alpha}R\cos\alpha\,.$$

In this case,

$$\beta_\rho^\sigma = \frac{\partial\dot{q}^\sigma}{\partial v_*^\rho} = \begin{cases} 1\,, & \sigma = \rho\,, \\ 0\,, & \sigma \neq \rho\,, \end{cases}$$

and hence, the nonholonomic basis agrees with the holonomic basis $\mathbf{e}_1$, $\mathbf{e}_2$, $\mathbf{e}_3$.

In this example, system (11.2) consists of one equation, which can be written in the form

$$MW_1 = Y_1\,, \tag{11.10}$$

where

$$MW_1 = \frac{d}{dt}\frac{\partial T}{\partial\dot{\beta}} - \frac{\partial T}{\partial\beta}\,, \quad Y_1 = -Fa\sin\beta\,.$$

Substituting the expression for the kinetic energy into Eq. (11.10), this gives

$$M(k^2 + b^2)\ddot{\beta} + Mb\frac{d}{dt}(\dot{y}\cos\beta - \dot{x}\sin\beta) +$$
$$+Mb(\dot{y}\dot{\beta}\sin\beta + \dot{x}\dot{\beta}\cos\beta) = -Fa\sin\beta\,. \tag{11.11}$$

To get a closed system, this equation need to be augmented with the equation of the control constraint

$$x = R\cos\alpha\,, \quad y = R\sin\alpha\,, \tag{11.12}$$

and the equation of the motion program

$$\beta = \alpha - \pi/2\,. \tag{11.13}$$

From (11.12), (11.13) it follows that

$$\dot{x} = -\dot{\beta}R\cos\beta, \quad \dot{y} = -\dot{\beta}R\sin\beta,$$

and hence,

$$\dot{y}\cos\beta - \dot{x}\sin\beta = 0, \quad \dot{y}\dot{\beta}\sin\beta + \dot{x}\dot{\beta}\cos\beta = -R\dot{\beta}^2. \tag{11.14}$$

In view of (11.14), (11.13) we may write Eq. (11.11) as an equation in the control parameter α:

$$M(k^2 + b^2)\ddot{\alpha} - MbR\dot{\alpha}^2 - Fa\cos\alpha = 0. \tag{11.15}$$

Taking into account that

$$\ddot{\alpha} = \frac{d\dot{\alpha}}{dt} = \frac{d\dot{\alpha}}{d\alpha}\frac{d\alpha}{dt} = \frac{1}{2}\frac{d(\dot{\alpha}^2)}{d\alpha},$$

we get a linear equation in $\dot{\alpha}^2$

$$\frac{d(\dot{\alpha}^2)}{d\alpha} - \lambda\dot{\alpha}^2 = \mu\cos\alpha, \tag{11.16}$$

where

$$\lambda = \frac{2bR}{k^2 + b^2}, \quad \mu = \frac{2Fa}{M(k^2 + b^2)}.$$

The partial solution to Eq. (11.16) should be sought in the form

$$\dot{\alpha}^2 = A\cos\alpha + B\sin\alpha.$$

Substituting this expression into Eq. (11.16), we find that

$$A = -\frac{\lambda\mu}{1 + \lambda^2}, \quad B = \frac{\mu}{1 + \lambda^2}.$$

The general solution of Eq. (11.16) is such that

$$\dot{\alpha}^2 = C_1 e^{\lambda\alpha} + \frac{\mu}{1 + \lambda^2}\sin\alpha - \frac{\lambda\mu}{1 + \lambda^2}\cos\alpha,$$

where C_1 is an arbitrary constant, which can be found from the initial conditions. Assuming for simplicity that

$$t_0 = 0, \quad \alpha(0) = \alpha_0 = 0, \quad \dot{\alpha}(0) = \dot{\alpha}_0 > 0,$$

we see that

$$\dot{\alpha} = \sqrt{f(\alpha)}, \quad f(\alpha) > 0,$$

where

$$f(\alpha) = \dot{\alpha}_0^2 e^{\lambda \alpha} + \frac{\mu}{1 + \lambda^2}(\lambda(e^{\lambda \alpha} - \cos \alpha) + \sin \alpha).$$

If the function $f(\alpha)$ is such that $f(\alpha) > 0$ for any α, then the sought-for function $\alpha = \alpha(t)$ can be found from the integral equation

$$t = \int_0^\alpha \frac{d\alpha}{\sqrt{f(\alpha)}}. \qquad (11.17)$$

Clearly, the function $f(\alpha)$ is positive for all α, provided that the second term, which is proportional to the function $f_1(\alpha) = \lambda(e^{\lambda \alpha} - \cos \alpha) + \sin \alpha$, is positive for all $\alpha > 0$.

Since $e^{\lambda \alpha} - \cos \alpha > 0$ for $\alpha > 0$ and since $\sin \alpha > 0$ for $0 < \alpha < \pi$, it follows that the function $f_1(\alpha)$, as defined on the interval $[0, 2\pi]$, can have real roots only in the interval $[\pi, 2\pi]$.

If λ is so small that $\lambda(e^{\lambda \alpha} - \cos \alpha) < 1$ for $\pi < \alpha < 2\pi$, then the function $f_1(\alpha)$ on the interval $[\pi, 2\pi]$ has two roots. And vice versa, if the parameter λ is so large that $\lambda(e^{\lambda \alpha} - \cos \alpha) > 1$ for $\alpha > \pi$, then the function $f_1(\alpha)$ for $\alpha > 0$ has no real roots.

The value $\lambda = \lambda_*$, for which the function $f_1(\alpha)$ has a double root $\alpha = \alpha_*$ in the interval $[\pi, 2\pi]$ is found from the conditions

$$f_1(\alpha_*)\Big|_{\lambda=\lambda_*} = 0, \quad \frac{\partial f_1}{\partial \alpha}\Big|_{a=a_*, \lambda=\lambda_*} = 0,$$

or in the expanded form

$$\lambda_*(e^{\lambda_* \alpha_*} - \cos \alpha_*) + \sin \alpha_* = 0,$$
$$\lambda_*(\lambda_* e^{\lambda_* \alpha_*} + \sin \alpha_*) + \cos \alpha_* = 0.$$

Multiplying the first equation by $(-\lambda_*)$ and adding the result with the second equation, this gives

$$(\lambda_*^2 + 1) \cos \alpha_* = 0.$$

It follows that $\alpha_* = 3\pi/2$, while λ_* satisfies the equation

$$\lambda_* e^{3\pi \lambda_*/2} = 1,$$

solving which we find that $\lambda_* = 0.2744$.

Fig. 22 Control of a ball on a plane

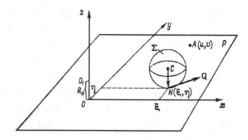

With $\lambda > \lambda_*$ the sought-for function $\alpha = \alpha(t)$ is given by integral (11.17) for any $\dot{\alpha}_0 > 0$ and μ. The problem becomes much more involved if $\lambda < \lambda_*$ and if $\dot{\alpha}_0$ and μ are such that $f(\alpha) < 0$ with some α. We shall not consider this more difficult case.

Assuming that $\lambda > \lambda_*$, we find the moment L, which needs to be applied to the disk $\sum_1$ in order that it would rotate by the law (11.17).

Under the condition $\beta = \alpha - \pi/2$ the disk–plate system can be looked upon as a whole rigid body, which rotates about the axis passing through the point O. The external forces acting on this system is the moment L and the force F; hence the motion equation of this system can be written as

$$(J_1 + M(k^2 + b^2 + R^2))\,\ddot{\alpha} = L - (R\sin\alpha - a\cos\alpha)F\,,$$

where J_1 is the moment of inertia of the disk with respect to the point O.

The function $\alpha = \alpha(t)$ satisfies Eq. (11.15), which can be put in the form (11.16), and hence

$$\ddot{\alpha} = \frac{1}{2}\lambda\dot{\alpha}^2 + \frac{1}{2}\mu\cos\alpha = \frac{1}{2}\lambda f(\alpha) + \frac{1}{2}\mu\cos\alpha\,.$$

Thus, the sought-for moment L *qua* a function of the angle α with the chosen initial conditions is given by

$$L = \frac{1}{2}(J_1 + M(k^2 + b^2 + R^2))\,(\lambda f(\alpha) + \mu\cos\alpha) + (R\sin\alpha - a\cos\alpha)\,F\,.$$

Example 15 *Control of a ball on the plane.* A material plane P slides forwards along an otherwise fixed horizontal Oxy-plane. A homogeneous ball $\sum$ of radius R_0 rolls without slipping along the Oxy-plane. The motion of the plane P is controlled automatically in order that the ball center would move uniformly with the velocity $v = \omega$ with respect to the fixed system $Oxyz$ along a circle of radius a with center on the z-axis (see Fig. 22).

In this problem, the controlled object is the ball, whose center may move only in the plane parallel to the Oxy. It has five degrees of freedom. As generalized coordinates, one may take the coordinates ξ and η of the ball center, with axes directed along the x- and y-axes, respectively, and the Euler angles ψ, θ and φ, which define its orientation.

The motion of the ball center in a given program is effected because of the constraint between the ball $\sum$ and the plane P. By the assumption, the ball rolls along the plane P without slipping, and hence, the material elements of the ball and plane that agree at a given time must have equal velocities. This constraint between the motion of the ball and the plain is the control constraint in the above problem. By the assumption the plane P may move only parallel to itself, and hence its motion depends only on two (control) parameters. Let u and v be the coordinates of a point A on the plane P with respect to the x- and y-axes. It is convenient to take the quantities u and v for the control parameters.

The reaction force applied to the ball $\sum$ from the plane P has the component (pointing in the z-axis; this component is balanced by the gravity force of the gravity force of the ball) and the component $\mathbf{Q}$ lying in the Oxy-plane. The control force, when considered as a vector of the tangent space, was denoted earlier by $\mathbf{R}$. In this problem, the vector $\mathbf{R}$ lies in the five-dimensional tangent space. In the real space, to this vector there corresponds the force $\mathbf{Q}$. To emphasize the difference between the vectors lying in different spaces, we shall denote these vectors differently.

The motion of the ball can be obtained by first translating the ball, and then rotating it about its center. The rotation is completely defined by three Euler angles qua functions of time. However, if in defining the rotation of the ball one is confined only with specifying the vector Ω of instantaneous angular velocity of the ball with the components p, q and r along the x-, y- and z-axes, then the problem becomes much more simple. Instead of the dynamic equations written for the tangent space, we shall use the motion equation for the center of mass

$$M\mathbf{w}_c = \mathbf{Q},$$
$$\mathbf{w}_c = \ddot{\xi}\mathbf{i} + \ddot{\eta}\mathbf{j}, \quad \mathbf{Q} = Q_x\mathbf{i} + Q_y\mathbf{j}, \tag{11.18}$$

where M is the mass of the ball, and employ the moment equations with respect to the centre of the ball

$$\frac{d\mathbf{l}}{dt} = \mathbf{L}. \tag{11.19}$$

Here, $\mathbf{L}$ is the moment of the control force $\mathbf{Q}$ with respect to the centre of the ball

$$\mathbf{L} = \overrightarrow{CN} \times \mathbf{Q} = \begin{vmatrix} \mathbf{i} & \mathbf{j} & \mathbf{k} \\ 0 & 0 & -R_0 \\ Q_x & Q_y & 0 \end{vmatrix} = Q_y R_0 \mathbf{i} - Q_x R_0 \mathbf{j}.$$

The vector of the angular momentum of the ball is as follows

$$\mathbf{l} = J\Omega = J(p\mathbf{i} + q\mathbf{j} + r\mathbf{k}), \quad J = \frac{2}{5}MR_0^2.$$

The velocity of the ball lower point, which touches at a given time the plane P, is as follows:

$$\mathbf{v}_N = \mathbf{v}_C + \Omega \times \overrightarrow{CN} = \dot{\xi}\mathbf{i} + \dot{\eta}\mathbf{j} + \begin{vmatrix} \mathbf{i} & \mathbf{j} & \mathbf{k} \\ p & q & r \\ 0 & 0 & -R_0 \end{vmatrix} = (\dot{\xi} - R_0 q)\mathbf{i} + (\dot{\eta} + R_0 p)\mathbf{j}.$$

At the same time, the velocity of an arbitrary point on the plane P is equal to

$$\mathbf{v} = \dot{u}\mathbf{i} + \dot{v}\mathbf{j},$$

and hence, the condition of the ball rolling without slipping can be described by the equation

$$\mathbf{v}_N = \mathbf{v}. \tag{11.20}$$

By the condition of the problem, one should find a motion of the plane P for which the ball center will execute a motion along a circular of radius a centered at a fixed z-axis with the velocity $v = \omega a$. Suppose that at $t = 0$ the coordinate η of the point C is zero. Then the equations of program motion can be put in the form

$$\xi = a\cos\omega t, \quad \eta = a\sin\omega t. \tag{11.21}$$

Thus, the complete system of equations, which describes the controlled motion of the ball, consists of the dynamic Eqs. (11.18), (11.19), the control constraint equation (11.20), and the motion program Eq. (11.21). This system of equations can be written in the scalar form as follows:

the dynamic equations:

$$M\ddot{\xi} = Q_x, \quad M\ddot{\eta} = Q_y, \quad J\dot{p} = Q_y R_0, \quad J\dot{q} = -Q_x R_0, \quad J\dot{r} = 0;$$

the control constraint equation

$$\dot{\xi} - R_0 q = \dot{u}, \quad \dot{\eta} + R_0 p = \dot{v};$$

the equations of program motion

$$\xi = a\cos\omega t, \quad \eta = a\sin\omega t.$$

Altogether, there are nine equations, and respectively, nine unknowns ξ, η, p, q, r, u, v, Q_x, Q_y.

Excluding Q_x and Q_y from the dynamic equations, this gives

$$\frac{2}{5}R_0\dot{p} = \ddot{\eta}, \quad -\frac{2}{5}R_0\dot{q} = \ddot{\xi}, \quad \dot{r} = 0. \tag{11.22}$$

These equations correspond to the equations, which in the general case were written as (11.2).

Differentiating the control constraint equations in time, this gives

$$\ddot{\xi} - \dot{q}R_0 = \ddot{u}, \quad \ddot{\eta} + \dot{p}R_0 = \ddot{v}.$$

Hence, taking into account the Eq. (11.22), we find that

$$2\ddot{u} = 7\ddot{\xi}, \quad 2\ddot{v} = 7\ddot{\eta}. \tag{11.23}$$

The relation between the acceleration of the plane P and the ball center hold for any program of motion.

Integrating Eq. (11.23) under the condition that the motion program is given by Eq. (11.21), this establishes

$$u = u_0 + \dot{u}_0 t - \frac{7a}{2}\left(1 - \cos \omega t\right),$$

$$v = v_0 + \dot{v}_0 t - \frac{7a}{2}\left(\omega t - \sin \omega t\right),$$

where u_0, v_0, $\dot{u}_0$, $\dot{v}_0$ are, respectively, the coordinates and the components of the velocity of the point A at the initial moment $t = 0$. Assuming for simplicity that the point A on the plane P is initially at the origin (that is, it agrees with the point O and has zero velocity), we find

$$u = -\frac{7a}{2}\left(1 - \cos \omega t\right),$$

$$v = -\frac{7a}{2}\left(\omega t - \sin \omega t\right).$$

This is the equation of the cycloid in the parametric form.

So, in order that the ball center would move along a circle, the point A on the plane P should travel along the cycloid.

The coordinates x_1, y_1 of the point C in the system of coordinate Ax_1y_1 fixed within the plane P are as follows:

$$x_1 = \xi - u = \frac{7a}{2}\left(1 - \frac{5}{7}\cos \omega t\right),$$

$$y_1 = \eta - v = \frac{7a}{2}\left(\omega t - \frac{5}{7}\sin \omega t\right).$$

This is the equation of a curtate cycloid. Hence, ball center executes a motion with respect to the observer, which moves together with the plane P along the curtate cycloid.

Let us go back to the dynamic Eq. (11.22). Integrating them, this gives

$$p = p_0 - \frac{5}{2}\frac{a\omega}{R_0}(1 - \cos\omega t),$$

$$q = q_0 + \frac{5}{2}\frac{a\omega}{R_0}\sin\omega t,$$

$$r = r_0,$$

where p_0, q_0, r_0 are the projections of the instantaneous angular velocity vector Ω to the fixed axes with $t = 0$.

This being so, using this simple algorithm one may easily find both the motion of the plane and the motion of the ball.

Chapter 7
Small Oscillations of Systems

P. E. Tovstik and M. P. Yushkov ⓘ

Small oscillations of mechanical systems are considered. The corresponding linear differential equations with constant coefficients are derived using the two methods: the linearization of the initial nonlinear system of equations or preliminary reducing of the expressions for kinetic and potential energies to quadratic forms with constant coefficients. General solutions to equations of small oscillation are found, the notions of natural frequencies and normal modes of oscillation are introduced, their properties are studied. The case of zero frequency and the case when several natural frequencies coincide are examined with the help of normal coordinates. The Rayleigh theorem and the Courant theorem are proved. Free small oscillations in the presence of resistance are considered. The Thomson and Tait theorems on the influence of dissipative and gyroscopic forces on the stability of state of equilibrium are presented. Forced oscillations under the action of arbitrary and periodic forces are considered. The relationship between the impulse transient function and the transfer function is established.

1 Differential Equations of Small Motion and Their Integration

Consider a certain holonomic system. In accordance with formula (5.1) of Chap. 5 the conditions of equilibrium read as

$$Q_\rho = 0, \quad \rho = \overline{1, s}. \tag{1.1}$$

We shall consider the generalized coordinates from this position. Equalities (1.1) will be rewritten as

$$(Q_\rho)_0 = 0, \quad \rho = \overline{1, s}. \tag{1.2}$$

The system starts moving when given initial deviation and initial velocities. Its motion is described by the Lagrange equations of the second kind, which in the stationary

© Springer Nature Switzerland AG 2021

N. N. Polyakhov et al., *Rational and Applied Mechanics*,
Foundations of Engineering Mechanics,
https://doi.org/10.1007/978-3-030-64061-3_7

case read in the extended form as

$$M\left(\sum_{\sigma=1}^{s} g_{\rho\sigma}\ddot{q}^{\sigma} + \sum_{\sigma,\tau=1}^{s} \Gamma_{\rho,\sigma\tau}\dot{q}^{\sigma}\dot{q}^{\tau}\right) = Q_{\rho}, \quad \rho = \overline{1,s}. \tag{1.3}$$

Here, we used the rule of summation over recurring indices before in writing the Lagrange equations of the second kind. In this chapter we shall not use this rule anywhere, for the cases are possible here when the index recurs, but the summation should not be done over it. Suppose that the velocities $\dot{q}^{\sigma}$ are so small that we can neglect the summands proportional to the products $\dot{q}^{\sigma}\dot{q}^{\tau}$ in comparison to the remaining of the summands in Eqs. (1.3). Besides, let in the expansions

$$g_{\rho\sigma}(q) = g_{\rho\sigma}(0) + \sum_{\tau=1}^{s}\left(\frac{\partial g_{\rho\sigma}}{\partial q^{\tau}}\right)_{0} q^{\tau} + \cdots$$

the second and following summands be small in comparison with the first one, so that we can assume $g_{\rho\sigma} \approx g_{\rho\sigma}(0)$.

We shall also suppose that the generalized forces Q_{ρ} represent analytic functions of generalized coordinates and generalized velocities. Limiting ourselves to the first degrees of expansion we can write

$$Q_{\rho}(q,\dot{q}) = (Q_{\rho})_{0} + \sum_{\sigma=1}^{s}\left(\frac{\partial Q_{\rho}}{\partial q^{\sigma}}\right)_{0} q^{\sigma} + \sum_{\sigma=1}^{s}\left(\frac{\partial Q_{\rho}}{\partial q^{\sigma}}\right)_{0} \dot{q}^{\sigma}, \tag{1.4}$$

where $(Q_{\rho})_{0}$ correspond to the state of equilibrium and should be taken to be zero by (1.2).

In view of the above Eqs. (1.3) can be written as

$$\sum_{\sigma=1}^{s}(a_{\rho\sigma}\ddot{q}^{\sigma} + b_{\rho\sigma}\dot{q}^{\sigma} + c_{\rho\sigma}q^{\sigma}) = 0, \quad \rho = \overline{1,s}, \tag{1.5}$$

where

$$a_{\rho\sigma} = Mg_{\rho\sigma}(0), \quad b_{\rho\sigma} = -\left(\frac{\partial Q_{\rho}}{\partial \dot{q}^{\sigma}}\right)_{0}, \quad c_{\rho\sigma} = \left(\frac{\partial Q_{\rho}}{\partial q^{\sigma}}\right)_{0};$$

these quantities are constants because so are $g_{\rho\sigma}(0)$.

In the matrix form the system of equations (1.5) appears as

$$\mathbf{A}\ddot{\mathbf{q}} + \mathbf{B}\dot{\mathbf{q}} + \mathbf{C}\mathbf{q} = 0, \tag{1.6}$$

where $\mathbf{A}$, $\mathbf{B}$, $\mathbf{C}$ are square matrices, $\mathbf{q}$ is a column matrix.

The system of equations of motion (1.5), or the matrix equation (1.6), represents a system of linear homogeneous equations of second order with constant coefficients. Such a system is integrated by representing the unknown partial solutions in the form

$$q^\sigma = H_\sigma e^{\lambda t},\tag{1.7}$$

where σ varies in the range $1, 2, \ldots, s$, and the quantities H_σ and λ are constants. Substituting expressions (1.7) for q^σ into Eqs. (1.5) we get the following system of equations that contain $s + 1$ unknowns $H_1, H_2, \ldots, H_s$ and λ:

$$\sum_{\sigma=1}^{s}(a_{\rho\sigma}\lambda^2 + b_{\rho\sigma}\lambda + c_{\rho\sigma})H_\sigma = 0, \quad \rho = \overline{1, s}.\tag{1.8}$$

This system represents a system of s linear homogeneous equations with respect to H_σ. Such systems are satisfied by the trivial solution

$$H_\sigma = 0, \quad \sigma = \overline{1, s}\tag{1.9}$$

and have nonzero solutions. For the existence of these solutions it is necessary for the determinant formed from the coefficients at H_σ to vanish. In this case the solution of the system is not unique. Equating to zero the determinant of the system equal, we get

$$\Delta(\lambda) = \begin{vmatrix} a_{11}\lambda^2 + b_{11}\lambda + c_{11}, \ldots, a_{1s}\lambda^2 + b_{1s}\lambda + c_{1s} \\ \ldots\ldots\ldots\ldots\ldots\ldots\ldots\ldots\ldots\ldots\ldots\ldots\ldots\ldots\ldots\ldots\ldots \\ a_{s1}\lambda^2 + b_{s1}\lambda + c_{s1}, \ldots, a_{ss}\lambda^2 + b_{ss}\lambda + c_{ss} \end{vmatrix} = 0,\tag{1.10}$$

or in the abbreviated form

$$\Delta(\lambda) = |a_{\rho\sigma}\lambda^2 + b_{\rho\sigma}\lambda + c_{\rho\sigma}| = 0.$$

Expanding the determinant, we have an algebraic equation of degree $2s$ with respect to λ:

$$\lambda^{2s} + h_1\lambda^{2s-1} + \cdots + h_{2s-1}\lambda + h_{2s} = 0.\tag{1.11}$$

The latter equation is called the *characteristic (eigenvalue) equation*, and the determinant (1.10) is called the *characteristic determinant*.

We denote solutions of the characteristic equation by $\lambda_1, \lambda_2, \ldots, \lambda_{2s}$ and shall assume that among these roots there are no multiple and zero ones. Substituting one of the indicated roots λ_k, $k = \overline{1, 2s}$, into formulas (1.8) we arrive at the following system of homogeneous equations for the unknowns $H_\sigma^{(k)}$:

$$\sum_{\sigma=1}^{s}(a_{\rho\sigma}\lambda_k^2 + b_{\rho\sigma}\lambda_k + c_{\rho\sigma})H_\sigma^{(k)} = 0, \quad \rho = \overline{1, s}.\tag{1.12}$$

The determinant of this system is the determinant (1.10) vanishing when $\lambda = \lambda_k$. This means that equations of system (1.12) are not independent. If all the roots λ_k are different, then only one equation is a corollary of the others, the remaining

$s - 1$ equations are independent, and hence, the rank of the matrix composed of the coefficients of these equations is $s - 1$.

Assume for definiteness that the equation which is a corollary of the others equations is the last equation of system (1.12) (with $\rho = s$). Then the system of independent equations can be written as

$$\sum_{\sigma=1}^{s-1} G_{\rho\sigma}(\lambda_k) \frac{H_\sigma^{(k)}}{H_s^{(k)}} = -G_{\rho s}(\lambda_k), \quad \rho = \overline{1, s - 1}, \tag{1.13}$$

where $G_{\rho\sigma}(\lambda_k) = a_{\rho\sigma}\lambda_k^2 + b_{\rho\sigma}\lambda_k + c_{\rho\sigma}$, and it is supposed that $H_s^{(k)} \neq 0$.

Solving the inhomogeneous system (1.13) by Cramer's rule, we have

$$\frac{H_\sigma^{(k)}}{H_s^{(k)}} = \frac{\Delta_\sigma(\lambda_k)}{\Delta_s(\lambda_k)}, \quad \sigma = \overline{1, s - 1}. \tag{1.14}$$

Here we the denote algebraic complements of elements $G_{s\sigma}$ of the last line of determinant (1.10) by $\Delta_\sigma(\lambda_k)$, $\sigma = \overline{1, s}$; in other words, $\Delta_\sigma(\lambda_k)$ is equal to the product of $(-1)^{s+\sigma}$ by the determinant, which we obtain from the determinant $\Delta(\lambda_k)$ by removing the last row and the column with index σ.

Formulas (1.14) can be rewritten as

$$\frac{H_1^{(k)}}{\Delta_1(\lambda_k)} = \frac{H_2^{(k)}}{\Delta_2(\lambda_k)} = \cdots = \frac{H_s^{(k)}}{\Delta_s(\lambda_k)} = C_k. \tag{1.15}$$

The forgoing implies that according to formulas (1.14) we can get the strictly defined numerical values only for the relations $H_\sigma^{(k)}/H_s^{(k)}$, $\sigma = \overline{1, s - 1}$. The solution of system (1.12) can be found accurate to an arbitrary factor C_k by the formulas

$$H_\sigma^{(k)} = C_k \Delta_\sigma(\lambda_k), \quad \sigma = \overline{1, s}, \tag{1.16}$$

obtained on the basis of (1.15). This implies that the partial solution (1.7) corresponding to the root λ_k can be represented as

$$q_k^\sigma = C_k \Delta_\sigma(\lambda_k) e^{\lambda_k t}, \quad \sigma = \overline{1, s}, \tag{1.17}$$

or in the vector form

$$\mathbf{q}_k = C_k \mathbf{H}^{(k)} e^{\lambda_k t}, \quad \mathbf{H}^{(k)} = (\Delta_1(\lambda_1), \ldots, \Delta_s(\lambda_s))^T, \tag{1.17'}$$

where $\mathbf{H}^{(k)}$ is the amplitude vector corresponding to the root λ_k of Eq. (1.11).

A general solution to the above linear system of differential equations (1.5) can now be written as a linear combination of partial solutions (1.17),

$$q^\sigma = \sum_{k=1}^{2s} q_k^\sigma = \sum_{k=1}^{2s} C_k \Delta_\sigma(\lambda_k) e^{\lambda_k t}, \quad \sigma = \overline{1, s}, \tag{1.18}$$

or

$$\mathbf{q} = \sum_{k=1}^{2s} \mathbf{q}_k = \sum_{k=1}^{2s} C_k \mathbf{H}^{(k)} e^{\lambda_k t}. \tag{1.18'}$$

It contains $2s$ arbitrary constants C_k; this corresponds to the number of arbitrary constants obtained when integrating s linear differential second-order equations. The values of these arbitrary constants should be defined by setting the initial values $q^\sigma(0)$ and $\dot{q}^\sigma(0)$, $\sigma = \overline{1, s}$.

Note that the trivial solution (1.9) generates a trivial solution of system (1.5) $q^\sigma(t) = 0$, $\sigma = \overline{1, s}$, corresponding to the state of rest of a mechanical system. This solution is discarded, because it corresponds to the solution of the Cauchy problem only under zero initial conditions. The general solution (1.18) of system (1.5), in which instead of formulas (1.9) we use expressions (1.16), makes it possible to solve the Cauchy problem under any initial conditions, including zero ones.

Coming back to the general solution of system (1.5), we note that its form (1.18) is easy-to-use only in the case when all λ_k, $k = \overline{1, 2s}$, are real numbers.

If some of roots of the characteristic Eq. (1.11) turn out to be complex numbers, then these numbers are complex conjugate, that is why their number is always even and equal to $2l$. For definiteness, we shall give them the first $2l$ numbers. We denote the complex conjugate roots by

$$\lambda_\varkappa = \alpha_\varkappa + i\beta_\varkappa, \quad \lambda_{\varkappa+l} = \alpha_\varkappa - i\beta_\varkappa, \quad i = \sqrt{-1}, \quad \varkappa = \overline{1, l}. \tag{1.19}$$

Recall that the expressions $H_\sigma^{(\varkappa)}$ and $H_\sigma^{(\varkappa+l)}$ corresponding to them are defined according to formulas (1.15):

$$\begin{aligned} \frac{H_1^{(\varkappa)}}{\Delta_1(\lambda_\varkappa)} &= \frac{H_2^{(\varkappa)}}{\Delta_2(\lambda_\varkappa)} = \cdots = \frac{H_s^{(\varkappa)}}{\Delta_s(\lambda_\varkappa)} = C_\varkappa, \\ \frac{H_1^{(\varkappa+l)}}{\Delta_1(\lambda_{\varkappa+l})} &= \frac{H_2^{(\varkappa+l)}}{\Delta_2(\lambda_{\varkappa+l})} = \cdots = \frac{H_s^{(\varkappa+l)}}{\Delta_s(\lambda_{\varkappa+l})} = C_{\varkappa+l}. \end{aligned} \tag{1.20}$$

As in the polynomial $\Delta_\sigma(\lambda)$ the coefficients fr all degrees of λ are real numbers, then $\Delta_\sigma(\lambda_\varkappa)$ is complex conjugate to $\Delta_\sigma(\lambda_{\varkappa+l})$.

Writing the complex number $\Delta_\sigma(\lambda_\varkappa)$ in the exponential form

$$\Delta_\sigma(\lambda_\varkappa) = |\Delta_\sigma(\lambda_\varkappa)| e^{i\varphi_{\sigma\varkappa}}, \tag{1.21}$$

we find that

$$\Delta_\sigma(\lambda_{\varkappa+l}) = |\Delta_\sigma(\lambda_\varkappa)| e^{-i\varphi_{\sigma\varkappa}}. \tag{1.22}$$

Here $|\Delta_\sigma(\lambda_\varkappa)|$ and $\varphi_{\sigma\varkappa}$ are known real numbers, which agree, respectively, with the absolute value of the complex number $\Delta_\sigma(\lambda_\varkappa)$ and its argument.

With the help of formulas (1.19)–(1.22) we can rewrite the sum

$$\sum_{\varkappa=1}^{l}(q_\varkappa^\sigma + q_{\varkappa+l}^\sigma) = \sum_{\varkappa=1}^{l}\left(C_\varkappa\Delta_\sigma(\lambda_\varkappa)e^{\lambda_\varkappa t} + C_{\varkappa+l}\Delta_\sigma(\lambda_{\varkappa+l})e^{\lambda_{\varkappa+l}t}\right) \qquad (1.23)$$

which is of interest to us, as

$$\sum_{\varkappa=1}^{l}(q_\varkappa^\sigma + q_{\varkappa+l}^\sigma) = \sum_{\varkappa=1}^{l}e^{\alpha_\varkappa t}|\Delta_\sigma(\lambda_\varkappa)|\left(C_\varkappa e^{i(\beta_\varkappa t+\varphi_{\sigma\varkappa})} + C_{\varkappa+l}e^{-i(\beta_\varkappa t+\varphi_{\sigma\varkappa})}\right).$$

We split the complex arbitrary constants into the real and imaginary parts. Then, using Euler's formulas, we obtain

$$\sum_{\varkappa=1}^{l}(q_\varkappa^\sigma + q_{\varkappa+l}^\sigma) = \sum_{\varkappa=1}^{l}e^{\alpha_\varkappa t}|\Delta_\sigma(\lambda_\varkappa)|\Big((\mathrm{Re}C_\varkappa + i\mathrm{Im}C_\varkappa)\big(\cos(\beta_\varkappa t + \varphi_{\sigma\varkappa})+$$
$$+i\sin(\beta_\varkappa t + \varphi_{\sigma\varkappa})\big) + (\mathrm{Re}C_{\varkappa+l} + i\mathrm{Im}C_{\varkappa+l})\big(\cos(\beta_\varkappa t + \varphi_{\sigma\varkappa})-$$
$$-i\sin(\beta_\varkappa t + \varphi_{\sigma\varkappa})\big)\Big).$$

Collecting similar terms, the latter expression appears as

$$\sum_{\varkappa=1}^{l}(q_\varkappa^\sigma + q_{\varkappa+l}^\sigma) = \sum_{\varkappa=1}^{l}e^{\alpha_\varkappa t}|\Delta_\sigma(\lambda_\varkappa)|\big((\mathrm{Re}C_\varkappa + \mathrm{Re}C_{\varkappa+l})\cos(\beta_\varkappa t + \varphi_{\sigma\varkappa})+$$
$$+ (\mathrm{Im}C_{\varkappa+l} - i\mathrm{Im}C_\varkappa)\sin(\beta_\varkappa t + \varphi_{\sigma\varkappa})\big)+$$
$$+i\sum_{\varkappa=1}^{l}e^{\alpha_\varkappa t}|\Delta_\sigma(\lambda_\varkappa)|\big((\mathrm{Im}C_\varkappa + \mathrm{Im}C_{\varkappa+l})\cos(\beta_\varkappa t + \varphi_{\sigma\varkappa})+$$
$$+ (\mathrm{Re}C_\varkappa - \mathrm{Re}C_{\varkappa+l})\sin(\beta_\varkappa t + \varphi_{\sigma\varkappa})\big). \qquad (1.24)$$

If the complex solution $\sum_{\varkappa=1}^{l}(q_\varkappa^\sigma + q_{\varkappa+l}^\sigma)$ satisfies the system of differential equations (1.5) with real coefficients, then both the real and imaginary components of solution (1.24) are solutions. That is why, instead of the sum (1.23), we can substitute the expression

$$\sum_{\varkappa=1}^{l}(q_\varkappa^\sigma + q_{\varkappa+l}^\sigma) = \sum_{\varkappa=1}^{l}e^{\alpha_\varkappa t}|\Delta_\sigma(\lambda_\varkappa)|\big(D_\varkappa\cos(\beta_\varkappa t + \varphi_{\sigma\varkappa})+$$
$$+ D_{\varkappa+l}\sin(\beta_\varkappa t + \varphi_{\sigma\varkappa})\big), \qquad (1.25)$$

into the general solution (1.18); here the real arbitrary constants $D_\varkappa$ and $D_{\varkappa+l}$ are connected to the arbitrary constants $C_\varkappa$ and $C_{\varkappa+l}$ by the formulas

$$D_\varkappa = \mathrm{Re}\,C_\varkappa + \mathrm{Re}\,C_{\varkappa+l}\,, \quad D_{\varkappa+l} = \mathrm{Im}\,C_{\varkappa+l} - \mathrm{Im}\,C_\varkappa\,,$$

or

$$D_\varkappa = \mathrm{Im}\,C_\varkappa + \mathrm{Im}\,C_{\varkappa+l}\,, \quad D_{\varkappa+l} = \mathrm{Re}\,C_\varkappa - \mathrm{Re}\,C_{\varkappa+l}\,,$$

depending on whether the real or imaginary component of solution (1.24) is used when writing the solution in the form (1.25). Instead of arbitrary constants $D_\varkappa$ and $D_{\varkappa+l}$ the new arbitrary constants $a_\varkappa$ and $\varepsilon_\varkappa$ are often introduced by formulas

$$D_\varkappa = a_\varkappa \sin \varepsilon_\varkappa\,, \quad D_{\varkappa+l} = a_\varkappa \cos \varepsilon_\varkappa\,.$$

In this case solution (1.25) can be rewritten as

$$\sum_{\varkappa=1}^{l} (q_\varkappa^\sigma + q_{\varkappa+l}^\sigma) = \sum_{\varkappa=1}^{l} a_\varkappa e^{\alpha_\varkappa t} |\Delta_\sigma(\lambda_\varkappa)| \sin (\beta_\varkappa t + \varphi_{\sigma\varkappa} + \varepsilon_\varkappa)\,. \tag{1.26}$$

If among the complex roots $\lambda_\varkappa$ there are purely imaginary ones, i.e., for some $\varkappa$ formulas (1.19) appear as

$$\lambda_\varkappa = i\beta_\varkappa\,, \quad \lambda_{\varkappa+l} = -i\beta_\varkappa\,,$$

then in expression (1.26) the factors $e^{\alpha_\varkappa t}$ are equal to 1 for the corresponding $\varkappa$.

2 Investigation of the Small Motion Pattern of Systems

The method of linearization and integration for equations of motion of holonomic systems with s degrees of freedom presented in the previous subsection is widely used when solving many practical problems. A simple form of a general solution of the problem in a linear formulation makes it possible to investigate the motion pattern both qualitatively and quantitatively.

Some conclusions on the motion pattern can be derived directly from the structure of differential equations (1.5) obtained on the basis of linearization of the Lagrange equations of the second kind.

First summands of equations (1.5) correspond to the summands $\frac{d}{dt}\frac{\partial T}{\partial \dot{q}^\rho}$ of the Lagrange equations under the condition that the kinetic energy T is approximately represented as

$$T = \frac{1}{2} \sum_{\rho,\sigma=1}^{s} a_{\rho\sigma} \dot{q}^\rho \dot{q}^\sigma\,, \quad a_{\rho\sigma} = M g_{\rho\sigma}(0)\,, \tag{2.1}$$

which implies that the matrix $\mathbf{A}$ is symmetric and positive definite.

The matrix $\mathbf{C}$ in equation (1.6) is generally nonsymmetric. Although this fact does not influence the way of integration of system of equations (1.5), but it makes more difficult the qualitative investigation of the required motion patterns. This is why we consider a simpler case, when the matrix $\mathbf{C}$ is symmetric, i.e., when the following condition

$$c_{\rho\sigma} = c_{\sigma\rho}, \quad \rho, \sigma = \overline{1, s},$$

is satisfied.

In this case one can state that the linear combinations $\sum_{\sigma=1}^{s} c_{\rho\sigma} q^{\sigma}$ represent the derivatives of the quadratic form

$$\Pi = \frac{1}{2} \sum_{\rho,\sigma=1}^{s} c_{\rho\sigma} q^{\rho} q^{\sigma}, \tag{2.2}$$

i.e.,

$$\frac{\partial \Pi}{\partial q^{\rho}} = \sum_{\sigma=1}^{s} c_{\rho\sigma} q^{\sigma}. \tag{2.3}$$

This quadratic form can be regarded as an expression for the potential energy of the system. Previously we defined the potential energy of a force field in curvilinear coordinates as the function $\Pi(t, q)$, its derivatives with respect to the coordinates q^{ρ} with the opposite sign representing the corresponding components of generalized forces. On the basis of this definition the summands $\sum_{\sigma=1}^{s} c_{\rho\sigma} q^{\sigma}$ in expression (1.5) reversed in sign can be considered as generalized forces $Q_{\rho 1}$ corresponding to the potential energy Π.

In just the same way, supposing that the matrix $\mathbf{B}$ is symmetric, i.e., $b_{\rho\sigma} = b_{\sigma\rho}, \rho, \sigma = \overline{1, s}$, we can state that the linear combinations $\sum_{\sigma=1}^{s} b_{\rho\sigma} \dot{q}^{\sigma}$ are derivatives of the quadratic form

$$D = \frac{1}{2} \sum_{\rho,\sigma=1}^{s} b_{\rho\sigma} \dot{q}^{\rho} \dot{q}^{\sigma}. \tag{2.4}$$

We introduce the following notation

$$Q_{\rho 2} = -\frac{\partial D}{\partial \dot{q}^{\rho}}, \quad \rho = \overline{1, s}. \tag{2.5}$$

If the function D is a positive definite form, then the forces $Q_{\rho 2}$ are called *dissipative (non-conservative) forces*, and the function D is called a *dissipation function*.[1] If the form D is positive definite then the *dissipation* is *complete*, otherwise it is *incomplete*. These names appear from the fact that the indicated forces serve to dissipate the total mechanical energy of a system, and so the energy integral does not exist in this case.

---

[1] From Latin *dissipare*, meaning "dissipate".

Actually, the Lagrange equations with both the potential and dissipation forces have the form

$$\frac{d}{dt}\frac{\partial T}{\partial \dot{q}^\sigma} - \frac{\partial T}{\partial q^\sigma} = -\frac{\partial \Pi}{\partial q^\sigma} - \frac{\partial D}{\partial \dot{q}^\sigma}, \quad \sigma = \overline{1, s}. \tag{2.6}$$

Multiplying these equations by $\dot{q}^\sigma$ and summing, we get

$$\sum_{\sigma=1}^{s}\left(\frac{d}{dt}\frac{\partial T}{\partial \dot{q}^\sigma} - \frac{\partial T}{\partial q^\sigma}\right)\dot{q}^\sigma = -\sum_{\sigma=1}^{s}\left(\frac{\partial \Pi}{\partial q^\sigma} + \frac{\partial D}{\partial \dot{q}^\sigma}\right)\dot{q}^\sigma,$$

which implies that

$$\frac{d}{dt}\left(\sum_{\sigma=1}^{s}\frac{\partial T}{\partial \dot{q}^\sigma}\dot{q}^\sigma\right) - \left(\sum_{\sigma=1}^{s}\frac{\partial T}{\partial \dot{q}^\sigma}\ddot{q}^\sigma + \sum_{\sigma=1}^{s}\frac{\partial T}{\partial q^\sigma}\dot{q}^\sigma\right) + \sum_{\sigma=1}^{s}\frac{\partial \Pi}{\partial q^\sigma}\dot{q}^\sigma =$$

$$= -\sum_{\sigma=1}^{s}\frac{\partial D}{\partial \dot{q}^\sigma}\dot{q}^\sigma. \tag{2.7}$$

The bracketed expression on the left part of equality (2.7) represents the total derivative with respect to time $dT(q, \dot{q})/dt$. One can say the same about the sum

$$\sum_{\sigma=1}^{s}\frac{\partial \Pi}{\partial q^\sigma}\dot{q}^\sigma = \frac{d\Pi}{dt}.$$

As the functions T and D are quadratic forms with respect to $\dot{q}^\sigma$, then by the Euler theorem on homogeneous functions we can write

$$\sum_{\sigma=1}^{s}\frac{\partial T}{\partial \dot{q}^\sigma}\dot{q}^\sigma = 2T, \quad \sum_{\sigma=1}^{s}\frac{\partial D}{\partial \dot{q}^\sigma}\dot{q}^\sigma = 2D,$$

and hence formula (2.7) can be expressed as

$$\frac{d}{dt}(2T) - \frac{d}{dt}(T - \Pi) = -2D,$$

or finally in the form

$$\frac{d(T + \Pi)}{dt} = -2D < 0. \tag{2.8}$$

Thus, in the process of motion the total mechanical energy $E = T + \Pi$ decreases, i.e., dissipates.

Let us consider the case when the potential energy Π represents a positive definite quadratic form. In this case for any solution of the system of equations (1.5) we have

$$0 \leqslant \Pi \leqslant E, \quad 0 \leqslant T \leqslant E \tag{2.9}$$

as a result of the fact that the kinetic energy can be only a positive definite quadratic form.

If $D > 0$ when $\dot{q} \neq 0$, then by (2.8) the total mechanical energy E decreases monotonically with time. This and inequalities (2.9) imply that in the case considered all the real values of $\lambda_{\varkappa}$ can be only negative in the general solution (1.18). If $\lambda_{\varkappa}$ and $\lambda_{\varkappa+l}$ are complex numbers (1.19), then we can conclude from the form of real solutions (1.26) and inequalities (2.8), (2.9) that the real parts $\alpha_{\varkappa}$ of the roots of Eq. (1.11) are negative.

The case just considered is of great practical importance, because to it there correspond oscillation of elastic systems with s degrees of freedom relative to the stable equilibrium position. In this case during oscillations the total mechanical energy E dissipates under the action of forces of internal friction appearing in elastic elements, as well as under the action of those forces of resistance to the motion of an element of an elastic structure that is caused by bodies which do not belong to the elastic system, i.e., under the action of forces of external friction. The both forces can be taken into consideration by means of the dissipation function D.

Note that the presence of resistance forces makes studying the motion of a system a little bit more complicated. This is why in the next subsection we shall consider oscillation of a system without resistance force. The analysis of a such simplest case makes it possible to establish some basic properties of the mechanical system of interest.

3 Small Oscillation of a System in the Absence of Resistance Forces

If in expressions (1.4) for generalized forces Q_ρ the coefficients at $\dot{q}^\sigma$ are equal to zero, then the system of differential equations for small oscillation (1.5) reads as

$$\sum_{\sigma=1}^{s} (a_{\rho\sigma}\ddot{q}^\sigma + c_{\rho\sigma}q^\sigma) = 0, \quad \rho = \overline{1, s}. \tag{3.1}$$

These equations are obtained by linearizing system (1.3).

Now we consider another method of composition of linear differential equations (3.1). To this point, we transform preliminary the expressions for the kinetic energy T and the potential energy Π.

If the kinetic energy does not depend explicitly on time, then it has the form

$$T = \frac{M}{2} \sum_{\sigma,\tau=1}^{s} g_{\sigma\tau}(q)\, \dot{q}^\sigma \dot{q}^\tau .$$

Expanding the coefficients $g_{\sigma\tau}(q)$ into the Maclaurin series,

$$g_{\sigma\tau}(q) = (g_{\sigma\tau})_0 + \sum_{\rho=1}^{s} \left(\frac{\partial g_{\sigma\tau}}{\partial q^\rho}\right)_0 q^\rho + \cdots . \tag{3.2}$$

It is obvious that if one needs to get linear differential equations from the Lagrange equations of the second kind, then in expansions (3.2) one should keep only the first terms. In this case, the kinetic energy appears as

$$T = \frac{1}{2} \sum_{\sigma,\tau=1}^{s} a_{\sigma\tau} \dot{q}^\sigma \dot{q}^\tau , \tag{3.3}$$

where $a_{\sigma\tau} = M(g_{\sigma\tau})_0$, and that is why the left-hand side of Eqs. (2.6) contains only linear summands.

We now turn our attention to the potential energy of elastic strains of the system

$$\Pi = \Pi(q) . \tag{3.4}$$

As is known, it is always defined with an accuracy up to the constant summand. For definiteness, we shall calculate the value of the potential energy from the stable equilibrium state of the system, in the vicinity of which the oscillations are studied, i.e., we assume that

$$\Pi(0) = 0 . \tag{3.5}$$

Expanding function (3.4) into the Maclaurin series and taking into account expression (3.5), we find that

$$\Pi(q) = \sum_{\sigma=1}^{s} \left(\frac{\partial \Pi}{\partial q^\sigma}\right)_0 q^\sigma + \frac{1}{2} \sum_{\sigma,\tau=1}^{s} \left(\frac{\partial^2 \Pi}{\partial q^\sigma \partial q^\tau}\right)_0 q^\sigma q^\tau + \cdots . \tag{3.6}$$

The system is in equilibrium when $q = 0$ by the assumption, and hence by formulas (1.2) we have

$$\left(\frac{\partial \Pi}{\partial q^\sigma}\right)_0 = 0 , \quad \sigma = \overline{1, s} .$$

To obtain the linear differential equations from the Lagrange equations (2.6) one should limit oneself to the second-order small quantities in expansion (3.6) Then the potential energy appears as

$$\Pi = \frac{1}{2} \sum_{\sigma,\tau=1}^{s} c_{\sigma\tau} q^\sigma q^\tau , \tag{3.7}$$

where $c_{\sigma\tau}$ are the constants $c_{\sigma\tau} = \left(\frac{\partial^2 \Pi}{\partial q^\sigma \partial q^\tau}\right)_0$. Because of the continuity of the second derivatives of the potential energy we have $c_{\tau\sigma} = c_{\sigma\tau}$.

It is worth noting that the expressions for the kinetic and potential energies (3.3), (3.7) as quadratic forms with constant coefficients coincide with representations (2.1), (2.2) obtained earlier. Note that the kinetic energy must be a positive definite quadratic form by definition.

Now from the Lagrange equations of the second kind (2.6) we actually obtain differential equations of small oscillations of the system in the form (3.1).

We shall seek a partial solution of the differential equations (3.1) corresponding to vibration of a mechanical system in the form

$$q^\sigma = H_\sigma \sin(\omega t + \varepsilon), \quad \sigma = \overline{1, s}. \tag{3.8}$$

After substituting (3.8) into (3.1) we have the following system of algebraic linear homogeneous equations for the unknowns $H_1, H_2, \ldots, H_s$

$$\sum_{\sigma=1}^{s} (c_{\rho\sigma} - \omega^2 a_{\rho\sigma})H_\sigma = 0, \quad \rho = \overline{1, s}, \tag{3.9}$$

or

$$(\mathbf{C} - \omega^2 \mathbf{A})\mathbf{H} = 0. \tag{3.9'}$$

For the existence of nontrivial solution of this system the determinant of its coefficients should vanish

$$\Delta(\omega^2) = \begin{vmatrix} c_{11} - \omega^2 a_{11}, \ldots, c_{1s} - \omega^2 a_{1s} \\ \cdots\cdots\cdots\cdots\cdots\cdots\cdots\cdots\cdots\cdots\cdots\cdots \\ c_{s1} - \omega^2 a_{s1}, \ldots, c_{ss} - \omega^2 a_{ss} \end{vmatrix} = 0. \tag{3.10}$$

Representing the determinant in expanded form we get the *frequency equation*

$$(\omega^2)^s + h_1(\omega^2)^{s-1} + \cdots + h_{s-1}\omega^2 + h_s = 0, \tag{3.11}$$

from which s roots can be found

$$\omega_1^2, \omega_2^2, \ldots, \omega_s^2. \tag{3.12}$$

Note that representation of the determinant (3.10) in the form of polynomial (3.11), when s is sufficiently large, presents some challenge; this is why in higher algebra a great number of methods for representing the characteristic determinant in the polynomial form is developed. The most efficient methods are applied to the determinants in which ω^2 is only in diagonal terms. We get such a determinant when the system of differential equations (3.1) is solvable with respect to the generalized accelerations (the so-called *direct form of equations of mechanical oscillation*).

Such a form of the system of differential equations is possible if the quadratic form (3.3) representing the expression for the kinetic energy of the system is preliminary reduced to the canonical form. Analogous reduction (simplification) of calculation is performed if the quadratic form (3.7) is reduced to the canonical form. In this case the equations of oscillation are solvable with respect to generalized coordinates (*inverse form of oscillation differential equations*), and a characteristic determinant is reduced to the determinant containing the term $1/\omega^2$ in the diagonal only.

We now turn our attention to solving the algebraic system (3.9). For any of values (3.12) ω_k^2, $k = \overline{1, s}$, the determinant (3.10) vanishes. We assume that all the roots of Eq. (3.11) are different. In this case one of the Eqs. (3.9) is a consequence of the others. Let the last one be such an equation. We omit it. Then the rest $s - 1$ equations of the system can be regarded as an inhomogeneous system with respect to the unknowns $H_1^{(k)}/H_s^{(k)}$, $H_2^{(k)}/H_s^{(k)}$, ..., $H_{s-1}^{(k)}/H_s^{(k)}$:

$$\sum_{\sigma=1}^{s-1}(c_{\rho\sigma} - \omega_k^2 a_{\rho\sigma})\frac{H_\sigma^{(k)}}{H_s^{(k)}} = \omega_k^2 a_{\rho s} - c_{\rho s}, \quad \rho = \overline{1, s-1}. \tag{3.13}$$

Here the subscript k means that a solution of system (3.9) is sought under the condition that a concrete value ω_k^2 is substituted instead of ω^2. As the values (3.12) are different, then the determinant of inhomogeneous system (3.13) is not equal to zero, and hence there exists a unique solution which by Cramer's rule can be represented as

$$\frac{H_\sigma^{(k)}}{H_s^{(k)}} = \frac{\Delta_\sigma(\omega_k^2)}{\Delta_s(\omega_k^2)}, \quad \sigma = \overline{1, s-1}. \tag{3.14}$$

It can be easily checked that the expression $\Delta_\sigma(\omega_k^2)$ is an algebraic complement of the element of the last row of the determinant (3.10) from the σ column.

Relations (3.14) demonstrate that the solution of system (3.9) can be obtained with the accuracy up to a common constant multiplier. If one rewrites solution (3.14) in the symmetric form

$$\frac{H_1^{(k)}}{\Delta_1(\omega_k^2)} = \frac{H_2^{(k)}}{\Delta_2(\omega_k^2)} = \cdots = \frac{H_s^{(k)}}{\Delta_s(\omega_k^2)},$$

then any of the relations

$$\frac{H_1^{(k)}}{\Delta_1(\omega_k^2)} = \frac{H_2^{(k)}}{\Delta_2(\omega_k^2)} = \cdots = \frac{H_s^{(k)}}{\Delta_s(\omega_k^2)} = C_k \tag{3.15}$$

is equal to an arbitrary constant C_k.

So, when substituting the values ω_k^2 into system (3.9) its solution can be represented as

$$H_\sigma^{(k)} = C_k\Delta_\sigma(\omega_k^2), \quad \sigma = \overline{1, s}. \tag{3.16}$$

Hence the partial solution (3.8) reads as

$$q_k^\sigma = C_k \Delta_\sigma(\omega_k^2) \sin(\omega_k t + \varepsilon_k), \quad \sigma = \overline{1, s}. \tag{3.17}$$

A general solution of system (3.1) can be obtained as a linear combination of such partial solutions,

$$q^\sigma = \sum_{k=1}^{s} C_k \Delta_\sigma(\omega_k^2) \sin(\omega_k t + \varepsilon_k), \quad \sigma = \overline{1, s}, \tag{3.18}$$

or

$$\mathbf{q} = \sum_{k=1}^{s} C_k \mathbf{H}^{(k)} \sin(\omega_k t + \varepsilon_k), \quad \mathbf{H}^{(k)} = (\Delta_\sigma(\omega_1^2), \dots, \Delta_s(\omega_s^2))^T. \tag{3.18'}$$

The values of arbitrary constants C_k and ε_k, $k = \overline{1, s}$, can be obtained from the initial conditions. As a result of an arbitrary choice of constants C_k and ε_k, it makes no difference what sign before the radical is when one takes the square root of expressions (3.12). The quantities ω_k are supposed to be positive numbers.

One gets a general solution (3.18) as a linear combination of partial solutions (3.17). A mechanical system can be easily made to perform motion according to any of partial solutions (3.17) by means of trying (fitting) initial conditions. In this case all the coordinates q^σ oscillate with the same frequency ω_k and have the same initial phase ε_k, the amplitudes of these oscillations $H_\sigma^{(k)}$ being related as in (3.15). Such *oscillations* are called *natural* or *normal*, and relations between the amplitudes define a *normal (natural) mode of vibration*. The frequencies $\omega_k, k = \overline{1, s}$, are called *natural frequencies of a system*. Any other motion of the system can be obtained as a linear combination of s natural vibrations.

Now let us show that in the assumption that the potential energy (3.7) is positive definite, the squared natural frequencies (3.12) turn out to be positive numbers. It is this fact that makes it possible to find the partial solutions in the form (3.8). To this purpose, we multiply every equation of system (3.9) by $H_\rho^{(k)}$ for a concrete value of ω_k. Summation all the resulting expressions, we have

$$\sum_{\rho,\sigma=1}^{s} (c_{\rho\sigma} - \omega_k^2 a_{\rho\sigma}) H_\rho^{(k)} H_\sigma^{(k)} = 0,$$

or

$$\omega_k^2 = \sum_{\rho,\sigma=1}^{s} c_{\rho\sigma} H_\rho^{(k)} H_\sigma^{(k)} \bigg/ \sum_{\rho,\sigma=1}^{s} a_{\rho\sigma} H_\rho^{(k)} H_\sigma^{(k)}. \tag{3.19}$$

Here the numerator and the denominator are quadratic forms having the coefficients of double potential and kinetic energies as the factors, respectively, in this case they are positive definite forms. Thus, according to (3.19) all the frequencies squared ω_k^2, $k = \overline{1, s}$, are actually the positive numbers.

Consider the two principal modes corresponding to natural frequencies

$$\omega_k \neq \omega_l . \tag{3.20}$$

It is obvious that these modes should satisfy the equations

$$\sum_{\rho,\sigma=1}^{s} (c_{\rho\sigma} - \omega_k^2 a_{\rho\sigma}) H_\sigma^{(k)} = 0 , \quad \rho = \overline{1, s} , \tag{3.21}$$

$$\sum_{\rho,\sigma=1}^{s} (c_{\rho\sigma} - \omega_l^2 a_{\rho\sigma}) H_\sigma^{(l)} = 0 , \quad \rho = \overline{1, s} , \tag{3.22}$$

respectively. Multiplying the equations of system (3.21) by $H_\rho^{(l)}$ and summing over ρ, we get

$$\sum_{\rho,\sigma=1}^{s} (c_{\rho\sigma} - \omega_k^2 a_{\rho\sigma}) H_\sigma^{(k)} H_\rho^{(l)} = 0 . \tag{3.23}$$

In a similar way one can derive the equation

$$\sum_{\rho,\sigma=1}^{s} (c_{\rho\sigma} - \omega_l^2 a_{\rho\sigma}) H_\sigma^{(l)} H_\rho^{(k)} = 0 \tag{3.24}$$

from system (3.22). If we subtract the previous equation from this one, we get

$$\sum_{\rho,\sigma=1}^{s} (\omega_k^2 - \omega_l^2) a_{\rho\sigma} H_\sigma^{(k)} H_\rho^{(l)} = 0 .$$

On the basis of inequality (3.20) this implies that

$$\sum_{\rho,\sigma=1}^{s} a_{\rho\sigma} H_\sigma^{(k)} H_\rho^{(l)} = 0 , \quad k \neq l . \tag{3.25}$$

It is obvious that as a result of this relation, invoking Eq. (3.23) or (3.24) we can also establish the validity of the following equality

$$\sum_{\rho,\sigma=1}^{s} c_{\rho\sigma} H_\sigma^{(k)} H_\rho^{(l)} = 0 , \quad k \neq l . \tag{3.26}$$

Formulas (3.25) and (3.26) express the *orthogonality of normal modes of vibration*. In the matrix form these formulas appear as

$$\mathbf{H}^{(k)^T} \mathbf{A} \mathbf{H}^{(l)} = 0, \quad \mathbf{H}^{(k)^T} \mathbf{C} \mathbf{H}^{(l)} = 0, \quad \omega_k^2 \neq \omega_l^2. \tag{3.27}$$

4 Principal Coordinates

In the previous subsection we have obtained a general solution for the system of differential equations of form (3.18), which shows that all generalized coordinates describing the state of a mechanical system are linear combinations of the same harmonic functions $C_k \sin(\omega_k t + \varepsilon_k)$, $k = \overline{1, s}$. It would be interesting whether it is possible to find a system of generalized coordinates θ^1, θ^2, ... , θ^s, instead of the generalized coordinates q^1, q^2, ... , q^s, such that any of the coordinates should be equal to a single corresponding harmonic function, but not to the superposition of the above harmonic functions, i.e.,

$$\theta^k = C_k \sin(\omega_k t + \varepsilon_k), \quad k = \overline{1, s}. \tag{4.1}$$

The generalized coordinates possessing such a property are called the *principal (master)* or *normal coordinates* of a mechanical system.

It should be noted that sometimes it is rather difficult to choose ab initio generalized coordinates so that they become the principal ones. However, if a problem of vibrations of a mechanical system is solved in some generalized coordinates q^σ, $\sigma = \overline{1, s}$, and if the general solution (3.18) is obtained, then its structure makes it possible to easily find a system of principal coordinates θ^ρ, $\rho = \overline{1, s}$. Actually, the normal coordinates should vary by law (4.1), and hence, according to the form of the general solution (3.18) the they and the generalized coordinates q^1, q^2, ... , q^s should be connected by the relation

$$q^\sigma = \sum_{k=1}^{s} \Delta_\sigma(\omega_k^2) \theta^k, \quad \sigma = \overline{1, s}. \tag{4.2}$$

Relations (4.2) are nothing but formulas of transformation of the principal θ^k coordinates to the generalized q^σ coordinates. For convenience we denote

$$\Delta_{\sigma k} = \Delta_\sigma(\omega_k^2), \quad \sigma, k = \overline{1, s} \tag{4.3}$$

and rewrite formulas (4.2) as

$$q^\sigma = \sum_{k=1}^{s} \Delta_{\sigma k} \theta^k, \quad \sigma = \overline{1, s}. \tag{4.4}$$

It is very important that in principal coordinates the quadratic forms expressing kinetic and potential energies are transformed to a canonical form simultaneously, i.e., to the form containing only the members with squared variables (sometimes it is this property of the coordinates θ^1, θ^2, ... , θ^s that is taken as a basis for the definition of the principal coordinates of a system). Indeed, formulas (3.3), (3.7) imply that

$$T = \frac{1}{2} \sum_{\sigma,\tau=1}^{s} \sum_{\mu,\nu=1}^{s} a_{\sigma\tau} \Delta_{\sigma\mu} \Delta_{\tau\nu} \dot{\theta}^\mu \dot{\theta}^\nu ,$$

$$\Pi = \frac{1}{2} \sum_{\sigma,\tau=1}^{s} \sum_{\mu,\nu=1}^{s} c_{\sigma\tau} \Delta_{\sigma\mu} \Delta_{\tau\nu} \theta^\mu \theta^\nu .$$

(4.5)

However, according to relations (3.16) and notation (4.3), the conditions for orthogonality of normal modes of vibration (3.25), (3.26) can be rewritten as

$$\sum_{\sigma,\tau=1}^{s} a_{\sigma\tau} \Delta_{\sigma\mu} \Delta_{\tau\nu} = 0 , \quad \sum_{\sigma,\tau=1}^{s} c_{\sigma\tau} \Delta_{\sigma\mu} \Delta_{\tau\nu} = 0 , \quad \mu \neq \nu .$$

This is why in double sums relative to indices μ and ν (4.5) only the members containing the squares of variables remain

$$T = \frac{1}{2} \sum_{\nu=1}^{s} a_\nu (\dot{\theta}^\nu)^2 , \quad \Pi = \frac{1}{2} \sum_{\nu=1}^{s} c_\nu (\theta^\nu)^2 ,$$

(4.6)

where

$$a_\nu = \sum_{\sigma,\tau=1}^{s} a_{\sigma\tau} \Delta_{\sigma\nu} \Delta_{\tau\nu} , \quad c_\nu = \sum_{\sigma,\tau=1}^{s} c_{\sigma\tau} \Delta_{\sigma\nu} \Delta_{\tau\nu} , \quad \nu = \overline{1,s} .$$

(4.7)

It follows from expressions (4.6) that system of the Lagrange equations of the second kind (2.6) in principal coordinates reads as

$$\ddot{\theta}^k + \frac{c_k}{a_k} \theta^k = 0 , \quad k = \overline{1,s} ,$$

(4.8)

where in accordance with formulas (4.7), (4.3), (3.16), and (3.19)

$$\frac{c_k}{a_k} = \frac{\sum_{\sigma,\tau=1}^{s} c_{\sigma\tau} \Delta_{\sigma k} \Delta_{\tau k}}{\sum_{\sigma,\tau=1}^{s} a_{\sigma\tau} \Delta_{\sigma k} \Delta_{\tau k}} = \frac{\sum_{\sigma,\tau=1}^{s} c_{\sigma\tau} H_\sigma^{(k)} H_\tau^{(k)}}{\sum_{\sigma,\tau=1}^{s} a_{\sigma\tau} H_\sigma^{(k)} H_\tau^{(k)}} = \omega_k^2 .$$

It is natural that solving the independent Eqs. (4.8) creates no difficulties contrary to solving the system of differential equations (3.1). The solution is set by simple formulas (4.1). However, as noted above the change of generalized coordinates by principal ones with the help of transformation (4.2) requires a preliminary solution of the problem of vibration. However, instead of this problem, one may consider the problem of finding a transformation that reduces the expressions for kinetic and potential energies to the sums of squares at the same time. As is known, such a problem is always solvable if one of the quadratic forms is of fixed sign. In this case this condition is fulfilled, as the kinetic energy is a positive definite quadratic form by definition. It is also worth noting that the procedure of simultaneous reduction of a quadratic forms to the canonical form actually repeats the procedures performed when solving a system of differential equations in direct or inverse forms.

Principal coordinates play a key role. When the number of degrees of freedom is increased unboundedly, i.e., when considering the elastic vibration of continua (a string, a rod, a plate, etc.), such a way of finding the natural modes of vibration makes it possible to reduce the problem to a system of independent equations. If the number of degrees of freedom is finite, then the principal coordinates can also be used in theoretical researches. Let us apply the principal coordinates to the analysis of cases when the characteristic Eq. (3.11) has a zero and multiple roots.

As follows from system (4.8), the equation $\ddot{\theta}^1 = 0$ corresponds to the root $\omega_1^2 = 0$, a general solution of the equation having the form $\theta^1 = C_1 + C_0 t$. From formulas (4.4) it follows that system (3.1) has a partial solution such that some coordinates q^σ are linear functions of time. When considering the vibration of elastic systems, a translational motion with the constant velocity can correspond, in particular, to this fact.

The structure of the general solution implies that when $\omega_1 = 0$ the generalized coordinates can increase unrestrictedly with time. Let us show that in such a case in the linear formulation of the problem the potential energy Π has no isolated minimum at the origin.[2] Actually, when $\omega_1 = 0$ the characteristic determinant (3.10) has the form

$$\Delta(0) = \begin{vmatrix} c_{11} & \cdots & c_{1s} \\ \cdots & \cdots & \cdots \\ c_{s1} & \cdots & c_{ss} \end{vmatrix} = 0,$$

which secures the existence of such a solution of the algebraic system

$$\sum_{\sigma=1}^{s} c_{\rho\sigma} q_*^\sigma = 0, \quad \rho = \overline{1,s}, \tag{4.9}$$

in which $q_*^\sigma \neq 0$ simultaneously for all σ. Multiplying Eqs. (4.9) by q_*^ρ and summing, we get

$$2\Pi = \sum_{\rho,\sigma=1}^{s} c_{\rho\sigma} q_*^\rho q_*^\sigma = 0,$$

---

[2] Recall that by the Lagrange theorem the sufficiency condition of the stability of a position is the existence of an isolated minimum of the potential energy.

i.e., the potential energy vanishes in the vicinity of the state of equilibrium of the system. Hence, for $\omega_1 = 0$ the potential energy of the system actually has no isolated minimum for $q^\sigma = 0$, $\sigma = \overline{1, s}$.

Consider now the case when two (or more) natural modes of the system are equal to each other: $\omega_\nu = \omega_{\nu+1} = \omega$. Here we have two identical equations in system (4.8):

$$\ddot{\theta}^\nu + \omega^2 \theta^\nu = 0, \quad \ddot{\theta}^{\nu+1} + \omega^2 \theta^{\nu+1} = 0.$$

The solutions of these equations,

$$\theta^\nu = C_\nu \sin(\omega t + \varepsilon_\nu), \quad \theta^{\nu+1} = C_{\nu+1} \sin(\omega t + \varepsilon_{\nu+1}),$$

show that the νth and $(\nu + 1)$st vibrations have the same frequency.

When applied to the system of equations (3.9), the existence of r equal squares of frequencies ω_ν^2 among the numbers (3.12) means that when substituting this value to system (3.9) the rank of the matrix of system coefficients is equal to $s - r$. Therefore, $s - r$ unknowns $H_\sigma^{(\nu)}$, $\sigma = \overline{1, s - r}$, can be expressed in terms of the rest r arbitrary quantities $H_\rho^{(\nu)}$, $\rho = s - r + 1, s - r + 2, \ldots, s$. Making r different combinations of the above arbitrary constants, one can compose r linearly independent solutions corresponding to the frequency ω_ν of r-multiplicity. Usually for the convenience of calculation these modes are chosen to orthogonal to each other, providing the validity of equalities (3.25) and (3.26).

When studying small oscillation of systems in the vicinity of the state of stable equilibrium, the case of multiple frequencies was not investigated for a long time. The originator of analytic mechanics Lagrange had been turning back to this problem over and over again. Assuming that in the case of equal frequencies one should introduce integer powers of time t as multipliers to obtain linearly independent solutions, as is usually done in differential equations, Lagrange fully realized at the same time that "these variables will not be able to contain the time t outside the sine and cosine functions, because then they could increase infinitely".[3] It was only a hundred years later when independently Weierstrass and Professor of Saint Petersburg University, Academician O.I. Somov managed to explain the validity of the structure of the general solution with the help of quadratic forms of kinetic and potential energies.

5 Minimum-Maximum Properties of Natural Frequencies

The Rayleigh function. Rayleigh's theorem. Consider the *Rayleigh function* that is equal to the fraction of two quadratic forms

---

[3] *J.L. Lagrange*. Analytic mechanics. 264 p. Translated and edited by August Boissonnade and Victor N. Vagliente. Kluwer Academic Publishers (Dordrecht. Boston. London). 1997. The Netherlands.

$$R(H) = R(H_1, H_2, \ldots, H_s) = \frac{\widehat{\Pi}(H)}{\widehat{T}(H)},$$

$$\widehat{\Pi}(H) = \sum_{\sigma,\tau=1}^{s} c_{\sigma\tau} H_\sigma H_\tau, \quad \widehat{T}(H) = \sum_{\sigma,\tau=1}^{s} a_{\sigma\tau} H_\sigma H_\tau, \tag{5.1}$$

where $H_\sigma, \sigma = \overline{1, s}$, are the oscillation amplitudes introduced in Sect. 3, and the quadratic forms $\widehat{\Pi}(H)$ and $\widehat{T}(H)$ relate to the maximum values of the potential and kinetic energies for harmonic oscillation with the frequency ω as

$$\max_t \Pi = 2\,\widehat{\Pi}(H), \quad \max_t T = 2\omega^2\,\widehat{T}(H).$$

Let us find the minimum in the variables H_σ of the Rayleigh function (5.1). It is convenient for its calculation to seek the minimum of the function $\widehat{\Pi}(H)$ provided that $\widehat{T}(H) = 1$. Introducing the Lagrange multiplier λ, we seek the minimum of the function $\widehat{\Pi}(H) - \lambda(\widehat{T}(H) - 1)$. Differentiating this function with respect to the variables H_ρ we get the system of equations

$$\sum_{\sigma=1}^{s} (c_{\rho\sigma} - \lambda a_{\rho\sigma}) H_\sigma = 0, \quad \rho = \overline{1, s}, \tag{5.2}$$

which coincides with system (3.9) for $\lambda = \omega^2$. The roots of determinant (3.10) of this system represent the frequencies of natural vibration. (3.12). That is why it is natural that the least of these roots represents the minimum of the Rayleigh function. Put the natural frequencies in increasing order

$$\omega_1^2 \leqslant \omega_2^2 \leqslant \cdots \leqslant \omega_s^2, \tag{5.3}$$

the case of multiple roots is not excluded from consideration.

Then one can formulate *Rayleigh's theorem*: the square of the least natural frequency agrees with the minimum of the Rayleigh function,

$$\omega_1^2 = \min_H R(H), \quad H = \{H_\sigma\}. \tag{5.4}$$

Rayleigh's theorem can be used to approximately determine the quantity ω_1^2 if one takes the amplitude vector H in an arbitrary way. In this case in view of formula (5.4) one gets the upper estimate for the quantity ω_1^2.

The process of constructing approximate estimates can be also continued for higher frequencies. Suppose that the amplitude vector $H^{(1)} = \{H_\sigma^{(1)}\}$ corresponding to the frequency ω_1 is known. Then the quantity ω_2^2 is equal to the minimum of the Rayleigh function if the orthogonality condition (3.25) of the sought-for vector H and the vector $H^{(1)}$ is satisfied:

$$\omega_2^2 = \min_{H,\,(H,H^{(1)})=0} R(H)\,, \quad \text{where} \quad (H, H^{(1)}) = \sum_{\sigma,\tau=1}^{s} a_{\sigma\tau}\, H_\sigma H_\tau^{(1)}\,. \tag{5.5}$$

In the same way, analogously to formula (5.5), the quantity ω_3^2 is equal to the minimum of the Rayleigh function if the two conditions are satisfied: $(H, H^{(1)}) = 0$ and $(H, H^{(2)}) = 0$, etc.

The process of constructing the frequencies described above has a grave disadvantage to the effect that in order to determine the quantity ω_{k+1}^2 one should know the amplitude vectors $H^{(1)},\ H^{(2)}, \ldots, H^{(k)}$. The technique presented below is free of this disadvantage.

Courant's Theorem *We shall find the minimum of the Rayleigh function (5.1) providing that k linear homogeneous constraints*

$$L^{(\varkappa)} H = 0 \quad \text{or} \quad \sum_{\sigma=1}^{s} l_\sigma^{(\varkappa)} H_\sigma = 0\,, \quad \varkappa = \overline{1,k} \tag{5.6}$$

are imposed on the amplitude vector H.
Courant's theorem states that

$$\omega_{k+1}^2 = \max_{L^{(\varkappa)},\,\varkappa=\overline{1,k}} \min_{H} R(H)\,, \quad L^{(\varkappa)} H = 0\,, \quad \varkappa = \overline{1,k}\,. \tag{5.7}$$

From this theorem the quantity ω_{k+1}^2 can be readily found. Here the maximum over the constraints $L^{(\varkappa)}$ means that we seek the maximum over all coefficients $l_\sigma^{(\varkappa)}$ entering (5.6).

To prove the theorem, we relate, as in formulas (4.2), the quantities H_σ to the amplitudes U_k of principal coordinates as

$$H_\sigma = \sum_{k=1}^{s} \Delta_{\sigma k} U_k\,, \quad \sigma = \overline{1,s}\,. \tag{5.8}$$

Now using formulas (5.8), we transform the expressions for $\widehat{\Pi}$ and $\widehat{T}$. Hence, analogously to formula (4.6), these quadratic forms become the sums of squares

$$\widehat{\Pi} = \sum_{\nu=1}^{s} c_\nu U_\nu^2\,, \quad \widehat{T} = \sum_{\nu=1}^{s} a_\nu U_\nu^2\,, \tag{5.9}$$

where the coefficients c_ν and a_ν are determined by formulas (4.7). We make one more change of variables $V_\nu = U_\nu \sqrt{a_\nu}$, $\nu = \overline{1,s}$. Hence, in view of the expressions (5.9) the Rayleigh function (5.1) reads as

$$R(V) = \frac{\omega_1^2 V_1^2 + \omega_2^2 V_2^2 + \cdots + \omega_s^2 V_s^2}{V_1^2 + V_2^2 + \cdots + V_s^2}\,, \quad V = \{V_\sigma\}\,. \tag{5.10}$$

Rayleigh's theorem is obvious from representation (5.10), because by inequalities (5.3) we have

$$\min_V R(V) = \omega_1^2 + \min_V \frac{(\omega_2^2 - \omega_1^2)V_2^2 + \cdots + (\omega_s^2 - \omega_1^2)V_s^2}{V_1^2 + V_2^2 + \cdots + V_s^2} = \omega_1^2.$$

Rewriting formula (5.7) in the principal coordinates and denoting the right-hand side by Z we have

$$\omega_{k+1}^2 = \max_{L^{(\varkappa)}, \, \varkappa = \overline{1,k}} \min_V R(V) = Z, \qquad L^{(\varkappa)}V = 0, \quad \varkappa = \overline{1,k}. \tag{5.11}$$

Let us now prove equality (5.11). Consider the partial case of constraints $V_1 = V_2 = \cdots = V_k = 0$. Then from formula (5.11) we obtain

$$\min_V R(V) = \min_V \frac{\omega_{k+1}^2 V_{k+1}^2 + \cdots + \omega_s^2 V_s^2}{V_{k+1}^2 + \cdots + V_s^2} = \omega_{k+1}^2,$$

which implies that

$$\omega_{k+1}^2 \geqslant Z, \tag{5.12}$$

because the maximum is not smaller than one of the possible values of the quantity in question.
Consider the system of the equations of constraints

$$\sum_{\sigma=1}^{s} l_\sigma^{(\varkappa)} V_\sigma = 0, \quad \varkappa = \overline{1,k}, \tag{5.13}$$

and take its partial solution, for which $V_{k+2} = \cdots = V_s = 0$. Then Eqs. (5.13) can be rewritten as

$$\sum_{\sigma=1}^{s} l_\sigma^{(\varkappa)} V_\sigma = 0, \quad \varkappa = \overline{1,k}. \tag{5.14}$$

The system of algebraic equations (5.14) is linear and homogeneous. The number of its equations being greater then the number of variables, it always has a nonlinear solution. For a chosen vector V we have

$$\max_{L^{(\varkappa)}, \, \varkappa = \overline{1,k}} \frac{\omega_1^2 V_1^2 + \cdots + \omega_{k+1}^2 V_{k+1}^2}{V_1^2 + \cdots + V_{k+1}^2} = \omega_{k+1}^2,$$

which implies that

$$\omega_{k+1}^2 \leqslant Z, \tag{5.15}$$

because the minimum is not greater than one of the possible values of the quantity of interest.

Courant's theorem is now proved by simultaneous realization of inequalities (5.12) and (5.15).

The case of small oscillation for which both quadratic forms $\widehat{\Pi}(H)$ and $\widehat{T}(H)$ in (5.1) are positively defined will be considered below. For the sake of validity of what was said above in this subsection it is only sufficient to assume that the form $\widehat{T}(H)$ is positively defined and the form $\widehat{\Pi}(H)$ can be arbitrary. In this case, the roots λ of the determinant of system (5.2) can be zero or negative. In these cases the state of equilibrium is unstable. For zero roots the solution can grow linearly with time, and for $\lambda < 0$ the exponential growth of the solution of a kind $e^{\sqrt{-\lambda}t}$ is possible. Nevertheless, when ordering the roots $\lambda_\sigma = \omega_\sigma^2$ by formula (5.3) the Rayleigh theorem and the Courant theorem keep their form taking into account that the quantities ω_σ^2 can be either zero or negative.

Corollaries to Courant's Theorem *Rayleigh's theorem can be used for approximate evaluation of the frequencies of vibrations, but the Courant theorem is not well suited for this purpose, since the procedure of enumeration of possible constraints is nonconstructive. However, there are two important corollaries to Courant's theorem that will be considered below.*

Corollary 1 *Dependence of the frequencies of vibrations on the masses and stiffnesses of a system. Let two conservative mechanical systems with the same number of degrees of freedom be considered. The quadratic forms of these systems are as follows:*

$$\widehat{\Pi}^{(m)}(H) = \sum_{\sigma,\tau=1}^{s} c_{\sigma\tau}^{(m)} H_\sigma H_\tau, \quad \widehat{T}^{(m)}(H) = \sum_{\sigma,\tau=1}^{n} a_{\sigma\tau}^{(m)} H_\sigma H_\tau, \quad m = 1,2.$$

If for any H the inequalities

$$\widehat{\Pi}^{(2)}(H) \geqslant \widehat{\Pi}^{(1)}(H), \quad \widehat{T}^{(2)}(H) \leqslant \widehat{T}^{(1)}(H) \tag{5.16}$$

are satisfied simultaneously, then all the frequencies of the second system are not less than the corresponding frequencies of the first system:

$$\left(\omega_\nu^{(2)}\right)^2 \geqslant \left(\omega_\nu^{(1)}\right)^2, \quad \nu = \overline{1,s}. \tag{5.17}$$

In fact, due to (5.16) for any H the inequality $R^{(2)}(H) \geqslant R^{(1)}(H)$ takes place and now inequalities (5.17) follow form Courant's theorem.

In particular, if a system consists of masses joined by springs, then increasing the stiffness of one of the springs leads to (nonstrict) increase of all frequencies, and increasing one of the masses leads to (nonstrict) decrease of all the frequencies.

Corollary 2 *Dependence of the frequencies on the constraints imposed.*

Consider a system with s degrees of freedom and with natural frequencies satisfying inequalities (5.3).

Let the linear constraint

$$L_* H = \sum_{\sigma=1}^{s} l_{\sigma*} H_\sigma = 0$$

be imposed on the system. As a result, we shall get a system with $s - 1$ degrees of freedom and with the frequencies of natural vibration $\bar{\omega}_\sigma$, $\sigma = \overline{1, s - 1}$. Then the frequencies of vibrations of the new system will lie between the frequencies of the initial system:

$$\omega_\sigma^2 \leqslant \bar{\omega}_\sigma^2 \leqslant \omega_{\sigma+1}^2, \quad \sigma = \overline{1, s - 1}. \tag{5.18}$$

Actually, by Courant's theorem, inequalities (5.18) follow immediately from the chain of relations:

$$\omega_k^2 = \max_{L^{(\varkappa)}, \, \varkappa = \overline{1, k-1}} \min_H R(H) \leqslant$$

$$\text{(for } L^{(\varkappa)} H = 0, \varkappa = \overline{1, k-1})$$

$$\leqslant \bar{\omega}_k^2 = \max_{L^{(\varkappa)}, \, \varkappa = \overline{1, k-1}} \min_H R(H) \leqslant \tag{5.19}$$

$$\text{(for } L^{(\varkappa)} H = 0, \varkappa = \overline{1, k-1}, L_* H = 0)$$

$$\leqslant \omega_{k+1}^2 = \max_{L^{(\varkappa)}, \, \varkappa = \overline{1, k-1}, L_*} \min_H R(H)$$

$$\text{(for } L^{(\varkappa)} H = 0, \varkappa = \overline{1, k-1}, L_* H = 0).$$

The inequality $\omega_k^2 \leqslant \bar{\omega}_k^2$ in (5.18) is connected with the fact that the minimum for $\bar{\omega}_k^2$ is sought over a more narrow manifold of values of H, and hence it is (unstrictly) greater. In turn, the inequality $\bar{\omega}_k^2 \leqslant \omega_{k+1}^2$ in (5.18) is connected with the fact that the maximum for ω_{k+1}^2 is sought over a more narrow manifold of constraints, and hence it is (unstrictly) greater.

Corollary 2 assumes a generalization to the effect that if r independent constraints are imposed on a system, then we obtain a system with $s - r$ degrees of freedom, inequalities (5.18) being replaced by

$$\omega_k^2 \leqslant \bar{\omega}_k^2 \leqslant \omega_{k+r}^2, \quad k = \overline{1, s - r}. \tag{5.20}$$

Extension to the case of systems with distributed parameters. Rayleigh's and Courant's theorems assume can be extended to systems with distributed parameters. For instance, consider the natural vibrations of a Bernoulli–Euler beam with constant cross section. These vibrations are described by the equation

$$EJ \frac{\partial^4 w}{\partial x^4} + \rho S \frac{\partial^2 w}{\partial t^2} = 0,$$

where $w(x, t)$ is the deflection, EJ is the bending stiffness, ρS is the mass of the unit length of a beam. We separate the variables: $w(x, t) = \varphi(x) \sin(\omega t)$. The Rayleigh function for a beam has the form

$$R(\varphi) = \frac{\int_0^a EJ \left(\frac{d^2 \varphi}{dx^2} \right)^2 dx}{\int_0^a \rho S \varphi^2 dx},$$

where a is the length of the beam.

Consider the beam with free ends. This beam has two zero frequencies $\omega_1 = \omega_2 = 0$, to which there correspond the displacements of the beam as a rigid solid (the natural functions have the form $\varphi_1(x) = 1$, $\varphi_2(x) = x$), and the infinite spectrum of nonzero frequencies $\omega_3 < \omega_4 < \cdots$, to which the bending modes of vibration correspond. If the beam is supported at two points, then the displacements of the beam as a rigid solid become impossible, and it will have the least natural frequency $\bar{\omega}_1 > 0$. We formulate the problem: find the position of the points x_1 and x_2 supporting at which the frequency $\bar{\omega}_1$ is the greatest. The supports at the points can be considered as two constraints $\varphi(x_1) = 0$ and $\varphi(x_2) = 0$. Using the Corollary 2 to Courant's theorem in formulation (5.20) for $k = 1$, $r = 2$, we obtain

$$0 \leqslant \bar{\omega}_1 \leqslant \omega_3,$$

where ω_3 is the first nonzero frequency of natural vibration of the beam. Hence, the best what can be expected is $\bar{\omega}_1 = \omega_3$. The first mode of vibration $\varphi_3(x)$ has two nodal points, at which $\varphi_3(x) = 0$. Suppose that they are x_1 and x_2. If the beam is supported at these points, then the supports will not preclude the vibration of the mode $\varphi_3(x)$, and we get the sought-for result $\bar{\omega}_1 = \omega_3$. The problem of constructing the function $\varphi_3(x)$ is not difficult, but it is beyond the scope of our consideration here.

6   Small Oscillation in the Presence of Resistance Forces and Gyroscopic Forces

In a general case the investigation of linear oscillation in the presence of resistance forces reduces to the consideration of a system of differential equations (1.5) or (1.6). We shall make a remark concerning the resistance forces. In the analysis of the pattern of motion (see Sect. 2) the matrix $\mathbf{B}$ of Eq. (1.6) was supposed to be symmetric for the sake of simplicity. It is natural that such a restriction is not always valid. Usually if the elements of some matrix are not symmetric $b_{\rho\sigma} \neq b_{\sigma\rho}$, then it should be partitioned into two matrices $\mathbf{B} = \mathbf{B}' + \mathbf{B}''$, the first one of which is symmetric, and the second one is asymmetric. The elements of such matrices are the numbers

$$b'_{\rho\sigma} = b'_{\sigma\rho} = \frac{1}{2}(b_{\rho\sigma} + b_{\sigma\rho}), \quad b''_{\rho\sigma} = -b''_{\sigma\rho} = \frac{1}{2}(b_{\rho\sigma} - b_{\sigma\rho}).$$

So, suppose, $\mathbf{B} = \mathbf{B}' + \mathbf{B}''$, where $\mathbf{B}'$ and $\mathbf{B}''$ are the symmetric and asymmetric parts of the matrix $\mathbf{B}$. As a result, the system of equations (1.5) can be represented as

$$\mathbf{A\ddot{q}} + \mathbf{B'\dot{q}} + \mathbf{B''\dot{q}} + \mathbf{Cq} = 0. \tag{6.1}$$

As before, the function D can be associated with the summand $\mathbf{B'\dot{q}}$ (see formulas (2.4), (2.5)). The summand $\mathbf{B''\dot{q}}$ describes the *gyroscopic forces*. Such forces can be present, in particular, in systems containing gyroscopes. It is easy to check that inequality (2.8) is also valid if there exist gyroscopic forces.

In the case of gyroscopic forces, i.e., when the matrix $\mathbf{B}$ is asymmetric, the general solution of system (6.1) can be constructed by the method used in Sect. 1.

Let us study the possibility of using the principal coordinates in the case when the matrix $\mathbf{B}$ is symmetric.

First of all we note that in a general case three quadratic forms T, Π, D cannot be simultaneously reduced to the canonical form.

Actually, changing from the generalized coordinates $q^1, q^2, \ldots, q^s$ to the coordinates $\theta^1, \theta^2, \ldots, \theta^s$ by formulas (4.4) we get

$$T = \frac{1}{2} \sum_{\nu=1}^{s} a_\nu (\dot{\theta}^\nu)^2, \quad \Pi = \frac{1}{2} \sum_{\nu=1}^{s} c_\nu (\theta^\nu)^2, \tag{6.2}$$

$$D = \frac{1}{2} \sum_{\mu,\nu=1}^{s} b^*_{\mu\nu} \dot{\theta}^\mu \dot{\theta}^\nu, \tag{6.3}$$

where

$$b^*_{\mu\nu} = \sum_{\rho,\sigma=1}^{s} b_{\rho\sigma} \Delta_{\rho\mu} \Delta_{\sigma\nu}. \tag{6.4}$$

Expressions (6.2), (6.3) imply that Eqs. (2.6) in the coordinates θ^ρ read as

$$a_\rho \ddot{\theta}^\rho + \sum_{\sigma=1}^{*s} b^*_{\rho\sigma} \dot{\theta}^\sigma + c_\rho \theta^\rho = 0, \quad \rho = \overline{1, s}. \tag{6.5}$$

Generally speaking, this system does not fall into s independent equations. But if there exists the following dependence between the coefficients of matrices of the kinetic and potential energies and the dissipation function

$$b_{\rho\sigma} = aa_{\rho\sigma} + cc_{\rho\sigma}, \quad \rho, \sigma = \overline{1, s}, \tag{6.6}$$

where $a = \text{const}$, $c = \text{const}$, then Eqs. (6.5) are independent. In this case, in accordance with expressions (4.7) formulas (6.4) appear as

$$b^*_{\mu\nu} = \sum_{\rho,\sigma=1}^{s} (aa_{\rho\sigma}\Delta_{\rho\mu}\Delta_{\sigma\nu} + cc_{\rho\sigma}\Delta_{\rho\mu}\Delta_{\sigma\nu}) = \begin{cases} aa_\nu + cc_\nu, & \mu = \nu, \\ 0, & \mu \neq \nu, \end{cases}$$

and, hence, Eqs. (6.5) can be written as

$$a_\rho\ddot{\theta}^\rho + b_\rho\dot{\theta}^\rho + c_\rho\theta^\rho = 0, \quad \rho = \overline{1, s}, \tag{6.7}$$

where $b_\rho = aa_\rho + cc_\rho$.

Let us find under which conditions relations (6.6) can be assumed.

Suppose that the forces of air resistance acting on each particle of the system depend linearly on the velocities of their motion

$$\mathbf{F}_i = -k_i\mathbf{v}_i, \quad i = \overline{1, n}.$$

In this case, the generalized forces read as

$$Q_\rho = \sum_{i=1}^{n} \mathbf{F}_i \cdot \frac{\partial \mathbf{r}_i}{\partial q^\rho} = -\sum_{i=1}^{n} k_i\dot{\mathbf{r}}_i \cdot \frac{\partial \mathbf{r}_i}{\partial q^\rho}. \tag{6.8}$$

We have

$$\dot{\mathbf{r}}_i = \sum_{\sigma=1}^{n} \frac{\partial \mathbf{r}_i}{\partial q^\rho}\dot{q}^\sigma, \quad \text{therefore} \quad \frac{\partial \dot{\mathbf{r}}_i}{\partial \dot{q}^\rho} = \frac{\partial \mathbf{r}_i}{\partial q^\rho},$$

and hence formula (6.8) can be rewritten as

$$Q_\rho = -\sum_{i=1}^{n} k_i\dot{\mathbf{r}}_i \cdot \frac{\partial \dot{\mathbf{r}}_i}{\partial \dot{q}^\rho} = -\frac{\partial}{\partial \dot{q}^\rho} \sum_{i=1}^{n} \frac{k_i v_i^2}{2}.$$

Supposing that the resistance coefficients k_i are proportional to the masses m_i we get

$$Q_\rho = -\frac{\partial}{\partial \dot{q}^\rho} \sum_{i=1}^{n} \frac{am_i v_i^2}{2} = -a\frac{\partial T}{\partial \dot{q}^\rho} = -a\sum_{\sigma=1}^{s} a_{\rho\sigma}\dot{q}^\sigma = -\sum_{\sigma=1}^{s} b_{\rho\sigma}\dot{q}^\sigma.$$

This implies that $b_{\rho\sigma} = aa_{\rho\sigma}$. Thus, the first summand in formula (6.6) takes into account the resistance forces that are proportional to the masses m_i by the assumption.

The dissipation of energy in oscillations of an elastic system is connected not only with the air friction of its elements (*external friction*) but also with the friction that accompanies strains (*internal friction*). Suppose that a system consists of n elastic elements with stiffnesses c_k. In this case the strain potential energy is

$$\Pi = \sum_{k=1}^{n} \frac{c_k \Delta_k^2}{2} = \frac{1}{2} \sum_{\rho,\sigma=1}^{s} c_{\rho\sigma} q^\rho q^\sigma \,,$$

where Δ_k is the strain (compression or extension) of the kth elastic element. A restoring force of a separate elastic element is $c_k \Delta_k$. The force of internal resistance in it can be approximately assumed to be proportional to the rate $\dot{\Delta}_k$ and to the stiffness c_k. The dissipation function characterizing the energy loss in this element is

$$D_k = c c_k \dot{\Delta}_k / 2 \,.$$

Supposing that the proportionality factor c is the same for all elastic elements we obtain

$$D = \sum_{k=1}^{n} D_k = c \sum_{k=1}^{n} \frac{c_k \dot{\Delta}_k^2}{2} = c \sum_{\rho,\sigma=1}^{s} \frac{c_{\rho\sigma} \dot{q}^\rho \dot{q}^\sigma}{2} \,.$$

Therefore,

$$Q_\rho = -\frac{\partial D}{\partial \dot{q}^\rho} = -c \sum_{\sigma=1}^{s} c_{\rho\sigma} \dot{q}^\sigma = -\sum_{\sigma=1}^{s} b_{\rho\sigma} \dot{q}^\sigma \,.$$

Thus, the second summand in formula (6.6) deals with the forces of internal friction which are proportional to the strain rate and the stiffness of the elastic system given.

According to formulas (7.1), (7.5) of Chap. 4 the general solutions of Eq. (6.7) have the form

$$\theta^\rho = e^{-\frac{b_{\rho} t}{2a_\rho}} \left(E_\rho \cos \sqrt{\frac{c_\rho}{a_\rho} - \frac{b_\rho^2}{4a_\rho^2}} \, t + F_\rho \sin \sqrt{\frac{c_\rho}{a_\rho} - \frac{b_\rho^2}{4a_\rho^2}} \, t \right) , \qquad (6.9)$$

$$\rho = \overline{1, s} \,.$$

Here E_ρ and F_ρ are arbitrary constants.

Representation of a general solution in generalized coordinates $q^1, q^2, \ldots, q^s$ can be performed with the help of linear transformation (4.2).

Thomson's and Tait's theorems on the influence of dissipative and gyroscopic forces on the stability of a state of equilibrium of a mechanical system. Let us go back to the system of equations (6.1) and suppose that the matrix $\mathbf{A}$ is symmetric and positive definite, the matrix $\mathbf{B}'$ is nonstrictly positive, the matrix $\mathbf{C}$ is symmetric. We shall be concerned with the roots of the eigenvalue equation

$$|\lambda^2 \mathbf{A} + \lambda(\mathbf{B}' + \mathbf{B}'') + \mathbf{C}| = 0 \,. \qquad (6.10)$$

If only one root with a positive real part exists, this means instability. We present the four Thomson's and Tait's theorems, which summarize the results of the analysis.[4]

---

[4] For details, see, for example, the book by *D.R. Merkin*. Introduction to the theory of stability (Texts in Applied Mathematics, Vol. 24). Springer Science and Business Media. 2012. 320 p.

Theorem 1 *Let* **C** *be a positive definite matrix. Then if there exist no dissipative and gyroscopic forces (when* **B** $= 0$), *then all the roots of Eq. (6.10) are purely imaginary and the state of equilibrium is stable. Addition of dissipative and gyroscopic forces does not disturb the stability of a state of equilibrium.*

Theorem 2 *If under the conditions of Theorem 1 the matrix* **B**′ *is positive definite (i.e., the resistance forces are with complete dissipation), then all the roots of Eq. (6.10) have negative real parts and the state of equilibrium is asymptotically stable.*

Before formulating Theorems 3 and 4, we first introduce the notion of the *degree of instability*. Consider the equation

$$|\mathbf{C} - \mu\mathbf{A}| = 0. \tag{6.11}$$

All the roots μ of this equation are real. The purely imaginary values of $\lambda = \pm\sqrt{-\mu}$ correspond to the positive roots μ, and the values of λ, one of which is positive, correspond the negative μ; in this case the solution unboundedly increases with time. The degree k of instability of a system is called the number of negative roots of Eq. (6.11). In this case, it is assumed that the equation has no zero roots.

Theorem 3 *If the degree of instability is even, then the gyroscopic forces exist, a state of equilibrium becoming stable (all the roots of Eq. (6.10) become purely imaginary) when these forces are added. If the degree of instability is odd, then the addition of gyroscopic forces is not able to influence the instability of an equilibrium state.*

Theorem 4 *If under the conditions of Theorem 3 when the degree of instability is even and gyroscopic forces are added, the stability is achieved, then the addition of resistance forces with complete dissipation changes again the state of equilibrium into instable.*

In this connection, we introduce the notion of *temporal stability* for the stability that is achieved by means of addition of gyroscopic forces. This stability is broken by the resistance forces.

As an example demonstrating Theorems 3 and 4, let us consider the motion of a gyroscope spinning in the field of gravity about a vertical axis. We have the system of equations

$$
\begin{aligned}
A\ddot{q}_1 + n\dot{q}_1 + C\omega\dot{q}_2 - PLq_1 &= 0, \\
A\ddot{q}_2 + n\dot{q}_2 - C\omega\dot{q}_1 - PLq_2 &= 0,
\end{aligned} \tag{6.12}
$$

where q_1, q_2 are the deviation angles of the gyroscope axis from the vertical, A, C are the inertia moments, ω is the angular velocity of a gyroscope, P is the gyroscope weight, L is the distance from the center of mass to the support point, and n is the resistance coefficient.

The characteristic equation of system (6.12) has the form

$$(A\lambda^2 + n\lambda - PL)^2 + C^2\omega^2\lambda^2 + 0. \tag{6.13}$$

With $\omega = n = 0$ Eq. (6.11) has a negative two-fold root $\mu = -PL/A$, that is why the degree of instability is two. With $n = 0$, Eq. (6.13) has only purely imaginary roots if $C^2\omega^2 > 4APL$, i.e., when the angular velocity of a gyroscope is sufficiently great.

Consider Eq. (6.13) in the presence of the resistance ($n > 0$). Expanding the polynomial (6.13) in powers of λ, we find that the coefficient at λ of the first degree is equal to $-2APL$. Therefore, the necessary condition for stability, according to which all the polynomial coefficients should be positive, is violated.

7 Forced Vibration of a Mechanical System

Consider vibration of a mechanical system in which along with conservative forces and resistance forces there are forces depending only on time. In this case the occurring oscillations are called *forced*. When composing the Lagrange equations of the second kind we get the following system of differential equations

$$\sum_{\tau=1}^{s}(a_{\sigma\tau}\ddot{q}^\tau + b_{\sigma\tau}\dot{q}^\tau + c_{\sigma\tau}q^\tau) = Q_\sigma(t), \quad \sigma = \overline{1, s}. \tag{7.1}$$

A general solution of this nonhomogeneous linear system with constant coefficients is a sum of the general solution (1.18) of the homogeneous system (1.5) and any partial solution of the nonhomogeneous system (7.1). The latter can be determined by the method of variation of constants.

Now we consider in more detail the case when the coefficients $b_{\sigma\tau}$ can be represented in form (6.6).

In Chap. 6 it was shown that if the Lagrange equations of the second kind, as written in the coordinates q^σ, are given, then in order to obtain the equations in new coordinates $\tilde{q}^\tau$ the initial equations should be multiplied by the transformation coefficients $\partial q^\sigma / \partial \tilde{q}^\tau$ and summed over σ. This implies that we can assume that Eqs. (6.7) are obtained as a result of the transformation

$$\sum_{\sigma=1}^{s}\left(\sum_{\tau=1}^{s}(a_{\sigma\tau}\ddot{q}^\tau + b_{\sigma\tau}\dot{q}^\tau + c_{\sigma\tau}q^\tau)\right)\frac{\partial q^\sigma}{\partial\theta^\rho} = a_\rho\ddot{\theta}^\rho + b_\rho\dot{\theta}^\rho + c_\rho\theta^\rho = 0, \tag{7.2}$$

$$\rho = \overline{1, s},$$

where in accordance with formulas (4.4) $\partial q^\sigma / \partial\theta^\rho = \Delta_{\sigma\rho}$. Equality (7.2) shows that if we multiply Eqs. (7.1) by $\Delta_{\sigma\rho}$ and sum over σ we get

$$a_\rho\ddot{\theta}^\rho + b_\rho\dot{\theta}^\rho + c_\rho\theta^\rho = \Theta_\rho(t), \tag{7.3}$$

where

$$\Theta_\rho = \sum_{\sigma=1}^{s} \Delta_{\sigma\rho} Q_\sigma . \tag{7.4}$$

The general solution of any of Eqs. (7.3) in accordance with formula (7.20) of Chap. 4 is

$$\theta^\rho = e^{-\frac{b_\rho t}{2a_\rho}} \left(E_\rho \cos\sqrt{\frac{c_\rho}{a_\rho} - \frac{b_\rho^2}{4a_\rho^2}}\, t + F_\rho \sin\sqrt{\frac{c_\rho}{a_\rho} - \frac{b_\rho^2}{4a_\rho^2}}\, t \right) +$$

$$+ \frac{1}{a_\rho\sqrt{\frac{c_\rho}{a_\rho} - \frac{b_\rho^2}{4a_\rho^2}}} \int_0^t \Theta_\rho(\xi) e^{-\frac{b_\rho(t-\xi)}{2a_\rho}} \sin\left(\sqrt{\frac{c_\rho}{a_\rho} - \frac{b_\rho^2}{4a_\rho^2}}\,(t-\xi) \right) d\xi, \tag{7.5}$$

$$\rho = \overline{1, s} .$$

Here E_ρ, F_ρ, $\rho = \overline{1, s}$, are arbitrary constants. If the case of no resistance, this solution assumes the form

$$\theta^\rho = E_\rho \cos\sqrt{\frac{c_\rho}{a_\rho}}\, t + F_\rho \sin\sqrt{\frac{c_\rho}{a_\rho}}\, t +$$

$$+ \frac{1}{\sqrt{a_\rho c_\rho}} \int_0^t \Theta_\rho(\xi) \sin\left(\sqrt{\frac{c_\rho}{a_\rho}}\,(t-\xi) \right) d\xi, \quad \rho = \overline{1, s} . \tag{7.6}$$

The values of arbitrary constants can be easily determined from the initial conditions $\theta^\rho(0) = \theta_0^\rho$, $\dot\theta^\rho(0) = \dot\theta_0^\rho$, $\rho = \overline{1, s}$, related to the initial conditions q_0^σ, $\dot q_0^\sigma$, $\sigma = \overline{1, s}$, by formulas (4.4).

Consider the case when all the generalized forces vary by harmonic laws with the same frequency and initial phase

$$Q_\sigma = B_\sigma \sin(pt + \gamma), \quad \sigma = \overline{1, s} . \tag{7.7}$$

Then according to formulas (7.4) and (7.7) the generalized forces read as

$$\Theta_\rho = H_\rho \sin(pt + \gamma), \quad \rho = \overline{1, s} ,$$

where

$$H_\rho = \sum_{\sigma=1}^{s} \Delta_{\sigma\rho} B_\sigma . \tag{7.8}$$

Now in solution (7.6) we can separate the following summand

$$H_\rho \sin(pt + \gamma)/(c_\rho - a_\rho p^2), \quad p \neq (c_\rho/a_\rho)^{1/2} ,$$

or

$$-t H_\rho \cos(pt + \gamma)/(2 p a_\rho), \quad p = (c_\rho/a_\rho)^{1/2} .$$

The last term describes the *resonance condition* in which the frequency of a disturbing force coincides with one of the natural frequencies of the system. In this case the deviations of the system proportional to time t can occur as great as possible. Taking into account the resistance, when the solution is given by formulas (7.5), deviations of the system are limited for any frequency p.

Note that if the coefficients B_σ of the generalized forces (7.7) satisfy the equation

$$\sum_{\sigma=1}^{s} \Delta_{\sigma\rho} B_\sigma = 0$$

for some ρ, then in accordance with formulas (7.8) we have $H_\rho = 0$. The latter means that the vibration mode corresponding to the frequency ω_ρ is not excited.

Let us go now to the general case when one cannot split the system into separate equations by transiting to principal coordinates. We write system (7.1) in the matrix notation

$$\mathbf{A\ddot{q}} + \mathbf{B\dot{q}} + \mathbf{Cq} = \mathbf{Q}(t) \tag{7.9}$$

and represent its general solution as

$$\mathbf{q}(t) = \sum_{k=1}^{2s} C_k \mathbf{U}_k(t) + \mathbf{q}_*(t), \tag{7.10}$$

where $\mathbf{U}_k(t)$ are linearly independent partial solutions of the homogeneous system (7.9) constructed above, C_k are arbitrary constants, $\mathbf{q}_*(t)$ is a partial solution of the inhomogeneous matrix equation.

Let us construct a partial solution $\mathbf{q}_*(t)$ satisfying the zero initial conditions

$$\mathbf{q}_*(0) = \dot{\mathbf{q}}_*(0) = 0 \tag{7.11}$$

on the assumption that $|\mathbf{A}| \neq 0$. We shall not yet impose any restrictions on the constant matrices $\mathbf{B}$ and $\mathbf{C}$. The solution of the above problem can be represented as

$$\mathbf{q}_*(t) = \int_0^t \mathbf{G}(t - \tau) \mathbf{Q}(\tau) d\tau, \quad \mathbf{G}(t - \tau) = \{g_{ij}(t - \tau)\}_{i,j=1,\dots,s}. \tag{7.12}$$

Here $\mathbf{G}(t - \tau)$ is the *matrix of impulse transition functions*, and its elements $g_{ij}(t - \tau)$ are reaction forces of the ith generalized coordinate on the unit impulse applied instead of the jth generalized force. Formula (7.12) is a generalization of the well-known *Duhamel integral* to the multidimensional setting.

Consider an important partial case when external forces are periodic time functions

$$\mathbf{Q}(t) = \mathbf{Q}_0 e^{i\omega t}, \tag{7.13}$$

where $\mathbf{Q}_0$ is a given constant vector and ω is the disturbing frequency. We shall seek a periodic partial solution in the form

$$\mathbf{q}_*(t) = \mathbf{q}_0 e^{i\omega t} \,. \tag{7.14}$$

Substitution into Eq. (7.9) gives

$$\mathbf{q}_0 = \mathbf{S}(\omega)\mathbf{Q}_0, \quad \mathbf{S}(\omega) = (\mathbf{A}(i\omega)^2 + \mathbf{B}i\omega + \mathbf{C})^{-1} \,. \tag{7.15}$$

Here $\mathbf{S}(\omega)$ is the matrix of *transfer functions* of the system.

Solution (7.14) makes sense only in the case when free oscillations satisfying the homogeneous system (7.9) dies out with time. This will take place if all the three matrices $\mathbf{A}$, $\mathbf{B}$, and $\mathbf{C}$ are positive definite, which we shall assume. Solution (7.14), (7.15) is complex. To obtain real solutions, one should separate the real and imaginary parts. Consider the motion by the kth generalized coordinate

$$q_{*k}(t) = |q_{0k}(\omega)| \cos(\omega t + \alpha_k) \,,$$

where α_k is the argument of quantity q_{0k}, and

$$|q_{0k}(\omega)| = |\mathbf{S}(\omega)\mathbf{Q}_0|_k = |(\mathbf{A}(i\omega)^2 + \mathbf{B}i\omega + \mathbf{C})^{-1}\mathbf{Q}_0|_k = A_k(\omega) \tag{7.16}$$

gives the *amplitude-frequency characteristic* $A_k(\omega)$ for the kth generalized coordinate.

If the resistance is absent ($\mathbf{B} = 0$), then for the disturbing frequency ω coinciding with one of the frequencies of natural vibration $\omega = \omega_j$, we shall have $A_k(\omega) = \infty$. In the case of small resistance due to formula (7.16) around $\omega = \omega_j$ the function $A_k(\omega)$ will have a local maximum (this phenomenon is called *resonance*).

Relationship between the impulse transition function and the transfer function. On the assumption that matrices $\mathbf{A}$, $\mathbf{B}$ and $\mathbf{C}$ are positive definite, we consider solution (7.12) in the case of periodic disturbance (7.13). In contrast to (7.11) we shall set initial conditions at time $t = -\infty$. Then the initial disturbance is damped out and we get the periodic solution (7.14)

$$\int_{-\infty}^{t} \mathbf{G}(t - \tau)\mathbf{Q}_0 e^{i\omega\tau} d\tau = \mathbf{S}(\omega)\mathbf{Q}_0 e^{i\omega t} \,. \tag{7.17}$$

Due to arbitrariness of the vector $\mathbf{Q}_0$ this gives

$$\int_{-\infty}^{t} \mathbf{G}(t - \tau)e^{i\omega\tau} d\tau = \mathbf{S}(\omega)e^{i\omega t} \,. \tag{7.18}$$

Changing the variables $t - \tau = u$, we find that

$$\mathbf{S}(\omega) = \int_0^\infty \mathbf{G}(u)e^{-i\omega u}\,du\,. \tag{7.19}$$

The inverse Fourier transformation

$$\mathbf{G}(u) = \frac{1}{2\pi}\int_{-\infty}^\infty \mathbf{S}(\omega)e^{i\omega u}\,d\omega \tag{7.20}$$

can be used for calculation of the function $\mathbf{G}(u)$.

Chapter 8
Dynamics of the Rigid Solid

M. P. Yushkov

The momentum, angular momentum, kinetic energy, and the inertia tensor of a rigid solid are studied. The dynamic Euler equations are derived from the angular impulse-momentum principle. Transformations of the force systems applied to a rigid solid are investigated, the equations of statics are considered as a particular case of the dynamic equations. The Euler and Lagrange cases of rotation of a heavy rigid solid around a fixed point are presented. The elementary theory of gyroscopes is built on the basis of the results obtained for the pseudoregular precession of a heavy gyroscope. A new special form of differential equations for motion of a system of rigid solids which is comfortable for using numerical methods is given.

1 Dynamic Characteristics of a Rigid Solid

Density of the body. The momentum of a rigid solid. Suppose that we have a system of mass points, the distance between any two of which is constant. Such a system is called a *rigid*. When studying the body motion we shall consider the body as a rigid system, the points of which form a continuum. This model of a real body is usually called an *absolutely rigid body* or a *rigid solid*. As was shown in Chap. 2 this model makes it possible to specify the position of a body from six parameters, i.e., to assume that a body has six degrees of freedom. A perception about a rigid body as a continuum can be obtained by mentally increasing the number of points in a rigid system and assuming them to fill entirely some volume.

The mass M of a rigid solid is considered to be continuously distributed in the volume τ which is taken by the body. Therefore, in any elementary volume of a rigid solid $\Delta\tau$, there is a nonzero mass Δm. The above assumption allows one to introduce the concept of density of a body at a point A. For the mathematical description of this

© Springer Nature Switzerland AG 2021
N. N. Polyakhov et al., *Rational and Applied Mechanics*,
Foundations of Engineering Mechanics,
https://doi.org/10.1007/978-3-030-64061-3_8

concept, we mentally cut out from the body some volume $\Delta\tau$ with mass Δm around a point A and circumscribe some closed surface around it which does not go out from the body borders. Changing the shape of this surface in such a way that all points tend to coincide with the point A that always stays inside the surface, we consider the limit of the relation $\mu(x, y, z) = \lim\limits_{\Delta\tau \to 0} (\Delta m/\Delta\tau)$. We shall suppose that the given limit exists and is a function of the coordinates of a point A. The magnitude $\mu(x, y, z)$ is called the *density of a body* at a given point. Introducing the concept of density as a function of the coordinates of points in a solid, we are moving ourselves from considering the molecular structure of a substance. In practice, the value μ, which can be computed or measured, represents the average density in a neighborhood of A point A, the size of this neighborhood being small in comparison with that of the body and is large in comparison with the distance between molecules.

Let us study the motion of rigid solids, supposing that all the results established for systems of mass points are valid for it. Here, we mean: the principle of linear momentum

$$\mathbf{K} - \mathbf{K}_0 = \int\limits_{t_0}^{t} \mathbf{F} \, dt \,;$$

the angular impulse-momentum principle

$$d\mathbf{l}/dt = \mathbf{L} \,;$$

the work-energy principle

$$T - T_0 = A;$$

the principle of motion of the center of mass

$$M\frac{d\mathbf{v}_c}{dt} = \mathbf{F} \,.$$

The momentum of the system $\mathbf{K}$, its angular momentum $\mathbf{l}$ and its kinetic energy T, as well as the velocity of the center of mass $\mathbf{v}_c$, which are principal dynamic characteristics, appear in the statements of the above theorems. Hence, for using these theorems in the framework of a rigid solid, it is necessary, first of all, to investigate its principal dynamic characteristics.

For a system consisting of a finite number of points, we have

$$\mathbf{K} = \sum_{\nu} m_\nu \mathbf{v}_\nu \,. \tag{1.1}$$

When extending the concept of momentum to the case when we have continuously distributed masses, the sum in expression (1.1) should be replaced by the integral of the function $\mathbf{v}$ with respect to the measure dm,

$$\mathbf{K} = \int_\tau \mathbf{v} \, dm. \qquad (1.2)$$

The integral is taken over the volume τ, which is occupied by the bodies entering the system. For using expression (1.2) as applied to a concrete system of bodies, it is necessary to explain more in detail what we understand as a measure dm in the given case. First of all, note that a concept of the measure of a set is closely connected with a concept of distribution of masses in space. The mass dm is a characteristic measure of the element $d\tau$ in the three-dimensional Euclidean space. It is a matter of convenience to denote further the volume of the element considered by $dxdydz$.

If the element $d\tau$ contains a concentrated mass m_ν, then its measure is mass m_ν. Such a measure is used to be called the Dirac measure. If there are only concentrated masses in the whole volume τ then we write the integral in expression (1.2) as the sum in (1.1). If the mass is continuously distributed in the elementary volume $d\tau$ then we can write

$$dm = \mu \, d\tau = \mu \, dxdydz,$$

where μ is the density.

Therefore, if a continuous rigid solid takes the volume $d\tau$, then expression (1.2) transforms into a common three-dimensional integral

$$\mathbf{K} = \int_\tau \mathbf{v}\mu \, dxdydz. \qquad (1.3)$$

If it consists of masses that are continuously distributed over the surface S, then the elementary mass dm is represented as

$$dm = \mu_s \, dS,$$

where dS is the area element of the surface S; μ_S is *the surface density of mass distribution*. In the case under consideration the vector of the momentum is expressed by the surface integral:

$$\mathbf{K} = \int_S \mathbf{v}\mu_s \, dS. \qquad (1.4)$$

For instance, the momentum of a thin shell can be written in such a form.

The mass of a thin bar can be considered as distributed along the line. Then the measure dm can be written in the form

$$dm = \mu_l \, dl,$$

where dl is the linear element of the line l; μ_l is the linear density. In this case the vector $\mathbf{K}$ is written as a contour integral

$$\mathbf{K} = \int_l \mathbf{v}\mu_l \, dl \, . \tag{1.5}$$

Formulas (1.3)–(1.5) give a method for computing the momentum of the system $\mathbf{K}$ depending on the introduction of the concepts of mass, surface, and linear densities of mass distribution. This makes it possible to visualize the concept of a measure.

The vector of the momentum is closely connected with the concept of the center of mass of a system. We recall that if the number of concentrated masses is finite, then the radius-vector of the center of mass is determined by the formula

$$\mathbf{r}_c = \frac{\sum\limits_{\nu} \mu_{\nu} \mathbf{r}_{\nu}}{\sum\limits_{\nu} \mu_{\nu}} = \frac{\sum\limits_{\nu} \mu_{\nu} \mathbf{r}_{\nu}}{M} \, ,$$

where M is the mass of the whole system. When changing to systems containing continuously distributed masses, by analogy with above finite sums are transformed to the integrals

$$\mathbf{r}_c = \frac{\int\limits_{\tau} \mathbf{r} \, dm}{\int\limits_{\tau} dm} = \frac{\int\limits_{\tau} \mu \mathbf{r} \, d\tau}{\int\limits_{\tau} \mu \, d\tau} = \frac{\int\limits_{\tau} \mu \mathbf{r} \, d\tau}{M} \, . \tag{1.6}$$

This is why the vector of the system momentum can be also written as

$$\mathbf{K} = \int_{\tau} \mathbf{v} \, dm = M\mathbf{v}_c \, ,$$

where $\mathbf{v}_c = d\mathbf{r}_c/dt$ is the velocity of the center of mass.

The angular momentum and the kinetic energy of a rigid solid. It was established in kinematics of rigid solids that if we take a point O of a body for a pole, then the velocity $\mathbf{v}$ of any other its point N is connected with that of the pole $\mathbf{v}_0$ by the formula

$$\mathbf{v} = \mathbf{v}_0 + \boldsymbol{\omega} \times \mathbf{r} \, , \tag{1.7}$$

where $\mathbf{r}$ is the radius-vector drawn from the pole to the point N, $\boldsymbol{\omega}$ is the instant angular velocity (see Fig. 1).

The velocities $\mathbf{v}$ and $\mathbf{v}_0$ in formula (1.7) are velocities of two points of a body in the fixed frame of reference $O_1\xi\eta\zeta$. The velocity $\boldsymbol{\omega} \times \mathbf{r}$ is the velocity of the point N with respect to the reference system $O\xi'\eta'\zeta'$ which is moving translationally together with the pole.

Let us introduce one more reference system $Oxyz$, which is rigidly connected with the body. We denote the angular momentum and kinetic energy of a rigid solid in the reference systems $O_1\xi\eta\zeta$ and $O\xi'\eta'\zeta'$ as $\mathbf{l}_{o_1}$, T_{o_1} and $\mathbf{l}$, T, respectively.

If a rigid solid with density μ occupies volume τ, then we have

Fig. 1 Three systems of
coordinate for a rigid solid

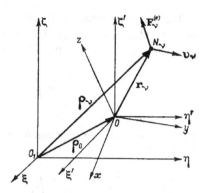

$$\mathbf{l}_{o_1} = \int_\tau \boldsymbol{\rho} \times (\mathbf{v}_0 + \boldsymbol{\omega} \times \mathbf{r})\, \mu d\tau \,,\quad \mathbf{l} = \int_\tau \mathbf{r} \times (\boldsymbol{\omega} \times \mathbf{r})\, \mu d\tau \,,$$

$$T_{o_1} = \tfrac{1}{2}\int_\tau (\mathbf{v}_0 + \boldsymbol{\omega} \times \mathbf{r})^2\, \mu d\tau \,,\quad T = \tfrac{1}{2}\int_\tau (\boldsymbol{\omega} \times \mathbf{r})^2\, \mu d\tau \,.$$

Here $\boldsymbol{\rho}$ is the radius-vector connecting the points O_1 and N.

We consider the motion of the reference system $O\xi'\eta'\zeta'$ as a bulk motion. As this takes place, T_{o_1} is the kinetic energy of the body in its absolute motion and T is its kinetic energy in the relative motion. An example of a relative motion of a rigid solid is its rotation around the point O. Since $(\mathbf{v}_0 + \boldsymbol{\omega} \times \mathbf{r})^2 = \mathbf{v}_0^2 + 2\,\mathbf{v}_0 \cdot (\boldsymbol{\omega} \times \mathbf{r}) + (\boldsymbol{\omega} \times \mathbf{r})^2$ and $\int_\tau \mathbf{r}\mu\, d\tau = M\mathbf{r}_c$, the magnitudes T_{o_1} and T are connected by the relation

$$T_{o_1} = \frac{1}{2}M(\mathbf{v}_0^2 + 2\,\mathbf{v}_0 \cdot (\boldsymbol{\omega} \times \mathbf{r}_c)) + T\,. \tag{1.8}$$

It is convenient to use formula (1.8), in particular, when computing the kinetic energy of manipulator links, whose motion is described on the whole by the Lagrange equations of the second kind.

Let us establish the connection between the vectors $\mathbf{l}_{o_1}$ and $\mathbf{l}$. Taking into account that $\boldsymbol{\rho} = \boldsymbol{\rho}_0 + \mathbf{r}$, $\boldsymbol{\rho}_0 = \overrightarrow{O_1 O}$ we can write

$$\mathbf{l}_{o_1} = \int_\tau (\boldsymbol{\rho}_0 + \mathbf{r}) \times \mu(\mathbf{v}_0 + \boldsymbol{\omega} \times \mathbf{r})\, d\tau = \int_\tau (\boldsymbol{\rho}_0 \times \mu \mathbf{v}_0)\, d\tau +$$

$$+ \int_\tau \boldsymbol{\rho}_0 \times (\boldsymbol{\omega} \times \mu \mathbf{r})\, d\tau + \int_\tau \mathbf{r} \times \mu \mathbf{v}_0\, d\tau + \int_\tau \mathbf{r} \times (\boldsymbol{\omega} \times \mu \mathbf{r})\, d\tau =$$

$$= \boldsymbol{\rho}_0 \times M\mathbf{v}_0 + [\boldsymbol{\rho}_0 \times (\boldsymbol{\omega} \times \int_\tau \mu \mathbf{r}\, d\tau)] + \int_\tau \mu \mathbf{r}\, d\tau \times \mathbf{v}_0 + \mathbf{l}\,.$$

In accordance with (1.6), for the radius vector of center of mass in the frame of reference $Oxyz$ we obtain

$$\mathbf{l}_{o_1} = \boldsymbol{\rho}_0 \times M\mathbf{v}_0 + \boldsymbol{\rho}_0 \times (\boldsymbol{\omega} \times M\mathbf{r}_c) + \mathbf{r}_c \times M\mathbf{v}_0 + \mathbf{l}\,.$$

The vector $\mathbf{l}$ represents the *angular momentum of a rigid solid with respect to the point O as it rotates around this point*. Using the known property of the double vector product, it can be written in the form

$$\mathbf{l} = \boldsymbol{\omega} \int_\tau r^2 \mu \, d\tau - \int_\tau \mathbf{r}(\mathbf{r} \cdot \boldsymbol{\omega}) \mu \, d\tau . \tag{1.9}$$

It follows from this formula that the projection of the vector $\mathbf{l}$ onto the direction $\boldsymbol{\omega}$ is expressed as

$$l_\omega = \omega \int_\tau r^2 \mu \, d\tau - \omega \int_\tau r^2 \cos^2 \alpha \mu \, d\tau = \omega \int_\tau p^2 \mu \, d\tau = J_\omega \omega , \tag{1.10}$$

where α is the angle between the vectors $\mathbf{r}$ and $\boldsymbol{\omega}$; $p = r \sin \alpha$ is the smallest distance from the point N to the axis $\boldsymbol{\omega}$. The magnitude $J_\omega = \int_\tau p^2 \mu \, d\tau$ is called the *moment of inertia of a rigid solid with respect to the axis with direction* $\boldsymbol{\omega}$.

Taking into account that $\omega^2 p^2 = (\boldsymbol{\omega} \times \mathbf{r})^2$, we have

$$T = \frac{\omega^2}{2} \int_\tau p^2 \mu \, d\tau = J_\omega \frac{\omega^2}{2} . \tag{1.11}$$

Using formulas (1.10) and (1.11), this gives

$$\mathbf{l} \cdot \boldsymbol{\omega} = J_\omega \omega^2 = 2T . \tag{1.12}$$

For the sake of simplicity, we suggest in what follows to sum over recurring indices and denote $x = x_1$, $y = x_2$, $z = x_3$, $\mathbf{i} = \mathbf{i}_1$, $\mathbf{j} = \mathbf{i}_2$, $\mathbf{k} = \mathbf{i}_3$, $\boldsymbol{\omega} \cdot \mathbf{i}_k = \omega_k$, $\mathbf{l} \cdot \mathbf{i}_k = l_k$, $k = 1, 2, 3$. Using (1.9), this gives

$$\begin{aligned} l_1 &= \omega_1 \int_\tau (x_1^2 + x_2^2 + x_3^2) \mu \, d\tau - \int_\tau \mu x_1 (\omega_1 x_1 + \omega_2 x_2 + \omega_3 x_3) \, d\tau = \\ &= J_{11} \omega_1 + J_{12} \omega_2 + J_{13} \omega_3 = J_{1k} \omega_k , \end{aligned} \tag{1.13}$$

where

$$J_{11} = \int_\tau \mu(x_2^2 + x_3^2) \, d\tau , \quad J_{12} = -\int_\tau \mu x_1 x_2 \, d\tau , \quad J_{13} = -\int_\tau \mu x_1 x_3 \, d\tau .$$

In a similar way we find

$$l_2 = J_{21} \omega_1 + J_{22} \omega_2 + J_{23} \omega_3 = J_{2k} \omega_k , \tag{1.14}$$

$$l_3 = J_{31} \omega_1 + J_{32} \omega_2 + J_{33} \omega_3 = J_{3k} \omega_k , \tag{1.15}$$

where

$$J_{22} = \int_\tau \mu(x_3^2 + x_1^2)\,d\tau\,, \qquad J_{21} = J_{12} = -\int_\tau \mu x_2 x_1\,d\tau\,,$$
$$J_{31} = J_{13} = -\int_\tau \mu x_3 x_1\,d\tau\,, \quad J_{32} = J_{23} = -\int_\tau \mu x_3 x_2\,d\tau\,, \qquad (1.16)$$
$$J_{33} = \int_\tau (x_1^2 + x_2^2)\mu\,d\tau\,.$$

When summing over recurring indice, formulas (1.13)–(1.15) can be written as

$$l_j = J_{jk}\,\omega_k\,, \quad j, k = 1, 2, 3\,. \qquad (1.17)$$

The magnitudes J_{11}, J_{22}, J_{33} are called *moments of inertia* of a rigid solid *with respect to the axes* x_1, x_2, x_3, respectively, and the minus magnitudes J_{12}, J_{23}, J_{31} sign are called *products of inertia*.

The tensor of inertia. If we introduce the symmetric matrix

$$\widehat{J} = (J_{jk}) = \begin{pmatrix} J_{11} & J_{12} & J_{13} \\ J_{21} & J_{22} & J_{23} \\ J_{31} & J_{32} & J_{33} \end{pmatrix}\,,$$

then the set of expressions (1.17) can represented in the form

$$\mathbf{l} = \widehat{J}\boldsymbol{\omega} = l_j \mathbf{i}_j = J_{jk}\,\omega_k \mathbf{i}_j\,. \qquad (1.18)$$

In accordance with formulas (1.12) and (1.17) the doubled kinetic energy reads as

$$2T = \mathbf{l} \cdot \boldsymbol{\omega} = \widehat{J}\boldsymbol{\omega} \cdot \boldsymbol{\omega} = J_{jk}\,\omega_j \omega_k =$$
$$= J_{11}\,\omega_1^2 + J_{22}\,\omega_2^2 + J_{33}\,\omega_3^2 + 2J_{12}\,\omega_1\omega_2 + 2J_{23}\,\omega_2\omega_3 + 2J_{31}\,\omega_3\omega_1\,. \qquad (1.19)$$

If we change from the frame of reference $Ox_1x_2x_3$ to the frame of reference $Ox_1'x_2'x_3'$ then we have in the latter

$$l_m' = J_{mn}'\,\omega_n'\,,$$
$$2T' = J_{mn}'\,\omega_m'\,\omega_n' = J_{jk}\omega_j\omega_k = 2T = \text{inv}\,. \qquad (1.20)$$

The magnitudes ω_j and ω_k are connected here with ω_m' and ω_n' by the transition formulas

$$\omega_j = \omega_m' \cos\alpha_{jm}\,, \qquad \omega_k = \omega_n' \cos\alpha_{kn}\,, \qquad (1.21)$$

where α_{jm} is the angle between the axes x_j and x_m'. Formulas (1.20) and (1.21) imply that when changing from the frame of reference $Ox_1x_2x_3$ to the frame $Ox_1'x_2'x_3'$ the elements of matrix $\widehat{J}$ are transformed by formulas

$$J_{mn}' = J_{jk} \cos\alpha_{jm} \cos\alpha_{kn}\,. \qquad (1.22)$$

In the general case, the matrix (a_{ij}), whose the elements are transformed by formulas

$$a'_{mn} = a_{jk} \cos \alpha_{jm} \cos \alpha_{kn},$$

when changing from one orthogonal frame of reference to another, is called a *second-order tensor*. It follows from formula (1.22) that the symmetric matrix $\widehat{J}$ is a second-order tensor. It is called *the inertia tensor*.

From the point of view of linear operators, the relation $\mathbf{l} = \widehat{J}\boldsymbol{\omega}$ establishes the correspondence between a vector $\boldsymbol{\omega}$ of the Euclidean space $\mathbb{R}^3$ and the vector $\mathbf{l}$ of the same space. Hence, the tensor $\widehat{J}$ is a linear operator.

Let us explain the physical sense of the inertia tensor. For simplicity, we shall assume that an arbitrary point O in the body is fixed. Suppose that the body rotates with angular velocity $\boldsymbol{\omega}$ around an axis which goes through the given point. The vector $\mathbf{l}$, corresponding to the rotation around this axis, is given by formula (1.18). It is worth noting that in spite of the arbitrariness of the direction of the vector $\boldsymbol{\omega}$, for calculation of the vector $\mathbf{l}$ it is enough to know only six parameters J_{jk}, which are called the *components of the tensor of inertia* in the chosen frame of reference $Ox_1x_2x_3$.

In a rotation of a rigid solid around a fixed point, the field of velocities of all points in the solid is entirely determined by setting the vector of instant angular velocity $\boldsymbol{\omega}$. Substituting expression (1.18) into the angular impulse-momentum principle, we obtain

$$\frac{d}{dt}(\widehat{J}\boldsymbol{\omega}) = \boldsymbol{L}.$$

Comparing this formula with the standard formula from the Newton second law, we can see that in this case the motion is characterized by the vector $\boldsymbol{\omega}$, the inertia tensor playing the role of the measure of inertia.

Principal axes representation of inertia tensor. The ellipsoid of inertia. We can see from formula (1.18) that the vectors $\boldsymbol{\omega}$ and $\mathbf{l}$ are not collinear in general. The natural question is under which conditions the above vectors lie on one line. In this case we have

$$\widehat{J}\boldsymbol{\omega} = \lambda\boldsymbol{\omega}, \tag{1.23}$$

where λ is an unknown scalar multiplier. The above equation can be also written in the form

$$(\widehat{J} - \lambda\widehat{E})\boldsymbol{\omega} = 0,$$

where $\widehat{E} = \begin{pmatrix} 1 & 0 & 0 \\ 0 & 1 & 0 \\ 0 & 0 & 1 \end{pmatrix}$ is a unit matrix.

In projections onto the axes of coordinates, the vector relation (1.23) reads as

$$J_{jk}\omega_k - \lambda\omega_j = 0, \quad j = 1, 2, 3. \tag{1.24}$$

Note that solution to the homogeneous linear algebraic system of equations (1.24) can be found with accuracy up to an arbitrary constant multiplier. Therefore the solutions of this system do not determine the angular velocities themselves, which satisfy condition (1.23), but are responsible for the directions of the rotation axes.

The necessary and sufficient condition of the existence of nonzero solutions of system (1.24) is that the determinant be zero:

$$|\widehat{J} - \lambda \widehat{E}| = \begin{vmatrix} J_{11} - \lambda & J_{12} & J_{13} \\ J_{21} & J_{22} - \lambda & J_{23} \\ J_{31} & J_{32} & J_{33} - \lambda \end{vmatrix} = 0 .$$

Calculating the determinant, we obtain the characteristic equation

$$\lambda^3 + A_1\lambda^2 + A_2\lambda + A_3 = 0 ,$$

where A_1, A_2, A_3 are real parameters.

Suppose that all roots of the characteristic equation are different; let λ_ν, $\nu = 1, 2, 3$ be one of them. Substituting this root into Eq. (1.23), we find $\widehat{J}\boldsymbol{\omega}_\nu = \lambda_\nu\boldsymbol{\omega}_\nu$. Multiplying both sides of the given relation by $\boldsymbol{\omega}_\nu$, we have by (1.19)

$$\lambda_\nu |\boldsymbol{\omega}_\nu|^2 = \widehat{J}\boldsymbol{\omega}_\nu \cdot \boldsymbol{\omega}_\nu = 2T .$$

Since $2T$ and $|\boldsymbol{\omega}_\nu|^2$ are real magnitudes, the root $\lambda_\nu > 0$ is also real.

We claim that if λ_ν and λ_μ are two different roots, then the corresponding vectors $\boldsymbol{\omega}_\nu$ and $\boldsymbol{\omega}_\mu$ are orthogonal. Indeed, since $\widehat{J}\boldsymbol{\omega}_\nu = \lambda_\nu\boldsymbol{\omega}_\nu$ and $\widehat{J}\boldsymbol{\omega}_\mu = \lambda_\mu\boldsymbol{\omega}_\mu$, we have

$$\widehat{J}\boldsymbol{\omega}_\mu \cdot \boldsymbol{\omega}_\nu - \widehat{J}\boldsymbol{\omega}_\nu \cdot \boldsymbol{\omega}_\mu = (\lambda_\mu - \lambda_\nu)(\boldsymbol{\omega}_\mu \cdot \boldsymbol{\omega}_\nu) . \tag{1.25}$$

Setting $\boldsymbol{\omega}_\mu = \omega_{\mu k} \mathbf{i}_k$, we have by (1.18) $\widehat{J}\boldsymbol{\omega}_\mu = J_{jk}\omega_{\mu k} \mathbf{i}_j$. This implies that

$$\widehat{J}\boldsymbol{\omega}_\mu \cdot \boldsymbol{\omega}_\nu = J_{jk}\omega_{\mu k}\omega_{\nu j} , \quad \widehat{J}\boldsymbol{\omega}_\nu \cdot \boldsymbol{\omega}_\mu = J_{jk}\omega_{\nu k}\omega_{\mu j} .$$

Changing the indices of summation j and k in the first sum and taking into account that $J_{jk} = J_{kj}$, we find

$$\widehat{J}\boldsymbol{\omega}_\mu \cdot \boldsymbol{\omega}_\nu = J_{jk}\omega_{\mu j}\omega_{\nu k} = J_{jk}\omega_{\nu k}\omega_{\mu j} = \widehat{J}\boldsymbol{\omega}_\nu \cdot \boldsymbol{\omega}_\mu .$$

Therefore, it follows from equality (1.25) that

$$(\lambda_\mu - \lambda_\nu)(\boldsymbol{\omega}_\mu \cdot \boldsymbol{\omega}_\nu) = 0 . \tag{1.26}$$

Since by the assumption $\lambda_\mu \neq \lambda_\nu$ and all the roots λ_ν are different, the vectors $\boldsymbol{\omega}_\nu$ are orthogonal.

If $\lambda_1 = \lambda_2 = \lambda_3$ we have $\boldsymbol{\omega}_1 \cdot \boldsymbol{\omega}_3 = 0$ and $\boldsymbol{\omega}_2 \cdot \boldsymbol{\omega}_3 = 0$ by (1.26). But the inner product $\boldsymbol{\omega}_1 \cdot \boldsymbol{\omega}_2$ can be any (zero or nonzero). This means that $\boldsymbol{\omega}_1$ and $\boldsymbol{\omega}_2$ are two arbitrary vectors which are orthogonal to the third vector $\boldsymbol{\omega}_3$. Let us choose them to be orthogonal to each other; that is, we suppose that $\boldsymbol{\omega}_1 \cdot \boldsymbol{\omega}_2 = 0$. If $\lambda_1 = \lambda_2 = \lambda_3$, then all three vectors $\boldsymbol{\omega}_\nu$ can be arbitrary (either orthogonal or not). We shall assume that they are orthogonal.

From the frame of reference $Ox_1x_2x_3$ we change to a new one $Ox_1'x_2'x_3'$, which is also body-fixed. Suppose that its basis vectors $\mathbf{i}_\nu'$ are collinear to the vectors $\boldsymbol{\omega}_\nu$. The axes of this frame of reference are called *principal inertia axes* of the body at a given point O. If the point O coincides with the center of mass, then the axes are called *principal central inertia axes*.

The basis vectors $\mathbf{i}_j'$ satisfy the equations

$$\widehat{J}\mathbf{i}_j' = \lambda_j \mathbf{i}_j', \quad j = 1, 2, 3,$$

and hence in accordance with formula (1.18) we have

$$\mathbf{l} = J_{jk}'\omega_k'\mathbf{i}_j' = \widehat{J}(\omega_j'\mathbf{i}_j') = \omega_j'\widehat{J}\mathbf{i}_j' = \lambda_j\omega_j'\mathbf{i}_j'.$$

Hence $J_{jk}' = 0$ for $j \neq k$, $J_{jj}' = \lambda_j$, and therefore, the inertia tensor in principal axes reads as

$$\widehat{J} = \begin{pmatrix} \lambda_1 & 0 & 0 \\ 0 & \lambda_2 & 0 \\ 0 & 0 & \lambda_3 \end{pmatrix}.$$

Note that the diagonalization of the inertia tensor is closely connected with the well-known algebraic problem of finding eigenvalues and eigenvectors of a symmetric matrix. In our case, the symmetric matrix is composed of components of the inertia tensor, and its eigenvalues and eigenvectors are λ_ν and $\boldsymbol{\omega}_\nu$, respectively, $\nu = 1, 2, 3$.

The initial equations (1.24), which make it possible to determine the values λ_ν and vectors $\boldsymbol{\omega}_\nu$, $\nu = 1, 2, 3$, were obtained earlier from the condition of collinearity of the vectors $\mathbf{l}$ and $\boldsymbol{\omega}$. Let us show that these equations reflect extreme properties of the kinetic energy $T = (1/2)J_{jk}\omega_j\omega_k$.

We shall consider the angular velocity $\boldsymbol{\omega}$ of constant absolute value. Let us establish in what directions of the vector $\boldsymbol{\omega}$, the magnitude T has extreme values; i.e., we find the extrema of the function T under the condition $2\widetilde{T} = \omega^2 = \omega_1^2 + \omega_2^2 + \omega_3^2 =$ const.

It is known that equations making it possible to solve the given problem of conditional extremum have the form

$$\frac{\partial(T - \lambda\widetilde{T})}{\partial\omega_j} = 0, \quad j = 1, 2, 3,$$

where λ is a Lagrange multiplier. These equations coincide with system (1.24), since

$$\frac{\partial T}{\partial \omega_j} = J_{jk}\omega_k, \quad \text{and} \quad \frac{\partial \tilde{T}}{\partial \omega_j} = \omega_j.$$

It follows from formula (1.11) that $2T = J_\omega \omega^2$. By the assumption $\omega^2 = \text{const}$, and therefore, in the directions $\boldsymbol{\omega}_\nu$ the magnitude J_ω has its extreme values, which are equal to λ_ν, respectively. According to formulas (1.12), (1.19) we have

$$J_\omega \omega^2 = J_{jk}\omega_j\omega_k,$$

which implies that

$$J_\omega = J_{jk}\cos\beta_j\cos\beta_k, \tag{1.27}$$

where $\cos\beta_j = \omega_j/\omega$, $j = 1, 2, 3$ are the directing cosines of the axis, with respect to which the inertia moment J_ω is calculated. Along this axis, we introduce the vector

$$\mathbf{x} = \frac{\cos\beta_j}{\sqrt{J_\omega}}\mathbf{i}_j = x_j\mathbf{i}_j$$

with the length $1/\sqrt{J_\omega}$. By (1.27) the end-point of this vector described the surface with the equation

$$J_{jk}x_j x_k = 1. \tag{1.28}$$

This surface is an ellipsoid, because $|\mathbf{x}| = 1/\sqrt{J_\omega} \neq \infty$. From the distance from the ellipsoid center to its surface we determine the inertia moment $J_\omega = 1/\mathbf{x}^2$. Hence the given surface, which characterizes the distribution of inertia moments with respect to the axes passing through the point O, is called the *ellipsoid of inertia*.

Equation (1.28) in the frame of reference $Ox_1'x_2'x_3'$, the axes of which are principal inertia axes, has the form

$$\lambda_1(x_1')^2 + \lambda_2(x_2')^2 + \lambda_3(x_3')^2 = 1.$$

If among the roots of the characteristic equation there are multiple roots, for example, $\lambda_1 = \lambda_2$, then the ellipsoid of inertia is an ellipsoid of revolution. If $\lambda_1 = \lambda_2 = \lambda_3$, then the ellipsoid of inertia is a sphere.

The ellipsoid of inertia built for the center of mass of a body is called *central*. Let us show that it is larger than the ellipsoids built for other points of a rigid solid. For this purpose, let us compare the moments of inertia with respect to two parallel axes, one of them passes through the point C, and another one passes through an arbitrary point O. We introduce two coordinate systems $Oxyz$ and $Cx'y'z'$, with parallel z- and z'-axes. By definition

$$J_{zz} = \int_\tau (x^2 + y^2)\mu\,d\tau, \quad J_{z'z'} = \int_\tau [(x')^2 + (y')^2]\mu\,d\tau.$$

Fig. 2 Huygens–Steiner theorem

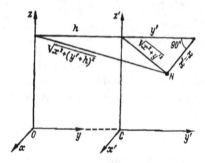

The inertia moment J_{zz} is independent of the choice of the axes x and y and of the position of a point O on the axis z, Hence, for the sake of simplicity, we shall consider the axes y and y' to lie on one line, and assume that $y = y' + h$, where h is the distance between the axes z and z' (Fig. 2). For the given choice of coordinate axes the moment of inertia J_{zz} can be represented as

$$J_{zz} = \int_\tau [x^2 + (y' + h)^2] \mu \, d\tau = J_{z'z'} + h^2 \int_\tau \mu \, d\tau + 2h \int_\tau y' \mu \, d\tau . \qquad (1.29)$$

However, the integral $\int_\tau \mu \, d\tau = M$, while the integral $\int_\tau y' \mu \, d\tau = M y_c' = 0$ vanishes, because the origin of coordinates coincides with the center of mass. Therefore, formula (1.29) assumes the form

$$J_{zz} = J_{z'z'} + Mh^2 . \qquad (1.30)$$

This equality is an analytic expression of the Huygens–Steiner theorem.

It follows from (1.30) that

$$|\mathbf{x}| = 1/\sqrt{J_{zz}} < 1/\sqrt{J_{z'z'}} .$$

Fig. 3 Ellipsoids of inertia

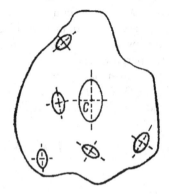

Therefore, the central ellipsoid of inertia is always larger than any other possible ellipsoids of inertia, built for a given body (Fig. 3).

2 Differential Equations of Motion of a Free Rigid Solid

The position of a rigid solid with respect to a fixed reference system $O_1 \xi \eta \zeta$ (see Fig. 1) is determined by setting the coordinates of the pole O and the Euler angles ψ, θ, φ, which characterize the orientation of the body-fixed axes x, y, z with respect to the fixed axes ξ, η, ζ. If the motion of a body is subject to no constraints, i.e., when it moves free (unconstrained), then those six parameters can assume any values. Taking the generalized Lagrange coordinates for these parameters and expressing with the help of them the kinetic and potential energies and the generalized forces, we arrive at six Lagrange equations of the second kind. Such a system of six second-order equations is equivalent to a system of twelve first-order equations. In particular, if a rigid solid moves under the action of potential forces, then these twelve equations can be written in the system of canonical equations (6.10) of Chap. 4. For a rigid solid, this way of composing the equations of motion proves to be more difficult, than immediately composition of such equations from the principle of motion of the center of mass and the angular impulse-momentum principle

$$M \frac{dv_c}{dt} = \sum_{\nu=1}^{n} \mathbf{F}_\nu^{(e)} \equiv \mathbf{F}, \qquad \frac{d\mathbf{l}_{O_1}}{dt} = \sum_{\nu=1}^{n} \boldsymbol{\rho}_\nu \times \mathbf{F}_\nu^{(e)} \equiv \mathbf{L}_{O_1}, \tag{2.1}$$

where n is the number of forces acting on a body. Each of these vector-valued equations reduces to three scalar second-order equations. It can be shown that the given six scalar equations are equivalent to the system of Lagrange equations of the second kind.

Denoting by ξ, η, ζ the coordinates of the center of mass with respect to a fixed reference system, we obtain in accordance with the principle of motion of the center of mass

$$M \ddot{\xi} = F_\xi, \quad M \ddot{\eta} = F_\eta, \quad M \ddot{\zeta} = F_\zeta.$$

These three equations are equivalent to the six first-order equations

$$\dot{\xi} = v_\xi, \qquad \dot{\eta} = v_\eta, \qquad \dot{\zeta} = v_\zeta, \\ \dot{v}_\xi = F_\xi/M, \ \dot{v}_\eta = F_\eta/M, \ \dot{v}_\zeta = F_\zeta/M. \tag{2.2}$$

In this case, it is expedient to write down the angular impulse-momentum principle with respect to the reference system $C\xi'\eta'\zeta'$, which moves translationally together with the center of mass. By formula (3.9) of Chap. 5, the angular impulse-momentum principle in the context of a rigid solid can be written as

$$\frac{d\mathbf{l}}{dt} = \mathbf{L}, \quad \mathbf{L} = \sum_{\nu=1}^{n} \mathbf{r}_\nu \times \mathbf{F}_\nu^{(e)}. \tag{2.3}$$

We assume that axes of the body-fixed coordinate system $Oxyz$ are principal axes of inertia. Then the vector $\mathbf{l}$ can be represented as

$$\mathbf{l} = Ap\,\mathbf{i} + Bq\,\mathbf{j} + Cr\,\mathbf{k}, \tag{2.4}$$

where

$$J_{11} = J_{xx} = A, \ J_{22} = J_{yy} = B, \ J_{33} = J_{zz} = C,$$
$$\omega_1 = \omega_x = p, \quad \omega_2 = \omega_y = q, \quad \omega_3 = \omega_z = r.$$

The possibility of writing the vector $\mathbf{l}$ in principal axes in a simple form of (2.4) indicates that it is expedient to project the derivative $d\mathbf{l}/dt$ onto the moving principal axes x, y, z. According to formula (1.7) of Chap. 3, we have

$$\frac{d\mathbf{l}}{dt} = \frac{d^*\mathbf{l}}{dt} + \boldsymbol{\omega} \times \mathbf{l}, \quad \frac{d^*\mathbf{l}}{dt} = A\dot{p}\,\mathbf{i} + B\dot{q}\,\mathbf{j} + C\dot{r}\,\mathbf{k}. \tag{2.5}$$

This shows that Eq. (2.3) can be written in projections onto the principal axes x, y, z in the form

$$\begin{aligned}
A\dot{p} + (C - B)qr &= L_x, \\
B\dot{q} + (A - C)rp &= L_y, \\
C\dot{r} + (B - A)pq &= L_z.
\end{aligned} \tag{2.6}$$

These equations are called the *dynamic Euler equations*. According to formula (4.6) of Chap. 2, the projections p, q, r of the angular velocity ω are connected with the Euler angles by the formulas

$$\begin{aligned}
p &= \dot{\psi} \sin\theta \sin\varphi + \dot{\theta} \cos\varphi, \\
q &= \dot{\psi} \sin\theta \cos\varphi - \dot{\theta} \sin\varphi, \\
r &= \dot{\psi} \cos\theta + \dot{\varphi}.
\end{aligned} \tag{2.7}$$

As a result, we have

$$\begin{aligned}
\dot{\psi} &= (p \sin\varphi + q \cos\varphi)/\sin\theta = f_1(\psi, \theta, \varphi, p, q, r), \\
\dot{\varphi} &= r - \cot\theta(p \sin\varphi + q \cos\varphi) = f_2(\psi, \theta, \varphi, p, q, r), \\
\dot{\theta} &= p \cos\varphi - q \sin\varphi = f_3(\psi, \theta, \varphi, p, q, r).
\end{aligned} \tag{2.8}$$

Relations (2.7) are called the *kinematic Euler equations*.

The system of equations (2.2), (2.6), and (2.8) represents twelve differential first-order equations solved with respect to the derivatives $\dot{p}, \dot{q}, \dot{r}, \dot{v}_\xi, \dot{v}_\eta, \dot{v}_\zeta, \dot{\xi}, \dot{\eta}, \dot{\zeta}, \dot{\psi}, \dot{\varphi},$ and $\dot{\theta}$. This system can be integrated in general only numerically.

3   Transformation of Force Systems Applied to a Perfectly Rigid Solid[1]

Equations of motion of a free rigid solid (2.1) do not change if we translate the forces $\mathbf{F}_\nu^{(e)}$ along lines of their action, because in this case the resultant vector $\mathbf{F}$ and the resultant moment $\mathbf{L}_{O_1}$ remain constant. The vectors that can be translated along lines of their action without any effect on the body motion are called *sliding vectors*. Force systems are called *equivalent* if they are characterized by the same resultant vector $\mathbf{F}$ and the resultant moment $\mathbf{L}_{O_1}$.

It should be specially noted that a force can be considered as a sliding vector only when the strains which can appear in a real rigid solid as a result of force system action are not taken into consideration. For example, a train can be accelerated by placing the locomotive ahead and behind the train, because the translation of the force along the line of action does not change the character of motion of the system. However, its inner state depends on the point of force action: if the locomotive is ahead of the train, it causes tension of the latter, but if we locate it behind, the train squeezes from the back.

Consider a translation of the given force $\mathbf{F}_\nu$ which goes through the point N_ν to another point of the space M_ν (Fig. 4a). In so doing, for getting an equivalent force system at the point M_ν, we should apply an additional force system $\mathbf{F}_\nu$ and $(-\mathbf{F}_\nu)$, which is equivalent to zero. The new system of three forces is the system equivalent to the initial force $\mathbf{F}_\nu$ applied to the point N_ν, as it has the same resultant vector $\mathbf{F}_\nu$ and the resultant moment $\boldsymbol{\rho}_\nu \times \mathbf{F}_\nu$ with respect to the point O_1. Now we can assume that the force $\mathbf{F}_\nu$ is translated from the point N_ν to the point M_ν, however as this takes place, an additional force system appears consisting of two parallel and reverse directed vectors $\mathbf{F}_\nu$ and $(-\mathbf{F}_\nu)$, whose action lines $A_\nu B_\nu$ and $C_\nu D_\nu$ do not coincide. From now on, we shall call such a system of two forces a *force couple*.

Thus, the force couple $(\mathbf{F}_\nu, -\mathbf{F}_\nu)$ appears as the force $\mathbf{F}_\nu$ is translated from the point N_ν to the point M_ν.

Let us analyze some properties of the couple $(\mathbf{F}_\nu, -\mathbf{F}_\nu)$. The plane containing the vectors composing the couple is called the *couple plane*. The distance p_ν between the action lines of the forces of the couple is called the *arm of the couple* (Fig. 4b). The resultant vector of the force couple is equal to zero. The resultant moment of the couple does not depend on the choice of the point O_1. Indeed, from Fig. 4b it follows that

$$\overrightarrow{O_1 N_\nu} \times \mathbf{F}_\nu + \overrightarrow{O_1 M_\nu} \times (-\mathbf{F}_\nu) = (\overrightarrow{O_1 N_\nu} - \overrightarrow{O_1 M_\nu}) \times \mathbf{F}_\nu =$$
$$= \overrightarrow{M_\nu N_\nu} \times \mathbf{F}_\nu . \tag{3.1}$$

This implies that the resultant moment is a free vector. It is called the *moment (torque) of a couple*.

---

[1] More compactly these questions are considered in Chap. 10.

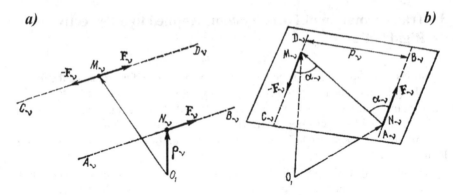

Fig. 4 Translation of a force and a force couple

Therefore, we can assertion that without disturbing the manner of the motion of a rigid solid we can translate the force $\mathbf{F}_\nu$ from the point of its application N_ν to the point M_ν by adding to this force a force couple whose moment is equal to the moment of the initial force $\mathbf{F}_\nu$ with respect to the new point M_ν.

The magnitude of the moment of a couple can be represented by the expression (see Fig. 4b)

$$F_\nu |\overrightarrow{M_\nu N_\nu}| \sin \alpha_\nu = F_\nu p_\nu ; \tag{3.2}$$

the direction of the moment is specified from the vector product (3.1). By formula (3.1) the moment of a couple is perpendicular to the couple plane on the side where the rotation caused by the couple is counterclockwise. In other words, the right-handed screw rule is used for finding the direction of the moment of a couple.

Note that the forces from a couple can be not only translated along their action lines, but also rotated with a change of their modulus F_ν and their arm p_ν in such a way that the magnitude (3.2) and the direction of the moment of the couple do not change. Besides, we can translate the couple plane parallel to itself. In any of these cases, we have the same resultant moment and the zero resultant. Consequently, we have the rule of couple summation, according to which the sum of the couples is a single couple whose moment vector is equal to the geometrical sum of the moments of the couples.

Let us consider the possibility of replacing a given force system $\mathbf{F}_1, \mathbf{F}_2, \ldots, \mathbf{F}_n$ by simplest equivalent systems.

If the action lines of the indicated force intersect at one point O, then the resultant moment with respect to this point is equal to zero and the system is equivalent to the resultant force $\mathbf{F}$. In this case, a force system is said to be reduced to the pure force (torque-free) resultant. Its action line passes through the point O.

If a force system $\mathbf{F}_1, \mathbf{F}_2, \ldots, \mathbf{F}_n$ is located in space in an arbitrary way, then by choosing some pole O we can translate all the forces to this point, therewith n couples appearing (Fig. 5a). As a result, the initial force system turns out to be equivalent to the resultant vector $\mathbf{F}$, which represents the geometric sum of all forces translated to

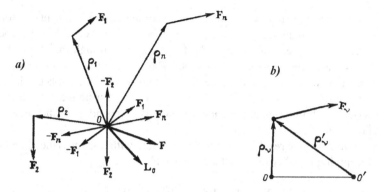

Fig. 5 Composition of a space force system

the pole O, and to the resultant moment $\mathbf{L}_o$, which is equal to a geometric sum of the moments of all appearing couples

$$\mathbf{F} = \sum_{\nu=1}^{n} \mathbf{F}_\nu , \tag{3.3}$$

$$\mathbf{L}_o = \sum_{\nu=1}^{n} \boldsymbol{\rho}_\nu \times \mathbf{F}_\nu . \tag{3.4}$$

Therefore, a force system located in space in arbitrary way is equivalent to a single force $\mathbf{F}$ the action line of which passes through the point O, and to a couple the moment of which is equal to $\mathbf{L}_o$. This transformation is called the *reduction of a force system to a single center*.

Let us find the effect of variation of the reduction center on the vectors $\mathbf{F}$ and $\mathbf{L}_o$. For the pole we take a point O' instead of O. We have $\mathbf{F}' = \sum_{\nu=1}^{n} \mathbf{F}_\nu$, which after comparison with (3.3) allows us to assert that the resultant of the force system does not change under variation of the reduction center.

For the resultant moment of a force system with respect to the point O' we have (Fig. 5b)

$$\mathbf{L}_{o'} = \sum_{\nu=1}^{n} \boldsymbol{\rho}'_\nu \times \mathbf{F}_\nu = \sum_{\nu=1}^{n} (\boldsymbol{\rho}_\nu + \overrightarrow{O'O}) \times \mathbf{F}_\nu =$$
$$= \sum_{\nu=1}^{n} \boldsymbol{\rho}_\nu \times \mathbf{F}_\nu + \overrightarrow{O'O} \times \sum_{\nu=1}^{n} \mathbf{F}_\nu ,$$

which, in accordance with formulas (3.3) and (3.4) implies that

$$\mathbf{L}_{o'} = \mathbf{L}_o + \overrightarrow{O'O} \times \mathbf{F} . \tag{3.5}$$

It follows that when choosing a new reduction center the resultant moment of a force system changes by the magnitude of the moment of the resultant vector **F** with respect to the new center O', the action line of the force passing through the point O. In particular, this means that if the resultant vector **F** is zero, then $\mathbf{L}_o = \mathbf{L}_{o'} = \mathbf{L}$, i.e., the resultant moment of the system is the same for all points of the space. The last statement also follows from the fact that when $\mathbf{F} = 0$ a force system can be reduced to a force couple, whose moment, as shown before, is independent of the choice of the reduction point, i.e., it is a free vector.

We have found that the resultant vector **F** is invariant with respect to the reduction center. Multiplying equality (3.5) by basis vector $\mathbf{F}^0$ we get

$$\mathbf{L}_{o'} \cdot \mathbf{F}^0 = \mathbf{L}_o \cdot \mathbf{F}^0, \qquad \mathbf{F}^0 = \mathbf{F}/|\mathbf{F}|,$$

because $(\overrightarrow{O'O} \times \mathbf{F}) \cdot \mathbf{F}^0 = 0$. Hence, the second invariant of the force system with respect to the reduction center is the projection of the resultant moment onto the direction of resultant vector.

Existence of the second invariant of a force system means that the resultant moment has the least modulus when its direction coincides with the resultant vector. Let us find the geometrical locus of reduction points which have this property.

In Sect. 3 of Chap. 3, we considered the summation of motions of a rigid solid. It was shown that the total angular velocity $\mathbf{\Omega} = \sum_{k=1}^{n} \boldsymbol{\omega}_k$ is independent of choice of a pole and that the vector **V**, which characterizes the resulting translational motion of a rigid solid, is determined from the choice of a pole. This dependence is expressed by formula (3.6) of Chap. 3. Comparing this with formula (3.5), we can see that a full analogy exists between the vectors $\mathbf{\Omega}, \mathbf{V}$ and $\mathbf{F}, \mathbf{L}_o$. Therefore, the corresponding material will be presented here briefly.

Suppose that at the point O the angle between the resultant vector **F** and the resultant moment $\mathbf{L}_o$ is α (Fig. 6). We represent the vector $\mathbf{L}_o$ as a sum of two mutually perpendicular components: $\mathbf{L}_o = \mathbf{L}'_o + \mathbf{L}''_o$, in which $\mathbf{L}'_o$ is collinear to the

Fig. 6 Wrench

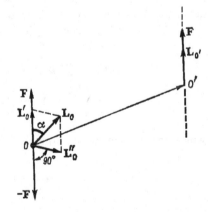

resultant vector $\mathbf{F}$. We construct the point O' with the position vector $\overrightarrow{OO'}$ satisfying the relation

$$\mathbf{L}''_o = \overrightarrow{OO'} \times \mathbf{F}. \tag{3.6}$$

If one now takes the point O' as a new reduction center, then the force system reduces to the resultant vector $\mathbf{F}$, which is invariant, and to the resultant moment, which is given by formula (3.5):

$$\mathbf{L}_{O'} = \mathbf{L}'_o + \mathbf{L}''_o + \overrightarrow{O'O} \times \mathbf{F}.$$

Hence, we get from (3.6)

$$\mathbf{L}_{O'} = \mathbf{L}'_o.$$

However, by construction the vector $\mathbf{L}'_o$ is collinear to the vector $\mathbf{F}$; this means that we found such a point of space for which the resultant moment and the resultant vector are collinear. Such a set of the resultant vector and the resultant moment is referred to as a *wrench*. It is obvious that the force system $\mathbf{F}_1, \mathbf{F}_2, \ldots, \mathbf{F}_n$ is reduced to a wrench if one locates the reduction points along the line passing through the point O' and parallely to the vector $\mathbf{F}$. This line is called the *central axis of a force system*. By formula (4.7) of Chap. 3 the equation for this axis assumes the form

$$\frac{L_x + F_y z - F_z y}{F_x} = \frac{L_y + F_z x - F_x z}{F_y} = \frac{L_z + F_x y - F_y x}{F_z}. \tag{3.7}$$

If the vectors $\mathbf{F}$ and $\mathbf{L}_o$ are orthogonal, then when reducing a given force system to the point O' the resultant moment is equal to zero. Hence, in this case Eq. (3.7) is the equation of a line along which the torque-free resultant of the initial force system is applied.

4 Equations of Statics of a Rigid Solid[2]

Let us suppose that a rigid solid is at rest or at a state of equilibrium if the velocities of all its points are identically zero during some time interval, that is,

$$\mathbf{v} \equiv 0. \tag{4.1}$$

Let us show for an equilibrium of a rigid solid it is necessary and sufficient that the resultant vector and resultant moment with respect to its center of mass be zero in this time interval and that the velocity of the center of mass and the angular velocity be zero at the initial moment, that is,

---

[2] A more detailed account of equations of statics is given in Chap. 10.

$$\mathbf{F} \equiv 0, \quad \mathbf{L} \equiv 0, \quad \mathbf{v}_c(0) = 0, \quad \boldsymbol{\omega}(0) = 0. \tag{4.2}$$

Consider the necessity of these conditions. In other words, assuming that the velocities of all the points of a rigid solid are all zero, let us verify equalities (4.2).

In accordance with definition (1.6) the velocity of the center of mass of a rigid solid is expressed by the formula

$$\mathbf{v}_c = \int_\tau \mu \mathbf{v} \, d\tau / M,$$

hence by conditions (4.1) we have $\mathbf{v}_c \equiv 0$, which shows that $\mathbf{v}_c(0) = 0$. From the principle of motion of the center of mass (2.1) we get $\mathbf{F} \equiv 0$. If one takes for the pole the center of mass, then the velocity of any point of the body can be represented as

$$\mathbf{v} = \mathbf{v}_c + \boldsymbol{\omega} \times \mathbf{r}, \tag{4.3}$$

where $\mathbf{r}$ is the position vector of the point in question with respect to the pole C. However, $\mathbf{v}$ and $\mathbf{v}_c$ are both zero, and hence $\boldsymbol{\omega} \times \mathbf{r} \equiv 0$. Since we consider an arbitrary point of the rigid solid, the last equality makes it possible to conclude that $\boldsymbol{\omega} \equiv 0$, i.e., we also have $\boldsymbol{\omega}(0) = 0$.

From the dynamic Euler equations (2.6) it follows that if the angular velocity $\boldsymbol{\omega}$ identically vanishes during some time interval, then $L_x = L_y = L_z \equiv 0$, i.e., $\mathbf{L} \equiv 0$. This proves the necessity of conditions (4.2).

The principle of motion of the center of mass (2.1) and the dynamic Euler equations (2.6) imply that if conditions (4.2) are satisfied, then

$$\mathbf{v}_c \equiv 0, \quad \boldsymbol{\omega} \equiv 0.$$

In turn, by (4.3) this means that relation (4.1) is fulfilled. This proves the sufficiency of conditions (4.2).

Formula (3.5) shows that under conditions (4.2) the resultant moment of external forces with respect to any point O which is taken for the pole is zero. Thus, at an equilibrium we have

$$\mathbf{F} = 0, \quad \mathbf{L}_o = 0,$$

or, in the projections onto the coordinate axes,

$$\sum_{\nu=1}^{n} F_{\nu x}^{(e)} = 0, \qquad \sum_{\nu=1}^{n} F_{\nu y}^{(e)} = 0, \qquad \sum_{\nu=1}^{n} F_{\nu z}^{(e)} = 0,$$
$$\sum_{\nu=1}^{n} (\mathbf{r}_\nu \times \mathbf{F}_\nu^{(e)})_x = 0, \ \sum_{\nu=1}^{n} (\mathbf{r}_\nu \times \mathbf{F}_\nu^{(e)})_y = 0, \ \sum_{\nu=1}^{n} (\mathbf{r}_\nu \times \mathbf{F}_\nu^{(e)})_z = 0. \tag{4.4}$$

Equations (4.4) are referred to as the *equations of statics of a rigid solid*.

If a body is subject to constraints, then according to the principle of releasability it can be considered as a free one, where the reactions are interpreted as constraints as external forces acting on the body. Equations (4.4) show that they allow us to find at most six unknowns, which characterize the reactions of constraints when the given system of external forces is applied to a rigid solid.

5 Rotation of a Rigid Solid Around a Fixed Axis

Rotation of a rigid solid around a fixed axis can be considered as a motion of a body under constraints. In this case, a rigid solid has one degree of freedom and the differential equation of its motion reads as (3.4) of Chap. 5.

Dynamic reaction forces in rotation of a rigid solid around a fixed axis. Suppose that a rigid solid is fixed at points A and A' on the z-axis (Fig. 7) and rotates about it by the law $\varphi = \varphi(t)$. A rigid solid is subject to the given force system $\mathbf{F}_1, \mathbf{F}_2, \ldots, \mathbf{F}_n$, for which the resultant vector and the resultant moment with respect to the point O lying on the axis of rotation are equal to $\mathbf{F}$ and $\mathbf{L}$, respectively. It is required to find the reaction forces $\mathbf{R}$ and $\mathbf{R}'$ at the fixed points A and A'.

By the principle of releasability from constraints in this case Eq. (2.1) reads as

$$M \frac{d\mathbf{v}_c}{dt} = \mathbf{F} + \mathbf{R} + \mathbf{R}',$$
$$\frac{d\mathbf{l}}{dt} = \mathbf{L} + \overrightarrow{OA} \times \mathbf{R} + \overrightarrow{OA'} \times \mathbf{R}'. \tag{5.1}$$

It is appropriate to project these equations onto the axes of a moving coordinate $Oxyz$-system. The z-axis is directed along the axis of rotation. By formula (1.7) of Chap. 3,

$$\frac{d\mathbf{v}_c}{dt} = \frac{d^*\mathbf{v}_c}{dt} + \boldsymbol{\omega} \times \mathbf{v}_c, \quad \frac{d^*\mathbf{v}_c}{dt} = \dot{v}_{c_x}\mathbf{i} + \dot{v}_{c_y}\mathbf{j} + \dot{v}_{c_z}\mathbf{k},$$
$$\frac{d\mathbf{l}}{dt} = \frac{d^*\mathbf{l}}{dt} + \boldsymbol{\omega} \times \mathbf{l}, \quad \frac{d^*\mathbf{l}}{dt} = \dot{l}_x\mathbf{i} + \dot{l}_y\mathbf{j} + \dot{l}_z\mathbf{k}.$$

Fig. 7 Rotation of a rigid solid around a fixed axis

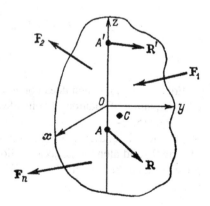

Next, we have

$$\mathbf{v}_c = \boldsymbol{\omega} \times \mathbf{r}_c = \begin{vmatrix} \mathbf{i} & \mathbf{j} & \mathbf{k} \\ 0 & 0 & \dot{\varphi} \\ x_c & y_c & z_c \end{vmatrix} = -\dot{\varphi} y_c \, \mathbf{i} + \dot{\varphi} \, x_c \mathbf{j} \,,$$

and so

$$\dot{v}_{c_x} = -\ddot{\varphi} y_c \,, \quad \dot{v}_{c_y} = \ddot{\varphi} x_c \,, \quad \dot{v}_{c_z} = 0 \,.$$

The vector $\mathbf{l}$ is specified by formulas (1.17). In the given case $\omega_1 = \omega_2 = 0$, and hence

$$l_x = l_1 = J_{13}\omega_3 \,, \quad l_y = l_2 = J_{23}\omega_3 \,, \quad l_z = l_3 = J_{33}\omega_3 \,.$$

The components of the tensor of inertia from this expression are evaluated by formulas (1.16). Setting

$$J_{13} = -J_{xz} \,, \quad J_{23} = -J_{yz} \,, \quad J_{33} = J_{zz} \,, \quad \omega_3 = \omega_z = \dot{\varphi} \,,$$

we obtain

$$l_x = -J_{xz}\dot{\varphi} \,, \quad l_y = -J_{yz}\dot{\varphi} \,, \quad l_z = J_{zz}\dot{\varphi} \,.$$

We also suppose that $\overrightarrow{OA} = -a\,\mathbf{k}$ and $\overrightarrow{OA'} = a'\mathbf{k}$.

The above formulas imply that Eq. (5.1) in the projections onto the moving axes x, y, z assume the form

$$- M(\ddot{\varphi}y_c + \dot{\varphi}^2 x_c) = F_x + R_x + R'_x \,, \tag{5.2}$$

$$M(\ddot{\varphi}x_c - \dot{\varphi}^2 y_c) = F_y + R_y + R'_y \,, \tag{5.3}$$

$$0 = F_z + R_z + R'_z \,, \tag{5.4}$$

$$- J_{xz}\ddot{\varphi} + J_{yz}\dot{\varphi}^2 = L_x + a\,R_y - a'\,R'_y \,, \tag{5.5}$$

$$- J_{yz}\ddot{\varphi} - J_{xz}\dot{\varphi}^2 = L_y - a\,R_x + a'\,R'_x \,, \tag{5.6}$$

$$J_{zz}\ddot{\varphi} = L_z \,. \tag{5.7}$$

Equation (5.7), which does not contain the unknown reaction forces, is a differential equation of rotation. It coincides with Eq. (3.4) of Chap. 5. The remaining of equations (5.2)–(5.6) allow one to find the unknown components of the reactions R_x, R_y, R'_x, R'_y and evaluate the sum $R_z + R'_z$ from Eq. (5.4). If the displacement of a rigid solid along the z-axis is allowed at the point A and is not allowed at the point A', then $R_z = 0$ and $R'_z = -F_z$. Here, the reactions $\mathbf{R}$ and $\mathbf{R}'$ are determined uniquely.

From Eqs. (5.2), (5.3) we can see that their left-hand sides vanish if the center of mass lies on the axis of rotation, i.e., when $x_c = y_c = 0$ (the case of *statical equilibrium of a rigid solid*). We shall call such an axis a *central axis*. From Eqs. (5.5), (5.6) it follows that if the z-axis is a principal inertia axis (and hence, $J_{xz} = J_{yz} = 0$), then the left-hand sides of these equations also vanish. So, if the axis of rotation is a principal central axis, then Eqs. (5.2)–(5.6) take the form

$$F_x + R_x + R'_x = 0,$$
$$F_y + R_y + R'_y = 0,$$
$$F_z + R_z + R'_z = 0,$$
$$L_x + a R_y - a' R'_y = 0,$$
$$L_y - a R_x + a' R'_x = 0.$$

These equations imply that in this case the rotation of a rigid solid, which is characterized by the angular velocity $\dot{\varphi}$ and the angular acceleration $\ddot{\varphi}$, has no effect on the magnitude of the reaction forces. A rigid solid which rotates about such an axis is in *dynamical equilibrium*.

Compound pendulum. Consider the case when a rigid solid rotates around the horizontal z-axis under the action of the gravity force only. The angle of rotation φ is measured from the position in which the center of gravity is at the lowest point. We denote the distance from the center of mass to the axis of rotation by a. The moment L_z tries to return the pendulum to the position in which $\varphi = 0$,

$$L_z = -Mga \sin \varphi.$$

When substituting this expression into Eq. (5.7) we get the following differential equation of motion for a compound pendulum:

$$\frac{d^2\varphi}{dt^2} + k^2 \sin \varphi = 0; \tag{5.8}$$

here $k^2 = Mga/J_{zz}$. It is worth noting that this nonlinear differential equation coincides with the equation of oscillation for a mathematical pendulum (2.15) of Chap. 6

$$\ddot{\varphi} + k_1^2 \sin \varphi = 0, \quad \text{where} \quad k_1^2 = g/l.$$

The length of a mathematical pendulum $l = J_{zz}/(Ma)$ corresponding to the condition $k_1 = k$ is called the *reduced length of a compound pendulum*. A mathematical pendulum of such a length oscillates synchronously with the initial compound pendulum (Fig. 8).

If deviations are small, then from Eq. (5.8) we get the linear differential equation

$$\ddot{\varphi} + \frac{Mga}{J_{zz}} \varphi = 0,$$

Fig. 8 Compound pendulum

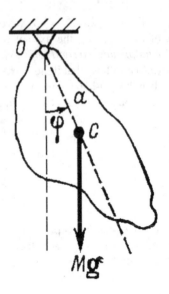

which implies that the moment of inertia $J_{zz} = MgaT^2/(4\pi^2)$ relates to the period of small oscillation

$$T = 2\pi \sqrt{\frac{J_{zz}}{Mga}} .$$

This formula is convenient for experimental evaluation of the axial moment of inertia. One can use this approach to find moments of inertia for bodies of complicated shape. Having defined the moment of inertia J_{zz}, we shall employ Huygens–Steiner's theorem (1.30) (the parallel axis theorem) to find the moment of inertia $J_{z'z'}$ with respect to the axis passing through the center of mass.

6   Rotation of a Rigid Solid Around a Fixed Point

When a rigid solid rotates about a fixed point O the differential equations of its motion can be obtained from the angular impulse-momentum principle (2.3) written with respect to the point O. We suppose that the body coordinate axes $Oxyz$ are principal axes of inertia. Here, Eq. (2.3) in its projections onto fixed axes can be written as (2.6). Augmenting equations (2.6) with the kinematic Euler equations we get a closed system of equations (2.6), (2.8). Its integration involves great difficulties. It is enough to say that even in the case of motion under the action of homogeneous gravity field a general solution of this system can be obtained only for classic cases (the Euler, Lagrange, and Kovalevskaya cases).

Euler case. The assumption is that the resultant moment of external forces acting on the body is zero. In particular, this can be realized if a rigid solid subject to only homogeneous gravity field is fixed at the center of mass (that is, the point of support coincides with the center of mass).

We start integration of the equations of motion with the simple case when $A = B$; this case corresponds to the ellipsoid of inertia which is the ellipsoid of revolution. This condition is satisfied for homogeneous bodies with the axis of symmetry. Now system (2.6), (2.8) can be integrated in elementary functions.

Since $\mathbf{L} = 0$ in this problem, on the basis of the angular impulse-momentum principle we can state that the vector of angular momentum does not change. We direct the fixed ζ-axis along the vector $\mathbf{l}$. Then $\mathbf{l} = l\mathbf{k}_0$, and hence,

$$\begin{aligned}
l_x &\equiv Ap = l \sin \theta \sin \varphi \,, \\
l_y &\equiv Aq = l \sin \theta \cos \varphi, \\
l_z &\equiv Cr = l \cos \theta \,.
\end{aligned} \tag{6.1}$$

Since $A = B$ and $L_z = 0$, from the third equation of system (2.6) we get $r = r_0 =$ const, and from the last equation of system (6.1) we have $\theta = \theta_0 =$ const. Hence from system (2.7) it follows that

$$p = \dot{\psi} \sin \theta_0 \sin \varphi \,,$$

and so the first equation of system (6.1) assumes the form

$$A\dot{\psi} \sin \theta_0 \sin \varphi = l \sin \theta_0 \sin \varphi \,,$$

i.e., the angular velocity of precession is constant $\dot{\psi} = \dot{\psi}_0 =$ const.

Further, from the last equation of system (2.7) we obtain $\dot{\varphi} = \dot{\varphi}_0 =$ const. By the above,

$$\varphi = \dot{\varphi}_0 t + \varphi_0 \,, \quad \psi = \dot{\psi}_0 t + \psi_0 \,, \quad \theta = \theta_0 \,.$$

Such a motion is called a *regular precession*. In particular, a gyroscope executes a similar motion. A homogeneous body that quickly rotates around an axis of symmetry is called a *gyroscope*. Technically one can make a gyroscope with a free axis using the Cardan suspension (Fig. 9).[3]

Let us consider a general case of motion of a rigid solid in the absence of external moments assuming that the moments of inertia are connected by the inequality $A > B > C$.

If $\mathbf{L} = 0$ the Euler equations (2.6) read as

---

[3] A detailed theory of gyroscopes can be found in the book *A.Yu.Ishlinskiy, V.I.Borzov, N.P.Stepanenko.* Lectures on the gyroscope theory. Moscow: MGU Publishers. 1983 [in Russian].

Fig. 9 Gyroscope

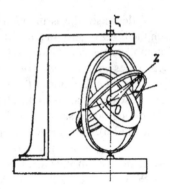

$$A\dot{p} + (C - B)qr = 0,$$
$$B\dot{q} + (A - C)rp = 0, \qquad (6.2)$$
$$C\dot{r} + (B - A)pq = 0.$$

Equation (6.2) should *a fortiori* have the integral of energy and the integral expressing the constancy of angular momentum of the system:

$$Ap^2 + Bq^2 + Cr^2 = h, \quad A^2p^2 + B^2q^2 + C^2r^2 = l^2. \qquad (6.3)$$

It is natural that integrals (6.3) can be obtained immediately from Eq. (6.2). For this purpose, it is enough to multiply these equations by p, q, r (or by Ap, Bq, Cr), respectively, to sum and integrate the resulting combinations.

Determining from Eq. (6.3) the values of p^2 and r^2, we have

$$p^2 = \frac{l^2 - Ch - Bq^2(B - C)}{A(A - C)}, \quad r^2 = \frac{Ah - l^2 - Bq^2(A - B)}{C(A - C)}.$$

Setting

$$\lambda_1^2 = \frac{l^2 - Ch}{B(B - C)}, \quad \lambda_2^2 = \frac{Ah - l^2}{B(A - B)}, \qquad (6.4)$$

one can rewrite the last expressions as

$$p^2 = \frac{B(B - C)}{A(A - C)}(\lambda_1^2 - q^2), \quad r^2 = \frac{B(A - B)}{C(A - C)}(\lambda_2^2 - q^2). \qquad (6.5)$$

Now we compose a differential equation for the function q. To this end, we substitute expressions (6.5) into the second equation of system (6.2). Then

$$\frac{dq}{dt} = \pm\sigma\sqrt{(\lambda_1^2 - q^2)(\lambda_2^2 - q^2)}, \quad \sigma = \sqrt{\frac{(A - B)(B - C)}{AC}}.$$

After separation of variables and integration we find

$$\pm (\sigma t + \alpha) = \int\limits_0^q \frac{dq}{\sqrt{(\lambda_1^2 - q^2)(\lambda_2^2 - q^2)}}, \qquad (6.6)$$

where α is an arbitrary constant. We claim that this quadrature can be reduced to the Jacobi elliptical integral of the first kind. Indeed, for definiteness we assume that

$$\lambda_1 > \lambda_2$$

. Then, changing to the variable $x = q/\lambda_2$, we get

$$\int\limits_0^q \frac{dq}{\sqrt{(\lambda_1^2 - q^2)(\lambda_2^2 - q^2)}} = \frac{1}{\lambda_1} \int\limits_0^x \frac{dx}{\sqrt{(1 - k^2 x^2)(1 - x^2)}},$$

where $k^2 = \lambda_2^2/\lambda_1^2 < 1$.

To write this elliptical integral in the Legendre form, another change is required, $x = \sin \Phi$. As a result, we have

$$\int\limits_0^q \frac{dq}{\sqrt{(\lambda_1^2 - q^2)(\lambda_2^2 - q^2)}} = \frac{1}{\lambda_1} \int\limits_0^\Phi \frac{d\Phi}{\sqrt{(1 - k^2 \sin^2 \Phi)}}.$$

The integral

$$u = \int\limits_0^\Phi \frac{d\Phi}{\sqrt{(1 - k^2 \sin^2 \Phi)}} \qquad (6.7)$$

is a function of the variables k and Φ: $u = F(k, \Phi)$. For $\Phi = \pi/2$ the elliptical integral is called *complete*:

$$K = \int\limits_0^{\pi/2} \frac{d\Phi}{\sqrt{(1 - k^2 \sin^2 \Phi)}}.$$

For function (6.7) we can find its inverse, which controls the dependence of Φ on u. This function is called an *amplitude* and is denoted by $\Phi = \text{am } u$.

Jacobi introduced three more *elliptic functions* of u:

$$\begin{aligned}
\text{sn } u &= \sin \Phi = \sin (\text{am } u), \\
\text{cn } u &= \cos \Phi = \cos (\text{am } u), \\
\text{dn } u &= \sqrt{1 - k^2 \sin^2 \Phi} = \sqrt{1 - k^2 \text{sn}^2 u}.
\end{aligned} \qquad (6.8)$$

Fig. 10 Jacobi elliptic
functions

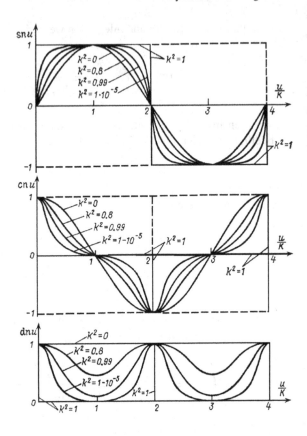

The character of variation of elliptic functions with different values of the parameter
k is shown in Fig. 10.

Definitions (6.8) imply the periodicity of the elliptic functions:

$$\operatorname{sn}(u + 4K) = \operatorname{sn} u, \quad \operatorname{cn}(u + 4K) = \operatorname{cn} u,$$
$$\operatorname{dn}(u + 2K) = \operatorname{dn} u, \quad \operatorname{sn}^2 u + \operatorname{cn}^2 u = 1. \tag{6.9}$$

Let us go back to relation (6.6) which can be now written in the form

$$u = \pm(\sigma \lambda_1 t + \alpha_1), \quad \alpha_1 = \alpha \lambda_1.$$

Then Jacobi function $\operatorname{sn} u$ now reads as

$$\operatorname{sn} u = \pm \operatorname{sn}(\sigma \lambda_1 t + \alpha_1).$$

But since $\operatorname{sn} u = \sin \Phi$ and according to the transition formulas $q = \lambda_2 \sin \Phi$, this
relation can be written as

$$q = \pm\lambda_2 \operatorname{sn}(\sigma\lambda_1 t + \alpha_1). \tag{6.10}$$

Using formulas (6.5), (6.8), and (6.9) for the two remaining projections of angular velocity, we obtain

$$\begin{aligned}
p &= \pm\lambda_1 \sqrt{\tfrac{B(B-C)}{A(A-C)}} \operatorname{dn}(\sigma\lambda_1 t + \alpha_1), \\
r &= \pm\lambda_2 \sqrt{\tfrac{B(A-B)}{C(A-C)}} \operatorname{cn}(\sigma\lambda_1 t + \alpha_1).
\end{aligned} \tag{6.11}$$

If $\lambda_1 < \lambda_2$, then we similarly find

$$\begin{aligned}
p &= \pm\lambda_1 \sqrt{\tfrac{B(B-C)}{A(A-C)}} \operatorname{cn}(\sigma\lambda_2 t + \alpha_2), \\
q &= \pm\lambda_1 \operatorname{sn}(\sigma\lambda_2 t + \alpha_2), \\
r &= \pm\lambda_2 \sqrt{\tfrac{B(A-B)}{C(A-C)}} \operatorname{dn}(\sigma\lambda_2 t + \alpha_2).
\end{aligned} \tag{6.12}$$

In these solutions λ_1, λ_2, and α_1 (or α_2) are arbitrary constants. The signs on their right-hand sides depend on the initial conditions. By properties of the elliptic functions (6.9) the above solutions have period $4K$. Note that the periodicity of angular velocity is established here with respect to the moving body, but not with respect to the absolute space.

Assume now that $\lambda_1 = \lambda_2 = \lambda$. In this case, by (6.5) the projections of the angular velocity p and r are connected by the linear relation

$$p = \pm\sqrt{\frac{C(B-C)}{A(A-B)}}\, r.$$

To find the function q we substitute the parameters p and r from (6.5) into the second equation of system (6.2) to obtain

$$\frac{dq}{dt} = \pm\sigma\,(\lambda^2 - q^2), \quad \sigma = \sqrt{\frac{(A-B)(B-C)}{AC}}.$$

We set $q = \lambda s$, where $0 \leqslant s \leqslant 1$. Then, separating the variables

$$\frac{ds}{1 - s^2} = \pm\sigma\lambda\,dt.$$

Integrating this equation, we obtain

$$\frac{1}{2}\ln\frac{1+s}{1-s} = \pm(\sigma\lambda t + \alpha),$$

where α is an arbitrary constant. Now one can find the function s:

$$s = \frac{e^{\pm 2(\sigma \lambda t + \alpha)} - 1}{e^{\pm 2(\sigma \lambda t + \alpha)} + 1} \equiv \pm \frac{e^{\sigma \lambda t + \alpha} - e^{-(\sigma \lambda t + \alpha)}}{e^{\sigma \lambda t + \alpha} + e^{-(\sigma \lambda t + \alpha)}}.$$

In other words, using the definition of the hyperbolic tangent, we can write

$$s = \pm \tanh (\sigma \lambda t + \alpha).$$

Let us now come back to the analysis of the function q. Using formulas (6.5) we get the following time law of time change of the projections of the angular velocity when $\lambda_1 = \lambda_2 = \lambda$:

$$q = \pm \lambda \, \mathrm{th} \, (\sigma \lambda t + \alpha),$$

$$p = \pm \sqrt{\frac{B(B-C)}{A(A-C)}} \, \frac{\lambda}{\mathrm{ch}(\sigma \lambda t + \alpha)},$$

$$r = \pm \sqrt{\frac{B(A-B)}{C(A-C)}} \, \frac{\lambda}{\mathrm{ch}(\sigma \lambda t + \alpha)}.$$

It is obvious that as $t \to \infty$ we have $q \to \pm \lambda$, $p \to 0$, $r \to 0$, that is a body rotates about the y-axis in the limit.

So far it was assumed that λ_1 and λ_2 are positive numbers. Let us investigate the case when one of them vanishes. We claim that in this case a body will rotate uniformly around one of its principal axes. We recall that the parameters λ_1 and λ_2 are connected by relations (6.4) with arbitrary constants l and h, which are determined from the initial velocities p_0, q_0, r_0 by formulas (6.3):

$$h = A p_0^2 + B q_0^2 + C r_0^2,$$
$$l^2 = A^2 p_0^2 + B^2 q_0^2 + C^2 r_0^2.$$

Let us analyze the quantity

$$\frac{l^2}{h} = \frac{A^2 p^2 + B^2 q^2 + C^2 r^2}{A p^2 + B q^2 + C r^2} = \frac{A(A p^2 + B \frac{B}{A} q^2 + C \frac{C}{A} r^2)}{A p^2 + B q^2 + C r^2} = \qquad (6.13)$$
$$= \frac{A(A p_0^2 + B \frac{B}{A} q_0^2 + C \frac{C}{A} r_0^2)}{A p_0^2 + B q_0^2 + C r_0^2}.$$

We can see from (6.4) that $\lambda_2 = 0$ when $Ah - l^2 = 0$. However, this expression vanishes only when $q = q_0 = 0$, $r = r_0 = 0$, $p = p_0 = \mathrm{const}$, as follows from (6.13), that is, when a body rotates uniformly around the x-axis.

A similar argument in the case $\lambda_1 = 0$ shows that

$$\frac{l^2}{h} = \frac{C(A \frac{A}{C} p^2 + B \frac{B}{C} q^2 + C r^2)}{A p^2 + B q^2 + C r^2} = \frac{C(A \frac{A}{C} p_0^2 + B \frac{B}{C} q_0^2 + C r_0^2)}{A p_0^2 + B q_0^2 + C r_0^2}.$$

As a result, $\lambda_1 = 0$ when $l^2 - Ch = 0$, that is when $p = p_0 = 0$, $q = q_0 = 0$, $r = r_0 = \mathrm{const}$, which corresponds to uniform rotation around the z-axis.

It can be proved case when the parameters λ_1 or λ_2 vanish both the necessary and sufficient condition for a uniform rotation of a body around the corresponding axes.

Note that under certain constraints on the initial (in particular, when $\lambda_1 = \lambda_2$, $\lambda_1 = 0$ or $\lambda_2 = 0$), the integration of the dynamic Euler equations is significantly simplified, and the parameters p, q, r can be expressed in terms of elementary functions.

Let us find now the Euler angles $\psi, \theta,$ and φ as time functions. Once the expressions for $p, q,$ and r are obtained, it is convenient to use relations (6.1) to determine the angles φ and θ

$$\tan \varphi = \frac{Ap}{Bq}, \quad \cos \theta = \frac{Cr}{l}. \tag{6.14}$$

Consider the first equation of system (2.8). Its right-hand side can be expressed in terms of the functions p and q. From (6.1) we have

$$Ap^2 + Bq^2 = l \sin \theta (p \sin \varphi + q \cos \varphi),$$
$$A^2 p^2 + B^2 q^2 = l^2 \sin^2 \theta,$$

and so

$$\dot{\psi} = \frac{p \sin \varphi + q \cos \varphi}{\sin \theta} = l \frac{Ap^2 + Bq^2}{A^2 p^2 + B^2 q^2}.$$

Using this expression, we can represent the angle ψ as the quadrature

$$\psi = l \int_0^t \frac{Ap^2 + Bq^2}{A^2 p^2 + B^2 q^2} dt + \psi_0, \tag{6.15}$$

which can be evaluated by substituting the functions p and q into the integration element. Recall that for $\lambda_1 = 0$, $\lambda_2 = 0$, and $\lambda_1 = \lambda_2$ they are expressed in terms of elementary functions. In the case $\lambda_1 > \lambda_2 \neq 0$ we obtain ($k^2 = \lambda_2^2/\lambda_1^2$) by formulas (6.10), (6.11)

$$\frac{Ap^2 + Bq^2}{A^2 p^2 + B^2 q^2} =$$
$$= \frac{\lambda_1^2 \frac{B(B-C)}{A-C}[1 - k^2 \sinh^2(\sigma \lambda_1 t + \alpha_1)] + B\lambda_2^2 \sinh^2(\sigma \lambda_1 t + \alpha_1)}{\lambda_1^2 \frac{AB(B-C)}{A-C}[1 - k^2 \sinh^2(\sigma \lambda_1 t + \alpha_1)] + B^2\lambda_2^2 \sinh^2(\sigma \lambda_1 t + \alpha_1)} =$$
$$= \frac{B - C + (A - B)k^2 \sinh^2(\sigma \lambda_1 t + \alpha_1)}{A(B - C) + C(A - B)k^2 \sinh^2(\sigma \lambda_1 t + \alpha_1)},$$

that makes it possible to express the angle ψ in terms of the integral containing the elliptic function $\operatorname{sn} u$ by formula (6.15). In a similar way, in the case $\lambda_2 > \lambda_1 \neq 0$ we have by formulas (6.12)

$$\frac{Ap^2 + Bq^2}{A^2 p^2 + B^2 q^2} = \frac{B - C + (A - B) \operatorname{sn}^2(\sigma \lambda_2 t + \alpha_2)}{A(B - C) + C(A - B) \operatorname{sn}^2(\sigma \lambda_2 t + \alpha_2)}.$$

Fig. 11 Lagrange case

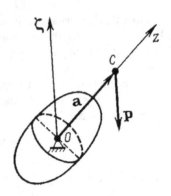

The general solution of the sixth-order system (2.6), (2.8) involves four arbitrary constants. For example, in the case $\lambda_1 > \lambda_2$ in formulas (6.10), (6.11), (6.14), (6.15) these constants are $\lambda_1, \lambda_2, \alpha_1, \psi_0$. The parameters ψ_0 and θ_0 do not appear in the solution, because the fixed axis was directed along the vector $\mathbf{l}$ for the sake of simplicity of consideration.

Lagrange case. Lagrange considered the case when the ellipsoid of inertia of a body corresponding to a fixed point is an ellipsoid of revolution and its center of mass C lies on its axis of dynamic symmetry (Fig. 11). These conditions are satisfied, for example, if a homogeneous body is fixed at one of the points of its axis of symmetry.

We denote the position vector of the center of mass with respect to the fixed point O by $\mathbf{a}$ and denote by $\gamma_1, \gamma_2, \gamma_3$ the directing cosines of the unit vector $\mathbf{k}_0$ of the vertical ζ-axis with respect to the moving axes x, y, z. In accordance with expressions (6.1) we have

$$
\begin{aligned}
\gamma_1 &= \sin\theta \sin\varphi\,, \\
\gamma_2 &= \sin\theta \cos\varphi\,, \\
\gamma_3 &= \cos\theta\,.
\end{aligned}
\tag{6.16}
$$

The moment of the gravity force $\mathbf{P}$ with respect to the fixed point can be represented as

$$
\mathbf{L} = \mathbf{a} \times \mathbf{P} = -P
\begin{vmatrix}
\mathbf{i} & \mathbf{j} & \mathbf{k} \\
0 & 0 & a \\
\gamma_1 & \gamma_2 & \gamma_3
\end{vmatrix}
= P
\begin{vmatrix}
\mathbf{i}_0 & \mathbf{j}_0 & \mathbf{k}_0 \\
a_\xi & a_\eta & a_\zeta \\
0 & 0 & -1
\end{vmatrix}\,,
\tag{6.17}
$$

where $\mathbf{i}_0, \mathbf{j}_0$, and $\mathbf{k}_0$ are the basis vectors of the coordinate system $O\xi\eta\zeta$.

According to Lagrange's assumption, the moments of inertia with respect to the x- and y-axes are equal to each other: $A = B$. So, the dynamic Euler equations (2.6) now read

$$
\begin{aligned}
A\dot{p} + (C - A)qr &= Pa\gamma_2\,, \\
A\dot{q} + (A - C)rp &= -Pa\gamma_1\,, \\
C\dot{r} &= 0\,.
\end{aligned}
\tag{6.18}
$$

From the third equation of this system we have

$$Cr = C_1 . \tag{6.19}$$

Since the gravity force acting on the body has a potential, the energy integral exists

$$T + \Pi = h , \tag{6.20}$$

where

$$\Pi = Pa\cos\theta , \tag{6.21}$$

$$2T = A(p^2 + q^2) + Cr^2 \equiv A(\dot{\psi}^2 \sin^2\theta + \dot{\theta}^2) + C(\dot{\psi}\cos\theta + \dot{\varphi})^2 . \tag{6.22}$$

The angular impulse-momentum principle (2.3) in the projection onto the fixed axis ζ has the form $dl_\zeta/dt = L_\zeta$. It follows from formula (6.17) that $L_\zeta = 0$, and hence

$$l_\zeta = \mathbf{l} \cdot \mathbf{k}_0 = (Ap\,\mathbf{i} + Aq\,\mathbf{j} + Cr\,\mathbf{k}) \cdot (\gamma_1\,\mathbf{i} + \gamma_2\,\mathbf{j} + \gamma_3\,\mathbf{k}) =$$
$$= A(p\gamma_1 + q\gamma_2) + Cr\gamma_3 = C_2 = \text{const} .$$

Using expressions (2.7) and (6.16), we find

$$A\dot{\psi}\sin^2\theta + C(\dot{\psi}\cos\theta + \dot{\varphi})\cos\theta = C_2 . \tag{6.23}$$

It is natural that integrals (6.20), (6.23) could be obtained immediately from the system of differential equations (6.18).

It is interesting to note that if instead of the system analyzed we compose the Lagrange equations of second kind, then integrals (6.19) and (6.23) can be obtained as cyclic integrals, because in this problem the Lagrange function $L = T - \Pi$ is independent of the generalized coordinates φ and ψ, and hence

$$\frac{\partial T}{\partial \dot{\varphi}} = \text{const} , \qquad \frac{\partial T}{\partial \dot{\psi}} = \text{const} .$$

Let us show that we can compose a differential equation with respect to the nutation angle θ from the integrals obtained. To this end we shall find the angular velocity $\dot{\psi}$ from formula (6.23) taking into account integral (6.19):

$$\dot{\psi} = \frac{C_2 - C_1\cos\theta}{A\sin^2\theta} . \tag{6.24}$$

Substituting this quantity into the energy integral (6.20) and taking into consideration relation (6.19) and expressions (6.21), (6.22), we have

$$A^2\dot{\theta}^2\sin^2\theta = A\sin^2\theta(h_1 - 2Pa\cos\theta) - (C_2 - C_1\cos\theta)^2 , \tag{6.25}$$

where

$$h_1 = 2h - C_1^2/C . \tag{6.26}$$

Fig. 12 Graph of the
function $f(u)$

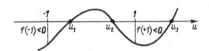

For convenience in calculations, we set

$$u = \cos\theta. \tag{6.27}$$

Then

$$\dot{u} = -\dot{\theta}\sin\theta,$$

and Eq. (6.25) can be rewritten in the form

$$A^2\dot{u}^2 = f(u), \tag{6.28}$$

where

$$f(u) = A(1 - u^2)(h_1 - 2Pau) - (C_2 - C_1u)^2, \tag{6.29}$$

whence

$$dt = \pm A\,du/\sqrt{f(u)}. \tag{6.30}$$

Here the sign is chosen depending on the sign of $\dot{u}$.

Consider the behavior of the function $f(u)$. From formula (6.29) we see that

$$f(u) \xrightarrow[u \to -\infty]{} -\infty, \qquad f(u) \xrightarrow[u \to +\infty]{} +\infty,$$
$$f(-1) = -(C_2 + C_1)^2 < 0, \quad f(+1) = -(C_2 - C_1)^2 < 0.$$

The variable u varies in the range $-1 \leqslant u \leqslant 1$, as follows from expression (6.27). Besides by formula (6.28) the function $f(u)$ should be positive in some closed subinterval of this interval. The character of variation of the function $f(u)$ is shown in Fig. 12. Here u_1, u_2, and u_3 denote the roots of the polynomial $f(u)$. The real motion, to which the inequalities $\dot{u}^2 > 0$ and $|u| \leqslant 1$ should correspond, occurs when the quantity u varies in the interval $[u_1, u_2]$. Hence, the angle θ varies in the range

$$\theta_2 \leqslant \theta \leqslant \theta_1, \tag{6.31}$$

where $\cos\theta_1 = u_1$, $\cos\theta_2 = u_2$.

Let us go back to Eq. (6.30). Factoring the polynomial $f(u)$ and taking into account that in the real motion u satisfies the conditions $u_1 \leqslant u \leqslant u_2$, we get

$$dt = \pm\frac{A\,du}{\sqrt{2PaA(u - u_1)(u_2 - u)(u_3 - u)}}.$$

We change the variable by the formula

$$u = u_1 + (u_2 - u_1)\, w^2, \quad -1 \leqslant w \leqslant 1. \tag{6.32}$$

Hence, we obtain

$$b\, dt = \pm \frac{dw}{\sqrt{(1 - w^2)(1 - k^2 w^2)}}, \tag{6.33}$$

where

$$b = \sqrt{\frac{Pa\,(u_3 - u_1)}{2A}}, \quad k^2 = \frac{u_2 - u_1}{u_3 - u_1} < 1.$$

Equation (6.28) is a completely analogous structure as Eq. (4.17) of Chap. 6, which describes the periodic motion. The periodicity of variation of the function $u(t)$ in the limits from u_1 to u_2 points to the expedience of introducing a more characteristic periodic function: the sine function. It was to this end that the variable u was changed to the new variable w which varies from (-1) to $(+1)$, as well as the sine function. If we introduce another change

$$w = \sin \Phi, \tag{6.34}$$

then we can assume that Φ increases with increasing t, and by changing from Eq. (6.33) to the equation connecting t and Φ, we obtain

$$b\, dt = \frac{d\Phi}{\sqrt{1 - k^2 \sin^2 \Phi}}. \tag{6.35}$$

For simplicity we shall assume that the moment $\Phi = 0$ corresponds to the moment $t = 0$. From formulas (6.34), (6.32), (6.27) this means that the time reading starts from the moment when $w = 0$, $u = u_1$, $\theta = \theta_1$. Besides, in accordance with Eq. (6.35) and expression (6.7) we have

$$bt = F(k, \Phi),$$

or $\Phi = \operatorname{am} bt$.

Thus, the variable u varies by the law

$$u = u_1 + (u_2 - u_1)\, \operatorname{sn}^2 bt,$$

and the nutation angle θ can be represented as a time function as follows

$$\theta = \arccos[u_1 + (u_2 - u_1)\, \operatorname{sn}^2 bt]. \tag{6.36}$$

The precession angle ψ and the angle of proper rotation φ can be written in terms of quadratures of elliptic functions. Indeed, from Eq. (6.24) for the angle ψ, we get

$$\psi = \frac{1}{A} \int \frac{C_2 - C_1[u_1 + (u_2 - u_1)\, \operatorname{sn}^2 bt]}{1 - [u_1 + (u_2 - u_1)\, \operatorname{sn}^2 bt]^2}\, dt + C_3. \tag{6.37}$$

The angular velocity of proper rotation $\dot{\varphi}$ can be found from integral (6.19), substituting in it the value of r from formulas (2.7). As a result we have

$$\dot{\varphi} = C_1/C - \dot{\psi} \cos \theta. \tag{6.38}$$

After substitution of expression (6.24) into this equation, we have

$$\dot{\varphi} = \frac{C_1}{C} - \frac{C_2 - C_1 \cos \theta}{A \sin^2 \theta} \cos \theta,$$

on the basis of which the angle in question can be represented as

$$\varphi = C_4 + \frac{C_1}{C} t - \frac{1}{A} \int \frac{C_2 - C_1[u_1 + (u_2 - u_1) \operatorname{sn}^2 bt]}{1 - [u_1 + (u_2 - u_1) \operatorname{sn}^2 bt]^2} [u_1 + (u_2 - u_1) \operatorname{sn}^2 bt] \, dt. \tag{6.39}$$

In the above solutions (6.36), (6.37), and (6.39) the quantities h, t_0, C_1, C_2, C_3, and C_4 are arbitrary constants. Recall that in view of (6.26) the quantity h_1, which controls the values of the roots u_1, u_2, and u_3, is expressed in terms of arbitrary constants h and C_1.

We now consider a qualitative pattern of rigid solid motion in the Lagrange case. The center of mass C of a system moves along a sphere of radius a, the center of which is located at a fixed point. The Oz-axis position of interest is characterized by the angles θ and ψ.

First, we examine the variation of the nutation angle θ. In a real motion, this angle varies in the limits specified by inequalities (6.31), and according to the law of its variation (6.36), this angle is a periodic time function with the period

$$T = 2K(k)/b,$$

where K is the complete elliptic integral of the first kind. As follows from the same representation (6.36), an increase of the angle from θ_2 to θ_1 and its decrease from θ_1 to θ_2 occur monotonically during the same time interval of length $K(k)/b$.

Thus, when a rigid solid moves, the point C describes on the sphere of radius a an undulating curve enclosed between two parallels formed by the intersection of the sphere with cones with conical angles $2\theta_2$ and $2\theta_1$ (Fig. 13). This point arrives in succession, after equal time intervals $K(k)/b$, on the upper and the lower parallels.

Let us now examine the variation of the precession angle ψ. The angular velocity $\dot{\psi}$ is given by formula (6.24), and so its sign depends on the value of $(C_2 - C_1 \cos \theta)$. For simplicity, we shall assume that $C_1 > 0$. Recall that by inequalities (6.31)

$$\cos \theta_1 \leqslant \cos \theta \leqslant \cos \theta_2. \tag{6.40}$$

Hence, if the choice of initial conditions secures the inequality

$$C_2 > C_1 \cos \theta_2, \tag{6.41}$$

Fig. 13 Possible motions of the point C on the sphere

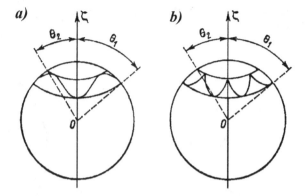

then in motion we have by inequalities (6.40)

$$C_2 > C_1 \cos\theta,$$

and hence, the angular velocity of the precession $\dot\psi$ is positive. This means that if $\theta = \theta_1$ or $\theta = \theta_2$, when $\dot\theta = 0$, then the quantity $\dot\psi$ does not vanish. This implies that the curve described by the point C touches the parallels shown in Fig. 13a. If the inequality

$$C_2 < C_1 \cos\theta_1$$

is satisfied instead of inequality (6.41), then in motion the angular velocity $\dot\psi$ is negative. In this case, the form of the trajectory of the point C does not change, but the body processes in the direction of decrease of the angle ψ.

If the equality is satisfied

$$C_2 = C_1 \cos\theta_2$$

instead of the last inequalities, then as the nutation angle reaches the value θ_2, then the quantity $\dot\theta$ and the quantity $\dot\psi$ both vanish. In this case, the trajectory of the point C is similar to a cycloid drawn on the sphere and supported by the upper parallel (Fig. 13b). One can show that the point C does not stop on the lower parallel.

Consider now the case when the inequalities are satisfied

$$C_1 \cos\theta_1 < C_2 < C_1 \cos\theta_2.$$

Then for the angles θ satisfying the conditions

$$\cos\theta_1 \leqslant \cos\theta < C_2/C_1,$$

the angular velocity $\dot\psi$ is positive, and when

$$C_2/C_1 < \cos\theta \leqslant \cos\theta_2$$

Fig. 14 Possible motion of
the point C on the sphere

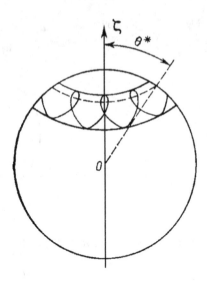

it is negative. Hence, if the angle θ^* satisfies the equation

$$\cos \theta^* = C_2/C_1 ,$$

then a change in the direction of precession motion of a rigid solid occurs. In this
case, the trajectory of the point C represents a loop-shaped spherical curve (Fig. 14)[4].

Pseudoregular precession of a heavy gyroscope. We now consider the case
when the first two roots of polynomial (6.29) are equal

$$u_1 = u_2 . \tag{6.42}$$

Then by formula (6.36) we have $\theta = \arccos u_1$, that is, in a motion the nutation angle
is constant and equal to the initial deviation,

$$\theta = \text{const} = \theta_0 , \tag{6.43}$$

and hence,

$$\dot{\theta} = \dot{\theta}_0 = 0 . \tag{6.44}$$

For the precession angle and the proper rotation angle, we obtain from formulas
(6.37), (6.39) under condition (6.42)

---

[4] As in the majority of university textbooks in Rational and Applied Mechanics, Figs. 13 and 14 give
for illustration purposes a large number of touchings and loops. For a real motion, the periodicity
of such motions is found to be close to one revolution with respect to the precession angle.

$$\psi = C_3 + \frac{C_2 - C_1 \cos \theta_0}{A \sin^2 \theta_0} t ,$$

$$\varphi = C_4 + \left(\frac{C_1}{C} - \frac{C_2 - C_1 \cos \theta_0}{A \sin^2 \theta_0} \right) t .$$

(6.45)

The arbitrary constants C_1, C_2, C_3, and C_4 can be determined by setting the initial conditions

$$t = 0, \quad \psi = \psi_0, \quad \dot{\psi} = \dot{\psi}_0, \quad \varphi = \varphi_0, \quad \dot{\varphi} = \dot{\varphi}_0 .$$

(6.46)

Then using formulas (2.7), (6.19), (6.23), (6.43)–(6.46), we obtain

$$C_1 = C(\dot{\psi}_0 \cos \theta_0 + \dot{\varphi}_0), \quad C_2 = A \dot{\psi}_0 \sin^2 \theta_0 + C(\dot{\psi}_0 \cos \theta_0 + \dot{\varphi}_0) \cos \theta_0 ,$$

$$C_3 = \psi_0, \quad C_4 = \varphi_0 .$$

Hence, by formula (6.24)

$$\frac{C_2 - C_1 \cos \theta_0}{A \sin^2 \theta_0} = \dot{\psi}_0 ,$$

and, therefore,

$$\frac{C_1}{C} - \frac{C_2 - C_1 \cos \theta_0}{A \sin^2 \theta_0} \cos \theta_0 = \dot{\varphi}_0 .$$

Hence, under conditions (6.42) the equations of motion have the form

$$\theta = \theta_0 = \text{const}, \quad \psi = \psi_0 + \dot{\psi}_0 t , \quad \varphi = \varphi_0 + \dot{\varphi}_0 t ,$$

that is the body executes a *regular precession*. In this case, the center of mass C of the body moves uniformly about the $O\zeta$-axis with angular velocity $\dot{\psi}_0$ describing a circle of radius $a \sin \theta_0$, while the body itself moves uniformly about its Oz-axis with angular velocity $\dot{\varphi}_0$.

Note that the regular precession can not take place in the Lagrange case under any initial conditions. A necessary and sufficient condition for the existence of a multiple first root of polynomial (6.29), and hence, the realization of equality (6.42), is the vanishing of the derivative of this polynomial with $u = u_1$,

$$f'(u_1) = 0 ,$$

or in an extended form,

$$2(C_1 C_2 - PaA) - 2u_1(C_1^2 + Ah_1) + 6PaAu_1^2 = 0 .$$

Substituting here the values of C_1 and C_2, taking into account formula (6.26), and canceling the expression obtained after transformations by the quantity $1 - u_1^2$, we get

$$(C - A)\dot{\psi}_0^2 \cos\theta_0 + C\dot{\psi}_0\dot{\varphi}_0 - Pa = 0. \tag{6.47}$$

Thus, for the existence of a regular precession in the Lagrange case the initial angular velocity of nutation should be equal to zero ($\dot{\theta}_0 = 0$), the angles ψ_0 and φ_0 can have any values, and the initial conditions θ_0, $\dot{\psi}_0$, $\dot{\varphi}_0$ should satisfy relation (6.47).

If the roots u_1 and u_2 are not equal, but are slightly different, then the parallels almost coincide in Fig. 13, and the point C describes a spherical curve between them. In this case that a body executes a *pseudoregular precession*. Here, the equations of gyroscope motion can be approximately integrated. Usually a gyroscope is made in such a way that the following inequality is valid

$$\frac{A}{C} = \frac{B}{C} < 1.$$

Let the initial conditions be of the form

$$t = 0, \quad \dot{\psi} = 0, \quad \dot{\theta} = 0, \quad \dot{\varphi} = \Omega, \\ \psi = \psi_0, \quad \theta = \theta_0, \quad \varphi = \varphi_0. \tag{6.48}$$

Hence, by formulas (6.19), (6.23), and (6.26), the differential equation (6.25) can be written as

$$\dot{\theta}^2 = \frac{2Pa}{A}(\cos\theta_0 - \cos\theta)\left[1 - \frac{C^2\Omega^2}{2PaA\sin^2\theta}(\cos\theta_0 - \cos\theta)\right]. \tag{6.49}$$

Note that in the process of motion the following condition should be satisfied

$$\cos\theta_0 \geqslant \cos\theta,$$

for otherwise the right-hand side of equation (6.49) becomes negative.

We assume that the required nutation angle θ deviates from the initial value θ_0 by a small magnitude ϑ:

$$\theta = \theta_0 + \vartheta. \tag{6.50}$$

Let us change from the variable θ to the variable ϑ in Eq. (6.49). From relation (6.50) we have

$$\frac{d\theta}{dt} = \frac{d\vartheta}{dt}.$$

Since

$$\cos\theta = \cos(\theta_0 + \vartheta) = \cos\theta_0\cos\vartheta - \sin\theta_0\sin\vartheta, \\ \sin\theta = \sin(\theta_0 + \vartheta) = \sin\theta_0\cos\vartheta + \cos\theta_0\sin\vartheta,$$

we have, for small ϑ,

$$\cos \theta \approx \cos \theta_0 - \vartheta \sin \theta_0 ,$$
$$\sin \theta \approx \sin \theta_0 + \vartheta \cos \theta_0,$$
$$\sin^2 \theta \approx \sin^2 \theta_0 + 2\vartheta \sin \theta_0 \cos \theta_0 ,$$

and hence,

$$\frac{cos\theta_0 - \cos \theta}{\sin^2 \theta} \approx \frac{\vartheta}{\sin \theta_0} .$$

Now we can write the following approximate differential equation instead of equation (6.49):

$$\dot{\vartheta}^2 = \frac{2Pa \sin \theta_0}{A} \vartheta \left(1 - \frac{C^2 \Omega^2}{2PaA \sin \theta_0} \vartheta \right) .$$

Setting $\alpha = C\Omega/A$, $\beta = 2PaA \sin \theta_0/(C^2\Omega^2)$, it can be written in the form

$$\dot{\vartheta}^2 = \alpha^2 \vartheta (\beta - \vartheta) ,$$

and hence, after taking the square root and separating the variables,

$$\frac{d\vartheta}{\sqrt{\beta\vartheta - \vartheta^2}} = \pm \alpha \, dt .$$

Integrating this equation and taking into account initial conditions (6.48), we have

$$\arccos \left(1 - \frac{2\vartheta}{\beta} \right) = \pm \alpha t ,$$

$$or \, \vartheta = \frac{Pa A \sin \theta_0}{C^2 \Omega^2} \left(1 - \cos \frac{C\Omega t}{A} \right) . \tag{6.51}$$

Hence we can see that an approximate solution in the form of (6.50) can be constructed under the condition

$$PaA \ll C^2\Omega^2 . \tag{6.52}$$

Besides, the nutation angle changes according to the law

$$\theta = \theta_0 + \frac{PaA \sin \theta_0}{C^2\Omega^2} - \frac{PaA \sin \theta_0}{C^2\Omega^2} \cos \frac{C\Omega t}{A} , \tag{6.53}$$

and the nutation angular velocity is a time function of the form

$$\dot{\theta} = \frac{Pa \sin \theta_0}{C\Omega} \sin \frac{C\Omega t}{A} . \tag{6.54}$$

For the quantity $\dot{\psi}$ we get from relation (6.24)

$$\dot{\psi} = \frac{C\Omega(\cos\theta_0 - \cos\theta)}{A\sin^2\theta} \tag{6.55}$$

under initial conditions (6.48); in other words, in view of (6.51) we can write approximately

$$\dot{\psi} = \frac{Pa}{C\Omega}\left(1 - \cos\frac{C\Omega t}{A}\right), \tag{6.56}$$

that is, the precession angular velocity $\dot{\psi}$ has the same sign as the quantity Ω. When $\theta_2 = \theta_0$ it vanishes, and for $\theta_1 = \theta_0 + 2PaA\sin\theta_0/(C^2\Omega^2)$ it assumes its maximum value. By the above, the center of mass C of a gyroscope moves along a spherical cycloid—this fact coincides with the assertion of the previous point (see Fig. 13b).

By formula (6.56) one can calculate the value of the mean angular precession velocity for the period of $2\pi A/(C\Omega)$

$$\tilde{\psi} = Pa/(C\Omega)$$

and the law of the precession angle change

$$\psi = \psi_0 + \frac{Pa}{C\Omega}t - \frac{PaA}{C^2\Omega^2}\sin\frac{C\Omega t}{A}. \tag{6.57}$$

If the angular momentum $C\Omega$ is such that inequality (6.52) is satisfied, then the third term in formula (6.57) can be neglected, that is why

$$\psi = \psi_0 + \frac{Pa}{C\Omega}t,$$

which corresponds to the uniform rotation with constant angular velocity $\tilde{\psi}$.

The expression for the angular velocity of proper rotation can be obtained from formula(6.38). Under the initial conditions (6.48) we have

$$C_1/C = r_0 = \Omega,$$

and hence,

$$\dot{\varphi} = \Omega - \dot{\psi}\cos\theta.$$

Substituting here expression (6.55) and supposing that $\sin\theta = \sin\theta_0$, $\cos\theta_0 - \cos\theta = \vartheta\sin\theta_0$, we can write

$$\dot{\varphi} = \Omega - \vartheta\frac{C\Omega}{A}\cot\theta_0,$$

and so, using the law (6.51),

$$\dot{\varphi} = \Omega - \frac{Pa\cos\theta_0}{C\Omega} + \frac{Pa\cos\theta_0}{C\Omega} \cos\frac{C\Omega t}{A}. \tag{6.58}$$

Integrating this differential equation under the initial conditions (6.48), this gives

$$\varphi = \varphi_0 + \left(\Omega - \frac{Pa\cos\theta_0}{C\Omega}\right)t + \frac{PaA\cos\theta_0}{C^2\Omega^2}\sin\frac{C\Omega t}{A}. \tag{6.59}$$

Thus, a pseudoregular precession is characterized by formulas (6.53), (6.57) and (6.59) in which the terms are neglected which contain the quantity $C\Omega$ in the power above two in the denominator. From this law of motion it follows that a pseudoregular precession can considered as a regular one specified by the formulas

$$\begin{aligned}
\theta' &= \theta_0 + \frac{PaA\sin\theta_0}{C^2\Omega^2}, \\
\psi' &= \psi_0 + \frac{Pa}{C\Omega}t, \\
\varphi' &= \varphi_0 + \left(\Omega - \frac{Pa\cos\theta_0}{C\Omega}\right)t,
\end{aligned} \tag{6.60}$$

with superimposed high-frequency harmonic vibrations with small amplitude

$$\begin{aligned}
\theta'' &= -\frac{PaA\sin\theta_0}{C^2\Omega^2}\cos\frac{C\Omega t}{A}, \\
\psi'' &= -\frac{PaA}{C^2\Omega^2}\sin\frac{C\Omega t}{A}, \\
\varphi'' &= \frac{PaA\cos\theta_0}{C^2\Omega^2}\sin\frac{C\Omega t}{A}.
\end{aligned}$$

The analysis of pseudoregular precession was conducted under the assumption that the initial conditions (6.48) were satisfied. Under more general initial conditions, when $\dot{\psi}_0, \dot{\theta}_0 \neq 0$, a pseudoregular precession is not executed with respect to the regular precession characterized by formulas (6.60), but rather with respect to some other regular precession.

The elementary theory of gyroscopes. Gyroscopes are widely useful in various engineering devices. This is explained by their special properties which make it possible to build different types of navigators (gyrocompasses, gyroscopic horizons, etc.). Let us give some elementary theory of gyroscopes.

An analysis of formulas (6.54), (6.56), and (6.58) shows that the angular velocities $\dot{\varphi}, \dot{\psi}, \dot{\theta}$ can be represented as

$$\dot{\varphi} = \Omega + O\left(\frac{Pa}{C\Omega}\right), \quad \dot{\psi} = O\left(\frac{Pa}{C\Omega}\right), \quad \dot{\theta} = O\left(\frac{Pa}{C\Omega}\right),$$

where the symbol $O(Pa/(C\Omega))$ stands for the quantity that has the order of smallness $Pa/(C\Omega)$. Hence by formulas (2.7)

$$p = O\left(\frac{Pa}{C\Omega}\right), \quad q = O\left(\frac{Pa}{C\Omega}\right), \quad r = \Omega + O\left(\frac{Pa}{C\Omega}\right).$$

This means that in accordance with formulas (6.1) the vector of angular momentum $\mathbf{l}$ can be given approximately in the form $\mathbf{l} = C\Omega\,\mathbf{k}$. In this case the angular impulse-momentum principle (2.3) can be written as

$$\frac{d\mathbf{l}}{dt} = C\Omega\,\frac{d\mathbf{k}}{dt} = \mathbf{L}\,.$$

Since $d\mathbf{k}/dt = \boldsymbol{\omega} \times \mathbf{k}$, we have

$$C\Omega\,\boldsymbol{\omega} \times \mathbf{k} = \mathbf{L}\,. \tag{6.61}$$

To define the vector $\boldsymbol{\omega}$ one should use formulas (6.60). Neglecting the high-frequency oscillation with small amplitude, i.e., supposing that $\varphi = \varphi'$, $\psi = \psi'$, $\theta = \theta'$, and approximately assuming that $\dot{\varphi} = \dot{\varphi}' = \Omega$, we have

$$\boldsymbol{\omega} = \dot{\varphi}\,\mathbf{k} + \dot{\psi}\,\mathbf{k}_0 = \Omega\,\mathbf{k} + \frac{Pa}{C\Omega}\,\mathbf{k}_0$$

in accordance with formula (2.5) of Chap. 2. Substituting this expression into formula (6.61), we get

$$\dot{\psi}\,\mathbf{k}_0 \times C\dot{\varphi}\,\mathbf{k} = \mathbf{L}\,. \tag{6.62}$$

Equation (6.62) underlies the elementary theory of gyroscopes. From this equation, it possible to find the precession angular velocity $\dot{\psi}$ with given moment $\mathbf{L}$ and angular velocity $\dot{\varphi}$. Vice versa, given angular velocities $\dot{\varphi}$ and $\dot{\psi}$, one can determine the moment $\mathbf{L}$ responsible for the precession in question. One should invoke equation (6.62) to solve such problems in the case when the angular velocity of proper rotation $\dot{\varphi}$ is essentially greater than the precession angular velocity $\dot{\psi}$.

Consider a rigid solid that rotates rapidly about bearings. We shall assume that it is a gyroscope subject to forces applied to it from the bearings. According to the law of action and reaction balance, the forces the moment of which is equal to $(-\mathbf{L})$ are applied to the bearings. The vector $\mathbf{L}_{gyr} = -\mathbf{L}$ is called the *gyroscopic moment*.

According to formula(6.62) we have

$$\mathbf{L}_{gyr} = C\dot{\varphi}\,\mathbf{k} \times \dot{\psi}\,\mathbf{k}_0\,. \tag{6.63}$$

Example 1 Let us apply the elementary theory of gyroscopes to solving the following problem. Suppose that a turbo-propelled plane is turning in the horizontal plane. The propeller and rapidly rotating parts of the turbine connected with it can be considered as a gyroscope. The angular momentum of the gyroscope $\mathbf{l} = C\dot{\varphi}\,\mathbf{k}$ is assumed to be oriented in the direction of motion of the plane. Besides, if the aeroplane is turning to the left in the horizontal plane, then the force pair $(\mathbf{F}_{gyr}, -\mathbf{F}_{gyr})$ applied from the gyroscope to the bearings is acting as shown in Fig. 15.

Fig. 15 Appearance of the gyroscopic moment in a ?? turbo-propelled plane

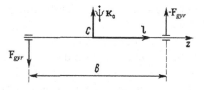

In the case when the aeroplane moves with velocity v along a circle of radius R, the precession angular velocity is $\dot{\psi} = v/R$, and the moment L_{gyr} can be represented as $L_{\text{gyr}} = F_{\text{gyr}}\, b$, here b is the distance between the bearings. Hence, formula (6.63) yields that each bearing of the turbine is subject to the force

$$F_{\text{gyr}} = C\dot{\varphi} v/(Rb).$$

A pair of such forces lifts the nose of the airplane (the so-called nosing-up). If under the same conditions the airplane is turning to the right in the horizontal plane, then its nose is going down (the so-called nosing-down). A turn of the airplane in the vertical plane causes it to yawn to the left or to the right. To avoid the gyroscopic moment, in comfortable liners with improved reliability, two propellers rotating in the opposite direction should be installed on each axis. Such a combination is also profitable from the point of view of aerodynamics because the second propeller uses the kinetic energy of a swirling slipstream produced by the front propeller.

7 Equations of Motion of a System of Rigid Solids in Redundant (Excess) Coordinates

In Chap. 6 it was shown that mathematical apparatus of theoretical mechanics, which was developed for a system of particles, can be applied to mechanical systems of rigid solids. This means that when studying the dynamics of systems of rigid solids one can use both the Lagrange equations of the first and second kind.

It is worth noting the fact that the Lagrange equations of the first kind composed for the system of finite number of particles will be of the same simplest structure in all coordinates if these coordinates are the Cartesian coordinates of the particle of the system (see Eq. (3.11) of Chap. 6). It is also essential that in the case in question the Lagrange equations of the first kind are solvable with respect to the second derivatives of the sought-for coordinates. This means that the equations under discussion have the form that is convenient for numerical solution. This raises the question of finding such a form of writing of the equations of motion for a system of rigid solids that is also convenient for numerical applications. The problem is that the equations of motion for a system of rigid solids (for a manipulator, for instance) written in the form of the Lagrange equations of the second kind are so complicated in the case of many solids that it is fairly difficult not only to integrate them, but even to write down.

The problem of finding the Lagrange equations of the first kind of simplest form for a system of rigid solids reduces to the search for a new form of the equations of motion of a single free rigid solid. Let us consider this question.

We introduce a fixed $O\xi\eta\zeta$-coordinate system and a moving $Cxyz$-coordinate system with unit vectors $\mathbf{i}, \mathbf{j}, \mathbf{k}$ directed along the principal central axes of inertia of the solid. Let $\boldsymbol{\rho} = \overrightarrow{OC}$.

The position of a rigid solid relative to the fixed system is controlled by defining the four vectors: $\boldsymbol{\rho}, \mathbf{i}, \mathbf{j}$, and $\mathbf{k}$. The assumption that a solid is not deformed in its motion, i.e., it is absolutely rigid, means that the vectors $\mathbf{i}, \mathbf{j}, \mathbf{k}$ remain orthogonal and unit during the motion of the body.

We shall consider the orthogonality and unity conditions for the vectors $\mathbf{i}, \mathbf{j}, \mathbf{k}$ as ideal constraints specified by the equations

$$f^1 = \mathbf{i}^2 - 1 = 0, \quad f^2 = \mathbf{j}^2 - 1 = 0, \quad f^3 = \mathbf{k}^2 - 1 = 0,$$
$$f^4 = \mathbf{i} \cdot \mathbf{j} = 0, \quad f^5 = \mathbf{j} \cdot \mathbf{k} = 0, \quad f^6 = \mathbf{k} \cdot \mathbf{i} = 0. \tag{7.1}$$

We shall take the projections of the vectors $\boldsymbol{\rho}, \mathbf{i}, \mathbf{j}$, and $\mathbf{k}$ onto the axes of a fixed $O\xi\eta\zeta$-coordinate system as the initial coordinates of a free rigid solid. The total number of these coordinates, which is twelve, is twice as much as the number of degrees of freedom of a rigid solid. This is why such coordinates are called *redundant* (*excess*, or *surplus*) coordinates. When applied to a mechanical system consisting of a finite number of particles, the Cartesian coordinates are redundant in the case of the presence of the holonomic constraints.

We note that in some courses in theoretical mechanics only the Eq. (3.11) of Chap. 6 written in Cartesian coordinates and containing the Lagrange multipliers are called the Lagrange equations of the first kind. Sometimes, the equations of motion for a holonomic mechanical system in generalized Lagrange coordinates in the case of their excess number are sometimes called the Lagrange equations with multipliers. In Sect. 3 of Chap. 6 it was shown that the Lagrange equations of the first and second kinds can be considered as two reciprocal systems of linear algebraic equations with respect to the constraint reaction forces. In accordance with this general approach, by the Lagrange equations of the first kind for any general mechanical system we shall mean the equations each of which contains all the Lagrange multipliers.

Let us now consider the question of obtaining the equations of motion of a free rigid solid in excess coordinates. We compose an expression for its kinetic energy. By definition we have

$$T = \frac{1}{2} \int_\tau v^2 \mu \, d\tau = \frac{1}{2} \int_\tau (\dot{\boldsymbol{\rho}} + x\mathbf{i} + y\mathbf{j} + z\mathbf{k})^2 \mu \, d\tau.$$

Here, as in Sect. 1, τ is the volume of a body, $d\tau = dx\,dy\,dz$, μ is its density, x, y, and z are the current coordinates of the rigid solid in the coordinate system $Cxyz$.

Since the axes Cx, Cy, and Cz are the principal axes of inertia of the body, we have

$$\int_\tau x\mu\,d\tau = \int_\tau y\mu\,d\tau = \int_\tau z\mu\,d\tau = \int_\tau xy\mu\,d\tau = \int_\tau yz\mu\,d\tau = \int_\tau zx\mu\,d\tau = 0\,,$$

and hence,

$$T = \frac{M\dot{\rho}^2}{2} + \frac{I_x \dot{\mathbf{i}}^2}{2} + \frac{I_y \dot{\mathbf{j}}^2}{2} + \frac{I_z \dot{\mathbf{k}}^2}{2}\,. \tag{7.2}$$

Here, M is the mass of the body, and

$$I_x = \int_\tau x^2\mu\,d\tau\,, \quad I_y = \int_\tau y^2\mu\,d\tau\,, \quad I_z = \int_\tau z^2\mu\,d\tau\,. \tag{7.3}$$

Suppose, the forces $\mathbf{F}_\nu$ are applied to the body at points $N_\nu = (x_\nu, y_\nu, z_\nu)$. The virtual elementary work of these forces can be written as

$$\delta A = \sum_\nu \mathbf{F}_\nu \cdot (\delta\rho + x_\nu\delta\mathbf{i} + y_\nu\delta\mathbf{j} + z_\nu\delta\mathbf{k}) =$$

$$= \mathbf{Q}_\rho \cdot \delta\rho + \mathbf{Q_i} \cdot \delta\mathbf{i} + \mathbf{Q_j} \cdot \delta\mathbf{j} + \mathbf{Q_k} \cdot \delta\mathbf{k}\,,$$

where

$$\mathbf{Q}_\rho = \sum_\nu \mathbf{F}_\nu\,, \quad \mathbf{Q_i} = \sum_\nu x_\nu\mathbf{F}_\nu\,,$$

$$\mathbf{Q_j} = \sum_\nu y_\nu\mathbf{F}_\nu\,, \quad \mathbf{Q_k} = \sum_\nu z_\nu\mathbf{F}_\nu\,. \tag{7.4}$$

Taking into consideration the expressions for the kinetic energy and the elementary work, as well as the form of the constraint equations (7.1), we see that it is expedient to represent the equations of motion corresponding to the three components of each of the vectors $\rho, \mathbf{i}, \mathbf{j}, \mathbf{k}$ as

$$\frac{d}{dt}\frac{\partial T}{\partial\dot{\rho}} - \frac{\partial T}{\partial\rho} = \mathbf{Q}_\rho\,, \quad \varkappa = \overline{1,k}\,,$$

$$\frac{d}{dt}\frac{\partial T}{\partial\dot{\mathbf{i}}} - \frac{\partial T}{\partial\mathbf{i}} = \mathbf{Q_i} + \Lambda_\varkappa\frac{\partial f^\varkappa}{\partial\mathbf{i}} \equiv \mathbf{Q_i} + 2\Lambda_1\mathbf{i} + \Lambda_4\mathbf{j} + \Lambda_6\mathbf{k}\,,$$

$$\frac{d}{dt}\frac{\partial T}{\partial\dot{\mathbf{j}}} - \frac{\partial T}{\partial\mathbf{j}} = \mathbf{Q_j} + \Lambda_\varkappa\frac{\partial f^\varkappa}{\partial\mathbf{j}} \equiv \mathbf{Q_j} + 2\Lambda_2\mathbf{j} + \Lambda_5\mathbf{k} + \Lambda_4\mathbf{i}\,, \tag{7.5}$$

$$\frac{d}{dt}\frac{\partial T}{\partial\dot{\mathbf{k}}} - \frac{\partial T}{\partial\mathbf{k}} = \mathbf{Q_k} + \Lambda_\varkappa\frac{\partial f^\varkappa}{\partial\mathbf{k}} \equiv \mathbf{Q_k} + 2\Lambda_3\mathbf{k} + \Lambda_6\mathbf{i} + \Lambda_5\mathbf{j}\,,$$

where we set

$$\frac{\partial}{\partial \mathbf{a}} = \frac{\partial}{\partial a_x}\mathbf{i} + \frac{\partial}{\partial a_y}\mathbf{j} + \frac{\partial}{\partial a_z}\mathbf{k},$$

$$\mathbf{a} = a_x\mathbf{i} + a_y\mathbf{j} + a_z\mathbf{k}.$$

Note that this notation is convenient, because when using it the vectors entering expressions (7.1) and (7.2) can be formally considered as scalar quantities. Taking into account this simple rule, as well as notation (7.4), Eq. (7.5) assume the form

$$M\ddot{\boldsymbol{\rho}} = \sum_{\nu} \mathbf{F}_{\nu}, \tag{7.6}$$

$$I_x\ddot{\mathbf{i}} = \sum_{\nu} x_{\nu}\mathbf{F}_{\nu} + 2\Lambda_1\mathbf{i} + \Lambda_4\mathbf{j} + \Lambda_6\mathbf{k},$$

$$I_y\ddot{\mathbf{j}} = \sum_{\nu} y_{\nu}\mathbf{F}_{\nu} + 2\Lambda_2\mathbf{j} + \Lambda_5\mathbf{k} + \Lambda_4\mathbf{i}, \tag{7.7}$$

$$I_z\ddot{\mathbf{k}} = \sum_{\nu} z_{\nu}\mathbf{F}_{\nu} + 2\Lambda_3\mathbf{k} + \Lambda_6\mathbf{i} + \Lambda_5\mathbf{j}.$$

Note that Eq. (7.6) is the equation of motion of the center of mass of a rigid solid.

Using the constraint equations, we shall eliminate the Lagrange multipliers from the Lagrange equations of the first kind (7.7). Differentiating twice the constraint equations of (7.1) with respect to time, we get

$$\ddot{\mathbf{i}}^2 = -\ddot{\mathbf{i}}\cdot\mathbf{i}, \quad \ddot{\mathbf{j}}^2 = -\ddot{\mathbf{j}}\cdot\mathbf{j}, \quad \dot{\mathbf{k}}^2 = -\ddot{\mathbf{k}}\cdot\mathbf{k},$$
$$2\dot{\mathbf{i}}\cdot\dot{\mathbf{j}} + \ddot{\mathbf{i}}\cdot\mathbf{j} + \mathbf{i}\cdot\ddot{\mathbf{j}} = 0, \quad 2\dot{\mathbf{j}}\cdot\dot{\mathbf{k}} + \ddot{\mathbf{j}}\cdot\mathbf{k} + \mathbf{j}\cdot\ddot{\mathbf{k}} = 0, \tag{7.8}$$
$$2\dot{\mathbf{k}}\cdot\dot{\mathbf{i}} + \ddot{\mathbf{k}}\cdot\mathbf{i} + \mathbf{k}\cdot\ddot{\mathbf{i}} = 0.$$

Multiplying the first equation of system (7.7) by the unit vector $\mathbf{i}$ and taking into account the first of the expressions (7.8), we arrive at

$$2\Lambda_1 = -\sum_{\nu} x_{\nu}\mathbf{F}_{\nu}\cdot\mathbf{i} - I_x\ddot{\mathbf{i}}^2.$$

Now we multiply the first of the equations of system (7.7) by $I_y\mathbf{j}$, multiply the second one by $I_x\mathbf{i}$, and then add them. Then, using the second of expressions (7.8), we have

$$(I_x + I_y)\Lambda_4 = -2I_xI_y\dot{\mathbf{i}}\cdot\dot{\mathbf{j}} - I_y\sum_{\nu} x_{\nu}\mathbf{F}_{\nu}\cdot\mathbf{j} - I_x\sum_{\nu} y_{\nu}\mathbf{F}_{\nu}\cdot\mathbf{i}.$$

As we can easily see, Eq. (7.7) and the constraint equations (7.1) assume circular permutation and hence, the formulas for Λ_2, Λ_3 and Λ_5, and Λ_6 can be obtained from the formulas written for Λ_1 and Λ_4 by means of circular permutation.

Substituting the expressions found for the Lagrange multipliers into the first equation of system (7.7) we get

$$I_x \ddot{\mathbf{i}} = \sum_\nu x_\nu \mathbf{F}_\nu - (I_x \dot{\mathbf{i}}^2) \mathbf{i} -$$
$$- \left(\sum_\nu x_\nu \mathbf{F}_\nu \cdot \mathbf{i} \right) \mathbf{i} - \frac{2 I_x I_y}{I_x + I_y} (\mathbf{i} \cdot \dot{\mathbf{j}}) \mathbf{j} - \frac{2 I_z I_x}{I_z + I_x} (\dot{\mathbf{k}} \cdot \mathbf{i}) \mathbf{k} -$$
$$- \frac{I_y}{I_x + I_y} \left(\sum_\nu x_\nu \mathbf{F}_\nu \cdot \mathbf{j} \right) \mathbf{j} - \frac{I_x}{I_x + I_y} \left(\sum_\nu y_\nu \mathbf{F}_\nu \cdot \mathbf{i} \right) \mathbf{j} -$$
$$- \frac{I_z}{I_z + I_x} \left(\sum_\nu x_\nu \mathbf{F}_\nu \cdot \mathbf{k} \right) \mathbf{k} - \frac{I_x}{I_z + I_x} \left(\sum_\nu z_\nu \mathbf{F}_\nu \cdot \mathbf{i} \right) \mathbf{k} .$$

We have

$$\mathbf{F}_\nu = (\mathbf{F}_\nu \cdot \mathbf{i}) \mathbf{i} + (\mathbf{F}_\nu \cdot \mathbf{j}) \mathbf{j} + (\mathbf{F}_\nu \cdot \mathbf{k}) \mathbf{k} ,$$
$$\frac{I_y}{I_x + I_y} = 1 - \frac{I_x}{I_x + I_y} , \ \frac{I_z}{I_z + I_x} = 1 - \frac{I_x}{I_z + I_x} ,$$
$$\sum_\nu (x_\nu \mathbf{F}_\nu \cdot \mathbf{j} - y_\nu \mathbf{F}_\nu \cdot \mathbf{i}) = L_z ,$$
$$\sum_\nu (z_\nu \mathbf{F}_\nu \cdot \mathbf{i} - x_\nu \mathbf{F}_\nu \cdot \mathbf{k}) = L_y ,$$

where as in formula (2.3), $\mathbf{L} = \sum_\nu (x_\nu \mathbf{i} + y_\nu \mathbf{j} + z_\nu \mathbf{k}) \times \mathbf{F}_\nu$. Consequently, we find that

$$\ddot{\mathbf{i}} = -\dot{\mathbf{i}}^2 \mathbf{i} - \frac{2 I_y}{I_x + I_y} (\mathbf{i} \cdot \dot{\mathbf{j}}) \mathbf{j} - \frac{2 I_z}{I_x + I_z} (\mathbf{i} \cdot \dot{\mathbf{k}}) \mathbf{k} + \frac{L_z}{I_x + I_y} \mathbf{j} - \frac{L_y}{I_x + I_z} \mathbf{k} . \quad (7.9)$$

The two other equations should be obtained by means of circular permutation.

In order to verify that Eq. (7.9) and analogous ones for unit vectors $\mathbf{j}$ and $\mathbf{k}$ are valid, we shall show that they reproduce the dynamic Euler equations (2.6).

First of all, one should note that the quantities $I_x, I_y, and I_z$ introduced by formulas (7.3) are connected to the moments of inertia relative to the axes $Cx, Cy,$ $and Cz$ by the formula

$$J_{xx} = A = I_y + I_z , \quad J_{yy} = B = I_z + I_x , \quad J_{zz} = C = I_x + I_y .$$

Therefore,

$$\frac{2 I_y}{I_x + I_y} = 1 + \frac{A - B}{C} , \quad \frac{2 I_z}{I_x + I_z} = 1 + \frac{A - C}{B} . \quad (7.10)$$

Further, the Euler formulas imply that

$$\dot{\mathbf{i}} = \omega \times \mathbf{i} = \begin{vmatrix} \mathbf{i} & \mathbf{j} & \mathbf{k} \\ p & q & r \\ 1 & 0 & 0 \end{vmatrix} = r \mathbf{j} - q \mathbf{k} .$$

Invoking a circular permutation, we have

$$\dot{\mathbf{j}} = p \mathbf{k} - r \mathbf{i} , \quad \dot{\mathbf{k}} = q \mathbf{i} - p \mathbf{j} .$$

This leads to

$$\mathbf{i}^2 = q^2 + r^2, \quad \mathbf{i} \cdot \mathbf{j} = -pq, \quad \mathbf{i} \cdot \mathbf{k} - pr,$$
$$\ddot{\mathbf{i}} = \dot{r}\mathbf{j} - \dot{q}\mathbf{k} + r(p\mathbf{k} - r\mathbf{i}) - q(q\mathbf{i} - p\mathbf{j}) = \qquad (7.11)$$
$$= \dot{r}\mathbf{j} - \dot{q}\mathbf{k} - \mathbf{i}^2\mathbf{i} + pq\mathbf{j} + rp\mathbf{k}.$$

Substituting expressions (7.10) and (7.11) into Eq. (7.9), we get

$$\dot{r}\mathbf{j} - \dot{q}\mathbf{k} = \frac{A-B}{C}pq\mathbf{j} + \frac{A-C}{B}rp\mathbf{k} + \frac{L_z}{C}\mathbf{j} - \frac{L_y}{B}\mathbf{k}.$$

Thus, the vector equation (7.9) is equivalent to the following two scalar equations

$$B\dot{q} + (A-C)rp = L_y, \quad C\dot{r} + (B-A)pq = L_z,$$

which are the second and third Euler equations in system (2.6). In a similar way, one can show that the equation for the vector $\mathbf{j}$ is equivalent to the first and third Euler equations, and the equation for the vector $\mathbf{k}$ is equivalent to the first and second Euler equations.

Consider now a system of rigid solids connected with each other by ball (spherical) joints. Let the number of joints s be equal to the number of movable solids. The friction in joints can be neglected, i.e., the constraints are considered to be ideal. Suppose that a joint with number σ connects a solid $\sigma - 1$ with the solid σ. In this case, the equations of constraints read as

$$\boldsymbol{\rho}_\sigma + x_\sigma^\sigma \mathbf{i}_\sigma + y_\sigma^\sigma \mathbf{j}_\sigma + z_\sigma^\sigma \mathbf{k}_\sigma -$$
$$-\boldsymbol{\rho}_{\sigma-1} - x_{\sigma-1}^\sigma \mathbf{i}_{\sigma-1} - y_{\sigma-1}^\sigma \mathbf{j}_{\sigma-1} - z_{\sigma-1}^\sigma \mathbf{k}_{\sigma-1} = 0, \qquad (7.12)$$
$$\sigma = \overline{1, s}.$$

Here the vectors $\boldsymbol{\rho}_\sigma, \mathbf{i}_\sigma, \mathbf{j}_\sigma, \mathbf{k}_\sigma$ corresponding to the σ-th solid have the same sense as before; $x_\rho^\sigma, y_\rho^\sigma, z_\rho^\sigma$ are the coordinates of the joint with number σ in the coordinate system $C_\rho x_\rho y_\rho z_\rho$. A fixed solid is considered to have number zero. We denote the force applied to the σth solid from the solid with number $\sigma - 1$ by means of a joint by $\mathbf{R}_\sigma$. We make use of the principle of releasability from constraints. Hence, the equations of motion of the σth solid, as written in the form of (7.6), (7.9), will contain the reaction forces $\mathbf{R}_\sigma$ and $\mathbf{R}_{\sigma+1}$. Note that the only reaction force $\mathbf{R}_s$ is applied to the sth body. We differentiate system (7.12) twice in time, and then eliminate the second derivatives using the new form of equations of motion found for each rigid solid. In this case, we get a system of s equations with respect to s unknown reaction forces $\mathbf{R}_\sigma$. The equation corresponding to any σ, which is not equal to 1 and s, contains the reaction forces $\mathbf{R}_{\sigma-1}, \mathbf{R}_\sigma$ and $\mathbf{R}_{\sigma+1}$. When $\sigma = 1$ and $\sigma = s$ we have equations with respect to $\mathbf{R}_1, \mathbf{R}_2$ and $\mathbf{R}_{s-1}, \mathbf{R}_s$, respectively. By the above, the given system of equations has the structure which is convenient for numerical solution by the method of sequential exclusion of unknown reaction forces. Having found the reactions and substituting them into the equations of motion, we obtain a system of differential equations of motion of the set of rigid solids under consideration. This system is solved with respect to the second derivatives, i.e., it is ready for numerical integration.

The theory of the vector Lagrange equations of first kind will be used in Chap. 3 of the second volume when studying the motion of a Stewart platform (in this chapter they will be called the *special form of the equations of motion of a rigid solid*).

Chapter 9
Variational Principles in Mechanics

N. N. Polyakhov, Sh. Kh. Soltakhanov, M. P. Yushkov⊙, and S. A. Zegzhda

Differential variational principles in mechanics for mechanical systems with a finite number of degrees of freedom under constraints are obtained from the corresponding scalar motion equations of these systems, as written for the tangent space to the manifold of all positions of the system which it may occupy at a given time. The concept of a virtual (possible) displacement of a system under holonomic constraints is introduced to formulate the d'Alembert–Lagrange principle, while for the derivation of the Suslov–Jourdain principle we need the concept of the virtual velocity of a mechanical system subject to nonholonomic first-order constraints. We shall discuss the Chetaev-type constraints and the relationship between the generalized d'Alembert–Lagrange and the Suslov–Jourdain principles. To formulate the Gauss principle, we introduce the concept of a virtual acceleration of a system due to linear second-order nonholonomic constraints. The differential variational principles obtained in this chapter are used to derive the principal forms of motion equations of constrained mechanical systems.

The integral variational Hamilton–Ostrogradskii and Lagrange principles, which reflect the extremal properties of the curves of motion under potential forces, are derived from the Hamilton principle of variable action. From this principle we shall also derive the Hamilton–Jacobi equation.

For the convenience of exposition the material of this chapter is split into two parts.

I) Differential Variational Principles in Mechanics

1 Classification of the Mechanical Principles. Virtual Displacements of Mechanical Systems

Classification of the mechanical principles. In developing any field of science one tries to formulate the basic provisions underlying it. Such provisions are called *principles*. In formulating the principles one tries not only to outline in the most brief form the foundations of the field of science on which the further research will depend, but also wishes that the principles would have a heuristic value, that is, they should be capable of promoting the comprehension of new phenomena in a particular field of science. Besides, new principles should also imply the previous fundamental results.

Finding new principles was always of utmost importance in mechanics. In this way, it proved possible not only to acquire new knowledge in mechanics, but also to develop new fields in different sciences and, first of all, in mathematics. A particularly clear example here is the extension of variational calculus, which is an important part of mathematical physics.

In mechanics, the principles subdivide into *differential* and *integral* ones. A principle formulated for a given time instant (respectively, a finite time interval) is called *differential* (respectively, *integral*). For example, the classical mechanics is based the Newton second law, which holds at any time instant. So, in our terms the Newton second law is a differential principle. A motion of a stationary conservative system is subject to the equation of conservation of the total mechanical energy, which is an underlying principle for the study of its motion. This principle is used on a finite time interval, and hence is an integral principle.

On the other hand, the mechanical principles can be subdivided into the *nonvariational* and *variational* ones. The above two principles are nonvariational. Unlike them, the d'Alembert–Lagrange and the Hamilton–Ostrogradskii principles, which will be considered below, are variational principles, because they prove instrumental in elucidating the properties peculiar to the actual motion of a mechanical system (in contrast with other virtual motions that are possible with the applied constraints). Furthermore, as we shall see below, the d'Alembert–Lagrange principle is a differential principle, while the Hamilton–Ostrogradskii principle is an integral one.

Thus, Newton's second law is a differential nonvariational principle, the law of conservation of the total mechanical energy is an integral nonvariational principle, the d'Alembert–Lagrange principle is a differential variational principle, and the Hamilton–Ostrogradskii principle is an integral variational principle.

Part *I* of this chapter will be concerned only with differential variational principles of mechanics. Holonomic, nonholonomic and second-order nonholonomic constraints will be denoted as $f_0^\varkappa$, $f_1^\varkappa$, $f_2^\varkappa$.

Virtual displacements of a material point. Let us first study the motion of one material point in the Cartesian coordinate system $Ox_1x_2x_3$. If a point is free, then, under the given initial conditions and when subject to the given forces, it will move along the path characterized by the radius vector

$$\mathbf{r}(t) = x_1(t)\,\mathbf{i}_1 + x_2(t)\,\mathbf{i}_2 + x_3(t)\,\mathbf{i}_3 .$$

The actual elementary displacement of a point in an infinitely small time interval dt is controlled by the vector

$$d\mathbf{r} = (dx_1, dx_2, dx_3), \quad dx_\sigma = \dot{x}_\sigma\, dt, \quad \sigma = 1, 2, 3.$$

The vector $d\mathbf{r}$ is directed along the tangent line to the path with this position of the point.

Let us now consider the case when the motion of a material point is subject to a holonomic constraint

$$f_0(t, x) = 0, \quad x = (x_1, x_2, x_3). \tag{1.1}$$

In this setting, the effect of this constraint can be assessed from the totality of all those elementary displacements that are possible for the point (without impairing the constraint (1.1)). These elementary displacements, unlike the elementary actual displacement $d\mathbf{r}$, are not actually executed over some infinitely small time interval, but can rather be looked at as the set of all conceivable elementary displacements that could be imposed on the point at a given time. These displacements depend only on the position of the point at a given time and on the constraint (1.1) imposed on the motion of the point.

Thus, we arrive at the following definition: any elementary displacement, which can be transmitted to a point from the position occupied by this point at a given time with conservation of all the constraints imposed on it, is called a *virtual (possible)*[1] *displacement*.

Unlike the true elementary displacement $d\mathbf{r}$ of a point, the above virtual displacements of a point will be denoted by

$$\delta\mathbf{r} = (\delta x_1, \delta x_2, \delta x_3) = \delta x_1\, \mathbf{i}_1 + \delta x_2\, \mathbf{i}_2 + \delta x_3\, \mathbf{i}_3. \tag{1.2}$$

Here, the notation δ reflects the fact that from the function $\mathbf{r} = \mathbf{r}(t)$ one takes the partial differential, in which the time t is assumed to be a fixed constant parameter. The components δx_σ, $\sigma = 1, 2, 3$, of the vector $\delta\mathbf{r}$ are also called the variations of coordinates x_σ, $\sigma = 1, 2, 3$.

Assume that the motion of a point is subject to the scleronomic (stationary) constraint

$$f_0(x) = 0. \tag{1.3}$$

If the point at a position $M(x)$ is given a virtual displacement $\delta\mathbf{r} = \overrightarrow{M\tilde{M}}$, then it will move to the position $\tilde{M}(x + \delta x)$. Since the motion of the system is subject to the nonreleasing holonomic constraint (1.3), the new coordinates of the point should satisfy the constraint equation

$$f_0(x + \delta x) = 0. \tag{1.4}$$

Expanding the function (1.4) in a Taylor series near the initial position M, we have

$$f_0(x + \delta x) \equiv f_0(x) + \frac{\partial f_0}{\partial x_\sigma} \delta x_\sigma + o(|\delta x_\sigma|) = 0, \tag{1.5}$$

---

[1] From the Latin '*virtualis*'.

where $o(|\delta x_\sigma|)$ collects the terms of the order higher than one. Taking into account that $f_0(x) = 0$ and abandoning the terms of the order higher then one, we have from expansion (1.5)

$$\frac{\partial f_0}{\partial x_\sigma} \delta x_\sigma = 0. \tag{1.6}$$

Thus, the variations of the coordinates under constraint (1.3) should satisfy condition (1.6). This means that in order to specify a virtual displacement of a point one may choose arbitrarily only two variations of the coordinates, while the third variation will be found from relation (1.6). It is worth pointing out that this corresponds to $l = 2$ degrees of freedom, which the point possesses when its motion is subject to one holonomic constraint (1.3).

Let us now consider the case when the motion of a point is subject to a rheonomic (non-stationary) holonomic constraint (1.1). Then, for a new virtual position $\widetilde{M}(x + \delta x)$, one should have

$$f_0(t, x + \delta x) \equiv f_0(t, x) + \frac{\partial f_0}{\partial x_\sigma} \delta x_\sigma + o(|\delta x_\sigma|) = 0,$$

whence we again have condition (1.6). So, the non-stationarity of the constraints imposes no additional conditions on the variations of the coordinates. This can be explained by the fact that the time is not varied when searching for the virtual displacement of a point.

Let us now discuss the true elementary displacement of a point $d\mathbf{r} = \overrightarrow{MM_1}$ under a scleronomic constraint (1.3). Again expanding the constraint in a Taylor series, we have

$$f_0(x + dx) \equiv f_0(x) + \frac{\partial f_0}{\partial x_\sigma} dx_\sigma + o(|dx_\sigma|) = 0,$$

and hence,

$$\frac{\partial f_0}{\partial x_\sigma} dx_\sigma = 0. \tag{1.7}$$

Comparing (1.7) and (1.6), we conclude that under the scleronomic constraint (1.3) the true elementary displacement lies in the class of virtual ones.

If the constraint is of form (1.1) (that is, it is rheonomic), then one should take into account the fact that to the actual elementary displacement at the point M_1 there corresponds the time $t + dt$, and hence, from the constraint equation we have

$$f_0(t + dt, x + dx) \equiv f_0(t, x) + \frac{\partial f_0}{\partial t} dt + \frac{\partial f_0}{\partial x_\sigma} dx_\sigma + o(|dt|, |dx_\sigma|) = 0,$$

which in turn gives

$$\frac{\partial f_0}{\partial t} dt + \frac{\partial f_0}{\partial x_\sigma} dx_\sigma = 0. \tag{1.8}$$

In other words, now the projections of the true elementary displacement satisfy the new relation (1.8), which is distinct from (1.6). Hence, the true elementary displacement under a rheonomic constraint is not virtual.

For a geometrical illustration of condition (1.6), we rewrite it using the Hamilton operator

$$\nabla f_0 \cdot \delta \mathbf{r} = 0, \quad \nabla f_0 = \frac{\partial f_0}{\partial x_\sigma} \mathbf{i}_\sigma. \tag{1.9}$$

The notation (1.9) shows that the virtual displacements of a point lie in the tangent plane $\mathbb{T}$ at the point M to the two-dimensional surface, which is given by equation (1.3). Note that the same plane contains the vector $d\mathbf{r}$, which is directed along the tangent line to the trajectory lying on the surface (1.3). So, for a scleronomic holonomic constraint, the vector of the actual displacement of a point indeed lies in the class of virtual displacements.

Under a rheonomic holonomic constraint (1.1), the vector ∇f_0 is directed along the normal vector to the surface $f_0(t, x) = 0$, and hence the vectors of virtual displacements $\delta \mathbf{r}$ again lie in the tangent plane $\mathbb{T}$ to this surface at the given point M, the end-point of the vector $d\mathbf{r} = \overrightarrow{MM_1}$ lying outside this tangent plane.

If the motion of a point is subject to two holonomic constraints

$$f_0^\varkappa(t, x) = 0, \quad \varkappa = 1, 2, \tag{1.10}$$

then in analogy with the above we get conditions of type (1.6)

$$\frac{\partial f_0^\varkappa}{\partial x_\sigma} \delta x_\sigma = 0, \quad \varkappa = 1, 2, \tag{1.11}$$

or of type (1.9)

$$\nabla f_0^\varkappa \cdot \delta \mathbf{r} = 0, \quad \varkappa = 1, 2. \tag{1.12}$$

Conditions (1.11) show that under constraints (1.10) there is only one free variation of the coordinate (which again corresponds to $l = 1$ degrees of freedom when the motion of a point is subject to two holonomic constraints (1.10)), the remaining ones can be expressed from it by formulas (1.11). In the case of scleronomic constraints, the point will move along a one-dimensional curve, which is the intersection of these surfaces. The vector $d\mathbf{r}$ is directed along the tangent line to the curve; this line also agrees with the one-dimensional domain $\mathbb{T}$, containing the vectors of virtual displacements of the point (they may differ only in the lengths and the direction). If the constraints are rheonomic, then the vector $d\mathbf{r}$ lies outside the domain $\mathbb{T}$.

The above difference between the vectors $d\mathbf{r}$ and $\delta \mathbf{r}$ will be explained on the example of a mathematical pendulum of variable length $l(t)$ (see Fig. 1).

The coordinates of a moving point satisfy the constraint equations

$$x_3 = 0, \quad x_1^2 + x_2^2 - l^2(t) = 0.$$

Fig. 1 Virtual displacement
of a point under a holonomic
rheonomic constraint

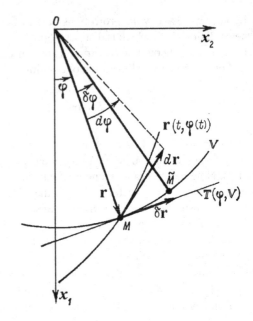

Fig. 1 Virtual displacement
of a point under a holonomic
rheonomic constraint

The set $\mathbb{V}$ of all virtual positions of the pendulum at a given time t is the circle of radius $l(t)$. Assume that to the position of the point M on this circle there corresponds the angle φ, while the angle $\varphi + \delta\varphi$ corresponds to the point $\widetilde{M}$. If the angle $\delta\varphi$ is sufficiently small, then one may approximately assume that the points M and $\widetilde{M}$ lie on one line $\mathbb{T}(\varphi, \mathbb{V})$, which touches the circle $\mathbb{V}$ at the point M.

The radius vector of the point M depends on t and φ, that is,

$$\mathbf{r} = \mathbf{r}(t, \varphi) \, .$$

A virtual displacement $\delta\mathbf{r}$ is the partial differential of this function, as calculated with a fixed t, and hence

$$\delta\mathbf{r} = \frac{\partial\mathbf{r}}{\partial\varphi} \delta\varphi \, .$$

This vector expresses the departure of the point $\widetilde{M}$ from M along the line $\mathbb{T}(\varphi, \mathbb{V})$, since the basis vector $\mathbf{e}_r$ of the polar coordinate system is equal to $\partial\mathbf{r}/\partial\varphi$.

The total differential

$$d\mathbf{r} = \frac{\partial\mathbf{r}}{\partial t} dt + \frac{\partial\mathbf{r}}{\partial\varphi} d\varphi = \left(\frac{\partial\mathbf{r}}{\partial t} + \frac{\partial\mathbf{r}}{\partial\varphi} \dot{\varphi} \right) dt = \dot{\mathbf{r}} \, dt \, ,$$

which corresponds to the oscillations of the pendulum by the law $\varphi = \varphi(t)$, is directed along the velocity vector $\mathbf{v} = \dot{\mathbf{r}}$, that is, along the tangent line to the path, and to a first-order approximation agrees with the actual displacement of the pendulum over time dt.

The difference between the virtual displacement $\delta \mathbf{r}$ of a point and the elementary actual displacement $d\mathbf{r}$ can be visualized from Fig. 1.

It would be interesting to consider the question on virtual displacements of a point from the viewpoint of the theory of curvilinear coordinates.

The motion of a free point can be described both in the Cartesian coordinates $x = (x_1, x_2, x_3)$ and in the curvilinear coordinates $q = (q^1, q^2, q^3)$. The formulas for changing between these coordinate systems are as follows:

$$q^\rho = q^\rho(t, x), \quad \rho = 1, 2, 3, \qquad (1.13)$$

$$x_\sigma = x_\sigma(t, q), \quad \sigma = 1, 2, 3. \qquad (1.14)$$

Let us introduce two systems of vectors defined by these transition formulas:

$$\mathbf{e}^\rho = \frac{\partial q^\rho}{\partial x_\tau} \mathbf{i}_\tau, \quad \mathbf{e}_\sigma = \frac{\partial x_{\tau_*}}{\partial q^\sigma} \mathbf{i}_{\tau_*}, \quad \rho, \sigma, \tau, \tau_* = 1, 2, 3. \qquad (1.15)$$

These two vector systems form the principal and the reciprocal bases of the above curvilinear coordinate system, because the following chain of equalities holds:

$$\mathbf{e}^\rho \cdot \mathbf{e}_\sigma = \frac{\partial q^\rho}{\partial x_\tau} \mathbf{i}_\tau \cdot \frac{\partial x_{\tau_*}}{\partial q^\sigma} \mathbf{i}_{\tau_*} = \frac{\partial q^\rho}{\partial x_\tau} \frac{\partial x_\tau}{\partial q^\sigma} = \delta_\sigma^\rho = \begin{cases} 0, & \sigma \neq \rho, \\ 1, & \sigma = \rho. \end{cases}$$

Assume now that the motion of a point satisfies the holonomic constraint (1.1). We write the transition formulas (1.13) in more detail as follows:

$$q^\lambda = f_*^\lambda(t, x), \quad \lambda = 1, 2, \quad q^3 = f_0(t, x). \qquad (1.16)$$

Here, the functions $f_*^\lambda(t, x)$, $\lambda = 1, 2$, are given by the researcher. By formulas (1.15) the vector $\mathbf{e}^3$ reads as

$$\mathbf{e}^3 = \frac{\partial q^3}{\partial x_\tau} \mathbf{i}_\tau = \frac{\partial f_0}{\partial x_\tau} \mathbf{i}_\tau = \nabla f_0,$$

the orthogonal vectors $\mathbf{e}_1$, $\mathbf{e}_2$ can be found from formulas (1.15), (1.16), (1.14).

Formulas (1.16) show that the variations δq^λ, $\lambda = 1, 2$, are arbitrary, δq^3 is zero by constraint (1.1), and besides

$$\delta q^3 \equiv \frac{\partial f_0}{\partial x_\sigma} \delta x_\sigma = 0 ;$$

this corresponds to the above condition on the variations of the Cartesian coordinates (1.6). Now the vector of virtual displacement of a point can be written in the form

$$\delta \mathbf{r}(t, q) = \frac{\partial \mathbf{r}}{\partial q^\lambda} \delta q^\lambda = \delta q^\lambda \, \mathbf{e}_\lambda \,. \tag{1.17}$$

This being so, the holonomic constraint (1.1) imposed on the motion defines the surface $\mathbb{V}$, which should contain the moving point, by formula (1.17) its virtual displacement lies in the plane $\mathbb{T}$ that is orthogonal to the vector $\mathbf{e}^3 = \nabla f_0$ (that is, in the tangent plane to the surface $\mathbb{V}$ at the point under consideration). It is worth pointing out that, according to the theory of Chap. 6, the introduction of a holonomic constraint (1.1) singles out, in the three-dimensional space, the one-dimensional $\mathbb{R}^K$-subspace with the basis vector $\mathbf{e}^3 = \nabla f_0$ and the orthogonal two-dimensional $\mathbb{R}_L$-subspace with the basis $\{\mathbf{e}_1, \mathbf{e}_2\}$. This means that the above tangent space $\mathbb{T}$, which contains the virtual displacements of a point, agrees with the $\mathbb{R}_L$-space.

If the motion of a point is subject to two holonomic constraints (1.10), then instead of the transition formulas (1.16) we take

$$q^1 = f_*(t, x) \,, \qquad q^{1+\varkappa} = f_0^\varkappa(t, x) \,, \quad \varkappa = 1, 2,$$

where the researcher only specifies the function $f_*(t, x)$. We thus have two vectors

$$\mathbf{e}^{1+\varkappa} = \nabla f_0^\varkappa \,, \quad \varkappa = 1, 2,$$

which single out the $\mathbb{R}^K$-subspace, and the orthogonal vector $\mathbf{e}_1$, which characterizes the $\mathbb{R}_L$-subspace.

The intersection of the surfaces corresponding to constraints (1.10) defines the curve (a one-dimensional $\mathbb{V}$-surface), which should contain a moving point, the vector of virtual displacement

$$\delta \mathbf{r} = \delta q^1 \, \mathbf{e}_1$$

being directed along the tangent line to it (in the $\mathbb{T}$-space, which agrees with the R_L-subspace).

Virtual displacements of a representative point. The above concepts can be carried over to the setting of the motion of a representative point (see Sect. 3 of Chap. 6). The vector of actual elementary displacement of a representative point $d\mathbf{y}$ can be written in the Cartesian coordinate system $Oy_1 \ldots y_{3n}$ in the form

$$d\mathbf{y} = dy_\mu \mathbf{j}_\mu \,, \quad dy_\mu = \dot{y}_\mu \, dt \,, \quad \mu = \overline{1, 3n} \,;$$

this vector is directed along the tangent line to the trajectory of the representative point.

Assume now that the motion of a representative point is subject to the holonomic constraints

$$f_0^\varkappa(t, y) = 0 \,, \quad \varkappa = \overline{1, k} \,, \quad y = (y_1, \ldots, y_{3n}) \,. \tag{1.18}$$

Then the virtual displacement of the representative point $\delta \mathbf{y} = \overrightarrow{M\tilde{M}}$, which is consistent with the constraints (1.18), is defined as the partial differential (calculated at a fixed t). The coordinates of the point $\tilde{M}(y + \delta y)$, as defined by this vector, should satisfy the constraint equations, and hence as in formulas (1.4)–(1.6), (1.11), (1.12) one may write

$$f_0^{\varkappa}(t, y + \delta y) = 0, \quad \varkappa = \overline{1, k}, \tag{1.19}$$

$$f_0^{\varkappa}(t, y + \delta y) \equiv f_0^{\varkappa}(t, y) + \frac{\partial f_0^{\varkappa}}{\partial y_{\mu}} \delta y_{\mu} + o(|\delta y_{\mu}|) = 0, \quad \varkappa = \overline{1, k}, \tag{1.20}$$

$$\frac{\partial f_0^{\varkappa}}{\partial y_{\mu}} \delta y_{\mu} = 0, \quad \varkappa = \overline{1, k}, \tag{1.21}$$

$$\nabla f_0^{\varkappa} \cdot \delta \mathbf{y} = 0, \quad \nabla f_0^{\varkappa} = \frac{\partial f_0^{\varkappa}}{\partial y_{\mu}} \mathbf{j}_{\mu}, \quad \varkappa = \overline{1, k}. \tag{1.22}$$

The constraints (1.18) define in the $3n$-dimensional Cartesian space the l-dimensional surface $\mathbb{V}(t, y)$, which at time t should contain the point $M(y_1, \ldots, y_{3n})$ (here, $l = 3n - k$ denotes the number of degrees of freedom of the representative point). If at this point one draws the tangent plane $\mathbb{T}(y, \mathbb{V})$ to the surface $\mathbb{V}(t, y)$, then in view of (1.19)–(1.22) it should contain the vectors of virtual displacement $\delta \mathbf{y} = \overrightarrow{M\tilde{M}}$.

Let us now change from the Cartesian coordinates $y = (y_1, \ldots, y_{3n})$ to the curvilinear coordinates $q = (q^1, \ldots, q^{3n})$ by the formulas

$$q^{\rho} = q^{\rho}(t, y), \quad y_{\mu} = y_{\mu}(t, q), \quad \mu, \rho = \overline{1, 3n}. \tag{1.23}$$

Formulas (1.23) define the vectors of the principal and reciprocal bases of this curvilinear coordinate system,

$$\mathbf{e}_{\sigma} = \frac{\partial y_{\mu}}{\partial q^{\sigma}} \mathbf{j}_{\mu}, \quad \mathbf{e}^{\rho} = \frac{\partial q^{\rho}}{\partial y_{\mu}} \mathbf{j}_{\mu}, \quad \mu, \rho, \sigma = \overline{1, 3n}. \tag{1.24}$$

If in the transition formulas (1.23) one takes into account the constraints (1.18) (the functions $f_*^{\lambda}(t, y)$ are defined by the researcher)

$$q^{\lambda} = f_*^{\lambda}(t, y), \quad \lambda = \overline{1, l}, \quad l = 3n - k, \quad q^{l+\varkappa} = f_0^{\varkappa}(t, y), \quad \varkappa = \overline{1, k},$$

then by (1.24) we get

$$\mathbf{e}^{l+\varkappa} = \nabla f_0^{\varkappa}, \quad \varkappa = \overline{1, k}. \tag{1.25}$$

Hence, the constraint equations singles out, in the $3n$-dimensional space, the k-dimensional subspace $\mathbb{R}^K$ with the reciprocal basis (1.25); in parallel we have the l-dimensional subspace $\mathbb{R}_L$ with the fundamental basis (see formulas (1.24))

$$\mathbf{e}_\lambda = \frac{\partial y_\mu}{\partial q^\lambda} \mathbf{j}_\mu, \quad \lambda = \overline{1, l}.$$

Since $\delta q^{l+\varkappa} = 0$, $\varkappa = \overline{1, k}$ by the constraints, we have the representation for the virtual displacement

$$\delta \mathbf{y} = \delta y_\mu \mathbf{j}_\mu = \frac{\partial y_\mu}{\partial q^\lambda} \delta q^\lambda \mathbf{j}_\mu = \delta q^\lambda \mathbf{e}_\lambda,$$

where the vectors $\mathbf{e}_\lambda$, $\lambda = \overline{1, l}$, lie in the plane $\mathbb{T}(y, \mathbb{V})$.

Virtual displacements of an arbitrary mechanical system. Out arguments in this paragraph resembles those of Sect. 7 of Chap. 6.

Assume that the position of a general mechanical system (not consisting of a finite number of material points) is described by the curvilinear coordinates $q = (q^1, \dots, q^s)$. Let us consider the differential manifold of all positions of the mechanical system under study, which it may occupy at a given time t. We shall fix some point with coordinates q^σ, $\sigma = \overline{1, s}$ on this manifold. In parallel with the above curvilinear coordinate system we shall consider a different system $q_* = (q_*^1, \dots, q_*^s)$, which is related with the first one by the formulas

$$q^\sigma = q^\sigma(t, q_*), \quad q_*^\rho = q_*^\rho(t, q), \quad \rho, \sigma = \overline{1, s}, \tag{1.26}$$

or in the differential form

$$\delta q^\sigma = \frac{\partial q^\sigma}{\partial q_*^\rho} \delta q_*^\rho, \quad \delta q_*^\rho = \frac{\partial q_*^\rho}{\partial q^\sigma} \delta q^\sigma, \quad \rho, \sigma = \overline{1, s}.$$

These quantities δq^σ and δq_*^ρ are called the contravariant components of the *tangent vector* $\delta \mathbf{y}$, the set of all vectors $\delta \mathbf{y}$ is called the *tangent space* to this manifold at a given point.[2] The components δq^σ, $\sigma = \overline{1, s}$, and δq_*^ρ, $\rho = \overline{1, s}$, are also called the *variations of coordinates*. It is expedient to write the vector $\delta \mathbf{y}$ in the form

$$\delta \mathbf{y} = \delta q^\sigma \mathbf{e}_\sigma = \delta q_*^{\sigma^*} \mathbf{e}_{\sigma^*}^*, \quad \sigma, \sigma^* = \overline{1, s}, \tag{1.27}$$

and consider the totalities of vectors $\mathbf{e}_\sigma$ and $\mathbf{e}_{\sigma^*}^*$ as the fundamental bases for the tangent space in the coordinates q and q_*. The fundamental tensors of the metric are defined by the matrices $(g_{\rho\sigma})$ and $(g_{\rho^*\sigma^*}^*)$, which are formed by the coefficients of the positive definite quadratic forms of the kinetic energy in the coordinates q and q_* (here, M denotes the mass of the whole system):

---

[2] See the book: *B.A. Dubrovin, A.T. Fomenko, and S.P. Novikov*, Modern geometry. Methods and applications, New York: Springer. 1984.

$$T^{(2)} = \frac{M}{2} g_{\rho\sigma} \dot{q}^\rho \dot{q}^\sigma, \quad \rho, \sigma = \overline{1, s},$$

$$T_*^{(2)} = \frac{M}{2} g_{\rho^*\sigma^*}^* \dot{q}_*^{\rho^*} \dot{q}_*^{\sigma^*}, \quad \rho^*, \sigma^* = \overline{1, s}. \tag{1.28}$$

We note in passing the concept of the variation of coordinates (formula (7.2) from Chap. 6) can be considered as a general rule for finding generalized forces: to this aim one should compose the elementary work δA on the virtual displacement of all active forces acting on the system; then the coefficients multiplying δq^σ, $\sigma = \overline{1, s}$, will be equal to the generalized forces Q_σ, $\sigma = \overline{1, s}$.

Example 1 Let us use this trick to find the generalized forces corresponding to the gravity forces of the first and second cylinders in Example 7 of Chap. 6. By the assumption, there are no other forces executing work with possible displacements of the system. We first fix φ_1 and φ_2. Following a displacement of the string by δs the gravity center of the first cylinder falls by $\sin \alpha \, \delta s$, and the gravity center of the second cylinder rises by $\sin \beta \, \delta s$. Hence, the total work is

$$\delta A_s = (P_1 \sin \alpha - P_2 \sin \beta) \, \delta s = Q_s \, \delta s,$$

which shows that $Q_s = P_1 \sin \alpha - P_2 \sin \beta$.

If, with fixed s and φ_2, the cylinder of weight P_1 is rotated by angle $\delta\varphi_1$, then its gravity center falls by $r_1 \sin \alpha \, \delta\varphi_1$. The following work is executed:

$$\delta A_{\varphi_1} = P_1 r_1 \sin \alpha \, \delta\varphi_1 = Q_{\varphi_1} \, \delta\varphi_1.$$

Hence $Q_{\varphi_1} = P_1 r_1 \sin \alpha$.

A similar analysis shows that $Q_{\varphi_2} = -P_2 r_2 \sin \beta$. The minus sign in this expression means that the second cylinder rises, not falls, when rotated by $\delta\varphi_2$.

Recall that by Sect. 7 of Chap. 6, to the above fundamental basis $\{\mathbf{e}_1, \dots, \mathbf{e}_s\}$ there corresponds the reciprocal basis $\{\mathbf{e}^1, \dots, \mathbf{e}^s\}$, whose vectors are defined by the relations

$$\mathbf{e}^\rho \cdot \mathbf{e}_\sigma = \delta_\sigma^\rho = \begin{cases} 0, & \rho \neq \sigma, \\ 1, & \rho = \sigma. \end{cases}$$

Let us now consider the case when the motion of a general mechanical system is subject to holonomic constraints

$$f_0^{\varkappa}(t, q) = 0, \quad \varkappa = \overline{1, k}. \tag{1.29}$$

For a virtual displacement of a system at a given time t from a position $q = (q^1, \dots, q^s)$ to the position $q + \delta q = (q^1 + \delta q^1, \dots, q^s + \delta q^s)$, the requirement that this new position should satisfy the system of constraint equations (1.29) implies the following conditions on the quantities $\delta q = (\delta q^1, \ldots, \delta q^s)$:

$$\frac{\partial f_0^\varkappa}{\partial q^\sigma} \delta q^\sigma = 0, \quad \varkappa = \overline{1, k}. \tag{1.30}$$

It shows that as soon as $l = s - k$ quantities (this number is equal to the number of degrees of freedom of a mechanical system) among $\delta q^1, \ldots, \delta q^s$ are independent, then the remaining ones will be expressible in rms of the first ones by formulas (1.30). This property could be also taken for the definition of the number of degrees of freedom.

Let us introduce the transformation (1.26) between coordinate systems in a special way, with due account of the constraint equations (1.29):

$$q_*^\lambda = f_*^\lambda(t, q), \quad \lambda = \overline{1, l}, \quad l = s - k, \quad q_*^{l+\varkappa} = f_0^\varkappa(t, q), \quad \varkappa = \overline{1, k}.$$

In this case, the constraint equations (1.29) imply that the variations of the coordinates $\delta q_*^{l+\varkappa}$, $\varkappa = \overline{1, k}$, are identical zeros, while the variations δq_*^λ, $\lambda = \overline{1, l}$, are arbitrary. Hence, the vector of virtual displacements assumes the form

$$\delta \mathbf{y} = \delta q_*^\lambda \mathbf{e}_\lambda^*. \tag{1.31}$$

By the theory of Sect. 7 of Chap. 6, in the case of holonomic constraints the vectors (7.13) of Chap. 6 form the $\mathbb{R}^K$-subspace, the orthogonal subspace $\mathbb{R}_L$ having the fundamental basis $\{\mathbf{e}_1^*, \ldots, \mathbf{e}_l^*\}$. Thus, it is the $\mathbb{R}_L$-space that contains the vectors (1.31) of virtual displacements of a general mechanical system.

2   The d'Alembert–Lagrange Principle

The d'Alembert–Lagrange principle. Let us consider the motion of holonomic mechanical system of general form with s degrees of freedom in the curvilinear coordinates $q = (q^1, \ldots, q^s)$ with the bases

$$\{\mathbf{e}_1, \ldots, \mathbf{e}_s\}, \quad \{\mathbf{e}^1, \ldots, \mathbf{e}^s\}. \tag{2.1}$$

According to Chap. 6, in this coordinate system under ideal holonomic constraints

$$f_0^\varkappa(t, q) = 0, \quad \varkappa = \overline{1, k}, \tag{2.2}$$

the motion is described by the first-order Lagrange equations in curvilinear coordinates (the second-order Lagrange equations with factors)

$$MW_\sigma \equiv \frac{d}{dt}\frac{\partial T}{\partial \dot{q}^\sigma} - \frac{\partial T}{\partial q^\sigma} = Q_\sigma + \Lambda_\varkappa \frac{\partial f_0^\varkappa}{\partial q^\sigma}, \quad \sigma = \overline{1,s}. \tag{2.3}$$

Let us consider a new system of curvilinear coordinates $q_* = (q_*^1, \dots, q_*^s)$ and introduce the transition formulas between two systems under consideration:

$$q_*^\rho = q_*^\rho(t,q), \quad q^\sigma = q^\sigma(t,q_*), \quad \rho, \sigma = \overline{1,s}. \tag{2.4}$$

With the help of formulas (2.4), we construct two systems of vectors

$$\mathbf{e}_*^\rho = \frac{\partial q_*^\rho}{\partial q^{\sigma_*}}\mathbf{e}^{\sigma_*}, \quad \mathbf{e}_\sigma^* = \frac{\partial q^\tau}{\partial q_*^\sigma}\mathbf{e}_\tau, \quad \rho, \sigma, \sigma_*, \tau = \overline{1,s}. \tag{2.5}$$

These vectors form the reciprocal and the inverse bases of the new coordinate system $q_* = (q_*^1, \dots, q_*^s)$.

As in Chap. 6, we shall change from the coordinate system $q = (q^1, \dots, q^s)$ to the system $q_* = (q_*^1, \dots, q_*^s)$ using the formulas that take into account the effective holonomic constraints (2.2):

$$\begin{aligned} q_*^\lambda &= q_*^\lambda(t,q), \quad \lambda = \overline{1,l}, \quad l = s-k, \\ q_*^{l+\varkappa} &= q_*^{l+\varkappa}(t,q) \equiv f_0^\varkappa(t,q), \quad \varkappa = \overline{1,k}. \end{aligned} \tag{2.6}$$

Hence, by formulas (2.5) and (2.6), the constraint equations (2.2) define k vectors

$$\mathbf{e}_*^{l+\varkappa} = \frac{\partial q_*^{l+\varkappa}}{\partial q^\sigma}\mathbf{e}^\sigma = \frac{\partial f_0^\varkappa}{\partial q^\sigma}\mathbf{e}^\sigma = \nabla f_0^\varkappa, \quad \varkappa = \overline{1,k},$$

which form the reciprocal basis of the K-space. By (2.5) the orthogonal L-space is formed by the vectors of the fundamental basis

$$\mathbf{e}_\lambda^* = \frac{\partial q^\tau}{\partial q_*^\lambda}\mathbf{e}_\tau, \quad \lambda = \overline{1,l}.$$

Now in the L-space the motion will be described by the system of second-order Lagrange equations

$$MW_\lambda^* \equiv \frac{d}{dt}\frac{\partial T^*}{\partial \dot{q}_*^\lambda} - \frac{\partial T^*}{\partial q_*^\lambda} = Q_\lambda^*, \quad \lambda = \overline{1,l}. \tag{2.7}$$

Here, T^* denotes the kinetic energy of the systems in the coordinates q_*.

It is worth noting that the form of the differential motion equations depends on the selected coordinate system. Let us find an expression equivalent to these equations and which is invariant with respect to the choice of a coordinate system.

Using formulas (2.4), given a fixed t, we write down the partial differentials (variations of the coordinates) δq_*^ρ and δq^σ:

$$\delta q_*^\rho = \frac{\partial q_*^\rho}{\partial q^\sigma}\, \delta q^\sigma\,,\quad \delta q^\sigma = \frac{\partial q^\sigma}{\partial q_*^\tau}\, \delta q_*^\tau\,,\qquad \rho,\sigma,\tau = \overline{1,s}\,. \tag{2.8}$$

In writing the conversion formulas (2.4) in the form (2.6), we take into account the constraints (2.2) and write formulas (2.8) as follows:

$$\delta q_*^\lambda = \frac{\partial q_*^\lambda}{\partial q^\sigma}\, \delta q^\sigma\,,\quad \delta q^\sigma = \frac{\partial q^\sigma}{\partial q_*^\lambda}\, \delta q_*^\lambda\,,\qquad \sigma = \overline{1,s}\,,\quad \lambda = \overline{1,l}\,, \tag{2.9}$$

where

$$\delta q_*^{l+\varkappa} = \frac{\partial f_0^\varkappa}{\partial q^\sigma}\, \delta q^\sigma = 0\,,\qquad \sigma = \overline{1,s}\,,\quad \varkappa = \overline{1,k}\,. \tag{2.10}$$

So, by formulas (2.9) and (2.10), the variations δq_*^λ, $\lambda = \overline{1,l}$, are free, the variations $\delta q_*^{l+\varkappa}$, $\varkappa = \overline{1,k}$, are zero, while the variations δq^σ, $\sigma = \overline{1,s}$, are related by formulas (2.10) (that is, among them one may specify l ones in an arbitrary way, the remaining k ones can be expressed from them by formulas (2.10)).

Let us now multiply equations (2.3) by the variations of the coordinates δq^σ and add the resulting expressions. We have

$$\left(\frac{d}{dt}\frac{\partial T}{\partial \dot q^\sigma} - \frac{\partial T}{\partial q^\sigma} - Q_\sigma - \Lambda_\varkappa \frac{\partial f_0^\varkappa}{\partial q^\sigma}\right) \delta q^\sigma = 0\,. \tag{2.11}$$

Consider the above double sum

$$\Lambda_\varkappa \frac{\partial f_0^\varkappa}{\partial q^\sigma}\, \delta q^\sigma\,. \tag{2.12}$$

Next, multiplying by $\Lambda_\varkappa$ the conditions (2.10) imposed on the variations and adding, we obtain

$$\Lambda_\varkappa \frac{\partial f_0^\varkappa}{\partial q^\sigma}\, \delta q^\sigma = 0\,. \tag{2.13}$$

This shows that the double sum (2.12) is zero, and hence equation (2.11) assumes the form

$$\left(\frac{d}{dt}\frac{\partial T}{\partial \dot q^\sigma} - \frac{\partial T}{\partial q^\sigma} - Q_\sigma\right) \delta q^\sigma = 0\,. \tag{2.14}$$

However, the resulting expression (2.14) is the scalar equation

$$\left(M\mathbf{W} - \mathbf{Y}\right)\cdot \delta\mathbf{y} = 0\,. \tag{2.15}$$

Here, the vector

$$\delta\mathbf{y} = \delta q^\sigma\, \mathbf{e}_\sigma \tag{2.16}$$

is a virtual displacement of the system if δq^σ, $\sigma = \overline{1,s}$, all satisfy conditions (2.10).

We now show that equations (2.7) can also be reduced to the same expression (2.15). Indeed, multiplying equations (2.7) by δq_*^λ and adding the resulting expressions, this establishes

$$(MW_\lambda^* - Q_\lambda^*)\,\delta q_*^\lambda = 0. \tag{2.17}$$

As a result, we again arrive at expression (2.15), in which now the virtual displacement is in the form

$$\delta \mathbf{y} = \delta q_*^\lambda\, \mathbf{e}_\lambda^*. \tag{2.18}$$

This being so, from the motion equations (2.3) and (2.7) we have obtained equation (2.15), which is independent of the choice of the curvilinear coordinate system.

Let us now prove the converse assertion that (2.15) implies the motion equations (2.3) and (2.7).

Indeed, rewriting the scalar product (2.15) as a sum of products of covariant and contravariant components of the vector factors and taking into account (2.18), we obtain formula (2.17). In turn, since δq_*^λ, $\lambda = \overline{1, l}$ are arbitrary, from (2.17) we see that the coefficients of these factors are zero, which means that the second-order Lagrange equations (2.7) are satisfied.

The derivation of the motion equations in the form (2.3) from formula (2.15) is more lengthy. We shall now employ representation (2.16) to expand the scalar product (2.15) of two vectors. Hence, expression (2.14) follows from formula (2.15). Subtracting from it the zero double sum (2.13), we arrive at expression (2.11). We split the s terms from (2.11) into two groups. In the first group, we place the terms with δq^λ, $\lambda = \overline{1, l}$, $l = s - k$, assuming that these variations are independent. The remaining (dependent) variations $\delta q^{l+\varkappa}$, $\varkappa = \overline{1, k}$, can be expressed from them using the conditions (2.10). In this second group of terms with $\delta q^{l+1}, \ldots, \delta q^s$ in the coefficients we shall take such $\Lambda_\varkappa$, $\varkappa = \overline{1, k}$, for which the coefficients multiplying these terms are zero; that is, we shall require that the relations

$$MW_{l+\varkappa} - Q_{l+\varkappa} - \Lambda_\varkappa \frac{\partial f_0^\varkappa}{\partial q^{l+\varkappa}} = 0, \quad \varkappa = \overline{1, k} \tag{2.19}$$

be satisfied. In the first l terms in the sum (2.11), the quantities $\delta q^1, \ldots, \delta q^l$ are independent, and hence this sum may vanish only if the coefficients of these terms are zero, that is, if

$$MW_\lambda - Q_\lambda - \Lambda_\varkappa \frac{\partial f_0^\varkappa}{\partial q^\lambda} = 0, \quad \lambda = \overline{1, l}. \tag{2.20}$$

Thus, we have obtained from equation (2.15) the first-order Lagrange equations in the curvilinear coordinates (2.19), (2.20).

This being so, from the principal forms of the equations of the dynamics of holonomic systems (2.3), (2.7), we have obtained, on one side, formula (2.15), and on the other side it turns out that equation (2.15) implies the nonvariational motion equations of the holonomic systems (2.3), (2.7). Hence, equation (2.15) can be taken for a principle mechanical characteristic of the motion of holonomic systems. This principle, which is called the *d'Alembert–Lagrange principle*, asserts that under ideal holonomic retaining constraints the work of the inertia forces and active forces exerted on a mechanical system is zero on a virtual displacement of the system.

The vector equation of constrained motion reads as

$$M\mathbf{W} = \mathbf{Y} + \mathbf{R},$$

and hence principle (2.15) can be written in the form

$$\mathbf{R} \cdot \delta\mathbf{y} = 0, \tag{2.21}$$

which reflects the orthogonality of the reaction $\mathbf{R}$ of the above ideal constraints (2.2) to the virtual displacement $\delta\mathbf{y}$. Comparing (2.21) with (2.13) shows that

$$\mathbf{R} = \Lambda_\varkappa \frac{\partial f_0^\varkappa}{\partial q^\sigma} \mathbf{e}^\sigma \equiv \Lambda_\varkappa \nabla f_0^\varkappa,$$

that is, under ideal constraints its reaction agrees with the component $\mathbf{R}^K$.

Principle (2.15) can be also extended to nonideal holonomic constraints (2.2), for which $\mathbf{R}_L \neq 0$. Assuming in this case that this vector is a known vector function of time and the generalized coordinates and velocities, it can be considered as an active force, thereby rewriting formula (2.15) in the form

$$\left(M\mathbf{W} - \mathbf{Y} - \mathbf{R}_L\right) \cdot \delta\mathbf{y} = 0. \tag{2.22}$$

The d'Alembert–Lagrange principle for system of material points. Using the concept of a representative point (see Chap. 6) for describing the motion of a system of material points, the d'Alembert–Lagrange principle (2.15) in the $3n$-dimensional Cartesian space can be written in the form

$$\left(M\ddot{y}_\mu - Y_\mu\right)\delta y_\mu = 0, \quad \mu = \overline{1, 3n}. \tag{2.23}$$

For the three-dimensional space, formula (2.23) assumes the form

$$\left(m_\mu\ddot{x}_\mu - X_\mu\right)\delta x_\mu = 0, \quad \mu = \overline{1, 3n}. \tag{2.24}$$

Now the holonomic constraints impose the following restrictions

$$f_0^\varkappa(t, x) = 0, \quad \varkappa = \overline{1, k}, \quad x = (x_1, \dots, x_{3n}) \tag{2.25}$$

on the coordinates of the n points under consideration, and hence the variations of the coordinates from formula (2.24) satisfy the equations

$$\frac{\partial f_0^{\varkappa}}{\partial x_{\mu}} \delta x_{\mu} = 0, \quad \mu = \overline{1, 3n}, \quad \varkappa = \overline{1, k}. \tag{2.26}$$

Changing in formula (2.24) to the numeration of points with masses m_{ν}, $\nu = \overline{1, n}$, in the vector form, we obtain the principle of virtual displacements for a system of material points:

$$\left(m_{\mu} \ddot{\mathbf{r}}_{\nu} - \mathbf{F}_{\nu} \right) \cdot \delta \mathbf{r}_{\nu} = 0, \quad \nu = \overline{1, n}. \tag{2.27}$$

In formula (2.27) the virtual displacements of material points are given by the formulas

$$\delta \mathbf{r}_{\nu} = \delta x_{\nu 1} \mathbf{i}_1 + \delta x_{\nu 2} \mathbf{i}_2 + \delta x_{\nu 3} \mathbf{i}_3,$$

where the variations $\delta x_{\nu j}$ ($\nu = \overline{1, n}$, $j = \overline{1, 3}$), satisfy conditions (2.26), inasmuch as

$$\delta x_{\mu} = \delta x_{\nu j}, \quad \mu = 3(\nu - 1) + j, \quad \nu = \overline{1, n}, \quad j = \overline{1, 3}$$

(see the definition of a representative point in Chap. 6).

Let us now dwell upon the concept of ideality of the holonomic constraints (2.2). Recall that these constraints in the case of motion of n material points can be put in the form (2.25). The ideality of the constraints implies the following chain of equalities (we abandon the use of "dummy indexes" and write the summation signs)

$$\mathbf{R} \cdot \delta \mathbf{y} = \sum_{\varkappa=1}^{k} \Lambda_{\varkappa} \nabla f_0^{\varkappa} \cdot \delta \mathbf{y} = \sum_{\varkappa=1}^{k} \Lambda_{\varkappa} \sum_{\mu=1}^{3n} \frac{\partial f_0^{\varkappa}}{\partial y_{\mu}} \delta y_{\mu} =$$

$$= \sum_{\varkappa=1}^{k} \Lambda_{\varkappa} \sum_{\nu=1}^{n} \left(\frac{\partial f_0^{\varkappa}}{\partial x_{\nu 1}} \delta x_{\nu 1} + \frac{\partial f_0^{\varkappa}}{\partial x_{\nu 2}} \delta x_{\nu 2} + \frac{\partial f_0^{\varkappa}}{\partial x_{\nu 3}} \delta x_{\nu 3} \right) = 0,$$

and hence,

$$\sum_{\nu=1}^{n} \mathbf{R}_{\nu} \cdot \delta \mathbf{r}_{\nu} = 0, \tag{2.28}$$

where $\mathbf{R}_{\nu}$ is the reaction force acting on the mass m_{ν} from the side of all constraints (2.25):

$$\mathbf{R}_{\nu} = \sum_{\varkappa=1}^{k} \Lambda_{\varkappa} \left(\frac{\partial f_0^{\varkappa}}{\partial x_{\nu 1}} \mathbf{i}_1 + \frac{\partial f_0^{\varkappa}}{\partial x_{\nu 2}} \mathbf{i}_2 + \frac{\partial f_0^{\varkappa}}{\partial x_{\nu 3}} \mathbf{i}_3 \right).$$

Relation (2.28) can be taken for the definition of the ideality of constraints (2.25). However, without the concept of a representative point this definition is puzzling and has an axiomatic character. In the actual fact, as was shown above, condition (2.28) is a simple extension of the standard concept of an ideal constraint for one point to the case of a representative point.

3 The Suslov–Jourdain Principle. Chetaev-Type Constraints. The Generalized d'Alembert–Lagrange Principle

The Suslov–Jourdain principle. Let us now consider the differential variational principle peculiar to nonholonomic systems. Our arguments will be similar to those of the previous subsection.

Assume now that the motion of a mechanical system (which is described, as above, in the curvilinear coordinates $q = (q^1, \ldots, q^s)$ with bases (2.1)) is subject to the ideal nonholonomic constraints

$$f_1^\varkappa(t, q, \dot{q}) = 0, \quad \varkappa = \overline{1, k}. \tag{3.1}$$

Let us relate the generalized velocities $\dot{q} = (\dot{q}^1, \ldots, \dot{q}^s)$ and the pseudo-velocities (quasi-velocities) $v_* = (v_*^1, \ldots, v_*^s)$ by the transformations:

$$v_*^\rho = v_*^\rho(t, q, \dot{q}), \quad \dot{q}^\sigma = \dot{q}^\sigma(t, q, v_*), \quad \rho, \sigma = \overline{1, s}; \tag{3.2}$$

here, t and q are considered as parameters. As was shown in Chap. 6, in this setting one may introduce the nonholonomic bases

$$\varepsilon^\rho = \frac{\partial v_*^\rho}{\partial \dot{q}^{\sigma_*}} \mathbf{e}^{\sigma_*}, \quad \varepsilon_\sigma = \frac{\partial \dot{q}^\tau}{\partial v_*^\sigma} \mathbf{e}_\tau, \quad \rho, \sigma, \sigma_*, \tau = \overline{1, s}.$$

Recall that, if in the transformations (3.2) one takes into account the nonholonomic constraints, then the tangent space decomposes into the direct sum of the K- and L-subspaces.

Let us write down the partial differentials $\delta' v_*^\rho$ and $\delta' \dot{q}^\sigma$ of transformations (3.2), assuming as before that t and q are parameters:

$$\delta' v_*^\rho = \frac{\partial v_*^\rho}{\partial \dot{q}^\sigma} \delta' \dot{q}^\sigma, \quad \delta' \dot{q}^\sigma = \frac{\partial \dot{q}^\sigma}{\partial v_*^\rho} \delta' v_*^\rho, \quad \rho, \sigma = \overline{1, s}. \tag{3.3}$$

If in transformations (3.2) the quasi-velocities $v_*^{l+\varkappa}$, $\varkappa = \overline{1, k}$, are equated to the functions $f_1^\varkappa$, $\varkappa = \overline{1, k}$, as given by the constraint equations (3.1), then formulas (3.3) assume the form

$$\delta' v_*^\lambda = \frac{\partial v_*^\lambda}{\partial \dot{q}^\sigma} \delta' \dot{q}^\sigma, \quad \delta' \dot{q}^\sigma = \frac{\partial \dot{q}^\sigma}{\partial v_*^\lambda} \delta' v_*^\lambda, \quad \sigma = \overline{1, s}, \quad \lambda = \overline{1, l}, \tag{3.4}$$

the conditions

$$\delta' v_*^{l+\varkappa} = \frac{\partial f_1^\varkappa}{\partial \dot{q}^\sigma} \delta' \dot{q}^\sigma = 0, \quad \sigma = \overline{1, s}, \quad \varkappa = \overline{1, k}, \tag{3.5}$$

being satisfied. Let us introduce the vector

$$\delta'\mathbf{V} = \delta'\dot{q}^\sigma \mathbf{e}_\sigma = \frac{\partial \dot{q}^\sigma}{\partial v_*^\lambda}\delta'v_*^\lambda \mathbf{e}_\sigma = \delta'v_*^\lambda \boldsymbol{\varepsilon}_\lambda \tag{3.6}$$

and construct (together with the vector $\mathbf{V}$, which is considered as a vector function of the variables $\dot{q} = (\dot{q}^1, \ldots, \dot{q}^s)$), the new vector

$$\widetilde{\mathbf{V}} = \mathbf{V} + \delta'\mathbf{V} = (\dot{q}^\sigma + \delta'\dot{q}^\sigma)\,\mathbf{e}_\sigma\,.$$

We next substitute the variables $\dot{q}^\sigma + \delta'\dot{q}^\sigma$ of the velocity vector $\widetilde{\mathbf{V}}$ into the constraint equations (3.1) and expand the functions $f_1^\varkappa$ (*qua* functions of only the variables $\dot{q}^\sigma$) into Taylor series near the point with coordinates $(\dot{q}^1, \ldots, \dot{q}^s)$, assuming that $q^1, \ldots, q^s, t$ are fixed parameters:

$$f_1^\varkappa(t, q, \dot{q} + \delta'\dot{q}) = f_1^\varkappa(t, q, \dot{q}) + \nabla'f_1^\varkappa \cdot \delta'\mathbf{V} + o(|\delta'\mathbf{V}|)\,, \qquad \varkappa = \overline{1, k}\,. \tag{3.7}$$

From these equalities we see that if, at some time t in an available position $(q^1, \ldots, q^s)$ with available generalized velocities $(\dot{q}^1, \ldots, \dot{q}^s)$, the velocity $\mathbf{V}$ is kinematically virtual, then, to a first order of approximation, the velocity $\widetilde{\mathbf{V}} = \mathbf{V} + \delta'\mathbf{V}$ is also kinematically virtual under the condition that

$$\nabla'f_1^\varkappa \cdot \delta'\mathbf{V} = 0\,, \qquad \varkappa = \overline{1, k}\,. \tag{3.8}$$

Thus, the set of vectors $\delta'\mathbf{V}$ satisfying equations (3.8), characterizes the kinematically virtual variations of the velocity $\mathbf{V}$, which are admitted by the nonholonomic constraints at a time t when the system is in the position $(q^1, \ldots, q^s)$. An arbitrary vector $\delta'\mathbf{V}$ satisfying (3.8), is called the *variation of the velocity* $\mathbf{V}$.

In Chap. 6 we derived the motion equations for nonholonomic systems in the form of Maggi equations

$$(MW_\sigma - Q_\sigma)\frac{\partial \dot{q}^\sigma}{\partial v_*^\lambda} = 0\,, \quad \lambda = \overline{1, l}\,, \quad l = s - k\,, \tag{3.9}$$

and in the form of first-order Lagrange equations in curvilinear coordinates for nonholonomic systems

$$\frac{d}{dt}\frac{\partial T}{\partial \dot{q}^\sigma} - \frac{\partial T}{\partial q^\sigma} = Q_\sigma + \Lambda_\varkappa \frac{\partial f_1^\varkappa}{\partial \dot{q}^\sigma}\,, \quad \sigma = \overline{1, s}\,. \tag{3.10}$$

These equations depend on the chosen system of curvilinear coordinates. Let us construct the corresponding expression, which is invariant with respect to the choice of the coordinate system.

We shall start with the Maggi equations (3.9). Multiplying each equation by $\delta'v_*^\lambda$ and adding, we find that the Maggi equations (3.9) are equivalent to the equation

$$\left(M W_\sigma - Q_\sigma\right)\frac{\partial \dot{q}^\sigma}{\partial v_*^\lambda}\,\delta' v_*^\lambda = 0\,, \tag{3.11}$$

which by formulas (3.4) can be put in the form

$$\left(M W_\sigma - Q_\sigma\right)\delta' \dot{q}^\sigma = 0\,. \tag{3.12}$$

Using the formula for the variation of velocity (3.6), the expression (3.12) can be written in the vector form

$$\left(M\mathbf{W} - \mathbf{Y}\right)\cdot \delta'\mathbf{V} = 0\,. \tag{3.13}$$

Let us now reduce equations (3.10) to the form (3.13). Multiplying these equations by the variations $\delta' \dot{q}^\sigma$ and adding, this establishes

$$\left(M W_\sigma - Q_\sigma - \Lambda_\varkappa \frac{\partial f_1^\varkappa}{\partial \dot{q}^\sigma}\right)\delta' \dot{q}^\sigma = 0\,. \tag{3.14}$$

Let us consider the double sum

$$\Lambda_\varkappa \frac{\partial f_1^\varkappa}{\partial \dot{q}^\sigma}\,\delta' \dot{q}^\sigma\,. \tag{3.15}$$

On multiplying the conditions by the variations $\delta' \dot{q}^\sigma$ from (3.5) by $\Lambda_\varkappa$ and adding, we find that

$$\Lambda_\varkappa \frac{\partial f_1^\varkappa}{\partial \dot{q}^\sigma}\,\delta' \dot{q}^\sigma = 0\,, \tag{3.16}$$

that is, expression (3.15) is zero, and thus formula (3.14) can be put in the form

$$\left(M W_\sigma - Q_\sigma\right)\delta' \dot{q}^\sigma = 0\,. \tag{3.17}$$

So, using the vector representation for the variation of the velocity (3.6), formula (3.17) can be put in the form (3.13), the result required.

This being so, from Eqs. (3.9) and (3.10), which express Newton's second law under the presence of ideal nonholonomic constraints (3.1), we derived expression (3.13). Let us now prove the converse assertion: expression (3.13) implies the motion equations (3.9) and (3.10). If we shall manage to do this, then expression (3.13) can be taken for the principle of mechanics which holds under ideal nonholonomic constraints.

Thus, let us take assertion (3.13) as a primitive principle of mechanics. Using the representation (3.6) for the variation of the velocity, the scalar product (3.13) can be written in the form (3.11). Since the variations $\delta' v_*^\lambda$, $\lambda = \overline{1, l}$, are independent, from this formula we arrive at the Maggi equations (3.9).

Let us now derive equations (3.10) from (3.13). Writing the scalar product (3.13) in the form (3.12) and subtracting from it the zero expression (3.16), we obtain (3.14).

We shall assume that the variations of the velocities $\delta' \dot{q}^\lambda$, $\lambda = \overline{1, l}$, are independent, and that the variations $\delta' \dot{q}^{l+\varkappa}$, $\varkappa = \overline{1, k}$, can be expressed from them by formulas (3.5). Hence, choosing $\Lambda_\varkappa$, $\varkappa = \overline{1, k}$, in the factors of the k last terms in a way that these coefficients vanish and equating to zero the coefficients in the first l terms with independent variations, we obtain Eq. (3.10).

As a result, formula (3.13) expresses the principle of mechanics that takes place under ideal nonholonomic constraints (3.1). We shall call it the *Suslov–Jourdain principle*.[3] This principle asserts that under ideal nonholonomic constraints on the motion of a system the scalar product of the inertia forces and the active forces exerted on a mechanical system by the vector of variation of the generalized velocity is zero.

Chetaev-type constraints. The generalized d'Alembert–Lagrange principle.
Many researchers, instead of deriving the principle in the Suslov–Jourdain form, were trying to extend the d'Alembert–Lagrange principle, which holds ab initio only for holonomic constraints, to the case of nonholonomic constraints. But for this purpose, it was required to introduce the concept of virtual displacements for nonholonomic systems.[4] This is why N.G. Chetaev[5] postulated that the above virtual displacements $\delta q = (\delta q^1, \dots, \delta q^s)$ under nonholonomic constraints (3.1) should obey the conditions

$$\frac{\partial f_1^\varkappa}{\partial \dot{q}^\sigma} \delta q^\sigma = 0, \quad \varkappa = \overline{1, k}. \tag{3.18}$$

The constraints satisfying these conditions were later called the *Chetaev-type constraints*. This question was addressed by many leading scientists of this time— for example, J. Papastavridis[6] calls this postulate the definition by Maurer–Appel–Chetaev–Hamel.

However, the Chetaev conditions (3.18) on the virtual displacements agree with the restrictions (3.5) imposed by the nonholonomic constraints (3.1) on the varia-

---

[3] This principle was put forward in 1908–1909 by Ph.E.B. Jourdain, but almost 10 years earlier, using different terminology, this principle was formulated by G.K. Suslov in his textbook. This is why the principle (3.13) should be properly called the Suslov–Jourdain principle (see the paper *N.N. Polyakhov, S.A. Zegzhda, M.P. Yushkov*, "The Suslov–Jourdain principle as a corollary of the dynamics equations," in Collection of Sci. and Method. Papers in Theoretical Mechanics, issue 12. Moscow, Vyshchaya shkola, 1982, pp. 72–79 [in Russian]).

[4] This is what V.I. Kirgetov, a famous mechanician, writes on this subject (see his paper "On 'virtual displacements' of material systems with linear differential constraints of the second order", Appl. Math. Mech. Vol. 23, issue 4, 1959, pp. 956–962 [in Russian]): "The concept of 'virtual displacement' of a system is, of course, a basic one in analytical mechanics. This is not just one of the concepts of analytical mechanics, but a concept on which the whole structure of analytical mechanics has been erected, a concept determining the character of analytical mechanics, the degree of its generality, the limits of its applications. Analytical mechanics extends only over those material systems for which the concept of 'virtual displacements' has been established or, in other words, the 'virtual displacements' have been defined".

[5] See the classical paper N.G. Chetaev: *N.G. Chetaev*, "On the Gauss principle", Izv. Fiz.-Matem. Obsch. Kazan Univ. Vol. 6, Ser. 3 (1932–1933), pp. 68–71 [in Russian].

[6] See the book *J.G. Papastavridis*, Analytical Mechanics. Oxford: University Press. 2002.

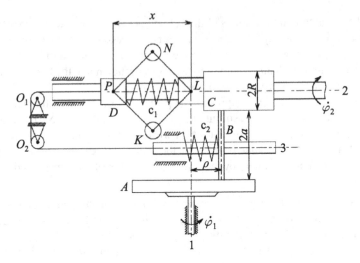

Fig. 2 Novoselov friction gear

tions of the velocity. Indeed,[7] if one divides equations (3.18) by an infinitely small time interval τ_0, as introduced by Herz, then the resulting factors $\delta q^\sigma / \tau_0$ can be regarded as virtual velocities $\delta' \dot{q}^\sigma$. Hence, the Suslov–Jourdain principle turns out to be equivalent to the d'Alembert–Lagrange principle (2.15), but in which the contravariant components of a virtual displacement $\delta \mathbf{y} = \delta q^\sigma \, \mathbf{e}_\sigma$ satisfy Eq. (3.18). The d'Alembert–Lagrange principle, as extended to nonholonomic systems in the case of Chetaev-type constraints, is called the *generalized d'Alembert–Lagrange principle*.

Example 2 *The motion equations of a Novoselov friction gear* (derivation of the motion equations using the Suslov–Jourdain principle). Let us derive the motion equations of a friction gear, which was first considered by V.S. Novoselov.[8] The friction gear (Fig. 2), which transmits the rotation from shaft *1* to shaft *2*, is composed of a disc *A*, with a rigidly fixed shaft *1*, a ring *B*, which rotates freely on shaft *3*, shaft2 with drum *C*, a centrifugal governor with masses *K* and *N* and a stiffener spring c_1. An operation of the governor sleeve *D* using a string which passes over fixed pulleys O_1 and O_2 and the stiffener spring c_2 moves the shaft *3* with the ring *B*, thereby changing the distance ρ between the middle circle of the ring *B* from the shaft axis *1*. The radius of the ring *B* is a; besides, $PN = NL = LK = KP = l$.

The position of the friction gear depends on the following generalized coordinates: the shaft rotation angles $q^1 = \varphi_1$ and $q^2 = \varphi_2$, and the distance $q^3 = x$ between the

---

[7] See the paper *N.N. Polyakhov*, "On differential principles of mechanics obtained from the motion equations of nonholonomic systems," Vest. Leningr. Univ. 1974. Issue 3, issue 13, pp. 106–116 [in Russian].

[8] See the paper *V.S. Novoselov*, "Example of non-Chetaev-type nonlinear nonholonomic constraints", Vestn. Leningr. Univ. 1957, issue 19, pp. 106–111 [in Russian].

governor sleeve D and the hinge joint L. Figure 2 shows that the distance ρ is related with x as follows:

$$x - \rho = c \equiv \text{const}.$$

The above system is subject to the nonholonomic constraint

$$f_1^1(t, q^1, q^2, q^3, \dot{q}^1, \dot{q}^2, \dot{q}^3) \equiv (x - c)\dot{\varphi}_1 - R\dot{\varphi}_2 = 0. \tag{3.19}$$

Under the no-slip condition, constraint (3.19) can be expressed from the condition that the circular velocities of points of contact of the ring B with the disc A and the drum C be equal.

The kinetic and potential energies are determined, respectively, from the expressions

$$T = \frac{1}{2}\left[J_A \dot{\varphi}_2^2 + J_C \dot{\varphi}_2^2 + m_D \dot{x}^2 + m_B \dot{\rho}^2 + J_B \frac{R^2}{a^2} \dot{\varphi}_2^2 + \right.$$

$$\left. + 2m_N \left(\left(l^2 - \frac{x^2}{4} \right) \dot{\varphi}_2^2 + \frac{l^2 \dot{x}^2}{4l^2 - x^2} \right) \right],$$

$$\Pi = \frac{1}{2} c_1 (\delta_1 + x - x_0)^2 + \frac{1}{2} c_2 (\delta_2 + x_0 - x)^2.$$

Here, δ_1, δ_2 are the static spring deformations of rigidities c_1, c_2, and x_0 is the static deflection of the governor sleeve D from the hinge joint L.

The Suslov–Jourdain principle for this system reads as

$$(MW_1 - Q_1)\,\delta'\dot{\varphi}_1 + (MW_2 - Q_2)\,\delta'\dot{\varphi}_2 + (MW_3 - Q_3)\,\delta'\dot{x} = 0. \tag{3.20}$$

The constraint between the variations of the velocities is as follows:

$$\frac{\partial f_1^1}{\partial \dot{\varphi}_1} \delta'\dot{\varphi}_1 + \frac{\partial f_1^1}{\partial \dot{\varphi}_2} \delta'\dot{\varphi}_2 = 0, \tag{3.21}$$

therefore, in equation (3.20) the variations $\delta'\dot{\varphi}_2$ and $\delta'\dot{x}$ are independent. Using (3.21) to express the variation $\delta'\dot{\varphi}_1$ in terms of $\delta'\dot{\varphi}_2$ and employing equation (3.20), this gives

$$(MW_1 - Q_1)\frac{R}{x - c} + (MW_2 - Q_2) = 0, \tag{3.22}$$

$$MW_3 - Q_3 = 0. \tag{3.23}$$

Here, $Q_1 = M_1$, $Q_2 = -M_2$ are the moments of force applied to shafts 1 and 2 respectively, and $Q_3 = -\partial\Pi/\partial x$.

As follows from the general theory, the equations thus obtained agree with the Maggi equations. It is worth noting that the second equation is the standard second-order Lagrange equation, because the coordinate x is holonomic.

We have

$$MW_\sigma = \frac{d}{dt}\frac{\partial T}{\partial \dot{q}^\sigma} - \frac{\partial T}{\partial q^\sigma}, \quad \sigma = \overline{1,3},$$

and hence Eqs. (3.22) and (3.23) can be put in the form

$$J_A \frac{R}{x-c}\ddot{\varphi}_1 + J(x)\ddot{\varphi}_2 - m_N x \dot{x}\dot{\varphi}_2 = M_1\frac{R}{x-c} - M_2,$$

$$m(x)\ddot{x} + \frac{1}{2}m_N x\dot{\varphi}_2^2 + \frac{2l^2 x}{(4l^2+x^2)^2}m_N \dot{x}^2 = c_1(-\delta_1 - x + x_0) + c_2(-x+x_0+\delta_2).$$

$$(3.24)$$

Here

$$J(x) = J_C + J_B \frac{R^2}{a^2} + \frac{1}{2}m_N(4l^2 - x^2),$$

$$m(x) = m_B + m_D + \frac{2m_N l^2}{4l^2 - x^2}.$$

$$(3.25)$$

The motion equations (3.24) together with the constraint equation (3.19) form a closed system for the functions $\varphi_1(t)$, $\varphi_2(t)$, and $x(t)$.

It should be noted that if in the first equation of system (3.24) one substitutes the time-differentiated constrained equation (3.19), then the equations will assume the form of Appel equation.[9]

4 The Gauss Principle

Let us now apply the ideas of the previous subsections to derive the Gauss principle, which holds under linear second-order ideal nonholonomic constraints

$$f_2^\varkappa(t,q,\dot{q},\ddot{q}) \equiv a_{2\sigma}^{l+\varkappa}(t,q,\dot{q})\ddot{q}^\sigma + a_{2,0}^{l+\varkappa}(t,q,\dot{q}) = 0, \quad \varkappa = \overline{1,k}, \quad (4.1)$$

in the study of the motion of a mechanical system in the curvilinear coordinate system $q = (q^1, \dots, q^s)$ with bases (2.1). In this setting we shall need the generalized Maggi equations

$$(MW_\sigma - Q_\sigma)\frac{\partial \ddot{q}^\sigma}{\partial w_*^\lambda} = 0, \quad \lambda = \overline{1,l}, \quad l = s-k, \quad (4.2)$$

and the generalized second-order Lagrange equations with the factors

$$MW_\sigma = Q_\sigma + \Lambda_\varkappa \frac{\partial f_2^\varkappa}{\partial \ddot{q}^\sigma}, \quad \sigma = \overline{1,s}, \quad (4.3)$$

which were obtained in Sect. 9 of Chap. 6.

---

[9] These equations were derived in this very form by A.I. Lur'e (see the book *A.I. Lur'e*, Analytical mechanics, Moscow: Izdat. "Fizmatgiz", 1961 [in Russian]).

Let us relate the generalized accelerations $\ddot{q}^{\,\sigma}$, $\sigma = \overline{1, s}$, with the pseudo-accelerations $w_* = (w_*^1, \ldots, w_*^s)$ by the transformations

$$w_*^\rho = w_*^\rho(t, q, \dot{q}, \ddot{q}), \quad \ddot{q}^{\,\sigma} = \ddot{q}^{\,\sigma}(t, q, \dot{q}, w_*), \quad \rho, \sigma = \overline{1, s}, \qquad (4.4)$$

which define the nonholonomic bases (see Sect. 9 of Chap. 6)

$$\varepsilon^\rho = \frac{\partial w_*^\rho}{\partial \ddot{q}^{\,\sigma_*}} \mathbf{e}^{\sigma_*}, \quad \varepsilon_\sigma = \frac{\partial \ddot{q}^{\,\tau}}{\partial w_*^\sigma} \mathbf{e}_\tau, \quad \rho, \sigma, \sigma_*, \tau = \overline{1, s}.$$

If the constraint equations (4.1) are taken into account in formulas (4.4),

$$\begin{aligned} w_*^\lambda &= w_*^\lambda(t, q, \dot{q}, \ddot{q}), \quad \lambda = \overline{1, l}, \quad l = s - k, \\ w_*^{l+\varkappa} &= w_*^{l+\varkappa}(t, q, \dot{q}, \ddot{q}) \equiv f_2^\varkappa(t, q, \dot{q}, \ddot{q}), \quad \varkappa = \overline{1, k}, \end{aligned} \qquad (4.5)$$

then the tangent space decomposes into a direct sum of K- and L-subspaces.

From transformations (4.4) we find the variations of the generalized accelerations and quasi-accelerations, assuming that $t, q, \dot{q}$ are parameters:

$$\delta'' w_*^\rho = \frac{\partial w_*^\rho}{\partial \ddot{q}^{\,\sigma}} \delta'' \ddot{q}^{\,\sigma}, \quad \delta'' \ddot{q}^{\,\sigma} = \frac{\partial \ddot{q}^{\,\sigma}}{\partial w_*^\rho} \delta'' w_*^\rho, \quad \rho, \sigma = \overline{1, s}. \qquad (4.6)$$

In the case of transformations (4.5), formulas (4.6) assume the form

$$\delta'' w_*^\lambda = \frac{\partial w_*^\lambda}{\partial \ddot{q}^{\,\sigma}} \delta'' \ddot{q}^{\,\sigma}, \quad \delta'' \ddot{q}^{\,\sigma} = \frac{\partial \ddot{q}^{\,\sigma}}{\partial w_*^\lambda} \delta'' w_*^\lambda, \quad \sigma = \overline{1, s}, \quad \lambda = \overline{1, l}, \qquad (4.7)$$

$$\delta'' w_*^{l+\varkappa} = \frac{\partial f_2^\varkappa}{\partial \ddot{q}^{\,\sigma}} \delta'' \ddot{q}^{\,\sigma} = 0, \quad \sigma = \overline{1, s}, \quad \varkappa = \overline{1, k}. \qquad (4.8)$$

Let us now introduce the vector

$$\delta'' \mathbf{W} = \delta'' \ddot{q}^{\,\sigma} \mathbf{e}_\sigma = \frac{\partial \ddot{q}^{\,\sigma}}{\partial w_*^\lambda} \delta'' w_*^\lambda \mathbf{e}_\sigma = \delta'' w_*^\lambda \varepsilon_\lambda \qquad (4.9)$$

and construct together with the vector $\mathbf{W}$ (considered as a vector function of the variables $\ddot{q} = (\ddot{q}^{\,1}, \ldots, \ddot{q}^{\,s})$) the new vector

$$\widetilde{\mathbf{W}} = \mathbf{W} + \delta'' \mathbf{W} = (\ddot{q}^{\,\sigma} + \delta'' \ddot{q}^{\,\sigma}) \mathbf{e}_\sigma.$$

In analogy with what we did for formula (3.7), we expand the constraint equations (4.1) into Taylor series near the point with coordinates $(\ddot{q}^{\,1}, \ldots, \ddot{q}^{\,s})$, assuming that $q^1, \ldots, q^s, \dot{q}^1, \ldots, \dot{q}^s, t$ are fixed parameters:

$$f_2^{\varkappa}(t, q, \dot{q}, \ddot{q} + \delta''\ddot{q}) = f_2^{\varkappa}(t, q, \dot{q}, \ddot{q}) + \nabla'' f_2^{\varkappa} \cdot \delta''\mathbf{W} + o(|\delta''\mathbf{W}|),$$

$$\varkappa = \overline{1, k}.$$ (4.10)

From equalities (4.10) it follows that if at a time t for a point with a phase state $(q^1, \ldots, q^s, \dot{q}^1, \ldots, \dot{q}^s)$, the acceleration $\mathbf{W}$ is kinematically virtual, then, to a first-order approximation, the acceleration $\widetilde{\mathbf{W}} = \mathbf{W} + \delta''\mathbf{W}$ is also kinematically virtual, provided that

$$\nabla'' f_2^{\varkappa} \cdot \delta''\mathbf{W} = 0, \qquad \varkappa = \overline{1, k}.$$ (4.11)

In other words, set of vectors $\delta''\mathbf{W}$, satisfying Eq. (4.11), characterizes the kinematically virtual variations of the acceleration $\mathbf{W}$, which are admitted by the non-holonomic constraints (4.1) at a time t, when the system is in the phase state $(q^1, \ldots, q^s, \dot{q}^1, \ldots, \dot{q}^s)$. An arbitrary vector $\delta''\mathbf{W}$ satisfying conditions (4.11) is called the *variation of the acceleration* $\mathbf{W}$.

Now, in a complete analogy with the arguments from the previous subsection, one easily shows that the expression

$$\left(M\mathbf{W} - \mathbf{Y}\right) \cdot \delta''\mathbf{W} = 0$$ (4.12)

can be obtained from Eqs. (4.2) and (4.3) in the vector representation of the variation of the generalized acceleration (4.9), provided that their contravariant components $\delta''w_*^{\lambda}$, $\lambda = \overline{1, l}$, and the components $\delta''\ddot{q}^1, \ldots, \delta''\ddot{q}^l$ are arbitrary, which are later used to determine from formulas (4.8) the remaining components $\delta''\ddot{q}^{l+1}, \ldots, \delta''\ddot{q}^s$ when the help of formulas (4.7). Vice versa, from the initial formula (4.12) one may obtain the motion equations (4.2) and (4.3).

Hence, assertion (4.12) can be taken for a variational differential principle of mechanics. This principle, which is called the *Gauss principle*, asserts that under ideal linear second-order nonholonomic constraints imposed on the motion of a mechanical system the scalar product of the inertia forces and the active forces acting on a system by the variation of the acceleration is zero.

This principle is usually written in the form

$$\delta''Z = 0,$$ (4.13)

where we set

$$Z = \frac{M}{2} \left(\mathbf{W} - \frac{\mathbf{Y}}{M}\right)^2.$$ (4.14)

The function Z from formulas (4.13), (4.14) is called the *Gauss constraint (compulsion)*.[10] Hence, the Gauss principle is sometimes called the *principle of least constraint(compulsion)*. It is worth pointing out that the idea of the constraint function was previously used by Gauss when he was creating the theory of errors.

---

[10] The letter "Z" comes from the German "*der Zwang*", meaning "enforcement", "constraint".

We recall that the two primes in formula (4.13) emphasize that only the second derivatives of the generalized coordinates are varied.

In Chap. 6 of the second volume, we shall outline the *generalized Gauss principle*, which holds under nonholonomic constraints of any order. This principle proves fairly efficient in solving one of the most important classes of problems of the control theory.

5   Uniform Vector Notation and Geometric Interpretation of Differential Variational Principles

Uniform vector notation of differential principles. Principle of virtual displacements. In previous subsections we obtained the d'Alembert–Lagrange (2.15), Suslov–Jourdain (3.13) and Gauss (4.12) differential variational principles. Taking into account their formal resemblance, these principles can be written in the following uniform vector notation:

$$(M\mathbf{W} - \mathbf{Y}) \cdot \delta\mathbf{x} = 0, \qquad (5.1)$$

where, in the case of holonomic constraints (2.2), the vector $\delta\mathbf{x}$ is the vector of virtual displacements in the form (2.16) or (2.18); under the nonholonomic constraints (3.1) this vector is the variation of velocity (3.6). With linear nonholonomic second-order constraints (4.1), this vector is the variation of acceleration (4.9). The importance of the contravariant structure of this vector will be pointed out in Sect. 9. Here the following points will be discussed.

The derivation of differential principles was chiefly based on the two-way relation between the Newton laws and the principle: the vector notation of the principle from two the most characteristic forms of notation of differential equations, which reflect the validity of the second Newton law under corresponding constraints, and then, the other way round, from this zero scalar product, which is now postulated, two original forms of constrained motion equations for a motion of a mechanical system were derived as a corollary.

Following the same logical approach, let us derive differential variational principles directly from the vector equation of constrained motion of a general mechanical system, as written in the tangent space to the manifold of all possible positions of the mechanical system which it can have at a given time (for more details, see Chap. 6):

$$M\mathbf{W} - \mathbf{Y} - \mathbf{R}^K - \mathbf{R}_L = 0. \qquad (5.2)$$

Under ideal constraints

$$\mathbf{R}_L \equiv 0; \qquad (5.3)$$

under the holonomic constraints (2.2) the vector $\mathbf{R}^K$ assumes the form

$$\mathbf{R}^K = \Lambda_\varkappa \frac{\partial f_0^\varkappa}{\partial q^\sigma} \mathbf{e}^\sigma \equiv \Lambda_\varkappa \nabla f_0^\varkappa,$$

under the first-order nonholonomic constraints (3.1) it assumes the form

$$\mathbf{R}^K = \Lambda_\varkappa \frac{\partial f_1^\varkappa}{\partial \dot{q}^\sigma} \mathbf{e}^\sigma \equiv \Lambda_\varkappa \nabla' f_1^\varkappa,$$

and under the second-order nonholonomic constraints (4.1) it reads as

$$\mathbf{R}^K = \Lambda_\varkappa \frac{\partial f_2^\varkappa}{\partial \ddot{q}^\sigma} \mathbf{e}^\sigma \equiv \Lambda_\varkappa \nabla'' f_2^\varkappa.$$

Multiplying the vector equation (5.2) by the variation of the vector $\mathbf{x}$, this gives the unified vector form of differential variational principles

$$(M\mathbf{W} - \mathbf{Y} - \mathbf{R}^K - \mathbf{R}_L) \cdot \delta \mathbf{x} = 0. \tag{5.4}$$

If now we assume ab initio that notation (5.4) holds, then from it we obtain the motion equation (5.2), since $\delta \mathbf{x}$ is arbitrary. Hence the zero scalar product (5.4) represents, depending on the choice of the vector $\delta \mathbf{x}$, the d'Alembert–Lagrange, Suslov–Jourdain, or Gauss differential variational principles.

In the case of nonideal constraints, the vector $\mathbf{R}_L$ is usually looked upon as an active force $\mathbf{Y}$; if the constraints are ideal, then condition (5.3) holds. Since $\mathbf{R}^K$ and $\delta \mathbf{x}$ are orthogonal, principle (5.4) can now be written in the standard form (5.1).

The convenience of the form (5.4) to represent differential variational principles is specially worth noting. Here, the scalar product is retained

$$\mathbf{R}^K \cdot \delta \mathbf{x} = 0;$$

this scalar product frequently arises in solving practical problems. As an example to illustrate this assertion, we write the *principle of virtual displacements*, which holds if a system has an equilibrium ($\mathbf{W} = 0$) under ideal holonomic constraints (2.2):

$$(\mathbf{Y} + \Lambda_\varkappa \nabla f_0^\varkappa) \cdot \delta \mathbf{y} = 0. \tag{5.5}$$

This principle is a necessary and sufficient condition that the system be in equilibrium. The principle of virtual displacements (5.5) is an underlying principle for the creation of the so-called *analytical statics* (for a more detailed account, see Chap. 10).

Another point is worth discussing. Earlier, in the derivation of principles we took into account either the dependencies between the contravariant components δx^σ, $\sigma = \overline{1, s}$ (from the effective forces) or the independent variations δx_*^λ, $\lambda = \overline{1, l}$, $l = s - k$. In this regard, this is why it proved possible to derive from principles, using the variations δx_*^λ, only the motion equations themselves; besides, it seemed that from them one cannot obtain second groups of Lagrange equations of the second kind, the Maggi equations, and the generalized Maggi equations:

$$MW_{l+\varkappa}^* = Q_{l+\varkappa}^* + \Lambda_\varkappa, \quad \varkappa = \overline{1,k},$$

$$(MW_\sigma - Q_\sigma)\frac{\partial \dot{q}^\sigma}{\partial v_*^{l+\varkappa}} = \Lambda_\varkappa, \quad \varkappa = \overline{1,k},$$

$$(MW_\sigma - Q_\sigma)\frac{\partial \ddot{q}^\sigma}{\partial w_*^{l+\varkappa}} = \Lambda_\varkappa, \quad \varkappa = \overline{1,k}, \tag{5.6}$$

$$MW_\sigma = \frac{d}{dt}\frac{\partial T}{\partial \dot{q}^\sigma} - \frac{\partial T}{\partial q^\sigma}, \quad \sigma = \overline{1,s}.$$

Taking this into account, let us go back to the vector motion equation (5.2). By the constraint-lose principle it can be considered as a motion of an unconstrained mechanical system under forces $\mathbf{Y}$, $\mathbf{R}^K$, $\mathbf{R}_L$. But if the system is unconstrained, then all variations δx^σ, $\sigma = \overline{1,s}$, are independent, and hence, since the coefficients multiplying $\delta x_*^{l+\varkappa}$, $\varkappa = \overline{1,k}$, are all zero, we obtain the desired equations (5.6), which contain Lagrange multipliers (the generalized reactions of ideal constraints). The knowledge of these quantities enables one to find out, in particular, whether a mechanical system can be relaxed from the effective holonomic constraints on its motion.

Geometrical interpretation of the d'Alembert–Lagrange principle. The d'Alembert–Lagrange principle can be written in the form (2.21)

$$\mathbf{R} \cdot \delta \mathbf{y} = 0. \tag{5.7}$$

Comparing formulas (5.7) and (2.13), we conclude that

$$\mathbf{R} = \Lambda_\varkappa \nabla f_0^\varkappa. \tag{5.8}$$

Formula (5.8) shows that the vector of reaction of ideal holonomic constraints can be expanded in the basis vectors $\mathbf{e}_*^{l+\varkappa} = \nabla f_0^\varkappa$, $\varkappa = \overline{1,k}$, of the K-space (under holonomic constraints, this space can be called the "subspace of reactions"). In the tangent space the holonomic constraint equations define the l-dimensional surface $\mathbb{V}(t,q)$, on which at a given time t must lie the point corresponding to the position of the system. To the curvilinear coordinate system $q_* = (q_*^1, \ldots, q_*^l)$ there corresponds the basis $\mathbf{e}_1^*, \ldots, \mathbf{e}_l^*$ lying in the plane $\mathbb{T}(q,V)$ which is tangential to the surface $\mathbb{V}(t,q)$. This plane contains the vectors $\delta \mathbf{y}$ of virtual displacements of the system (under holonomic constraints, this space may be called the "subspace of virtual displacements"). So, the d'Alembert–Lagrange principle in the form (5.7) asserts that for ideal holonomic constraints the subspace of reactions (the K-space) is orthogonal to the subspace of virtual displacements (the L-space).

The above arguments can be explained on an example of the motion of one material point with one holonomic constraint. At a given time t it lies on the surface, as given by the equation of holonomic constraint $f_0^1(t, y_1, y_2, y_3) = 0$, at the point M with the radius vector $\mathbf{y}$. Figure 3 shows the tangent plane to this surface at the point M (the surface itself is not shown in the figure; it is an analogue of the two-dimensional sur-

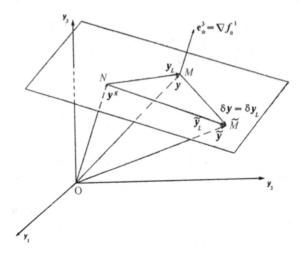

Fig. 3 Virtual displacement of a point under a holonomic constraint

face $\mathbb{V}(t, q)$, the tangent plane corresponds to the plane $\mathbb{T}(q, \mathbb{V})$). In our example, the Cartesian coordinates of the point $y = (y_1, y_2, y_3)$ play the role of the original curvilinear coordinate system $q = (q^1, q^2, q^3)$. Since the Cartesian coordinate system is orthonormal and has the unit coordinate vectors $\mathbf{i}_\sigma$, $\sigma = \overline{1, 3}$, we have $\mathbf{e}_\sigma = \mathbf{e}^\sigma = \mathbf{i}_\sigma$, $\sigma = \overline{1, 3}$. Changing from the original coordinate system $q = (q^1, q^2, q^3)$ to the new curvilinear coordinate system $q_* = (q_*^1, q_*^2, q_*^3)$ by the formulas

$$q_*^\lambda = q_*^\lambda(t, q), \quad \lambda = 1, 2, \quad q_*^3 = f_0^1(t, y_1, y_2, y_3), \tag{5.9}$$

where the functions $q_*^\lambda(t, q)$ are given by the researcher, we decompose the available three-dimensional space into a direct sum of the one-dimensional K-space with the vector $\mathbf{e}_*^3$ and the two-dimensional L-space with the fundamental basis $\{\mathbf{e}_1^*, \mathbf{e}_2^*\}$. The last two vectors depend on the choice by the researcher of functions $q_*^\lambda(t, q)$, $\lambda = 1, 2$, and are determined from the transform

$$q^\sigma = q^\sigma(t, q_*), \quad \sigma = \overline{1, 3},$$

which is inverse to (5.3), by the formulas

$$\mathbf{e}_\lambda^* = \frac{\partial q^\sigma}{\partial q_*^\lambda} \mathbf{e}_\sigma, \quad \lambda = 1, 2, \quad \sigma = \overline{1, 3}.$$

These vectors lie in the plane tangent to the surface $f_0^1(t, y_1, y_2, y_3) = 0$ at the point M (now shown in Fig. 3).

By Fig. 3 a material point, when is positioned at a point M, may, without breaking the constraint, obtain a virtual displacement $\delta \mathbf{y}$ and move at the position $\widetilde{M}$, which

Fig. 4 Velocities of a point
under a nonholonomic
constraint

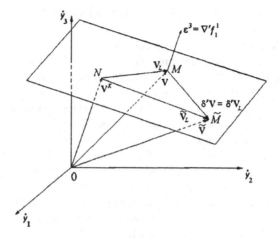

is characterized by the radius vector $\widetilde{\mathbf{y}}$. That the subspaces K and L are orthogonal is clear from the figure.

Geometrical interpretation of the Suslov–Jourdain principle. Since

$$M\mathbf{W} = \mathbf{Y} + \mathbf{R},$$

the Suslov–Jourdain principle (3.13) can be put in the form

$$\mathbf{R} \cdot \delta'\mathbf{V} = 0. \tag{5.10}$$

On the other hand, from formula (5.10) it follows by expressions (3.6) and (3.16) that under ideal nonholonomic constraints their reaction looks like

$$\mathbf{R} = \Lambda_{\varkappa} \nabla' f_1^{\varkappa}. \tag{5.11}$$

The constraint equations decompose the s-dimensional tangent space into a direct sum of two subspaces K and L of dimensions k and l. These spaces have the non-holonomic bases $\{\varepsilon^{l+1}, \ldots, \varepsilon^{s}\}$ and $\{\varepsilon_1, \ldots, \varepsilon_l\}$. Formula (5.11) shows that reaction of ideal nonholonomic constraints lies in the K-space.

To the constraint equations (3.1) in the space of velocities there corresponds the l-dimensional surface $\mathbb{V}(t, q, \dot{q})$, for which t and q are given parameters. Depending on the active forces and the initial conditions with the above t and q, the system may have various velocities $\mathbf{V}$, but in order to satisfy constraints (3.1) the end-points of these vectors should lie on the surface $\mathbb{V}(t, q, \dot{q})$. We let M denote the point corresponding to the available vector $\mathbf{V}$. At this point, we draw the tangent plane $\mathbb{T}(q, \mathbb{V})$ to the surface $\mathbb{V}(t, q, \dot{q})$. This plane contains the basis vectors $\{\varepsilon_1, \ldots, \varepsilon_l\}$, in which by formula (3.6) the vector of variation of the velocity is expanded. Besides, principle (3.13) asserts that the scalar product of the forces $(M\mathbf{W} - \mathbf{Y})$ by the variation of the velocity $\delta'\mathbf{V}$ is zero, and formula (5.10) shows the orthogonality of the reaction

$\mathbf{R} = \Lambda_{\varkappa} \nabla' f_1^{\varkappa}$ of the ideal nonholonomic constraints (3.1) to this variation of the velocity.

Figure 4 gives a geometrical illustration of the above on an example of the motion of one material point with one ideal nonholonomic constraint $f_1^1(t, q, \dot{q}) = 0$. As before, the role of the original curvilinear coordinate system $q = (q^1, q^2, q^3)$ is played by the Cartesian system $Oy_1y_2y_3$. At a given time t, the point is at a position $(q^1, q^2, q^3) \equiv (y_1, y_2, y_3)$ and has the velocity whose vector $\mathbf{V}$ ends at the point M, which lies on the surface given in the space of velocities $O\dot{y}_1\dot{y}_2\dot{y}_3$ by the equation of nonholonomic constraint $f_1^1 = 0$ (this surface, which was denoted above by $\mathbb{V}(t, q, \dot{q})$, is not shown in Fig. 4; t and q are assumed to be fixed parameters). The figure also shows the tangent plane to the surface under consideration at the point M. In the general case, this plane was denoted by $\mathbb{T}(t, \mathbb{V})$. This plane contains the vectors ε_1, ε_2 (not shown in the figure) of the principal nonholonomic basis for the L-space, in which by formula (3.6) the vector of the variation of the velocity $\delta' \mathbf{V}$ is expanded. The vector $\varepsilon^3 = \nabla' f_1^1$, which is the vector of the reciprocal basis for the K-space is orthogonal to this plane; along this vector there goes the reaction of the ideal nonholonomic constraint. Now the orthogonality of the spaces K and L is clear. This makes transparent the Suslov–Jourdain principle, as written in the form (5.10).

The component of the velocity $\mathbf{V}^K$ is specified by the expression for the nonholonomic constraint $f_1^1(t, q, \dot{q}) = 0$; in absolute value it is equal to the distance from the origin to the tangent plane. This component is not involved in the variation, and hence the virtual velocity $\tilde{\mathbf{V}}$, which is the sum $\mathbf{V} + \delta' \mathbf{V}$, is equal to the vector $\overrightarrow{OM}$, satisfying to the first order of approximation the nonholonomic constraint equation.

If the configuration of vectors is moved to the end-point of the vector $\mathbf{y}$, then the Chetaev's definition for a 'virtual displacements' of a nonholonomic system becomes transparent, when it is assumed that $\delta q^{\sigma} = \tau_0 \delta' \dot{q}^{\sigma}$, where τ_0 is an infinitely small time interval, which was introduced by Gauss.

Geometrical interpretation of the Gauss principle. The geometrical interpretation of the Gauss principle is similar. Now the equations of linear second-order nonholonomic constraints (4.1) decompose the tangent space into two orthogonal subspaces K and L with nonholonomic bases $\{\varepsilon^{l+1}, \ldots, \varepsilon^s\}$ and $\{\varepsilon_1, \ldots, \varepsilon_l\}$. The constraints in the space of accelerations specify the l-dimensional plane $\mathbb{T}(t, q, \dot{q}, \ddot{q})$, for which t, q and $\dot{q}$ are given parameters. This plane must contain the end-points of the acceleration vectors $\mathbf{W}$ of the mechanical system. It also contains the vectors ε_{λ}, $\lambda = \overline{1, l}$, into which the vector of variation of the acceleration $\delta'' \mathbf{W}$ is expanded by formula (4.9). From formulas (4.11) and (4.12), one easily obtains that the reaction of the ideal second-order nonholonomic constraints is $\mathbf{R} = \Lambda_{\varkappa} \nabla'' f_2^{\varkappa} \equiv \Lambda_{\varkappa} \varepsilon^{l+\varkappa}$; that is, it lies in the space K.

The Gauss principle in the form (4.13) shows that in the case of ideal linear second-order nonholonomic constraints their reaction $\mathbf{R}/M = \mathbf{W} - \mathbf{Y}/M$ "forces" the mechanical system to move with the smallest value of the magnitude of this reaction.

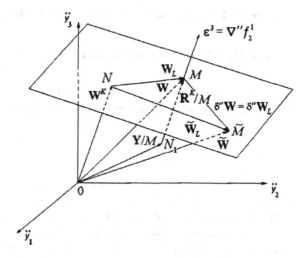

Fig. 5 Accelerations of the system with a second-order nonholonomic constraint

Figure 5 explains the above analysis in the case of a motion of one material point under an ideal nonholonomic constraint

$$f_2^1(t, q, \dot{q}, \ddot{q}) \equiv a_{2\sigma}^3(t, q, \dot{q}) \ddot{q}^\sigma + a_{2,0}^3(t, q, \dot{q}) = 0, \quad \sigma = \overline{1, 3}.$$

In extending the Chetaev conditions (3.18) to the second-order constraints (4.1) in the form[11]

$$\frac{\partial f_2^\varkappa}{\partial \ddot{q}^\sigma} \delta q^\sigma = 0, \quad \varkappa = \overline{1, k},$$

the "virtual displacements" can be looked upon as variations of the virtual generalized accelerations[12] with the factor $\tau_0^2/2$:

$$\delta q^\sigma = \frac{\tau_0^2}{2} \delta'' \ddot{q}^\sigma, \quad \sigma = \overline{1, s}. \tag{5.12}$$

Here, τ_0 is the Gauss infinitely small time interval. It is convenient to imagine that the "virtual displacements" of a nonholonomic system, as given by formulas (5.12), are applied to the end of the "radius vector" $\mathbf{y}$ of the mechanical system.

---

[11] For the case of one constraint, such conditions were first introduced by G. Hamel (see the paper G. Hamel, "Nichtholonome Systeme höherer Art," Sitzungsbererichte der Berliner Mathematischen Gesellschaft. 1938. Bd 37. S. 41–52.).

[12] See the paper N.N. Polyakhov, "On differential principles of mechanics obtained from the motion equations of nonholonomic systems," Vestn. Leningr. Univ. 1974. Issue 3, issue 13, pp. 106–116 [in Russian].

II) Integral Variational Principles of Mechanics

6 The Hamilton–Ostrogradskii Principle

Let us consider the motion equations of a holonomic system. We shall prove that these equations, as taken in the form of the Lagrange equation of the second kind,

$$\frac{d}{dt}\frac{\partial L}{\partial \dot{q}^\sigma} - \frac{\partial L}{\partial q^\sigma} = 0, \quad \sigma = \overline{1, s}, \tag{6.1}$$

provide a sufficient condition for an extremum of some function (the *action function*) in the case of a potential force field under certain additional assumptions.

For the purpose of mathematical description of extremal properties of motion paths of mechanical systems, as given by functions $q^\sigma(t)$, we shall compare them with the curves defined in terms of arbitrary smooth functions $\widetilde{q}^\sigma(t)$. Such curves will be called the *comparison curves*. It is appropriate to consider comparison curves as a family of smooth curves depending on the parameter α, assuming that the motion path of a system lies in this family (with $\alpha = 0$); that is,

$$\widetilde{q}^\sigma(t) = q^\sigma(t, \alpha), \quad q^\sigma(t) = q^\sigma(t, 0), \quad \sigma = \overline{1, s}.$$

In what follows, we shall introduce various quantities related to the comparison curves and hence depending on the parameter α.

By definition we assume that the variation of the scalar quantity S is as follows:

$$\delta S = \left.\frac{\partial S}{\partial \alpha}\right|_{\alpha=0} \alpha. \tag{6.2}$$

This is the partial differential of S, which describes in the linear approximation of the character of variation of this quantity when changing from the actual path to the comparison curve.

We expand definition (6.2) to the functions $q^\sigma(t, \alpha)$; that is, we assume that by definition the time functions

$$\delta q^\sigma(t) = \left.\frac{\partial q^\sigma}{\partial \alpha}\right|_{\alpha=0} \alpha \tag{6.3}$$

are variations of the functions $q^\sigma(t)$.

Definitions (6.2), (6.3) enable one to employ the standard rules for derivatives of composites in evaluating the variation δS. For example, if $S = S(t, q(t, \alpha))$, then

$$\delta S = \frac{\partial S}{\partial \alpha}\bigg|_{\alpha=0} \alpha = \left(\frac{\partial S}{\partial q^\sigma} \frac{\partial q^\sigma}{\partial \alpha}\right)_{\alpha=0} \alpha =$$

$$= \left(\frac{\partial S}{\partial q^\sigma}\right)_{\alpha=0} \left(\frac{\partial q^\sigma}{\partial \alpha}\right)_{\alpha=0} \alpha = \left(\frac{\partial S}{\partial q^\sigma}\right)_{\alpha=0} \delta q^\sigma ,$$

and hence,

$$\delta S = \frac{\partial S}{\partial q^\sigma} \delta q^\sigma .$$

Here for simplicity we do not indicate $\alpha = 0$ in the expressions like $\partial S/\partial q^\sigma$.

The introduction of the parameterizations and definitions (6.2), (6.3) becomes very instrumental if S is a real number associated according to some rule with the family of functions $\widetilde{q}^{\,1}(t), \widetilde{q}^{\,2}(t), \ldots, \widetilde{q}^{\,s}(t)$, defined on the interval $[t_0, t_1]$. This operation of obtaining the number S can be considered as a function defined on all possible families of functions $\widetilde{q}^{\,1}(t), \widetilde{q}^{\,2}(t), \ldots, \widetilde{q}^{\,s}(t)$. Such functions are called *functionals*; their arguments are functions. The definite integral is a typical example of a functional.

In what follows, we shall show that the principal properties of motion in a potential force field are determined by those of the functional

$$S = \int\limits_{t_0}^{t_1} L(t, q, \dot{q}) \, dt ,$$

which is called the *Hamilton action*.

Clearly, the definition of the partial differential δS of the functional S according to the same logical scheme for standard functions of real variable should naturally lead to the consideration of the differences of the functions $\widetilde{q}^{\,\sigma}(t) - q^\sigma(t)$, which are known as *variations of the functions* $q^\sigma(t)$ in modern courses in variational calculus.

By using the parametrization and definition (6.2), (6.3) one is able, without introducing new concepts from the functional analysis, to employ much more simple machinery in a fairly wide range of topics. Under this approach, the final results are written in a form useful in a more general framework.

Multiplying Eq. (6.1) by the variations of coordinates δq^σ and summing in σ, we obtain

$$\left(\frac{d}{dt} \frac{\partial L}{\partial \dot{q}^\sigma} - \frac{\partial L}{\partial q^\sigma}\right) \delta q^\sigma = 0 .$$

Pulling the functions $\delta q^\sigma(t)$ inside the derivative d/dt, we get

$$\frac{d}{dt}\left(\frac{\partial L}{\partial \dot{q}^\sigma} \delta q^\sigma\right) = \frac{\partial L}{\partial \dot{q}^\sigma} \frac{d(\delta q^\sigma)}{dt} + \frac{\partial L}{\partial q^\sigma} \delta q^\sigma . \tag{6.4}$$

Let us discuss the derivative $d(\delta q^\sigma)/dt$. Using formulas (6.2), (6.3), it can be written as

$$\frac{d}{dt}(\delta q^\sigma) = \frac{d}{dt}\left(\left.\frac{\partial q^\sigma}{\partial \alpha}\right|_{\alpha=0}\alpha\right) = \left.\frac{\partial \dot{q}^\sigma}{\partial \alpha}\right|_{\alpha=0}\alpha = \delta \dot{q}^\sigma .$$

So, we have

$$\frac{d}{dt}(\delta q^\sigma) = \delta \dot{q}^\sigma ,$$

that is, the differentiation commutes with the operation of varying.

In view of this relation, expression (6.4) assumes the form

$$\frac{d}{dt}\left(\frac{\partial L}{\partial \dot{q}^\sigma}\delta q^\sigma\right) = \frac{\partial L}{\partial \dot{q}^\sigma}\delta \dot{q}^\sigma + \frac{\partial L}{\partial q^\sigma}\delta q^\sigma .$$

The right-hand side of this equality is the partial differential of the function $L(t, q, \dot{q})$ with a fixed t; it corresponds to the increment of its arguments q^σ and $\dot{q}^\sigma$, respectively, by δq^σ and $\delta \dot{q}^\sigma$. By definitions (6.2), (6.3), this differential agrees with the variation δL of the function L. Hence,

$$d\left(\frac{\partial L}{\partial \dot{q}^\sigma}\delta q^\sigma\right) = \delta L\, dt ,$$

and so,

$$\int_{t_0}^{t_1} \delta L\, dt = \left(\frac{\partial L}{\partial \dot{q}^\sigma}\delta q^\sigma\right)_{t_1} - \left(\frac{\partial L}{\partial \dot{q}^\sigma}\delta q^\sigma\right)_{t_0} . \tag{6.5}$$

From elementary calculus, it is known that the derivative in α of the integral

$$S(\alpha) = \int_{t_0(\alpha)}^{t_1(\alpha)} L(t, q(t, \alpha), \dot{q}(t, \alpha))\, dt ,$$

is as follows:

$$\frac{dS}{d\alpha} = \int_{t_0(\alpha)}^{t_1(\alpha)} \frac{dL}{d\alpha}\, dt + L_1\frac{dt_1(\alpha)}{d\alpha} - L_0\frac{dt_0(\alpha)}{d\alpha} ,$$

where L_1 and L_0 are the values of the function L, respectively, with $t = t_1(\alpha)$ and $t = t_0(\alpha)$.

From this formula and definition (6.2), the variation δS of the functional

$$S = \int_{t_0}^{t_1} L\, dt \tag{6.6}$$

Fig. 6 Variation of a
function with fixed
end-points

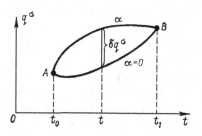

is found to be

$$\delta S = \int_{t_0}^{t_1} \delta L \, dt + L_1 \, \delta t_1 - L_0 \, \delta t_0 \, .$$

Here, we choose the functions $t_0(\alpha)$ and $t_1(\alpha)$ so as to have $t_0(0) = t_0$ and $t_1(0) = t_1$. Taking into account (6.5), this establishes

$$\delta S \equiv \delta \int_{t_0}^{t_1} L \, dt = \left(\frac{\partial L}{\partial \dot{q}^{\sigma}} \delta q^{\sigma} \right)_{t_1} - \left(\frac{\partial L}{\partial \dot{q}^{\sigma}} \delta q^{\sigma} \right)_{t_0} + L_1 \, \delta t_1 - L_0 \, \delta t_0 \, . \qquad (6.7)$$

Consider a particular case. Assume that t_0 and t_1 are fixed; the corresponding variations are as follows: $(\delta q^{\sigma})_{t_1} = (\delta q^{\sigma})_{t_0} = 0$. Geometrically, this case corresponds to fixed end-points (see Fig. 6).

Under the above assumptions, formula (6.7) assumes the form

$$\delta S = \delta \int_{t_0}^{t_1} L \, dt = 0 \, . \qquad (6.8)$$

As a result, the integral (functional) on the actual path assumes its stationary value in comparison with its values on the neighboring "paths". These neighboring "paths" are also called "devious paths" and "comparison trajectories".

It is worth pointing out that this result is obtained under the assumption

$$(\delta q^{\sigma})_{t_1} = (\delta q^{\sigma})_{t_0} = 0 \, , \quad \sigma = \overline{1, s} \, , \qquad (6.9)$$

while the functional (6.6) is calculated with fixed t_0 and t_1:

$$\delta t_1 = \delta t_0 = 0 \, . \qquad (6.10)$$

The condition $\delta S = 0$ is obtained as a consequence of the Lagrange equations (6.1). Hence, these equations are sufficient conditions for vanishing the variation δS under boundary conditions (6.9), (6.10). One can show that they are also necessary conditions for this, that is vanishing the variation δS for arbitrary functions $q^\sigma(t, \alpha)$ satisfying boundary conditions (6.9), (6.10) is possible only in the case, when functions $q^\sigma(t)$ are the solution of system of equations (6.1).

Actually, basing on the formulae introduced and derived before we can write the following sequence of equalities:

$$\delta S = \frac{\partial S}{\partial \alpha}\bigg|_{\alpha=0} \alpha = \left(\frac{\partial}{\partial \alpha} \int_{t_0}^{t_1} L(t, q(t, \alpha), \dot{q}(t, \alpha))\, dt\right)_{\alpha=0} \alpha =$$

$$= \int_{t_0}^{t_1} \left[\left(\frac{\partial L}{\partial q^\sigma}\right)_{\alpha=0} \left(\frac{\partial q^\sigma}{\partial \alpha}\right)_{\alpha=0} + \left(\frac{\partial L}{\partial \dot{q}^\sigma}\right)_{\alpha=0} \left(\frac{\partial \dot{q}^\sigma}{\partial \alpha}\right)_{\alpha=0}\right] \alpha\, dt =$$

$$= \int_{t_0}^{t_1} \left[\left(\frac{\partial L}{\partial q^\sigma}\right)_{\alpha=0} \delta q^\sigma + \left(\frac{\partial L}{\partial \dot{q}^\sigma}\right)_{\alpha=0} \delta \dot{q}^\sigma\right] dt =$$

$$= \int_{t_0}^{t_1} \left(\frac{\partial L}{\partial q^\sigma} \delta q^\sigma + \frac{\partial L}{\partial \dot{q}^\sigma} \frac{d}{dt}\delta q^\sigma\right) dt = 0.$$

Applying integration by parts, we obtain

$$\int_{t_0}^{t_1} \delta q^\sigma \left(\frac{\partial L}{\partial q^\sigma} - \frac{d}{dt}\frac{\partial L}{\partial \dot{q}^\sigma}\right) dt = 0.$$

But since the integration interval $[t_0, t_1]$ can be chosen arbitrarily, then this implies that the integrand should vanish

$$\delta q^\sigma \left(\frac{\partial L}{\partial q^\sigma} - \frac{d}{dt}\frac{\partial L}{\partial \dot{q}^\sigma}\right) = 0,$$

and due to arbitrariness of the variations δq^σ from the equality of the sum to zero we get

$$\frac{\partial L}{\partial q^\sigma} - \frac{d}{dt}\frac{\partial L}{\partial \dot{q}^\sigma} = 0, \qquad \sigma = \overline{1, s},$$

that is the Lagrange equations of the second kind are satisfied, which was to be proved.

Relations (6.8)–(6.10), which are taken axiomatically for the initial ones, express the *Hamilton–Ostrogradskii principle* that is formulated as follows: a real motion of a system from the position M_0 to the position M_1 differs from all the kinematically possible motions performed between the same positions during the same time interval by the fact that for the real motion the Hamilton action has a stationary value.[13] The principle is widely applied when composing differential equations of motion of mechanical systems including systems with distributed parameters, as a result of which the equations in partial derivatives are obtained.

The Hamilton–Ostrogradskii principle belongs to *integral principles*. The most general variational relation with respect to the Hamilton action is relation (6.7). It was obtained by Hamilton for the first time, and he called it *the principle of varying, or varied, action*. All the rest of integral principles will be further obtained from this relation.

7 The Lagrange Principle

Let us now consider the case when the end-points t_0 and t_1 are functions of the parameter α; that is, we consider the problem with variable end-points.

We use formula (6.7); for the sake of brevity we rewrite it in the following form:

$$\delta S \equiv \delta \int_{t_0}^{t_1} L \, dt = (p_\sigma \delta q^\sigma)_{t_0}^{t_1} + (L \delta t)_{t_0}^{t_1} \,,$$

$$p_\sigma = \frac{\partial T}{\partial \dot{q}^\sigma} = \frac{\partial L}{\partial \dot{q}^\sigma} \,.$$

Replacing under the integral sign the function L by its value from the Legendre transform $L = p_\sigma \dot{q}^\sigma - H$, we get

$$\delta \int_{t_0}^{t_1} (p_\sigma \dot{q}^\sigma - H) \, dt = (p_\sigma \delta q^\sigma)_{t_0}^{t_1} + (L \delta t)_{t_0}^{t_1} \,.$$

Let us now consider the stationary force field. We have (see Sect. 6 of Chap. 4)

$$p_\sigma \dot{q}^\sigma = 2T \,, \quad H = T + \Pi = \text{const} = h \,.$$

---

[13] For stationary constraints, this principle was set forth by Hamilton in 1834–1835s. Independently, in 1848 M.V. Ostrogradskii formulated this principle for non-stationary constraints, and hence it is now known as the *Hamilton–Ostrogradskii principle*.

Under these assumptions, we get

$$\delta \int_{t_0}^{t_1} 2T \, dt - \delta \left[h(t_1 - t_0) \right] = (p_\sigma \delta q^\sigma)_{t_0}^{t_1} + (L \delta t)_{t_0}^{t_1} \, .$$

We consider an additional condition to the effect that the constant h is the same on all possible paths (that is, it is independent of the parameter α, and hence, $\delta h = 0$). We have

$$\delta \int_{t_0}^{t_1} 2T \, dt = (p_\sigma \delta q^\sigma)_{t_0}^{t_1} + ((L + h) \, \delta t)_{t_0}^{t_1} \, .$$

However, $L + h = p_\sigma \dot{q}^\sigma$, and so

$$\delta \int_{t_0}^{t_1} 2T \, dt = (p_\sigma (\delta q^\sigma + \dot{q}^\sigma \delta t))_{t_0}^{t_1} \, . \tag{7.1}$$

For brevity, we set $\Delta q^\sigma = \delta q^\sigma + \dot{q}^\sigma \delta t$. The quantity Δq^σ will be called the *total non-isochronous variation* q^σ at the end-points t_0 and t_1, respectively. Note that, as distinct from the total variation Δq^σ, the above δq^σ is sometimes called the *isochronous variation*. Now we rewrite formula (7.1) as

$$\delta \int_{t_0}^{t_1} 2T \, dt = (p_\sigma \Delta q^\sigma)_{t_0}^{t_1} \, . \tag{7.2}$$

The actual motion of a system between the positions M_0 and M_1, which is described by the functions $q^\sigma(t)$ on the interval $[t_0, t_1]$, will be compared with other kinematically possible motions between the above positions. On all kinematically possible motions, which are defined by the functions $\widetilde{q}^\sigma(t) = q^\sigma(t, \alpha)$, the total mechanical energy h is assumed to be the same. The functions $\widetilde{q}^\sigma(t) = q^\sigma(t, \alpha)$ are supposed to be defined on the interval $[t_0(\alpha), t_1(\alpha)]$. Besides, at the times $t_0(\alpha)$ and $t_1(\alpha)$ the system is, respectively, at the positions M_0 and M_1; that is,

$$q^\sigma(t_0(\alpha), \alpha) = q_0^\sigma \, , \quad q^\sigma(t_1(\alpha), \alpha) = q_1^\sigma \, . \tag{7.3}$$

Expanding the functions on the left of these equalities into the Maclaurin series in α, we find that

$$q^\sigma(t_0(0), 0) + \frac{d}{d\alpha} q^\sigma(t_0(\alpha), \alpha)\bigg|_{\alpha=0} \alpha + O(\alpha^2) = q_0^\sigma,$$

$$q^\sigma(t_1(0), 0) + \frac{d}{d\alpha} q^\sigma(t_1(\alpha), \alpha)\bigg|_{\alpha=0} \alpha + O(\alpha^2) = q_1^\sigma. \tag{7.4}$$

Since for $\alpha = 0$ the kinematically possible motion becomes the actual motion, we have

$$q^\sigma(t_0(0), 0) = q_0^\sigma, \quad q^\sigma(t_1(0), 0) = q_1^\sigma.$$

Further,

$$\frac{d}{d\alpha} q^\sigma(t(\alpha), \alpha)\bigg|_{\alpha=0} \alpha = \frac{\partial q^\sigma}{\partial \alpha}\bigg|_{\alpha=0} \alpha + \dot{q}^\sigma \frac{dt}{d\alpha}\bigg|_{\alpha=0} \alpha = \delta q^\sigma + \dot{q}^\sigma \delta t = \Delta q^\sigma,$$

and so, using (7.4) we conclude that the corollary to the conditions that the end-points (7.3) shall not move is that the total non-isochronous variation Δq^σ should vanish at the end-points t_0 and t_1; that is,

$$(\Delta q^\sigma)_{t_0} = (\Delta q^\sigma)_{t_1} = 0.$$

From formula (7.2) follows that

$$\delta W \equiv \delta \int_{t_0}^{t_1} 2T \, dt = 0,$$

provided that the initial and final configurations of the system are invariant.

Thus, the actual motion path of the system between M_0 and M_1 differs from all kinematically possible trajectories between the same positions and with the same total mechanical energy h in that on this path the functional

$$W \equiv \int_{t_0}^{t_1} 2T \, dt,$$

representing the *Lagrange action*, has stationary value. This assertion is known as the *stationary action principle in the Lagrange form*. From this principle one may obtain the principal equations of mechanics under the above constraints.

8 Various Forms of the Lagrange Principle

If we introduce the representative point, then the Lagrange action W assumes the form

$$W = \int_{t_0}^{t_1} MV^2 dt = \int_{\smile M_0 M_1} MV \, ds \,, \tag{8.1}$$

where $ds = V dt = \sqrt{g_{\sigma\tau} \dot{q}^\sigma \dot{q}^\tau} \, dt > 0, dt > 0, M_0, M_1$ are points in $\mathbb{R}^{3n}$ corresponding to the position of the representative point at t_0 and t_1.

Expression (8.1) is known as the *Maupertuis–Euler action*. The *least action principle in the Maupertuis forms* is written as

$$\delta W = \delta \int_{\smile M_0 M_1} MV \, ds = 0 \,.$$

Here, M_0 and M_1 should be considered, when varying, as stationary points, all possible paths lying between them. Note that time intervals to travel the above paths are distinct.

Since the energy integral has the form $T = h - \Pi$, for the representative point we get

$$M^2 V^2 = 2M(h - \Pi) \,,$$

and so,

$$W = \int_{\smile M_0 M_1} MV \, ds = \int_{\smile M_0 M_1} \sqrt{2M(h - \Pi)} \, ds \,.$$

We set

$$(d\Sigma)^2 = 2M(h - \Pi) \, g_{\sigma\tau} dq^\sigma dq^\tau = \tilde{g}_{\sigma\tau} dq^\sigma dq^\tau \,, \tag{8.2}$$

where $\tilde{g}_{\sigma\tau} = 2M(h - \Pi) \, g_{\sigma\tau}$.

The metric coefficients $g_{\sigma\tau}$ for a free representative point correspond to the $3n$-dimensional Euclidean space. Multiplying by the coordinate functions $2M(h - \Pi)$ transforms these coefficients into new metric coefficients $\tilde{g}_{\sigma\tau}$, which determine in general a non-Euclidean metric. Besides, this transforms the Euclidean space into the Riemannian one. Mathematically, the latter is defined as a space equipped with the quadratic form at each point:

$$\tilde{g}_{\sigma\tau} dq^\sigma dq^\tau \,.$$

Since from geometrical considerations this form should express the squared length element, it should remain invariant under transformations of coordinates.

For a nonfree mechanical system, the input coefficients $g_{\sigma\tau}$, which are defined from the expression for the kinetic energy of a system, are in general metric coefficients of the Riemann metric

$$ds^2 = g_{\sigma\tau}dq^\sigma dq^\tau .$$

Clearly, the metric defined by the quadratic form (8.2) is also a Riemannian metric.

In view of the above, we have the following expression for the action W, which is called the *Jacobi form*:

$$W = \int_{\smile M_0 M_1} \sqrt{\tilde{g}_{\sigma\tau}dq^\sigma dq^\tau} = \int_{\smile M_0 M_1} d\Sigma .$$

The least action principle, which requires that

$$\delta W = \delta \int_{\smile M_0 M_1} d\Sigma = 0 ,$$

is equivalent to the requirement that the action of a representative point under forces of stationary potential field from the position M_0 to the position M_1 follow the geodesic line in the Riemannian space, whose metric is defined by the quadratic form (8.2).

9 On the Variational Principles of Mechanics

Differential and[14] integral principles of mechanics depend on the concept of variation and hence are called *variational principles*.

The differential principles differ in that each of them is concerned with variations of certain new quantities. Of special importance is that in all differential principles the variations are introduced using the general logical schema, which depends on the definition of a contravariant vector. We recall this definition.

Let X be a manifold on which a point x is defined by the quantities x^σ, $\sigma = \overline{1, s}$. We introduce the new variables x_*^ρ, which in total define the same point x. The contravariant vector $\delta\mathbf{x}$ is defined in terms of the quantities δx^σ and δx_*^ρ, which are related as follows:

$$\delta x^\sigma = \frac{\partial x^\sigma}{\partial x_*^\rho}\delta x_*^\rho , \quad \delta x_*^\rho = \frac{\partial x_*^\rho}{\partial x^\sigma}\delta x^\sigma , \quad \rho, \sigma = \overline{1, s} . \tag{9.1}$$

---

[14] Genesis and development of the variational principles of mechanics are studied in the books:
L.S. Polak, Variational principles of mechanics (Moscow: Izdat. "Fizmatgiz", 1960) [in Russian];
V.N. Shchelkachev, Variational principles of mechanics (Moscow, 1989) [in Russian].

In the d'Alembert–Lagrange principle, x governs the position of a system, which is given by the coordinates q^σ; in the Suslov–Jourdain principle, x is the velocity of a system defined by the generalized velocities $\dot{q}^\sigma$; and finally, in the Gauss principle, x is the acceleration of the system, which means in the present setting the set of accelerations $\ddot{q}^\sigma$. This also means that the above differential principles can be written in the unique form

$$(MW_\sigma - Q_\sigma)\,\delta x^\sigma = 0\,, \tag{9.2}$$

which reflects their logical unity.

The coefficients of the linear form (9.2) of the contravariant vector $\delta\mathbf{x}$ are the constraint forces $R_\sigma = MW_\sigma - Q_\sigma$. Since this form is invariant with respect to the origin on the manifold X, it follows that the totality R_σ defines the vector $\mathbf{R}$. The crux of the corresponding differential principle lies in the invariance of the linear form (9.2). Let us consider this question in more detail.

We introduce arbitrary new "coordinates" x_*^ρ. Using (9.2), (9.1), we have

$$R_\sigma\,\delta x^\sigma = R_\sigma\,\frac{\partial x^\sigma}{\partial x_*^\rho}\,\delta x_*^\rho = \Lambda_\rho^*\,\delta x_*^\rho = \Lambda_\rho^*\,\frac{\partial x_*^\rho}{\partial x^\sigma}\,\delta x^\sigma = 0\,.$$

It follows that the components R_σ and Λ_ρ^* of the covector $\mathbf{R}$ in the "coordinates" x^σ and x_*^ρ are related by as follows:

$$R_\sigma = \Lambda_\rho^*\,\frac{\partial x_*^\rho}{\partial x^\sigma}\,, \qquad \Lambda_\rho^* = R_\sigma\,\frac{\partial x^\sigma}{\partial x_*^\rho}\,. \tag{9.3}$$

Assume that the variables x^λ, $\lambda = \overline{1,l}$, $l < s$, are chosen arbitrarily, and the variables $x_*^{l+\varkappa}$, $\varkappa = \overline{1,k}$, $k = s - l$, are given. In the case under consideration, the differentials $\delta x_*^{l+\varkappa} = 0$, the linear form (9.2) in the variables x_*^ρ assumes the form

$$\Lambda_\lambda^*\,\delta x_*^\lambda = 0\,. \tag{9.4}$$

Since the differentials δx_*^λ of the arbitrarily chosen variables are linearly independent, from equation (9.4) it follows that $\Lambda_\lambda^* = 0$, $\lambda = \overline{1,l}$, and hence,

$$R_\sigma = \Lambda_{l+\varkappa}^*\,\frac{\partial x_*^{l+\varkappa}}{\partial x^\sigma}\,. \tag{9.5}$$

Formulas (9.3) can be looked upon as the definition of the generalized forces Λ_ρ^*, which correspond to the variables x_*^ρ. Recall that the variables x_*^ρ are quasi-velocities or quasi-accelerations.

The generalized forces Q_σ, corresponding to the generalized Lagrange coordinates q^σ, for a holonomic mechanical system were introduced in Sect. 3 of Chap. 6, where the following fundamental theorem was proved: the motion under which one of the generalized coordinates is a give function of time can be realized by introducing one additional generalized force corresponding to this coordinate. This theorem has the following direct corollary.

Assume that the variables $q_*^{l+\varkappa} = f_0^\varkappa(t, q)$, $\varkappa = \overline{1, k}$, $k = s - l$ are given time functions. Then to implement the above motion it suffices to apply the additional generalized forces $\Lambda_{l+\varkappa}^*$, which correspond to the variables $q_*^{l+\varkappa}$; in other words, the generalized forces R_σ, as follows from the general expression (9.5), can be written in the form

$$R_\sigma = \Lambda_{l+\varkappa}^* \frac{\partial f_0^\varkappa}{\partial q^\sigma} \, .$$

The Suslov–Jourdain and Gauss principles extend this assertion to the case when the given variables are, respectively, the quasi-velocities $v_*^{l+\varkappa} = f_1^\varkappa(t, q, \dot{q})$ and the quasi-accelerations $w_*^{l+\varkappa} = f_2^\varkappa(t, q, \dot{q}, \ddot{q})$.

So, the differential variational principles show that with each constraint equation one may associate a generalized force controlling this constraint.

The following fact is worth pointing out. The variations of coordinates, velocities, and accelerations, as introduced from the unified approach (9.1), are not logically related with the concept of kinematically possible trajectories between two positions of a mechanical system. The comparison of the actual motion with the kinematically possible one is performed on the basis of the integral variational principles. The introduction of the comparison curves leads to the concept of variations of coordinates *qua* time functions. These time functions are denoted in the same way as the variations of coordinates in the d'Alembert–Lagrange principle. So, on one side δq^σ are coordinates of the tangent vector, and on the other side, these are time functions, which by definition are given by formulas (6.3).

The use of the same symbol to denote different concepts can be explained as follows.

Assume that a system, whose position depends on the generalized coordinates $q^1, q^2, \ldots, q^s$, is subject to the holonomic constraints

$$f_0^\varkappa(t, q) = 0 \,, \quad \varkappa = \overline{1, k} \,.$$

Then it has l degrees of freedom ($l = s - k$). We denote by denoted by $\mathbb{V}$ (respectively, by $\mathbb{V}_*$) the set of all possible positions of a system subject to no constraints (subject to constraints). Let q_*^λ, $\lambda = \overline{1, l}$ be the generalized coordinates of a position $M \in \mathbb{V}_*$

The set $\mathbb{V}_*$ will be looked upon as a subset of the set $\mathbb{V}$. The tangent space $\mathbb{T}(q_*, \mathbb{V}_*)$ to the manifold $\mathbb{V}_*$ in this case can be considered as a subspace of the tangent space $\mathbb{T}(q, \mathbb{V})$ only if the independent q_*^λ and dependent q^σ coordinates describe the same position of the system. A possible deviation $\delta \mathbf{y}$ is by definition an arbitrary vector of the tangent space $\mathbb{T}(q_*, \mathbb{V}_*)$. However, the vector $\delta \mathbf{y}$ can be dealt with in the same way as the vector from $\mathbb{T}(q, \mathbb{V})$, which satisfies the relations

$$\nabla f_0^\varkappa \cdot \delta \mathbf{y} = 0 \,, \quad \varkappa = \overline{1, k} \,, \tag{9.6}$$

or, in the dependent coordinates,

$$\frac{\partial f_0^{\varkappa}}{\partial q^{\sigma}} \delta q^{\sigma} = 0 \,.$$

Assume that the actual motion of a system is described by the functions $q^{\sigma}(t)$, and the kinematically possible motions, by the functions $\widetilde{q}^{\sigma}(t) = q^{\sigma}(t, \alpha)$, which for $\alpha = 0$ go into $q^{\sigma}(t)$. The functions $\widetilde{q}^{\sigma}(t) = q^{\sigma}(t, \alpha)$ should satisfy the constraint equations, and hence,

$$f_0^{\varkappa}(t, q(t, \alpha)) = 0 \,.$$

Therefore, in accordance with the definitions (6.2), (6.3) we have

$$\delta f_0^{\varkappa} = \frac{\partial f_0^{\varkappa}}{\partial q^{\sigma}} \delta q^{\sigma} = 0 \,, \quad \varkappa = \overline{1, k} \,. \tag{9.7}$$

So, the functions $\delta q^{\sigma}(t)$ are not independent, but must satisfy relations (9.7). In turn, relations (9.7) agree with Eq. (9.6) for the coordinates δq^{σ} of the vector of possible deviation $\delta \mathbf{y}$. Hence, the values of the function $\delta q^{\sigma}(t)$ with fixed t can be looked upon as coordinates of the vector $\delta \mathbf{y}$. This explains why different mathematical concepts have the same notation.

To clarify this, we consider a mathematical pendulum. Under the no-binding condition, the set of possible positions $\mathbb{V}$ is the three-dimensional Euclidean space; for simplicity we set $q^{\sigma} = x_{\sigma}$, $\sigma = 1, 2, 3$. Assume that the suspension point of a pendulum of length l_0 is at the origin, $O x_1 x_2$ is the plane of motion of the pendulum. The constraint equations now assume the form

$$x_3 = 0 \,, \quad x_1^2 + x_2^2 - l_0^2 = 0 \,.$$

Under the above constraints, the set of possible positions $\mathbb{V}_*$ of the pendulum is the circle of radius l_0 (Fig. 7). The position M of the pendulum depends on the angle of deviation from the vertical $\varphi = q_*^1$.

The coordinates of the vector $\delta \mathbf{y}$ in the space $\mathbb{T}(q, \mathbb{V}) = \mathbb{R}^3$ are related with the coordinate $\delta \varphi$ of the same vector $\delta \mathbf{y}$ (but which is considered as a vector in the space $\mathbb{T}(\varphi, \mathbb{V}_*)$) by the following relations:

$$\delta x_1 = -l_0 \sin \varphi \, \delta \varphi \,, \quad \delta x_2 = l_0 \cos \varphi \, \delta \varphi \,, \quad \delta x_3 = 0 \,.$$

It is worth pointing out that the above relations may be obtained by varying the Cartesian coordinates $x_1 = l_0 \cos \varphi$, $x_2 = l_0 \sin \varphi$, $x_3 = 0$, as given as functions of the angle φ.

So, the coordinates of the vector $\delta \mathbf{y}$ can be obtained by varying in the sense of the variational calculus (that is, in accordance with definition (6.2)). This example shows that no contradiction appears when using the same symbol δ in two different definitions (1.27) and (6.3). For principally nonholonomic systems, when in the cor-

Fig. 7 Possible travel of a mathematical pendulum

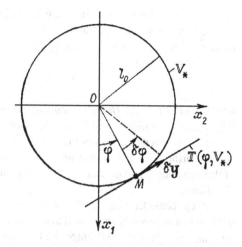

responding differential principle one has to use the symbol δ with the required number of primes, it proves impossible to relate the definition of variations by scheme (9.1) with definition (6.2).

All the integral principles follow from the principal variational relation (6.7), which shows how in the linear approximation the Hamiltonian action varies away from the chosen path of real motion. This variational relation (the principle of variable action) is closely related to the theory of integration of the motion equations (see Chap. 11).

10 The Hamilton–Jacobi Equation for the Function

Consider the action function S in the form

$$S = \int_{t_0}^{t_1} L(t, q, \dot{q})\, dt \,.$$

To evaluate the above integral one should first find L as a time function; that is, to ascertain the law of variation in time of the coordinate q^σ, $\sigma = \overline{1, s}$. In other words, one should find the law of motion obtained by integrating the Lagrange equations of the second kind.

According to the general theory of integration of the equations of mechanics, the coordinates q^σ, $\sigma = \overline{1, s}$, are functions of time t and $2s$ arbitrary integration constants. To find these constants one may specify s values of coordinates at the initial t_0 and finite t_1 times. So, we have

$$q_0^\sigma = q^\sigma(t_0, C_1, C_2, \ldots, C_{2s}) \,, \quad q_1^\sigma = q^\sigma(t_1, C_1, C_2, \ldots, C_{2s}) \,.$$

From these equations, assuming that the corresponding solvability conditions are satisfied, one may find $2s$ arbitrary constants C_1, C_2, ..., C_{2s}. They are expressed in terms of the constants q_0^σ and q_1^σ and the parameters t_0 and t_1. Hence S can be written as the function

$$S = S(t_0, t_1, q_0, q_1),$$

which is known as the *Hamilton's principal function*. As will be shown in Chap. 11, this function is fundamental in the theory of integration of the motion equations of mechanical systems for forces that have potentials.

The partial differential equation for the Hamilton's principal function as a function of the variables t_1 and q_1^σ can be obtained from the principle of variable action (6.7) as follows.

Let us consider the family of actual motions, as given by the functions $q^\sigma(t, \alpha)$. The times t_0 and t_1 will also be assumed to depend on the parameter α. In this case, the quantity S is composite function of the parameter α:

$$S(\alpha) = S(t_0(\alpha), t_1(\alpha), q^\sigma(t_0(\alpha), \alpha), q^\sigma(t_1(\alpha), \alpha)).$$

Hence,

$$\delta S = \left(\frac{dS}{d\alpha}\right)_{\alpha=0} \alpha = \frac{\partial S}{\partial t_1} \delta t_1 + \frac{\partial S}{\partial q_1^\sigma}(\delta q_1^\sigma +$$

$$+\dot{q}_1^\sigma \, \delta t_1) + \frac{\partial S}{\partial t_0} \delta t_0 + \frac{\partial S}{\partial q_0^\sigma}(\delta q_0^\sigma + \dot{q}_0^\sigma \, \delta t_0).$$

At the same time, in view of (6.7), the variation of the motion integral with variable limits is as follows:

$$\delta S = (p_\sigma \delta q^\sigma)_{t_1} - (p_\sigma \delta q^\sigma)_{t_0} + L_1 \, \delta t_1 - L_0 \, \delta t_0.$$

In the derivation of this formula, the comparison curves $q^\sigma(t, \alpha)$ were not considered as real paths — they were assumed to be arbitrary functions of time and the parameter α.

The curves of this family were assumed to satisfy only one constraint: for $\alpha = 0$ they should correspond to the actual path, with respect to which one calculates the variation δS. The above functions $q^\sigma(t, \alpha)$ both for $\alpha = 0$ and for $\alpha \neq 0$ correspond to actual paths. Hence, for them, in particular, the variation δS can be calculated by formula (6.7). This enables one to equate the coefficients of the independent δt_0, δt_1, δq_0^σ, δq_1^σ in two different equations for the variation δS. As a result, we have

$$\frac{\partial S}{\partial t_1} + \frac{\partial S}{\partial q_1^\sigma} \dot{q}_1^\sigma = L_1, \quad \frac{\partial S}{\partial t_0} + \frac{\partial S}{\partial q_0^\sigma} \dot{q}_0^\sigma = -L_0, \qquad (10.1)$$

$$\frac{\partial S}{\partial q_1^\sigma} = p_{\sigma 1}, \quad \frac{\partial S}{\partial q_0^\sigma} = -p_{\sigma 0}. \qquad (10.2)$$

In view of (10.2), we may write (10.1) as follows:

$$\frac{\partial S}{\partial t_1} = L_1 - p_{\sigma 1}\dot{q}_1^{\sigma} = -H_1, \quad \frac{\partial S}{\partial t_0} = p_{\sigma 0}\dot{q}_0^{\sigma} - L_0 = H_0, \tag{10.3}$$

where in accordance with (6.7) of Chap. 4 H_0 and H_1 are the values of the Hamilton function

$$H(t, q, p) = p_{\sigma}\dot{q}^{\sigma} - L(t, q, \dot{q})$$

at times t_0 and t_1.

We shall consider t_1 as the current time, and q_1^{σ}, as current coordinates. Writing t and q^{α}, respectively, in place of t_1 and q_1^{σ}, the first relation in (10.3) assumes the form

$$\frac{\partial S}{\partial t} + H\left(t, q, \frac{\partial S}{\partial q}\right) = 0, \tag{10.4}$$

where in view of (10.2) the current value of the generalized momentum p_{σ} is as follows:

$$p_{\sigma} = \frac{\partial S}{\partial q^{\sigma}}.$$

Hence, the action S, when considered as a function of time t and coordinates q^{σ}, $\sigma = \overline{1, s}$, satisfies the partial differential equation (10.4). The latter is known as the *Hamilton–Jacobi equation*. In Chap. 11 we shall show that the integration of this equation is equivalent to that of the system of the canonical equations of mechanics.

Let us show that the Hamilton–Jacobi equation (10.4) can also be obtained directly from the Newton equation for the representative point

$$M\mathbf{W} = \mathbf{Y}, \quad d(M\mathbf{V}) = \mathbf{Y}dt. \tag{10.5}$$

For simplicity, it may be assumed that the representative point is free and that the number of degrees of freedom s is $3n$, even though the subsequent analysis also apply to general mechanical systems with s degrees of freedom. Recall that under this general approach, the Newton equation (10.5) is considered valid for the tangent space to the manifold V_* of all possible positions of a given mechanical system.

We assume that the force field has the potential

$$\mathbf{Y} = -\nabla \Pi. \tag{10.6}$$

Integrating the Newton equation from t_0 to t along the path corresponding to the actual motion, this gives

$$M\mathbf{V}(q, t) = M\mathbf{V}_0(q_0, t_0) - \int_{t_0}^{t} \nabla \Pi \, dt. \tag{10.7}$$

Here, $\mathbf{V_0}(q_0, t_0)$ is the velocity of the representative point at time t_0 in position N_0 with coordinates q_0^σ. With given initial conditions, at time t the representative point occupies the position N with coordinates q^σ and has velocity $\mathbf{V}(q, t)$.

Consider the position N_0 as an arbitrary point of the set $\mathbb{V}_*$ of all possible positions of the mechanical system, we find information about the vector field on the manifold $\mathbb{V}_*$.

If one knows the motion paths corresponding to arbitrary initial conditions q_0 and $\mathbf{V_0}$ (that is, the general solution of the Newton equation (10.5)), then formula (10.7) enables one, from the velocity field at time t_0, to construct the velocity field at time t.

If $\mathbb{V}_* = \mathbb{R}^3$, then $\mathbf{V_0}(q_0, t_0)$ can be looked upon as a velocity field of some fluid at time t_0, and $\mathbf{V}(q, t)$, as the velocity field at time t. So, with the general solution of the Newton equation with $\mathbb{V}_* = \mathbb{R}^3$ one may associate the motion of some fluid.

The velocity field $\mathbf{V}(q, t)$ corresponding to the general solution to the motion equation will be searched in the form

$$M\mathbf{V}(q, t) = \nabla S, \tag{10.8}$$

where $S(q, t)$ is some unknown function.

Substituting (10.6), (10.8) into the Newton equation (10.5), this gives

$$\frac{d}{dt}(\nabla S) = -\nabla \Pi. \tag{10.9}$$

Since $\nabla S = \dfrac{\partial S}{\partial q^\sigma} \mathbf{e}^\sigma$, we have

$$\begin{aligned}
\frac{d}{dt}(\nabla S) &= \frac{\partial^2 S}{\partial t \partial q^\sigma} \mathbf{e}^\sigma + \frac{\partial^2 S}{\partial q^\tau \partial q^\sigma} \dot{q}^\tau \mathbf{e}^\sigma + \frac{\partial S}{\partial q^\sigma} \dot{\mathbf{e}}^\sigma = \\
&= \frac{\partial}{\partial q^\sigma}\left(\frac{\partial S}{\partial t} + \frac{\partial S}{\partial q^\tau} \dot{q}^\tau \right) \mathbf{e}^\sigma + \frac{\partial S}{\partial q^\sigma} \dot{\mathbf{e}}^\sigma.
\end{aligned} \tag{10.10}$$

At the same time,

$$M\mathbf{W} = MW_\sigma \mathbf{e}^\sigma = \left(\frac{d}{dt}\frac{\partial T}{\partial \dot{q}^\sigma} - \frac{\partial T}{\partial q^\sigma} \right) \mathbf{e}^\sigma. \tag{10.11}$$

However,

$$p_\sigma = \frac{\partial T}{\partial \dot{q}^\sigma} = \frac{\partial S}{\partial q^\sigma}, \quad M\mathbf{V} = p_\sigma \mathbf{e}^\sigma, \tag{10.12}$$

and hence,

$$M\mathbf{W} = \frac{d(M\mathbf{V})}{dt} = \dot{p}_\sigma \mathbf{e}^\sigma + p_\sigma \dot{\mathbf{e}}^\sigma. \tag{10.13}$$

Comparing (10.11), (10.13) and taking into account (10.12), we see that

$$\frac{\partial S}{\partial q^\sigma} \dot{\mathbf{e}}^\sigma = -\frac{\partial T}{\partial q^\sigma} \mathbf{e}^\sigma .$$

Substituting this expression into (10.10), we find that

$$\frac{d}{dt}(\nabla S) = \frac{\partial}{\partial q^\sigma} \left(\frac{\partial S}{\partial t} + p_\tau \dot{q}^\tau - T \right) \mathbf{e}^\sigma .$$

So, the Newton equation (10.9) can be written in the form

$$\frac{\partial}{\partial q^\sigma} \left(\frac{\partial S}{\partial t} + p_\tau \dot{q}^\tau - T + \Pi \right) \mathbf{e}^\sigma = 0 .$$

Further since

$$p_\tau \dot{q}^\tau - T + \Pi = p_\tau \dot{q}^\tau - L(t,q,\dot{q}) = H(t,q,p) = H\left(t,q,\frac{\partial S}{\partial q}\right),$$

we have

$$\frac{\partial}{\partial q^\sigma} \left(\frac{\partial S}{\partial t} + H\left(t,q,\frac{\partial S}{\partial q}\right) \right) \mathbf{e}^\sigma = 0 . \tag{10.14}$$

Since the required function S is a function of time t and coordinates q^σ, then the expression

$$\frac{\partial S}{\partial t} + H\left(t,q,\frac{\partial S}{\partial q}\right) \tag{10.15}$$

is also a function of the variables indicated.

From the Newton equation in the form (10.14) it follows that expression (10.15) is independent of q^σ, and so is a function of only time. Since the function S is given up to an arbitrary time function, it may be assumed without loss of generality that expression (10.15) is zero:

$$\frac{\partial S}{\partial t} + H\left(t,q,\frac{\partial S}{\partial q}\right) = 0 .$$

Hence, the velocity field $\mathbf{V}(q,t)$ corresponding to the general solution to the motion equations can be put in the form (10.8) if the function S satisfies the Hamilton–Jacobi equation (10.4).

Chapter 10
Statics

S. B. Filippov⊙

This chapter gives a brief outline of the statics of rigid solids and systems of rigid solids. Two definitions of the equivalence of systems of forces are given and shown to be equivalent. Setting up of equilibrium equations of mechanical systems by various methods is demonstrated by examples. Problems of determination of the position of the center of mass of a rigid solid, an equilibrium of a truss, and a flexible inextensible string are considered. Properties of friction forces are described, the solution to the Euler's equilibrium problem of a string wound on a cylinder is found.

1 Equivalent Systems of Forces

Statics is concerned with properties of forces and conditions for equilibrium in rigid solids subject to forces acting on them. The question of the equivalence of systems of forces acting on an absolutely rigid body was examined in Sect. 3 of Chap. 8. In this subsection, we give a more detailed account of the equivalence of systems of forces.

The vectors

$$\mathbf{F} = \sum_{\nu=1}^{n} \mathbf{F}_\nu, \quad \mathbf{L}_O = \sum_{\nu=1}^{n} \mathbf{r}_\nu \times \mathbf{F}_\nu,$$

where $\mathbf{r}_\nu$ is the vector pointing from the point O to the point M_ν of application of the force $\mathbf{F}_\nu$, are, respectively, the *resultant vector* and the *resultant torque* of a system of forces $(\mathbf{F}_\nu, M_\nu)$, $\nu = \overline{1, n}$. Let $\mathbf{F}'$ and $\mathbf{L}'_O$ denote, respectively, the resultant vector and resultant torque of a system of forces $(\mathbf{F}'_{\nu'}, M'_{\nu'})$, $\nu' = \overline{1, n'}$.

Definition 1 Systems of forces $(\mathbf{F}_\nu, M_\nu)$ and $(\mathbf{F}'_{\nu'}, M'_{\nu'})$ are called *equivalent* if $\mathbf{F} = \mathbf{F}'$, $\mathbf{L}_O = \mathbf{L}'_O$.

© Springer Nature Switzerland AG 2021
N. N. Polyakhov et al., *Rational and Applied Mechanics*,
Foundations of Engineering Mechanics,
https://doi.org/10.1007/978-3-030-64061-3_10

This definition of equivalence agrees with the definition from Sect. 3 of Chap. 8. In the same subsection, it was shown that equivalent systems of forces have equal resultant torques about any point P. Equivalent systems of forces act equally on a rigid solid.

In textbooks on mechanics, one may also find a different definition of the equivalence of systems of forces based on the notion of an elementary transformation. By definition, *elementary transformations* of a system of forces $(\mathbf{F}_\nu, M_\nu)$ are as follows:

(1) augmenting the system with a zero force or deleting a zero force from it;
(2) translation of the force $\mathbf{F}_\nu$ along the line of its action;
(3) replacement of two forces applied to one point by their vector sum;
(4) replacement of the force $\mathbf{F}_\nu$ by the sum of two ones applied to the same point.

Definition 2 Two systems of forces are called *equivalent* if one of them can be obtained from the other one by a finite number of elementary transformations.

Clearly, elementary transformations do not change the resultant vector of a system of forces. It is also obvious that the elementary transformations (1) and (2) do not change its resultant torque. The resultant torque is also invariant under the elementary transformations (3) and (4). Indeed, let $\mathbf{F}_\nu = \mathbf{F}_{\nu 1} + \mathbf{F}_{\nu 2}$; we assume that all three forces are applied at the point M_ν and the vector $\mathbf{r}_\nu$ is directed from the point O to the point M_ν. Hence,

$$\mathbf{L}_O(\mathbf{F}_\nu) = \mathbf{r}_\nu \times \mathbf{F}_\nu = \mathbf{r}_\nu \times (\mathbf{F}_{\nu 1} + \mathbf{F}_{\nu 2}) = \mathbf{L}_O(\mathbf{F}_{\nu 1}) + \mathbf{L}_O(\mathbf{F}_{\nu 2}).$$

Thus we have

Theorem 1 *A system of forces $(\mathbf{F}_\nu, M_\nu)$ and a system $(\mathbf{F}'_{\nu'}, M'_{\nu'})$, as obtained from it by a finite number of elementary transformations, have the same resultant vectors and resultant torques.*

From Theorem 1, it follows that if two systems of forces are equivalent in the sense of Definition 2, then they are equivalent in the sense of Definition 1. We now prove that the equivalence of systems of forces in the sense of Definition 1 implies their equivalence in the sense of Definition 2.

Theorem 2 *If two systems of forces have the same resultant vectors and resultant torques, then one of them can be obtained from the other one after a finite number of elementary transformations.*

Proof Our first aim is to show that any system of forces can be reduced by elementary transformations to a system of two forces. To do so, it suffices to show that any system of three forces $(\mathbf{F}_\nu, M_\nu)$, $\nu = 1, 2, 3$ is reduced to a system of two forces.

The case of parallel forces. Let us consider the case when all three forces are parallel to each other. Suppose that the forces $(\mathbf{F}_1, M_1)$ and $(\mathbf{F}_2, M_2)$ have the same direction. We claim that the system of two parallel forces $(\mathbf{F}_1, M_1)$ and $(\mathbf{F}_2, M_2)$ is reduced by elementary transformations to one force. By this, it will be shown that a system of three parallel forces is reduced to a system of two forces.

Fig. 1 System of two
parallel forces

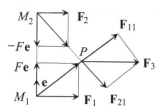

Let $\mathbf{e}$ be the unit vector directed along the interval $M_1 M_2$ (Fig. 1). We augment the system under consideration with two forces $F\mathbf{e}$ and $-F\mathbf{e}$ applied at the point M_1. We translate the force $-F\mathbf{e}$ to the point M_2 and add the forces applied at the points M_1 and M_2. This gives us a system of two forces $(\mathbf{F}_{11}, M_1)$, $(\mathbf{F}_{21}, M_2)$, where $\mathbf{F}_{11} = \mathbf{F}_1 + F\mathbf{e}$, $\mathbf{F}_{21} = \mathbf{F}_2 - F\mathbf{e}$. The lines of action of the forces $\mathbf{F}_{11}$ and $\mathbf{F}_{21}$ intersect at some point P. Moving these forces to the point P and adding them, we get the system $(\mathbf{F}_3, P)$ consisting of one force (Fig. 1), and besides $\mathbf{F}_3 = \mathbf{F}_1 + \mathbf{F}_2$.

In the same way, a system of two oppositely directed parallel forces can also be reduced to one force if $\mathbf{F}_1 \neq -\mathbf{F}_2$. In the case $\mathbf{F}_1 = -\mathbf{F}_2$ the lines of action of the forces $\mathbf{F}_{11}$ and $\mathbf{F}_{21}$ are parallel. A system of forces $(\mathbf{F}_1, M_1)$, $(-\mathbf{F}_1, M_2)$ is called a *couple of forces*.

The case of unparallel forces. We assume that the lines of action of three forces are not parallel. Suppose that the force $\mathbf{F}_3$ is not parallel to $\mathbf{F}_1$ and $\mathbf{F}_2$. We claim that in this case there exists a plane S passing through the line of action of the force $\mathbf{F}_3$ and intersecting the lines of action of the forces $\mathbf{F}_1$ and $\mathbf{F}_2$.

We introduce the fixed coordinate system with unit basis vectors $\mathbf{i}$, $\mathbf{j}$, $\mathbf{k}$, where the unit basis vector $\mathbf{k}$ is directed along the vector $\mathbf{F}_3$. We construct some vector $\mathbf{p}$ emerging from the origin of the force $\mathbf{F}_3$ and which is orthogonal to the unit basis vector $\mathbf{k}$. We consider the plane $\widetilde{S}$ passing through the vectors $\mathbf{F}_3$ and $\mathbf{p}$ and find out for which vectors $\mathbf{p}$ the plane $\widetilde{S}$ thus obtained will coincide with the sought-for plane S.

Assume that the angle between the vector $\mathbf{p}$ and the unit basis vector $\mathbf{i}$ is φ (see Fig. 2). Then the unit normal vector $\mathbf{n}$ to the plane $\widetilde{S}$, which is orthogonal to the vectors $\mathbf{p}$ and $\mathbf{k}$, can be expanded into the basis $\mathbf{i}$, $\mathbf{j}$, $\mathbf{k}$ as follows:

$$\mathbf{n} = \mathbf{i} \sin \varphi + \mathbf{j} \cos \varphi.$$

Fig. 2 The plane $\widetilde{S}$

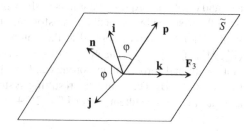

Fig. 3 The plane S

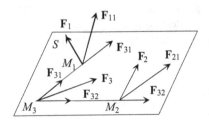

In turn, the vectors $\mathbf{F}_1$ and $\mathbf{F}_2$ have the following representations:

$$\mathbf{F}_k = a_k \mathbf{i} + b_k \mathbf{j} + c_k \mathbf{k}, \quad k = 1, 2.$$

The line of action of the force $\mathbf{F}_1$ does not intersect the plane $\widetilde{S}$ if the vector $\mathbf{F}_1$ lies in a plane parallel to $\widetilde{S}$. In this case, the vectors $\mathbf{n}$ and $\mathbf{F}_1$ are orthogonal, that is, $\mathbf{n} \cdot \mathbf{F}_1 = 0$. Substituting the vectors $\mathbf{n}$ and $\mathbf{F}_1$ into the last equality in the expansion, we get $a_1 \sin \varphi + b_1 \cos \varphi = 0$. If $a_1 = b_1 = 0$, then the equality $\mathbf{n} \cdot \mathbf{F}_1 = 0$ holds for any φ. However, in the case $a_1 = b_1 = 0$ the vector $\mathbf{F}_1$ is parallel to the vector $\mathbf{F}_3$, which contradicts the assumption that these vectors are not parallel. Hence, at least one of the numbers a_1 and b_1 should be nonzero, but then the equality $\mathbf{n} \cdot \mathbf{F}_1 = 0$ holds only for the angles φ for which $\tan \varphi = -b_1/a_1$. A similar analysis shows that the line of action of the force $\mathbf{F}_2$ does not meet the plane $\widetilde{S}$ if $\tan \varphi = -b_2/a_2$. Choosing the angle φ so that its tangent be different from $-b_1/a_1$ and $-b_2/a_2$, we get the requited plane S, which is intersected by the lines of action of both forces $\mathbf{F}_1$ and $\mathbf{F}_2$. Clearly, there are infinite number of such planes.

Suppose that the lines of action of the forces $\mathbf{F}_1$ and $\mathbf{F}_2$ intersect the plane S at the points M_1 and M_2 (Fig. 3). We replace the vector $\mathbf{F}_3$ by the sum of two forces directed along the intervals $M_3 M_1$ and $M_3 M_2$: $\mathbf{F}_3 = \mathbf{F}_{31} + \mathbf{F}_{32}$. We translate the forces $\mathbf{F}_{31}$ and $\mathbf{F}_{32}$ along the lines of their action to the points M_1 and M_2 and add with the forces $(\mathbf{F}_1, M_1)$ and $(\mathbf{F}_2, M_2)$, respectively. This gives us the system of two forces $(\mathbf{F}_{11}, M_1)$ and $(\mathbf{F}_{21}, M_2)$ (see Fig. 3), where

$$\mathbf{F}_{11} = \mathbf{F}_{31} + \mathbf{F}_1, \quad \mathbf{F}_{21} = \mathbf{F}_{32} + \mathbf{F}_2.$$

We come now to the final step of the proof of Theorem 2. We consider the systems of forces $(\mathbf{F}_\nu, M_\nu)$ and $(\mathbf{F}'_{\nu'}, M'_{\nu'})$, $\nu = \overline{1, n}$, $\nu' = \overline{1, n'}$, with the same resultant vectors and resultant torques. Our aim is to show that the second system can be obtained from the first one by elementary transformations. We augment the system $(\mathbf{F}_\nu, M_\nu)$ with n' nonzero forces applied at the points M'_ν and replace them by the system $(\mathbf{F}'_\nu, M'_\nu)$, $(-\mathbf{F}'_\nu, M'_\nu)$. The system of forces $(\mathbf{F}_\nu, M_\nu)$, $(-\mathbf{F}'_\nu, M'_\nu)$, with zero resultant vector and zero resultant torque can be reduced via elementary transformations to two forces. By Theorem 1 the resulting system of two forces $(\mathbf{F}''_1, M''_1)$, $(\mathbf{F}''_2, M''_2)$ also has the zero resultant vector $\mathbf{F}'' = \mathbf{F}''_1 + \mathbf{F}''_2$ and zero resultant torque $\mathbf{L}''_O$. Since the resultant vector is zero, it follows that $\mathbf{F}_2'' = -\mathbf{F}_1''$, and hence the resultant torque reads as

$$\mathbf{L}_O'' = \mathbf{r}_1 \times \mathbf{F}_1'' - \mathbf{r}_2 \times \mathbf{F}_1'' = (\mathbf{r}_1 - \mathbf{r}_2) \times \mathbf{F}_1'',$$

where $\mathbf{r}_1$ and $\mathbf{r}_2$ are the vectors pointing from some point O to the points M_1'' and M_2''. The resultant torque $\mathbf{L}_O''$ is zero if (1) $\mathbf{F}_1'' = 0$, (2) $\mathbf{r}_1 = \mathbf{r}_2$, (3) $(\mathbf{r}_1 - \mathbf{r}_2) \parallel \mathbf{F}_1''$. In the first case both forces are nonzero. Excluding the zero forces, we get the system of forces $(\mathbf{F}_\nu', M_\nu')$. In the second case the points coincide: $M_1'' = M_2''$. Adding the forces $\mathbf{F}_1''$ and $\mathbf{F}_2''$ applied at one point, we get the zero force, which can be excluded. Finally, in the third case the forces $\mathbf{F}_1''$ and $\mathbf{F}_2''$ lie on one line. Translating the force $\mathbf{F}_2''$ along the line of its action to the point M_1'' reduces the third case to the second one.

This proves Theorem 2 and therefore the equivalence of Definitions 1 and 2.

The definition of the equivalence based on the introduction of elementary transformations is more easily checked by experiments, but for theoretical studies it is more convenient to use the definition given in terms of the resultant vector and the resultant torque.

2 Systems of Parallel Forces. The Center of Mass

In the previous subsection, it was shown that any system of forces is equivalent to a system consisting of at most two forces. If the resultant vector $\mathbf{F}$ and the resultant torque $\mathbf{L}_O$ of a system of forces are orthogonal, then from the results of Sect. 3 of Chap. 8 it follows that the system is equivalent to one force $\mathbf{F}$, that is, the system is reduced to the *torque-free (pure force) resultant*. In particular, a system of forces converging at one point can be reduced to the *torque-free (pure force) resultant*, because the resultant torque about this point is zero. We shall show that a *system of parallel forces* with nonzero resultant vector can be reduced to the torque-free resultant $\mathbf{F}$ and find the line of action of the force $\mathbf{F}$.

The forces forming the system $(\mathbf{F}_\nu, M_\nu)$, $\nu = \overline{1, n}$, are parallel if $\mathbf{F}_\nu = F_\nu \mathbf{e}$, where $\mathbf{e}$ is the unit vector governing the direction of forces. We assume that the resultant vector is as follows:

$$\mathbf{F} = \sum_{\nu=1}^{n} \mathbf{F}_\nu = \mathbf{e} \sum_{\nu=1}^{n} F_{\nu e} \neq 0.$$

In this case we also have

$$F_e = \sum_{\nu=1}^{n} F_{\nu e} \neq 0.$$

If the point P lies on the line of action of $\mathbf{F}$, then the resultant torque $\mathbf{L}_P$ about the point P is zero.

Let $\mathbf{r}_\nu$, $\mathbf{p}_\nu$ and $\mathbf{p}$ be vectors pointing from the origin O to the point M_ν of application of the force $\mathbf{F}_\nu$, from the point P to the point M_ν, and from the point O to the point P, respectively. The vector

$$\mathbf{L}_P = \sum_{\nu=1}^{n}(\mathbf{p}_\nu \times \mathbf{F}_\nu) = \left[\sum_{\nu=1}^{n}(\mathbf{r}_\nu - \mathbf{p})F_{\nu e}\right] \times \mathbf{e} = \left[\sum_{\nu=1}^{n}(\mathbf{r}_\nu F_{\nu e}) - \mathbf{p}F_e\right] \times \mathbf{e}$$

will be zero if

$$\sum_{\nu=1}^{n}(\mathbf{r}_\nu F_{\nu e}) - \mathbf{p}F_e = \lambda \mathbf{e},$$

where λ is any real number. Hence, for the point P, the position of which is governed by the vector

$$\mathbf{p} = F_e^{-1}\left(\sum_{\nu=1}^{n}\mathbf{r}_\nu F_{\nu e} - \lambda \mathbf{e}\right),$$

we have $\mathbf{L}_P = 0$. There are infinite number of such points corresponding to all possible values of λ; they all lie on the line of action of the force $\mathbf{F}$.

So, we have proved that in the case $\mathbf{F} \neq 0$ the system of parallel forces is reduced to the pure force resultant. If $\mathbf{F} = 0$, then the system of parallel forces is equivalent to the couple of forces with moment equal to the resultant torque of this system of forces.

Assume that $\mathbf{L}_P = 0$ for the system of parallel forces $(\mathbf{F}_\nu, M_\nu)$, $\nu = \overline{1,n}$. We consider the system of parallel forces $(\mathbf{F}'_\nu, M_\nu)$, $\nu = \overline{1,n}$, obtained from the system $(\mathbf{F}_\nu, M_\nu)$ by rotating all forces by the same angle. For such a system we have

$$\mathbf{F}'_\nu = F_{\nu e}\mathbf{e}', \quad \nu = \overline{1,n}, \quad e' = 1.$$

The resultant torque of the system of forces $(\mathbf{F}'_\nu, M_\nu)$, $\nu = \overline{1,n}$, about the point P

$$\mathbf{L}'_P = \left[\sum_{\nu=1}^{n}(\mathbf{r}_\nu F_{\nu e}) - \mathbf{p}F_e\right] \times \mathbf{e}' = \lambda \mathbf{e} \times \mathbf{e}'$$

will be zero for any $\mathbf{e}'$ only in the case $\lambda = 0$. The point C corresponding to $\lambda = 0$ is called the *center of parallel forces*. The position of the point C is governed by the vector

$$\mathbf{r}_c = F_e^{-1}\left(\sum_{\nu=1}^{n}\mathbf{r}_\nu F_{\nu e}\right).$$

The center of parallel forces C is the unique point for which the resultant torque of the system of parallel forces remains zero for any rotation of all forces by the same angle.

We partition the region τ occupied by a rigid solid into n disjoint parts of volume $\Delta\tau_\nu$. We replace each part by a particle $M_\nu \in \tau_\nu$, the mass of which Δm_ν equals the mass of the corresponding part of the body. The resulting system of particles is an approximate model of a rigid solid.

In the homogeneous gravitational field, each of the points is subject to the force of gravity $\mathbf{P}_\nu = \Delta m_\nu \mathbf{g}$, where $\mathbf{g}$ is the gravitational acceleration vector. The center of parallel forces of the system $(\mathbf{P}_\nu, M_\nu)$, $\nu = \overline{1, n}$, is called the *center of gravity* of a system of particles. Its position is governed by the vector

$$\mathbf{r}_c = P^{-1}\left(\sum_{\nu=1}^{n}\mathbf{r}_\nu P_\nu\right), \quad P = \sum_{\nu=1}^{n}P_\nu,$$

where $P_\nu = \mathbf{P}_\nu \cdot \mathbf{e}$, $\mathbf{e} = \mathbf{g}/g$. Taking into account that $P_\nu = \Delta m_\nu g$, this gives

$$\mathbf{r}_c = M^{-1}\left(\sum_{\nu=1}^{n}\mathbf{r}_\nu \Delta m_\nu\right), \quad M = \sum_{\nu=1}^{n}\Delta m_\nu. \tag{2.1}$$

In view of this equality, the point C is also called the *center of mass* of a system of particles.

If n is increased and simultaneously the largest of the volumes $\Delta\tau = \max_\nu \Delta\tau_\nu$ is decreased, then the approximate model will describe more and more accurately a continuous rigid solid. Hence, the formula for the vector $\mathbf{r}_c$, which determines the position of the center of mass of a rigid solid, can be obtained from (2.1) by making $n \to \infty$, $\Delta\tau \to 0$, where the sums change into the volume integrals

$$\mathbf{r}_c = M^{-1}\int_\tau \mathbf{r}\,dm, \quad M = \int_\tau dm.$$

Here, M is the mass of the body, $dm = \mu\,d\tau$, μ is the density (see Sect. 3 of Chap. 8). The action of gravity forces on a rigid solid is equivalent to the action of the weight $\mathbf{P} = M\mathbf{g}$ as applied to the center of mass C.

The coordinates of the center of mass of a rigid solid are as follows:

$$x_c = M^{-1}\int_\tau x\,dm, \quad y_c = M^{-1}\int_\tau y\,dm, \quad z_c = M^{-1}\int_\tau x\,dm. \tag{2.2}$$

For a homogeneous body, the density μ is constant, and so formulas (2.2) assume the form

$$x_c = \tau^{-1}\int_\tau x\,d\tau, \quad y_c = \tau^{-1}\int_\tau y\,d\tau, \quad z_c = \tau^{-1}\int_\tau z\,d\tau. \tag{2.3}$$

If a body is composed of masses continuously distributed over the surface S, then the vector $\mathbf{r}_c$ is expressed in terms of the surface integrals

Fig. 4 Body in the form of a
rectangular parallelepiped

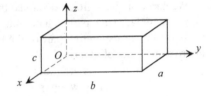

$$\mathbf{r}_c = M^{-1} \int_S \mathbf{r}\, dm\,, \quad M = \int_S dm\,, \quad dm = \mu_S\, dS\,,$$

where μ_S is the *surface density*. The mass of a thin rod can be assumed to be distributed over the line. In this case, $\mathbf{r}_c$ can be determined using the curvilinear integrals

$$\mathbf{r}_c = M^{-1} \int_l \mathbf{r}\, dm\,, \quad M = \int_l dm\,, \quad dm = \mu_l\, dl\,.$$

The function μ_l is the *linear mass density*.

For bodies of simple form, the coordinates of the center of mass can be easily obtained by direct calculation of the integrals.

Example 1 Assume that a homogeneous body is a rectangular parallelepiped of width a, length b, and height c (Fig. 4).

Since $\tau = abc$, it is found by formulas (2.3) that

$$x_c = \tau^{-1} \int_0^c \int_0^b \int_0^a x\, dx dy dz = \tau^{-1} \int_0^c \int_0^b \frac{a^2}{2}\, dy dz = \frac{a^2 bc}{2abc} = \frac{a}{2}\,, \quad y_c = \frac{b}{2}\,, \quad z_c = \frac{c}{2}\,.$$

If the height c of the rectangular parallelepiped is much smaller than its width and length, then in finding its center of mass the parallelepiped can be assumed to be a rectangle with $c = 0$. The center of mass of the rectangle with sides a and b has the coordinates $x_c = a/2$, $y_c = b/2$. In the case $a, c \ll b$, the rectangular parallelepiped can be looked upon as a thin rod of length b, its center of mass being in its midpoint ($y_c = b/2$).

Example 2 In order to calculate the coordinates of the center of mass of a homogeneous disc of radius R and area $S = \pi R^2$, we introduce the rectangular Cartesian coordinate system with origin at the disc center and consider the polar coordinates ρ and ψ. We have $x = \rho\cos\psi$, $y = \rho\sin\psi$, and hence,

$$x_c = S^{-1} \int_0^{2\pi} \int_0^R \rho^2 \cos\psi\, d\psi d\rho = 0\,, \quad y_c = S^{-1} \int_0^{2\pi} \int_0^R \rho^2 \sin\psi\, d\psi d\rho = 0\,,$$

that is, the center of mass is in the disc center.

Fig. 5 The center of mass of
the plane figure

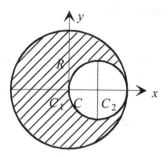

In the case when a body can be partitioned into parts of simple form, the center of
its mass can be found using the additivity of the integral. We split the region occupied
by a rigid solid into two disjoint parts of volumes τ_1, τ_2 and masses M_1 and M_2.
Assume that we know the positions of the centers of masses C_1 and C_2 of the parts
of the body:

$$\mathbf{r}_{c1} = M_1^{-1} \int_{\tau_1} \mathbf{r}\, dm\,, \quad \mathbf{r}_{c2} = M_2^{-1} \int_{\tau_2} \mathbf{r}\, dm\,.$$

In this case, the position of the center of mass C of the entire body can be found
from the formula

$$\mathbf{r}_c = \frac{\int_\tau \mathbf{r}\, dV}{M} = \frac{\int_{\tau_1} \mathbf{r}\, dm + \int_{\tau_2} \mathbf{r}\, dm}{M_1 + M_2} = \frac{\mathbf{r}_{c1} M_1 + \mathbf{r}_{c2} M_2}{M_1 + M_2}\,. \tag{2.4}$$

We note that the center of mass of a body lies on the interval $C_1 C_2$. This can be most
easily checked by taking the point C_1 as the origin. In this case, $\mathbf{r}_{c1} = 0$, the vector
$\mathbf{r}_{c2}$ is directed from the point C_1 to the point C_2, and hence, using (2.4),

$$\mathbf{r}_c = \alpha \mathbf{r}_{c2}\,, \quad \alpha = V_2/(V_1 + V_2)\,.$$

Since $0 \leqslant \alpha \leqslant 1$, the last equality yields that the point C lies in the interval $C_1 C_2$.

Example 3 Let us find the coordinates of the center of mass C_1 of the homogeneous
plane figure obtained by removing a disc of radius $R/2$ from a disc of radius R (see
Fig. 5).
   The center of mass C of the large disc lies in the interval $C_1 C_2$, where C_2 is the
center of mass of the small one. Hence, the point C_1 lies on the Ox-axis to the left
from the point C. Its coordinate x_{c1} is found from the equation

$$x_c = \frac{x_{c1} S_1 + x_{c2} S_2}{S_1 + S_2}\,,$$

which is obtained by projecting the vector equality (2.4) onto the x-axis and replacing
in it the masses M_k by the areas of the parts of the disc $S_k, k = 1, 2$. The solution of

the equation reads as

$$x_{c1} = (x_c S - x_{c2} S_2)/(S - S_2),$$

where $S = S_1 + S_2$ is the area of a disc of radius R. Taking into account that

$$x_c = 0, \quad x_{c2} = R/2, \quad S = \pi R^2, \quad S_2 = \pi R^2/4,$$

we find $x_{c1} = -R/6$.

3 Equilibrium Equations

Equations of statics in Cartesian coordinates. In Sect. 4 of Chap. 8 it was shown that a rigid solid may be in an equilibrium position under the action of a system of forces $(\mathbf{F}_\nu, M_\nu)$, $\nu = \overline{1, n}$, if the resultant vector and the resultant torque of the system of forces are both zero:

$$\mathbf{F} = \sum_{\nu=1}^{n} \mathbf{F}_\nu = 0, \quad \mathbf{L}_O = \sum_{\nu=1}^{n} \mathbf{r}_\nu \times \mathbf{F}_\nu = 0. \tag{3.1}$$

Here, $\mathbf{r}_\nu = x_\nu \mathbf{i} + y_\nu \mathbf{j} + z_\nu \mathbf{k}$ is the vector pointing from the origin of the rectangular Cartesian coordinate system $Oxyz$ to the point M_ν of application of the force $\mathbf{F}_\nu = F_{\nu x} \mathbf{i} + F_{\nu y} \mathbf{j} + F_{\nu z} \mathbf{k}$. Projecting equations (3.1) on the coordinate axes, we obtain six scalar equilibrium equations of a rigid solid:

$$\sum_{\nu=1}^{n} F_{\nu x} = 0, \quad \sum_{\nu=1}^{n} (y_\nu F_{\nu z} - z_\nu F_{\nu y}) = 0,$$

$$\sum_{\nu=1}^{n} F_{\nu y} = 0, \quad \sum_{\nu=1}^{n} (z_\nu F_{\nu x} - x_\nu F_{\nu z}) = 0, \tag{3.2}$$

$$\sum_{\nu=1}^{n} F_{\nu z} = 0, \quad \sum_{\nu=1}^{n} (x_\nu F_{\nu y} - y_\nu F_{\nu x}) = 0.$$

In solving problems in statics, it is convenient to use the concept of the *moment of the force about the axis*. We consider the axis $\overrightarrow{OL}$, whose position in the space is controlled by the unit vector $\mathbf{l}$ ($l = 1$). The moment of the force $\mathbf{F}$ about the axis $\overrightarrow{OL}$ is the scalar

$$m_l(\mathbf{F}) = \mathbf{m}_O(\mathbf{F}) \cdot \mathbf{l} = (\mathbf{r} \times \mathbf{F}) \cdot \mathbf{l},$$

which is the projection of the vector $\mathbf{m}_O(\mathbf{F})$ on the axis $\overrightarrow{OL}$ ($\mathbf{r}$ is the vector from the point O to the point M of the application of the force).

Fig. 6 Vectors **F**, **r**, **r′** and **r₀**

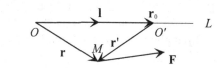

Fig. 7 The force **F** and its projections **F**ₗ and **F**ₛ

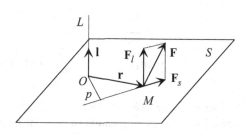

We claim that $m_l(\mathbf{F})$ does not depend on the choice of the point O on the axis $\overrightarrow{OL}$. Let O' be a different point on the axis $\overrightarrow{OL}$, let $\mathbf{r}'$ be the vector pointing from the point O' to the point M, and let $\mathbf{r}_0$ be the vector from the point O to the O', which is parallel to the vector **l** (Fig. 6).

We have

$$m_l'(\mathbf{F}) = (\mathbf{r}' \times \mathbf{F}) \cdot \mathbf{l} = [(\mathbf{r} - \mathbf{r}_0) \times \mathbf{F}] \cdot \mathbf{l} = (\mathbf{r} \times \mathbf{F}) \cdot \mathbf{l} = m_l(\mathbf{F}).$$

Let us consider the plane S perpendicular to the vector **l** and passing through the point of application of the force M (see Fig. 7).

We represent the vector **F** in the form $\mathbf{F} = \mathbf{F}_l + \mathbf{F}_s$, where $\mathbf{F}_l \parallel \mathbf{l}, \mathbf{F}_s \in S$. Let O be the point of intersection of the axis $\overrightarrow{OL}$ and the plane S and let **r** be the vector from the point O to the point M. We have

$$m_l(\mathbf{F}) = [\mathbf{r} \times (\mathbf{F}_l + \mathbf{F}_s)] \cdot \mathbf{l} = (\mathbf{r} \times \mathbf{F}_s) \cdot \mathbf{l},$$

and besides, $\mathbf{r} \times \mathbf{F}_s \parallel \mathbf{l}$. Hence,

$$|m_l(\mathbf{F})| = |\mathbf{m}_O(\mathbf{F}_s)| = p \, |\mathbf{F}_s|,$$

where p is called the *arm of the force* $\mathbf{F}_s$.

If $p = 0$ or $\mathbf{F}_s = 0$, then the moment of the force about the axis is zero. In the first case, the line of action of the force intersects the axis, and in the second case, the force **F** is parallel to the axis.

Thus, in order to calculate the absolute value of the moment of the force about the axis one needs to find the projection of this force to the plane S, which is perpendicular to the axis and then find the magnitude of the moment of this projection relative to the point of intersection of the axis and the plane. In order to find the sign of the moment of the force, one needs to look at the plane S from the end of the vector **l**.

If as a result the force $\mathbf{F}_s$ will try to rotate the body counterclockwise (respectively, clockwise), then $m_l(\mathbf{F}) > 0$ ($m_l(\mathbf{F}) < 0$).

Multiplying scalarly the equilibrium equations (3.1) by the unit basis vectors $\mathbf{i}$, $\mathbf{j}$ and $\mathbf{k}$ gives

$$\sum_{\nu=1}^{n} F_{\nu x} = 0, \quad \sum_{\nu=1}^{n} F_{\nu y} = 0, \quad \sum_{\nu=1}^{n} F_{\nu z} = 0,$$

$$\sum_{\nu=1}^{n} m_x(\mathbf{F}_\nu) = 0, \quad \sum_{\nu=1}^{n} m_y(\mathbf{F}_\nu) = 0, \quad \sum_{\nu=1}^{n} m_z(\mathbf{F}_\nu) = 0, \tag{3.3}$$

where $m_x(\mathbf{F}_\nu)$, $m_y(\mathbf{F}_\nu)$, $m_z(\mathbf{F}_\nu)$ are the moments of the force $\mathbf{F}_\nu$ about the coordinate axes. Since the multiplication of a vector by a unit basis vector is equivalent to its projection on the coordinate axis, equations (3.3) can be looked upon as different forms of the equilibrium equations (3.2).

If one considers a system of n rigid solids, then for each of them one may write down six equilibrium equations of the form (3.2) or (3.3). So, the number of scalar equilibrium equations for a system of n rigid solids is $6n$.

Equations of statics in curvilinear coordinates. We consider the motion of a system of material bodies. To describe its motion in curvilinear coordinates one may use the Lagrange equations from Chap. 6. The equations of statics derived in this way are subsumed into the so-called *analytical statics*.

Assume that the position of a mechanical system is given in the curvilinear coordinates $q = (q^1, \ldots, q^s)$. If there are holonomic constraints

$$f^{\varkappa}(t, q) = 0, \quad \varkappa = \overline{1, k}, \tag{3.4}$$

then the motion of such a mechanical system can be described by the Lagrange equations of first kind in curvilinear coordinates (the Lagrange equations of second kind with multipliers)

$$\frac{d}{dt}\frac{\partial T}{\partial \dot{q}^\sigma} - \frac{\partial T}{\partial q^\sigma} = Q_\sigma + \Lambda_\varkappa \frac{\partial f^\varkappa}{\partial q^\sigma}, \quad \sigma = \overline{1, s}. \tag{3.5}$$

We note that in our case the equations of holonomic constraints (3.4) will describe the constraints on the generalized displacements of a system of material bodies that provide for the possibility of its equilibrium position.

From the system of coordinates $q = (q^1, \ldots, q^s)$ one may frequently change to the new system of curvilinear coordinates $q_* = (q^1_*, \ldots, q^s_*)$. In the latter system, the first l coordinates ($l = s - k$) are the Lagrange coordinates, and their relation with the original system of coordinates $q = (q^1, \ldots, q^s)$ is specified by the researcher.

$$q^\lambda_* = f^\lambda_*(t, q), \quad \lambda = \overline{1, l}. \tag{3.6}$$

In turn, the last k coordinates are set to be equal to the equations of constraints (3.4)

$$q_*^{l+\varkappa} = f^{\varkappa}(t, q), \quad \varkappa = \overline{1, k}, \tag{3.7}$$

and hence, they are zero in the process of motion. In the new system of coordinates, there hold the Lagrange equations of the second kind, which is split into two groups of equations:

$$\frac{d}{dt} \frac{\partial T}{\partial \dot{q}_*^{\lambda}} - \frac{\partial T}{\partial q_*^{\lambda}} = Q_{\lambda}^*, \quad \lambda = \overline{1, l}, \tag{3.8}$$

$$\frac{d}{dt} \frac{\partial T}{\partial \dot{q}_*^{l+\varkappa}} - \frac{\partial T}{\partial q_*^{l+\varkappa}} = Q_{l+\varkappa}^* + \Lambda_{\varkappa}, \quad \varkappa = \overline{1, k}. \tag{3.9}$$

Considering the case of an equilibrium of a system of rigid solids, when the kinetic energy is identically equal to zero, from Eqs. (3.5) and (3.8), (3.9) one gets two kinds of *equations of statics in curvilinear coordinates*:

$$Q_{\sigma} + \Lambda_{\varkappa} \frac{\partial f^{\varkappa}}{\partial q^{\sigma}} = 0, \quad \sigma = \overline{1, s}, \tag{3.10}$$

$$Q_{\lambda}^* = 0, \quad \lambda = \overline{1, l}, \quad Q_{l+\varkappa}^* + \Lambda_{\varkappa} = 0, \quad \varkappa = \overline{1, k}. \tag{3.11}$$

Here, the Lagrange multipliers $\Lambda_{\varkappa}$, $\varkappa = \overline{1, k}$, are the *generalized reaction forces* of the ideal holonomic constraints (3.4), while the terms in the sum $\Lambda_{\varkappa} \dfrac{\partial f^{\varkappa}}{\partial q^{\sigma}}$ are the covariant components of the reactions of these constraints by using the system of curvilinear coordinates $q = (q^1, \ldots, q^s)$.

Application of the principle of virtual displacements. The derivation of the equilibrium equations in the form (3.10) and (3.11) can also be explained using the principle of virtual displacements (see Chap. 9).

This principle, when written in the tangential space, reads under the ideal constraints (3.4) as

$$(\mathbf{Y} + \Lambda_{\varkappa} \nabla f^{\varkappa}) \cdot \delta \mathbf{y} = 0, \tag{3.12}$$

where $\delta \mathbf{y}$ is the vector of virtual displacement of the system (see formula (5.5) of Chap. 9).

Together with the curvilinear frame $q = (q^1, \ldots, q^s)$ with the bases $\{\mathbf{e}_1, \ldots, \mathbf{e}_s\}$ and $\{\mathbf{e}^1, \ldots, \mathbf{e}^s\}$, the new frame $q_* = (q_*^1, \ldots, q_*^s)$ was introduced by formulas (3.6), (3.7). To this transformation there corresponds the inverse one

$$q^{\sigma} = q^{\sigma}(t, q_*), \quad \sigma = \overline{1, s}. \tag{3.13}$$

Transformations (3.6), (3.7), (3.13) define the fundamental and reciprocal bases in the new coordinate system

$$\mathbf{e}_{\tau}^* = \frac{\partial q^{\sigma}}{\partial q_*^{\tau}} \mathbf{e}_{\sigma}, \quad \mathbf{e}_*^{\rho} = \frac{\partial q_*^{\rho}}{\partial q^{\tau}} \mathbf{e}^{\tau}, \quad \rho, \sigma, \tau = \overline{1, s}. \tag{3.14}$$

It is important here that $\mathbf{e}_*^{l+\varkappa} = \nabla f^\varkappa$, $\varkappa = \overline{1, k}$; that is, constraints (3.4) form the K-"subspace of reaction forces". Now the virtual displacement, the active forces, and the reaction forces of ideal holonomic constraints can be written, depending on the frame under consideration, in the form

$$\delta \mathbf{y} = \delta q^\sigma \mathbf{e}_\sigma = \delta q_*^\tau \mathbf{e}_\tau^*, \quad \sigma, \tau = \overline{1, s}, \tag{3.15}$$

$$\mathbf{Y} = Q_\sigma \mathbf{e}^\sigma = Q_\tau^* \mathbf{e}_*^\tau, \quad \mathbf{R} = \mathbf{R}^K = \Lambda_\varkappa \frac{\partial f^\varkappa}{\partial q^\sigma} \mathbf{e}^\sigma = \Lambda_\varkappa \mathbf{e}_*^{l+\varkappa}, \tag{3.16}$$

$$\sigma, \tau = \overline{1, s}, \quad \varkappa = \overline{1, k}.$$

Using formulas (3.15), (3.16), principle (3.12) can be written in the form

$$\left(Q_\sigma + \Lambda_\varkappa \frac{\partial f^\varkappa}{\partial q^\sigma} \right) \delta q^\sigma = 0, \quad \sigma = \overline{1, s}, \tag{3.17}$$

or

$$Q_\lambda^* \delta q_*^\lambda + (Q_{l+\varkappa}^* + \Lambda_\varkappa) \delta q_*^{l+\varkappa} = 0, \quad l = s - k, \quad \lambda = \overline{1, l}, \quad \varkappa = \overline{1, k}. \tag{3.18}$$

We recall that when applying the *principle of releasability from constraints*, all variations in representations (3.15) are independent, and hence in order that formulas (3.17) and (3.18) be satisfied all the coefficients of these independent variations should vanish, which gives the statics equations (3.10) and (3.11).

In solving practical problems, one should choose a reference system, indicate in the mechanical system the active forces and momenta acting on the system, and find the sought-for forces and moments, which are the reactions of the imposed holonomic constraints. From the resulting system of $6n$ algebraic equilibrium equations of n bodies one finds $6n$ unknown components of reactions of the constraints. These reactions are responsible for the equilibrium of the system of mutually connected bodies.

Equilibrium equations may have infinite number of solutions if the number of unknown forces involved in these equations is larger than the number of equations. In this case, the problems are called *statically indeterminate*. Statically indeterminate problems cannot be solved completely in the framework of statics. In solving such problems one should add to the equilibrium equations additional equations, which may be taken from the theory of deformable solids.

Comparing the number of equilibrium equations with the number of unknown forces, one should take into account that for specific problems some of the equations may prove to be dependent. In particular, some equations may turn to be identities. In finding out whether a problem is statically indeterminate, one should compare the number of unknowns with the number of independent equilibrium equations.

In the next section, in solving some examples we shall discuss different kinds of reaction forces appearing under specific holonomic constraints.

4 Setting Up and Solving Equilibrium Equations

We consider two systems of forces for which there are at most three independent equations in the six equations of (3.3).

Concurrent system of forces. As the origin O we take the point of intersection of the lines of action of all the forces. Since in this case the resultant torque L_O is 0, the last three equilibrium equations become identities, and so there are three equations

$$\sum_{i=1}^{n} F_{ix} = 0, \quad \sum_{i=1}^{n} F_{iy} = 0, \quad \sum_{i=1}^{n} F_{iz} = 0. \tag{4.1}$$

for the projections of forces. In choosing as the origin the point P, as given by the vector $\mathbf{p} = p_x \mathbf{i} + p_y \mathbf{j} + p_z \mathbf{k}$ relative to the point O, we get six equilibrium equations. However, three of them, as obtained by projecting the principal torque $L_P = -\mathbf{p} \times \mathbf{F}$ on the coordinate axes, will be linear combinations of equations (4.1) with the coefficients p_x, p_y and p_z.

Planar system of forces. Assume that all forces lie in the plane S. Taking $O \in S$ as the origin, we direct the Oz-axis to be perpendicular to the plane S. Now since $F_{iz} = 0$, $z_i = 0$ the three scalar equilibrium equations become identities. The remaining three equations

$$\sum_{i=1}^{n} F_{ix} = 0, \quad \sum_{i=1}^{n} F_{iy} = 0, \quad \sum_{i=1}^{n} m_z(\mathbf{F}_i) = 0 \tag{4.2}$$

are used to solve problems in statics for a planar system of forces. For a planar concurrent system of forces, the third equation becomes an identity if as the original one takes the point of intersection of the action lines of forces.

Example 4 We consider the equilibrium problem for a homogeneous rod of length l which has its ends A and B on two absolutely smooth perpendicular planes (Fig. 8).

The rod lies in the plane Oxy. A string is attached to the point B, the second end of the string is fixed at the point O. The gravity force P acts on the rod midpoint. The magnitude of P and the angle $\alpha \neq 0$ between the intervals BO and BA are known. Find the tension T of the string and the reactions X_A and Y_B, which are directed perpendicular to the corresponding planes (no friction case). A string does

Fig. 8 Forces acting on the rod

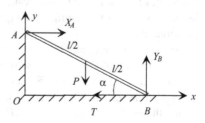

work only by extension, and hence it acts on the rod by the force T directed along the string from right to left.

For this problem, Eq. (4.2) assume the form

$$X_A - T = 0, \quad -P + Y_B = 0, \quad -X_A l \sin \alpha + P \frac{l}{2} \cos \alpha = 0.$$

The third equation is obtained from the condition $\mathbf{L}_B = 0$. The system of equations has the solution

$$X_A = T = \frac{P}{2} \cot \alpha, \quad Y_B = P.$$

Assume that there is another string attached to the rod end B; the second end of this string is fixed at some interior point of the interval OA. In this case, the system of equations will contain an additional unknown (the tension T' of the second string), the problem becoming statically indeterminate. In solving such a problem one needs to take into account the expansion of strings.

If a system of forces is neither concurrent nor planar, then in order to solve the problem one may need to have recourse to the six equilibrium equations.

Example 5 Let us now examine the equilibrium of a spatial system of forces. We consider the equilibrium problem of a homogeneous rectangular door with hinges at points A and B (see Fig. 9).

The hinge at the point B plays the role of a pin joint around the axis of which the body can rotate freely, but the joint hinders any displacement of the body in the direction perpendicular to its axis. Hence, the pin joint (the hinge B) produces an unknown horizontal reaction force, which will be sought for in the form of its projections X_B and Y_B.

As distinct from the upper hinge B, the lower hinge A (see Fig. 9) also inhibits the vertical displacement of the door. To this hinge one may associate a pin joint with foot (such a joint is called "foot-step bearing"; usually a body on the axis of rotation is fixed using a combination of a pin joint on the top and a foot-step bearing on the lower end). As a result, the hinge A may create a reaction of arbitrary magnitude and direction; this reaction will be sought for as its unknown projections X_A, Y_A, Z_A. Considering the door as a whole, one should note that the unknown vertical component Z_A was referred to the lower hinge only conventionally—depending on a particular installation of the door on hinges this load can be also born by the upper hinge. Moreover, for an exceptionally accurate suspension of the door, both the hinges may simultaneously restrain the door from a vertical displacement, and in this case, from the equilibrium equations, one may get the sum of their vertical components.

As an origin, we take the point A and direct the z-axis along the door axis of rotation. In addition to the gravity force $\mathbf{P}$, which is parallel to the Oz-axis, the door is subject to the forces $\mathbf{T}$ and $\mathbf{S}$, which lie in the planes perpendicular to the z-axis, and which are created by the ropes CD and EF. The magnitudes of the forces P, Q and the angles α, β are given: $AB = 2a$, $BC = 2b$.

Fig. 9 Forces acting on the door

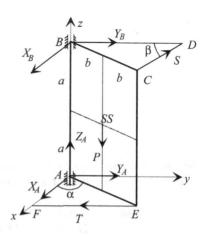

Fig. 10 Top view of the door

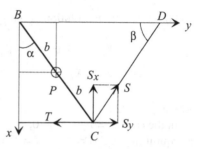

The reactions X_A, Y_A, Z_A, X_B, Y_B and the magnitude of the force T are unknown.

It is convenient to replace the force S by the sum $\mathbf{S}_x + \mathbf{S}_y$, where the forces $\mathbf{S}_x$ and $\mathbf{S}_y$ are parallel to the Ox- and Oy-axes,

$$S_x = S \sin \beta, \quad S_y = S \cos \beta$$

(see Fig. 10 for the top view of the door).

Projecting the resultant vector of the system of forces on the coordinate axes, we get the first three equilibrium equations:

$$X_A + X_B - S_x = 0, \quad Y_A + Y_B - T + S_y = 0, \quad Z_A - P = 0.$$

In developing the equilibrium equations involving the torques relative to the coordinate axes, it should be taken into account that in this problem several torques are zero. So, for example, the reacting torques X_A, Y_A and Z_A relative to all coordinate axes are zero, because each of such forces is parallel to one of the axes and intersects

Fig. 11 Equilibrium
position of two rods

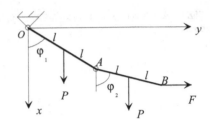

the other two ones. Equating to zero the sums of torques relative to the Ox-, Oy-,
and Oz-axes, we get the remaining three equilibrium equations:

$$-2Y_B a - Pb \sin\alpha - 2S_y a = 0, \quad 2X_B a + Pb \cos\alpha - 2S_x a = 0,$$
$$-2Tb \cos\alpha + 2S_x b \sin\alpha + 2S_y b \cos\alpha = 0.$$

For $\alpha \neq \pi/2$, the solution of the system of equilibrium equations reads as

$$X_A = \gamma P \cos\alpha, \quad Y_A = T + \gamma P \sin\alpha, \quad Z_A = P,$$
$$X_B = S_x - X_A, \quad Y_B = -S_y - \gamma P \sin\alpha, \quad T = S\frac{\cos(\alpha - \beta)}{\cos\alpha},$$

where $\gamma = b/2a$.

In the case $\alpha = \pi/2$, $\beta \neq 0$, $S \neq 0$ there are no solutions (a door may not be in
an equilibrium). If $\alpha = \pi/2$, $\beta = 0$, then the sixth equilibrium equation becomes an
identity, the problem becoming statically indeterminant.

Example 6 Let us find the equilibrium position of two equal homogeneous rods OA
and AB of weight P and length $2l$ with hinge at the point A. The point O is a hinged
end, the end B of the second rod is subject to a horizontal force of magnitude F (see
Fig. 11).

To solve the problem we employ equations (3.11). This system of two rods has
two degrees of freedom. As generalized coordinates, we take the rotation angles of
the rods, that is, we assume that $q^1_* = \varphi_1$ and $q^2_* = \varphi_2$. To find the generalized forces
corresponding to these generalized coordinates, we write an expression for the virtual
elementary work

$$\delta A = P\delta x_1 + P\delta x_2 + F\delta y_3. \tag{4.3}$$

Here, the coordinates of the points of applications of forces x_1, x_2 and y_3 are expressed
in terms of the generalized coordinates by the formulas

Fig. 12 Forces acting on the beam

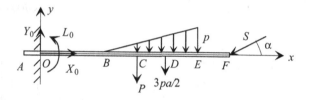

$$x_1 = l \cos \varphi_1 , \quad x_2 = 2l \cos \varphi_1 + l \cos \varphi_2 , \quad y_3 = 2l(\sin \varphi_1 + \sin \varphi_2) .$$

The virtual displacements are related to the variations of the generalized coordinates as follows:

$$\delta x_1 = -l \sin \varphi_1 \delta \varphi_1 , \quad \delta x_2 = -2l \sin \varphi_1 \delta \varphi_1 - l \sin \varphi_2 \delta \varphi_2 ,$$
$$\delta y_3 = 2l(\cos \varphi_1 \delta \varphi_1 + \cos \varphi_2 \delta \varphi_2) .$$

We substitute the expressions for the virtual displacements in (4.3). The coefficients of the variations of the generalized coordinates $\delta \varphi_1$ and $\delta \varphi_2$ are the generalized forces. Hence, in an equilibrium position

$$Q^*_{\varphi_1} = -3Pl \sin \varphi_1 + 2Fl \cos \varphi_1 = 0 , \quad Q^*_{\varphi_2} = -Pl \sin \varphi_2 + 2Fl \cos \varphi_2 = 0 .$$

The solution of the resulting system of equations

$$\tan \varphi_1 = \frac{2F}{3P} , \quad \tan \varphi_2 = \frac{2F}{P}$$

gives the angles φ_1 and φ_2 in an equilibrium position.

If one uses the equilibrium equation (3.3) for the solution of this problem, then one gets the system of six equations (three for each rod) for the determination of the six unknown: the angles φ_1 and φ_2, the reactions X_O, Y_O in the hinge O and two projections on the axes Ox and Oy of the rods exchange force. In the actual fact, in solving this system of equations one will get in addition the reaction force at the hinge O and the interacting (exchange) force of the rods in the hinge A.

Example 7 We consider an equilibrium of a horizontal homogeneous cantilever beam OF of length $6a$ (Fig. 12), the left end of which is fastened to a vertical wall. In this case, a beam is said to be embedded at the point O. The beam is subject to the gravity force **P** applied to its center, the active force **S** on the right end of the beam, and the vertical load, which is distributed on the segment BE of length $3a$ according to the linear law from zero to p N/m. The length of OB is $2a$.

Let us discuss how one may change the load distributed on the interval $[a, b]$ according to the law $f(x)$ by the pure force resultant (Fig. 13).

We partition the interval $[a, b]$ into n subintervals of length $\Delta x_1, \Delta x_2, ..., \Delta x_n$ and choose on each subinterval a point with coordinate x_i, $i = \overline{1, n}$ (in Fig. 13 $n = 5$).

Fig. 13 Distributed load

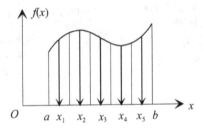

We replace the load distributed over the ith part of $[a, b]$ by the force directed vertically downwards, of magnitude $F_i = f(x_i)\Delta x_i$, and passing through the point with coordinate x_i. From the results of Sect. 2 it follows that the resulting system of parallel forces is equivalent to the pure force resultant

$$F' = \sum_{i=1}^{n} F_i = \sum_{i=1}^{n} f(x_i)\Delta x_i, \tag{4.4}$$

the point of application of which has the coordinate

$$x_c' = \frac{1}{F'} \sum_{i=1}^{n} x_i F_i = \frac{1}{F'} \sum_{i=1}^{n} x_i f(x_i)\Delta x_i. \tag{4.5}$$

The smaller the lengths of the intervals Δx_i the better the system of parallel forces F_i describes the action of the distributed load. For $\Delta x_i \to 0$ the sums on the right-hand sides of formulas (4.4) and (4.5) become the definite integrals. The limit values F and x_c of F' and x_c'

$$F = \int_a^b f(x)\,dx, \quad x_c = \frac{1}{F} \int_a^b x f(x)\,dx$$

give the magnitude of the resultant distributed load and the abscissa of the point of its application. Hence, the magnitude of the pure force resultant equals the area of the curvilinear trapezoid under the graph of the function $f(x)$, and the coordinate x_c of the point of application of this force agrees with the abscissa of the "center of gravity" of the curvilinear trapezoid.

For this example, we replace the distributed load acting on the beam by the force $3pa/2$ applied to the point D lying at distance $4a$ from the origin (see Fig. 12).

Let us now discuss the reaction force created by the embedding at the point O. The action at this point on the beam OF is created by the part AO of the beam embedded in the wall. Under external loads, it hinders the motion of the point O by creating the force with the projections X_O, Y_O and it does not allow the beam to rotate by creating the torque L_O (which is depicted in Fig. 12 by the curved arc, which tries to rotate the beam counterclockwise around the point O).

So, by the principle of system releasability from constraints one may assume that the beam (being free) is in equilibrium under the action of the generalized forces acting to it

$$P, \quad S, \quad 3pa/2, \quad X_O, \quad Y_O, \quad L_O. \tag{4.6}$$

The spatial position of the free beam in the vertical plane is determined by the coordinates x, y of the point O and the rotation of the beam around the point O by the angle φ relative to the horizontal axis Ox. According to the principle of virtual displacements, the work of the generalized forces (4.6) on any virtual displacement of the system

$$(\delta x, \delta y, \delta \varphi) \tag{4.7}$$

must be zero,

$$(X_O - S\cos\alpha)\delta x + (Y_O - S\sin\alpha - P - \frac{3}{2}pa)\delta y + (L_O - 3Pa - 6pa^2 - 6aS\sin\alpha)\delta\varphi = 0. \tag{4.8}$$

The virtual displacements (4.7) being independence, the coefficients multiplying by them in (4.8) should be zero:

$$X_O - S\cos\alpha = 0,$$

$$Y_O - S\sin\alpha - P - \frac{3}{2}pa,$$

$$L_O - 3Pa - 6pa^2 - 6aS\sin\alpha = 0. \tag{4.9}$$

The solution of system (4.9) reads as

$$X_O = S\cos\alpha, \quad Y_O = S\sin\alpha + P + \frac{3}{2}pa, \quad L_O = 3a(P + 2pa + 2S\sin\alpha).$$

It is worth pointing out that in setting up the equilibrium equations (4.9) with chosen coordinates of the system x, y, φ, we in fact obtained the equilibrium equations of a planar system of forces in the case of the Cartesian system of coordinates. The application of the principle of virtual displacements may be expedient when working in a more involved system of curvilinear coordinates.

Example 8 Figure 14 depicts a crank press as a rhombus consisting of four equal pin-connected rods of length[1] l.

The rhombus has the fixed upper point O, which is taken as the origin of the Cartesian coordinate system, its lower point C is in contact with the upper plate of the press DE. The joint points A and B of rods are subject to two oppositely directed horizontal forces of magnitude F. Assuming that the rhombus is in equilibrium, find the magnitude of the force P acting on the lower point of the rhombus C from the

---

[1] See Example 46.2 in the book: *I. V. Meshchersky*. Collection of Problems in Theoretical Mechanics (Moscow: Izdat. "Nauka", 1986) [in Russian].

Fig. 14 The scheme of
crank press

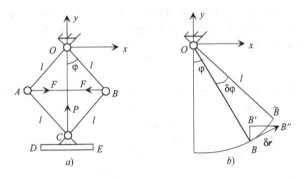

side of the upper plate of the press at the time when the angle at the rhombus vertex
is 2φ. According to Newton's third law, the press plate is subject to the force of
magnitude P directed vertically downwards.

To solve the problem, we employ the principle of virtual displacements (3.17).
As a generalized coordinate we choose the angle φ between the intervals OC and
OB. We have

$$x_1 = -x_2 = -l \sin \varphi, \quad y_1 = y_2 = -l \cos \varphi, \quad x_3 = 0, \quad y_3 = 2y_1,$$

where x_1, y_1, x_2, y_2 and x_3, y_3 are the coordinates of the points A, B and C, respectively. To calculate the virtual displacements δx_2 and δy_2 we'll use the Fig. 14b. Due
to the fact that $B\widetilde{B}$ and BB'' are equivalent infinitesimals, the equality $B\widetilde{B} = l\delta\varphi$
implies an approximate formula $BB'' \approx l\delta\varphi$. Taking into consideration that the angle
$B'B''B$ in the triangle $B'B''B$ is equal to φ we get

$$\delta x_2 = B'B'' = l\delta\varphi \cos \varphi, \quad \delta y_2 = BB' = l\delta\varphi \sin \varphi.$$

The same result can be obtained if one makes use of the the fact that calculating
δf for an arbitrary function f is analogous to determining a differential df. For
example, the formula $dx_2 = l d\varphi \cos \varphi$ implies that $\delta x_2 = l\delta\varphi \cos \varphi$. Hence,

$$\delta x_1 = -\delta x_2 = -l\delta\varphi \cos \varphi, \quad \delta y_1 = \delta y_2 = l\delta\varphi \sin \varphi, \quad \delta y_3 = 2\delta y_1.$$

Substituting the virtual displacements δx_1, δx_2 and δy_3 into the equality

$$F\delta x_1 - F\delta x_2 + P\delta y_3 = 0,$$

which follows from the principle of virtual displacements, we find

$$(-F \cos \varphi + P \sin \varphi)\delta\varphi = 0.$$

Since $\delta\varphi$ is arbitrary, the last equality implies that

$$P = F \cot \varphi.$$

One can solve the problem under discussion in another way expressing δx_2 and δy_3 in terms of δx_1 with the help of equations of constraints

$$x_2 = -x_1, \quad x_1^2 + y_1^2 = l^2, \quad y_3 = 2y_1.$$

It follows from these equations that

$$\delta x_2 = -\delta x_1, \quad x_1 \delta x_1 + y_1 \delta y_1 = 0, \quad \delta y_1 = -\frac{x_1}{y_1}\delta x_1 = -\operatorname{tg}\varphi \delta x_1, \quad \delta y_3 = -2\operatorname{tg}\varphi \delta x_1.$$

Substitution of these expressions for δx_2 and δy_3 into the principle of virtual displacements gives the equality

$$2F\delta x_1 - 2P \operatorname{tg}\varphi \delta x_1 = 0,$$

which implies the formula from above, since δx_1 is arbitrary,

$$P = F \cot \varphi.$$

5 Equilibrium of Trusses

A *truss* is a structure consisting of straight rods. The joint points of rods between each other and their free ends are known as *joints* of the truss. It will be assumed that at all joints of a truss there are hinges and that the external forces act only at the truss joints. If all rods of a truss lie in one plane, then such a truss is called a *plane truss*. A truss is called *invariable* if it does not change its shape under the action of any system of forces applied to it. Otherwise, a truss is called *variable*. A plane truss depicted in Fig. 15a is invariable, while the trusses from Fig. 15b and c are variable. Depending on the kind of the load applied to a truss, a variable truss either folds (Fig. 15b) or preserves its form (Fig. 15c).

We shall be concerned only with plane trusses. After adding one joint an invariable plane truss will remain invariable if the new joint is connected with the other joints by at least two rods. The simplest invariable truss consists of one rod and two joints on its ends. Any truss can be obtained from the simplest one by adding to it rods and

Fig. 15 Invariable and variable trusses

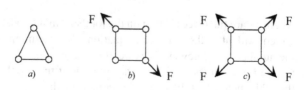

Fig. 16 Plane truss

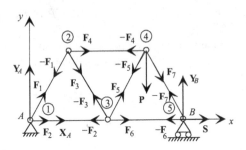

joints. Assume that a truss obtained in this way from the simplest truss is invariable and contains s rods and n joints. Then the number $s - 1$ of added rods should not be less than the doubled number $2(n - 2)$ of added joints, that is, $s - 1 \geqslant 2(n - 2)$, or, what is the same,

$$s \geqslant 2n - 3.$$

An invariable truss is called *simple* if it contains the least possible number of rods; that is, $s = 2n - 3$.

The problem of solving a truss is to determine, from the given external forces, the internal forces in the rods and the support reaction forces.

Example 9 Let us analyze the truss shown in Fig. 16. The truss is made of seven equal rods lying in the plane xy. This truss is simple, because the number of rods $s = 7$ and the number of joints $n = 5$ satisfy the equality $s = 2n - 3$. The magnitudes of the external forces P and S are given. The internal forces in rods F_i, $i = \overline{1, 7}$, and reactions X_A, Y_A, Y_B are unknowns. The right end of the truss is supported by a cart (a "roller support"). Under no friction condition, it may only create the reaction Y_B perpendicular to the direction of its possible motion.

The truss will be in equilibrium if all its rods and joints are in equilibrium. Joints will be regarded as points, their sizes being neglected. Each rod is subject to the action of two forces applied to its ends. A rod will be in equilibrium if these forces are equal in value, opposite in the direction, and directed along the rod. The actions of rods on the joints are equal in value and opposite in the direction to the actions of joints on the rods. The equilibrium equations for joints for this truss read as

$$\mathbf{F}_1 + \mathbf{F}_2 + \mathbf{X}_A + \mathbf{Y}_A = 0, \quad -\mathbf{F}_1 + \mathbf{F}_3 + \mathbf{F}_4 = 0,$$
$$-\mathbf{F}_2 - \mathbf{F}_3 + \mathbf{F}_5 + \mathbf{F}_6 = 0, \quad -\mathbf{F}_4 - \mathbf{F}_5 + \mathbf{F}_7 + \mathbf{P} = 0, \quad (5.1)$$
$$-\mathbf{F}_6 - \mathbf{F}_7 + \mathbf{S} + \mathbf{Y}_B = 0,$$

where $\mathbf{F}_i$ are the forces acting on the joints from the rods. The vector equations (5.1) are equivalent to the ten scalar equations. So, the number of equations agrees with the number of unknowns. The directions of the unknown forces $\mathbf{F}_i$ correspond to the tensile forces in rods. If, after the solution of the problem, it happens that $F_i < 0$, then this means that the i-th rod is compressed.

Summing up the equations in (5.1), we find that the resultant vector of the external forces (including support reactions) acting on the truss vanishes:

$$\mathbf{F} = \mathbf{X}_A + \mathbf{Y}_A + \mathbf{P} + \mathbf{S} + \mathbf{Y}_B = 0.$$

Assume that the positions of the joints relative to some point O are given by the vectors $\mathbf{r}_k$. We take the vector product of the k-th equation (5.1) and $\mathbf{r}_k$ and sum up the resulting equalities. The sum of the vector products involving the vector $\mathbf{F}_i$ is zero. Indeed,

$$\mathbf{r}_p \times \mathbf{F}_i - \mathbf{r}_q \times \mathbf{F}_i = (\mathbf{r}_p - \mathbf{r}_q) \times \mathbf{F}_i = 0,$$

inasmuch as $(\mathbf{r}_p - \mathbf{r}_q) \parallel \mathbf{F}_i$. Hence, the resultant torque of the external forces is also zero:

$$\mathbf{L}_O = \mathbf{r}_1 \times \mathbf{X}_A + \mathbf{r}_1 \times \mathbf{Y}_A + \mathbf{r}_4 \times \mathbf{P} + \mathbf{r}_5 \times \mathbf{S} + \mathbf{r}_5 \times \mathbf{Y}_B = 0.$$

Since $\mathbf{F} = 0$, for any point P we have $\mathbf{L}_P = 0$.

Clearly, the above analysis holds for any plane truss in an equilibrium position. Hence, a truss, when considered as a solid, is in equilibrium under the action of external forces. If the problem for a truss *qua* a rigid solid is statically determinate (that is, it contains three unknown external forces), then the total number of unknown external and internal forces is $s + 3$. The number of equilibrium equations for the joints is $2n$. If a truss is simple, then $2n = s + 3$, and the problem will be statically determinate in large.

Projecting equations (5.1) onto the x- and y-axes, we find that

$$
\begin{aligned}
X_A + \tfrac{1}{2}F_1 + F_2 &= 0, & Y_A + \tfrac{\sqrt{3}}{2}F_1 &= 0, \\
-\tfrac{1}{2}F_1 + \tfrac{1}{2}F_3 + F_4 &= 0, & -\tfrac{\sqrt{3}}{2}F_1 - \tfrac{\sqrt{3}}{2}F_3 &= 0, \\
-F_2 - \tfrac{1}{2}F_3 + \tfrac{1}{2}F_5 + F_6 &= 0, & \tfrac{\sqrt{3}}{2}F_3 + \tfrac{\sqrt{3}}{2}F_5 &= 0, & (5.2) \\
-F_4 - \tfrac{1}{2}F_5 + \tfrac{1}{2}F_7 &= 0, & -\tfrac{\sqrt{3}}{2}F_5 - \tfrac{\sqrt{3}}{2}F_7 - P &= 0, \\
-F_6 - \tfrac{1}{2}F_7 + S &= 0, & \tfrac{\sqrt{3}}{2}F_7 + Y_B &= 0.
\end{aligned}
$$

In solving system (5.2) it is convenient to use the equilibrium equations of a truss as a rigid solid,

$$X_A + S = 0, \quad Y_A + Y_b - P = 0, \quad -\frac{3}{2}Pl + 2lY_B = 0, \qquad (5.3)$$

where l is the rod length. The third equation (5.3) is obtained from the condition $\mathbf{L}_A = 0$. A system of equations equivalent to (5.3) can be obtained by excluding the unknowns $F_1, F_2, \ldots, F_7$ from system (5.2). The solution of system (5.3) reads as

$$X_A = -S, \quad Y_B = \frac{3}{4}P, \quad Y_A = \frac{1}{4}P.$$

Substituting these expressions into system (5.2)gives

$$F_1 = -\frac{1}{2\sqrt{3}}P, \quad F_2 = S + \frac{1}{4\sqrt{3}}P, \quad F_3 = -F_1,$$

$$F_4 = F_5 = F_1, \quad F_6 = S + \frac{\sqrt{3}}{4}P, \quad F_7 = 3F_1.$$

6 Equilibrium for Systems with Friction

We consider a body lying on a rough surface. The gravity force **P** acting on a body is balanced by the reaction force **N** (Fig. 17).

If the body is subject to a horizontal force **F** such that its magnitude is $F \leqslant kN$, where k is the *friction coefficient*, then the body will remain in equilibrium due to the emerging *static friction force* $\mathbf{F}_k = -\mathbf{F}$ (Fig. 17). The friction coefficient k, which depends on the material and the properties of the surfaces in contact, is determined experimentally. In the case $F > kN$ the body starts to slide along the plane. For a moving body, the dry friction force is directed opposite to the body velocity, its magnitude is determined by the equation $F_k = k_s N$. The friction coefficient k_s depends on the motion velocity, but in many problems this dependence can be ignored assuming that $k_s = k$.

We note that in an equilibrium position the direction and magnitude of the static friction force depend on the magnitude and direction of the force **F**, and besides $0 \leqslant F_k \leqslant kN$. However, if a body moves, then the magnitude of the dry friction force $F_k = kN$ will be independent of the force **F** if we ignore any dependence of the friction coefficient on the velocity.

A cone of semi-angle $\alpha = \arctan k$, whose axis is directed along the normal pressure vector **N** is called the *friction cone* (Fig. 17). If the reaction $\mathbf{R} = \mathbf{N} + \mathbf{F}$ lies inside the friction cone, then the body is in equilibrium. Indeed, in this case $F/N \leqslant \tan \alpha = k$.

The Euler formula. We consider a string touching the surface of a circular cylinder along an arc with central angle α (Fig. 18a).

The friction coefficient of the string at the cylinder is k. Assume that one end of the string is subject to the force $\mathbf{T}_1$. Let us find the force $\mathbf{T}_0$ of smallest magnitude,

Fig. 17 Friction cone

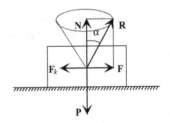

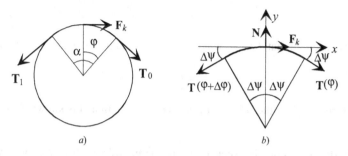

Fig. 18 String at the surface of a circular cylinder

which needs to be applied to the other end of the string in order that the string be in equilibrium. In the absence of the friction force the string would be in equilibrium only for $T_0 = T_1$. In the presence of the friction force an equilibrium is also possible for $T_0 \neq T_1$. If, for example, $T_1 > T_0$, then between the string and the cylinder there appears the friction force, which prevents the counterclockwise motion of the string (see Fig. 18a). With a given value of T_1, the value of T_0 will be the smallest in the case when the friction force has the greatest magnitude.

As the coordinate of a point on the string we shall use the angle φ. Let $\mathbf{T}(\varphi)$ denote the string tension force at the point with coordinate φ. If the string is in equilibrium, then the portion of the string between the angles φ and $\varphi + \Delta\varphi$ will also be in equilibrium. Hence, the resultant vector force acting on this portion of the string will be zero (Fig. 18b),

$$\mathbf{T}(\varphi + \Delta\varphi) + \mathbf{T}(\varphi) + \mathbf{N} + \mathbf{F}_k = 0, \tag{6.1}$$

where $\mathbf{N}$ and $\mathbf{F}_k$ are, respectively, the resultant reaction and friction forces acting on the highlighted segment of the string. Projecting (6.1) on the Ox- and Oy-axes, we find that

$$-(T + \Delta T)\cos \Delta\psi + T \cos \Delta\psi + F_k = 0,$$
$$-(T + \Delta T)\sin \Delta\psi - T \sin \Delta\psi + N = 0, \tag{6.2}$$

where $\Delta\psi = \Delta\varphi/2$, $\Delta T = T(\varphi + \Delta\varphi) - T(\varphi)$. Replacing in (6.2) the magnitude of the friction force F_k by its maximum value kN and excluding N from (6.2) gives

$$\Delta T \cos \Delta\psi = k(2T + \Delta T)\sin \Delta\psi.$$

Dividing this equality by $\Delta\varphi$ and making $\Delta\varphi \to 0$, we find an equation for the function $T(\varphi)$

$$\frac{dT}{d\varphi} = kT,$$

the solution of which reads as

$$T = Ce^{k\varphi}.$$

Taking into account that $T = T_0$ when $\varphi = 0$, we find the constant $C = T_0$. Hence,

$$T = T_0 e^{k\varphi}, \quad T_1 = T_0 e^{k\alpha}, \quad T_0 = T_1 e^{-k\alpha}.$$

The last formula gives the solution of the above problem and is known as the *Euler formula*.

Example 10 For friction between a rope and a tree, we have $k = 0.5$. However, if a rope is wrapped twice about a wooden pile, then

$$T_0 = T_1 e^{-4\pi k} = T_1 e^{-2\pi} \approx 0.002 T_1.$$

Hence, in this case, the force $T_1 = 1000\,\text{N}$ can be balanced by the force $T_0 \approx 2\,\text{N}$.

7 Equilibrium of a String

We consider an absolutely flexible string with end-points at A and B in equilibrium under the action of a load distributed along the string. As the coordinate of a point M on the string axis, we take the length s of the arc AM (Fig. 19).

The portion of the string from the point M with coordinate s to the point M_1 with coordinate $s + \Delta s$ will also be in equilibrium. This portion is subject to the *tension forces* $\mathbf{T} = \mathbf{T}(s)$ and $\mathbf{T}_1 = -\mathbf{T}(s + \Delta s)$, which act at the points M and M_1, and the resultant of distributed forces $\Delta\mathbf{F}$ (see Fig. 19).

The vector $\mathbf{q}_* = \Delta\mathbf{F}/\Delta s$ is called the *mean distributed load* on the interval $[s, s + \Delta s]$, and the vector

$$\mathbf{q} = \lim_{\Delta s \to 0} \mathbf{q}_* = \lim_{\Delta s \to 0} \frac{\Delta\mathbf{F}}{\Delta s}$$

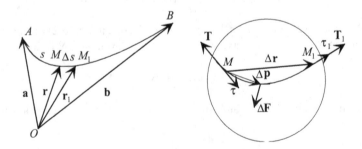

Fig. 19 Equilibrium of a string

is called the *distributed load* at the point M. In the case, the string is subject to the gravity forces

$$\Delta \mathbf{F} = \Delta m \mathbf{g} = S \Delta s \rho \mathbf{g}, \quad \mathbf{q} = S \rho \mathbf{g},$$

where S is the cross-sectional area of the string, ρ is the string density, $\mathbf{g}$ is the gravitational acceleration vector.

We choose a rectangular Cartesian system with origin at an arbitrary point O. Let $\mathbf{r} = \mathbf{r}(s)$ and $\mathbf{r}_1 = \mathbf{r}(s + \Delta s)$ be vectors from the point O to the points M and M_1, respectively.

Since the portion of the string MM_1 is in equilibrium, the resultant torque of the forces acting on the spring with respect to a point (including the point M) is zero:

$$\Delta \mathbf{r} \times \mathbf{T}_1 + \Delta \mathbf{p} \times \Delta \mathbf{F} = 0.$$

Here, $\Delta \mathbf{r} = \mathbf{r}(s + \Delta s) - \mathbf{r}(s)$, $\Delta \mathbf{p}$ is the vector from the point M to the point of application of the force $\Delta \mathbf{F}$ (see Fig. 19). We divide the last equality by Δs and make $\Delta s \to 0$. Hence, taking into account that

$$\frac{\Delta \mathbf{r}}{\Delta s} \to \frac{d\mathbf{r}}{ds} = \boldsymbol{\tau}, \quad \mathbf{T}_1 \to -\mathbf{T}, \quad \Delta \mathbf{p} \to 0, \quad \frac{\Delta \mathbf{F}}{\Delta s} \to \mathbf{q},$$

as $\Delta s \to 0$, where $\boldsymbol{\tau}$ is the unit tangential vector at the point M, we find that

$$\boldsymbol{\tau} \times \mathbf{T} = 0.$$

This means that the tension force is $\mathbf{T} = -T\boldsymbol{\tau}$, that is, it is directed along the tangential to the string axis, and besides $T > 0$, because the string cannot be in equilibrium under the action of the compression force.

The resultant vector of the forces acting on the portion MM_1 of the string is also zero,

$$\mathbf{T}_1 + \mathbf{T} + \Delta \mathbf{F} = 0,$$

where $\mathbf{T}_1 = T_1 \boldsymbol{\tau}_1$, $T_1 = T(s + \Delta s)$, $\boldsymbol{\tau}_1 = \boldsymbol{\tau}(s + \Delta s)$. We divide this equality by Δs and rewrite as

$$\frac{T_1 \boldsymbol{\tau}_1 - T\boldsymbol{\tau}}{\Delta s} + \frac{\Delta \mathbf{F}}{\Delta s} = 0.$$

Making $\Delta s \to 0$, we obtain

$$\frac{d}{ds}(T\boldsymbol{\tau}) + \mathbf{q} = 0.$$

Substituting $\boldsymbol{\tau} = d\mathbf{r}/ds$ into this equality, we get the *vector equilibrium equation of the string*

$$\frac{d}{ds}\left(T\frac{d\mathbf{r}}{ds}\right) + \mathbf{q} = 0.$$

Projecting this vector equation on the Ox-, Oy- and Oz-axes, we obtain three scalar equations

$$(Tx')' + q_x = 0, \quad (Ty')' + q_y = 0, \quad (Tz')' + q_z = 0,$$

where q_x, q_y, q_z are the projections of the distributed load on the coordinate axes, $x' = dx/ds$.

We have only three equations to determine the four unknown functions $x(s)$, $y(s)$, $z(s)$ and $T(s)$. The fourth equation

$$(x')^2 + (y')^2 + (z')^2 = 1$$

follows from the equality $\tau = 1$, because

$$\tau = x'\mathbf{i} + y'\mathbf{j} + z'\mathbf{k}.$$

The system of differential equilibrium equations of a string is of sixth order. Its general solution depends on six arbitrary constants. They are determined from the boundary conditions on the string ends. Assume that the left and right ends of the string A and B have the coordinates $s = s_1$ and $s = s_2$. Hence, the boundary conditions, from which the string ends positions can be found, read as

$$\mathbf{r}(s_1) = \mathbf{a}, \quad \mathbf{r}(s_2) = \mathbf{b},$$

where $\mathbf{a}$ and $\mathbf{b}$ the vectors from the point O to the points A and B (see Fig. 19). The last two vector equalities are equivalent to six scalar boundary conditions.

Let us consider the equilibrium problem for a string under the action of its weight. Assume that the string ends are fixed at the points A and B. As the origin, we take a point on the plane S passing through A and B and which is parallel to the gravitational acceleration vector $\mathbf{g}$. As unit directed vectors, we take the vectors $\mathbf{j} = -\mathbf{g}/g$, $\mathbf{k} \perp S$ and $\mathbf{i} = \mathbf{j} \times \mathbf{k}$ (Fig. 20).

Taking into account that the distributed load $\mathbf{q} = S\rho\mathbf{g} = -S\rho g\mathbf{j}$ has the projections $q_x = q_z = 0$, $q_y = -q$, where $q = S\rho g$, the *equilibrium equations of the string* read as

$$(Tx')' = 0, \quad (Ty')' = q, \quad (Tz')' = 0. \tag{7.1}$$

Fig. 20 String under the action of its weight

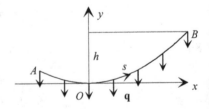

We have $z(s_1) = z(s_2) = 0$ at the string ends, and hence the function $z(s) = 0$ satisfies the equilibrium equations (7.1) and the boundary conditions. Hence, the string axis lies in the plane Oxy. We assume that $x(s_1) \neq x(s_2)$.

Integrating the first two equations (7.1), this gives

$$Tx' = C_1, \quad Ty' = qs + C_2. \tag{7.2}$$

Assume that $Ty' = 0$ at some point of the string. We take this point for the reference point of the s-coordinate. Hence, $C_2 = 0$. If $T(0) = 0$, then $C_1 = 0$ and either $T(s) = 0$ or $x' = 0$. The equality $T(s) = 0$ contradicts the equations, and the equality $x' = 0$ contradicts the condition $x(s_1) \neq x(s_2)$. Hence, $T(0) \neq 0$, and $y' = 0$ for $s = 0$. From the equation

$$(x')^2 + (y')^2 = 1 \tag{7.3}$$

it follows that $x' = 1$ for $s = 0$, and hence, $C_1 = T(0) \equiv T_0 > 0$. So, Eq. (7.2) can be written as

$$Tx' = T_0, \quad Ty' = qs. \tag{7.4}$$

The case when $Ty' \neq 0$ at any point of the string is not considered, because for it the solution of the problem becomes more involved.

Squaring and adding equations (7.4), we have, in view of equation (7.3),

$$T^2 = T_0^2 + q^2 s^2.$$

Hence,

$$x' = \frac{T_0}{T} = \frac{T_0}{\sqrt{T_0^2 + q^2 s^2}}.$$

Setting $\beta = T_0/q$, we rewrite this equality as

$$\frac{dx}{ds} = \frac{\beta}{\sqrt{\beta^2 + s^2}}.$$

We change the variables $s = \beta \sinh v$. Next, taking into account that

$$ds = \beta \cosh v \, dv, \quad \cosh^2 v - \sinh^2 v = 1,$$

we find that

$$dx = \beta dv, \quad x = \beta v + C_3.$$

As the origin O we take the point on the string with coordinate $s = 0$ (see Fig. 20). Hence, $C_3 = 0$,

$$s = \beta \sinh v = \beta \sinh(x/\beta).$$

Dividing the second equation of (7.4) by the first one, we find that

$$\frac{dy}{dx} = \frac{qs}{T_0} = \frac{s}{\beta} = \sinh(x/\beta), \quad y = \beta\cosh(x/\beta) + C_4.$$

From the condition $y = 0$ with $s = 0$ it follows that $C_4 = -\beta$. Hence,

$$y = \beta[\cosh(x/\beta) - 1],$$

that is, the string axis is a *catenary curve*.

Setting $x(s_k) = a_k$, $y(s_k) = b_k$, $k = 1, 2$, we have

$$s_k = \beta\sinh(a_k/\beta), \quad b_k = \beta[\cosh(a_k/\beta) - 1], \quad k = 1, 2. \tag{7.5}$$

Let us consider the case $b_1 = b_2 = h$. Then from (7.5) and the condition $a_1 \neq a_2$ it follows that

$$a_1 = -a, \quad a_2 = a, \quad s_1 = -l, \quad s_2 = l.$$

The string length is as follows

$$L = s_2 - s_1 = 2l = 2\beta\sinh(a/\beta),$$

and its maximum deflection is

$$h = \beta[\cosh(a/\beta) - 1].$$

Setting $\alpha = a/\beta$, we have

$$\frac{h}{a} = \frac{\cosh\alpha - 1}{\alpha}.$$

We assume that the relative maximum deflection h/a of the string is a small quantity. In this case, α is also small. Expanding $\cosh\alpha$ in a series and rejecting the terms which are small, we obtain the following approximative formula:

$$\frac{h}{a} = \frac{1}{\alpha}\left(1 + \frac{\alpha^2}{2} + \ldots - 1\right) \approx \frac{\alpha}{2}.$$

Hence,

$$h \approx \frac{qa^2}{2T_0}. \tag{7.6}$$

Example 11 A catenary wire along which the pantograph atop a commuter train moves can be looked upon as a flexible inextensible string. A considerable deflection of the catenary wire prevents the normal motion of the train. Let us find the tension force of the wire for which the maximum deflection is 0.1 m. Assume that the distance between poles is $2a = 50$ m and the distributed load is $q = 0.5$ N/m. Using (7.6), we find that

$$T_0 \approx \frac{qa^2}{2h} = \frac{0.5 \cdot 625}{2 \cdot 0.1} \approx 1560 \, \text{N}.$$

Chapter 11
Integration of Equations in Mechanics

N. N. Polyakhov

This chapter is concerned with the questions of integration theory of equations of motion of holonomic mechanical systems under the action of potential forces. The main attention is paid to those aspects of the theory, which make it possible to reveal general properties of motion of such systems.

1 Hamilton–Jacobi Theorem

The Jacobi theorem. In Chap. 9 it was shown that the Newton vector equation, which expresses the law of motion, is equivalent to both a system of canonical equations and a single equation in partial derivatives with respect to the action function S. Thus, there exists an intimate relationship between the system of ordinary differential equations and the equation in partial derivatives. Let us consider this question more in detail.

Suppose that a general solution of equations of motion is known, i.e., the functions

$$q^\sigma = f^\sigma(\tau, C_1, C_2, \dots, C_{2s}), \quad \sigma = \overline{1, s}, \tag{1.1}$$

are known, where τ is current time, σ is the number of degrees of freedom of a mechanical system, C_σ are arbitrary constants. Then, as was already shown, the Hamilton action integral

$$S = \int_{t_0}^{t} L(\tau, q, \dot{q}) d\tau$$

© Springer Nature Switzerland AG 2021
N. N. Polyakhov et al., *Rational and Applied Mechanics*,
Foundations of Engineering Mechanics,
https://doi.org/10.1007/978-3-030-64061-3_11

can be represented as the function of variables

$$S = S(t, q, t_0, q_0).$$ (1.2)

Here q_0^σ corresponds to the position of the system at time t_0, and q^σ corresponds to its position at time t. The function S, which is considered as a function of variables t and q^σ, satisfies the Hamilton–Jacobi equation (10.4) of Chap. 9

$$\frac{\partial S}{\partial t} + H\left(t, q, \frac{\partial S}{\partial q}\right) = 0.$$ (1.3)

Suppose now that we do not know the general solution (1.1), and hence we do not know a function S given in form (1.2) either. We shall find it from the equation in partial derivatives (1.3). Recall that the total integral of the Hamilton–Jacobi equation (1.3) is such a solution of this equation

$$S = S(t, q, \alpha) + \alpha^{s+1},$$ (1.4)

which depends on the $(s + 1)$st arbitrary constant α^ρ and satisfies the condition

$$\det\left[\frac{\partial^2 S}{\partial q^\tau \partial \alpha^\sigma}\right] \neq 0, \quad \sigma, \tau = \overline{1, s}.$$

The constant α^{s+1} enters the total integral additively. This is explained by the fact that Eq. (1.3) does not contain the unknown function S.

There is a close relationship between the total integral of Eq. (1.3) and the general solution of the equations of motion. It consists in the fact that one can find the total integral of Eq. (1.3) by the general solution of equations of motion, and vice versa, by the total integral of this equation one can construct the general solution of equations of motion. The equations of motion can be given both in the form of the Lagrange equations of the second kind and in the form of canonical equations, so the general solution (1.1) can be considered as a general solution of the Lagrange equations and as a general solution of the canonical equations.

As was already noted, specifying a function S satisfying Eq. (1.3) in the form (1.2) is a direct corollary of the general solution (1.1). It is easy to show that it is indeed a total integral. A converse proposition, which makes it possible to construct a general solution of the canonical equations from the total integral of Eq. (1.3), represents the *Jacobi theorem*, which reads as follows: *if the function $S(t, q, \alpha)$ is a total integral of of the Hamilton–Jacobi equation* (1.3), *then the general solution of the canonical equations*

$$\frac{dq^\sigma}{dt} = \frac{\partial H}{\partial p_\sigma}, \quad \frac{dp_\sigma}{dt} = -\frac{\partial H}{\partial q^\sigma}, \quad \sigma = \overline{1, s},$$ (1.5)

can be found from the following relations:

$$p_\sigma = \frac{\partial S}{\partial q^\sigma}, \quad \sigma = \overline{1,s}, \tag{1.6}$$

$$\beta_\sigma = \frac{\partial S}{\partial \alpha^\sigma}, \quad \sigma = \overline{1,s}, \tag{1.7}$$

where α^σ and β_σ are arbitrary constants.

The function S is a specified function of variables t, q^σ and α^σ, and hence relations (1.6) imply that the quantities p_σ and β_σ are functions of the form

$$p_\sigma = p_\sigma(t, q, \alpha), \quad \sigma = \overline{1,s}, \tag{1.8}$$

$$\beta_\sigma = \beta_\sigma(t, q, \alpha), \quad \sigma = \overline{1,s}. \tag{1.9}$$

According to the condition, the determinant

$$\det\left[\frac{\partial^2 S}{\partial q^\tau \partial \alpha^\sigma}\right] = \det\left[\frac{\partial \beta_\sigma}{\partial q^\tau}\right], \quad \sigma, \tau = \overline{1,s},$$

is nonzero, and so, Eq. (1.9) can be solved with respect to q^σ, i.e., we can write

$$q^\sigma = q^\sigma(t, \alpha, \beta), \quad \sigma = \overline{1,s}. \tag{1.10}$$

Substituting these equations into formulas (1.8) we get

$$p_\sigma = p_\sigma(t, \alpha, \beta), \quad \sigma = \overline{1,s}. \tag{1.11}$$

Thus, relations (1.6), (1.7) make it possible to represent canonical variables q^σ and p_σ in the form of functions of time t and $2s$ arbitrary constants α^σ and β_σ.

As is known, expressions (1.10), (1.11) are a general solution of the canonical equations (1.5) if for any α^σ and β_σ the functions q^σ and p_σ, which are considered as functions of time t, satisfy the above equations. To be more precise, we proceed as follows.

Differentiating expressions (1.7) in time, we find

$$0 \equiv \frac{d\beta_\sigma}{dt} = \frac{\partial^2 S}{\partial t \partial \alpha^\sigma} + \frac{\partial^2 S}{\partial q^\tau \partial \alpha^\sigma}\dot{q}^\tau, \quad \sigma, \tau = \overline{1,s}. \tag{1.12}$$

We have

$$\frac{\partial^2 S}{\partial t \partial \alpha^\sigma} = \frac{\partial^2 S}{\partial \alpha^\sigma \partial t}, \quad \frac{\partial^2 S}{\partial q^\tau \partial \alpha^\sigma} = \frac{\partial^2 S}{\partial \alpha^\sigma \partial q^\tau},$$

and hence equality (1.12) can be represented as

$$\frac{\partial^2 S}{\partial \alpha^\sigma \partial t} + \dot{q}^\tau \frac{\partial^2 S}{\partial \alpha^\sigma \partial q^\tau} = 0. \tag{1.13}$$

The function $S(t, q, \alpha)$ satisfies the Eq. (1.3), and so, when substituting the total integral (1.4) into the latter, this equation should be regarded as an identity. Differentiating this identity with respect to α^σ, this gives

$$\frac{\partial^2 S}{\partial \alpha^\sigma \partial t} + \frac{\partial H}{\partial p_\tau} \frac{\partial^2 S}{\partial \alpha^\sigma \partial q^\tau} = 0. \tag{1.14}$$

Comparing expressions (1.13), (1.14), we see that

$$\frac{dq^\tau}{dt} = \frac{\partial H}{\partial p_\tau}, \quad \tau = \overline{1, s}. \tag{1.15}$$

In a similar way, it can be proved that the function p_σ, which is specified by the relations (1.6), (1.7) in form (1.11), satisfies the equation

$$\frac{dp_\sigma}{dt} = -\frac{\partial H}{\partial q^\sigma}, \quad \sigma = \overline{1, s}. \tag{1.16}$$

In fact, differentiating expressions (1.6) in time, we get

$$\frac{dp_\sigma}{dt} = \frac{\partial^2 S}{\partial t \partial q^\sigma} + \frac{\partial^2 S}{\partial q^\tau \partial q^\sigma} \dot{q}^\tau. \tag{1.17}$$

Moreover, differentiating identity (1.3) with respect to q^σ, we have

$$\frac{\partial^2 S}{\partial q^\sigma \partial t} + \frac{\partial H}{\partial q^\sigma} + \frac{\partial H}{\partial p_\tau} \frac{\partial^2 S}{\partial q^\sigma \partial q^\tau} = 0.$$

Taking into account (1.15), this equality can be represented as

$$\frac{\partial^2 S}{\partial t \partial q^\sigma} + \frac{\partial^2 S}{\partial q^\tau \partial q^\sigma} \dot{q}^\tau = -\frac{\partial H}{\partial q^\sigma}. \tag{1.18}$$

Now comparing expressions (1.17), (1.18) we make sure that the function p_σ indeed satisfies Eq. (1.16). This proves the Jacobi theorem.

The technique for constructing a general solution of a canonical system of equations, based on the Jacobi theorem, is called the *Hamilton–Jacobi method*. It is of great importance for both the theory and applications, and makes it often possible to represent a solution in a convenient analytic form. In some cases, a total integral of the Hamilton–Jacobi equation can be found by separating the variables. This method is based on the fact that the solution is sought in the form

$$S = S_0(t) + S_1(q^1, t) + \ldots + S_s(q^s, t),$$

i.e., in the form of the sum of functions, of which each is a function of one variable q^σ, the time t, and of course, arbitrary constants α^σ or some of them. The possibility of separating the variables depends on both the character of the problem itself and the Lagrange coordinates chosen for its description.

It is worth noting that finding the total differential of the nonlinear partial differential equation (1.3) is in general no simpler than finding the general solution of the canonical equations. Moreover, it is known that in the cases, when one succeeds in finding the total integral of Eq. (1.3), the canonical system (1.5) can be directly integrated. However such a possibility occurs rarely. Usually, the equations involved and important problems of mechanics are nonintegrable, which raises the question about their approximate solution or numerical integration. The Hamilton–Jacobi is of great value in the development of approximate methods of integration of Eq. (1.5). Using this theory it proves possible to represent these methods in the most simple form. It is also worth noting that the above methods were developed, for the most part, in connection with the solution of problems of celestial mechanics.

As an example illustrating the Hamilton–Jacobi theory, we consider the problem of the harmonic oscillator and the Kepler problem.

The harmonic oscillator. As was already shown in Sect. 7 of Chap. 4, a load suspended on a spring performs harmonic oscillations when deviating from the state of equilibrium. Oscillations of such a type are widely spread in nature, and are typical for both elastic and electromechanical systems. Any system with one degree of freedom, the kinetic and potential energies of which are given in the form

$$T = \frac{m\dot{q}^2}{2}, \quad \Pi = \frac{cq^2}{2},$$

where m and c are positive constants, and q is a generalized Lagrange coordinate, is commonly called *harmonic oscillator*.

The system under discussion is conservative, and hence the Hamilton function H is equal to the total mechanical energy: $H = T + \Pi$. Expressing T in terms of the generalized momentum $p = \partial T / \partial \dot{q} = m\dot{q}$, we obtain

$$H = \frac{p^2}{2m} + \frac{cq^2}{2}.$$

Thus, in this case, the Hamilton–Jacobi equation reads as

$$\frac{\partial S}{\partial t} + \frac{1}{2m}\left(\frac{\partial S}{\partial q}\right)^2 + \frac{cq^2}{2} = 0.$$

Time t does not enter explicitly this equation, and so its solution can be represented as

$$S = -\alpha t + w(q, \alpha),$$

where α is an arbitrary constant, $w(q, \alpha)$ is a new unknown function, which should satisfy the equation

$$\frac{1}{2m}\left(\frac{\partial w}{\partial q}\right)^2 = \alpha - \frac{cq^2}{2}. \tag{1.19}$$

Note that the constant α is equal to the total energy of the system, because

$$H_0 = H = -\frac{\partial S}{\partial t} = \alpha.$$

The functions S and w can be determined up to an additive constant. For the sake of simplicity, we assume that $w = 0$ when $q = 0$. Integrating Eq. (1.19) under these initial conditions, we obtain

$$w = \pm\sqrt{mc}\int_0^q \sqrt{\frac{2\alpha}{c} - q^2}\,dq. \tag{1.20}$$

We find the general solution with respect to the function q in the accordance with the Jacobi theory from the relation

$$\beta = \frac{\partial S}{\partial \alpha} = -t + \frac{\partial w}{\partial \alpha},$$

where β is the second arbitrary constant. Substituting expression (1.20) into this formula, we have

$$t + \beta = \pm\sqrt{\frac{m}{c}}\int_0^q dq\bigg/\sqrt{\frac{2\alpha}{c} - q^2} = \pm\sqrt{\frac{m}{c}}\arcsin\left(q\sqrt{\frac{c}{2\alpha}}\right). \tag{1.21}$$

Since $\arcsin x = \pi/2 - \arccos x$, solution (1.21) can be represented as

$$q = \sqrt{\frac{2\alpha}{c}}\cos(\omega t + \gamma), \quad \omega = \sqrt{\frac{c}{m}}, \quad \gamma = \omega\beta \pm \frac{\pi}{2}.$$

The constants α and γ can be found from the initial conditions. The solution just constructed coincides with the usual formula for harmonic oscillations.

The Kepler problem. It is known that when a particle of mass m is attracted to a fixed center O, the trajectory of motion is a plane curve. Assuming that the motion is under the action of gravity force is occurring in the Oxy-plane, and if the polar coordinates r and φ are used, we represent the kinetic and the potential energies as

$$T = \frac{m}{2}(\dot{r}^2 + r^2\dot{\varphi}^2), \quad \Pi = -\frac{\mu m}{r},$$

where $\mu = \gamma M$ is the gravity parameter of the attracting center.

In accordance with formulas for the generalized momenta

$$\frac{\partial S}{\partial r} = p_r = \frac{\partial T}{\partial \dot{r}} = m\dot{r}, \quad \frac{\partial S}{\partial \varphi} = p_\varphi = \frac{\partial T}{\partial \dot{\varphi}} = mr^2\dot{\varphi}.$$

In this case, the Hamilton–Jacobi equation (1.3) reads as

$$\frac{\partial S}{\partial t} + \frac{1}{2m}\left[\left(\frac{\partial S}{\partial r}\right)^2 + \frac{1}{r^2}\left(\frac{\partial S}{\partial \varphi}\right)^2\right] - \frac{\mu m}{r} = 0.$$

The time t and the angle φ do not enter this equation explicitly, and hence, its total integral, depending on the two arbitrary constants α^1 and α^2, can be represented as

$$S = \alpha^1 t + \alpha^2 \varphi + w(r, \alpha^1, \alpha^2),$$

where the unknown function w satisfies the equation

$$\frac{\partial w}{\partial r} = \pm\frac{m}{r}\sqrt{2\mu r - \frac{2\alpha^1 r^2}{m} - \frac{(\alpha^2)^2}{m^2}}. \tag{1.22}$$

The radical is taken with the plus or minus sign depending on what sign the generalized momentum has with given r,

$$p_r = \frac{\partial S}{\partial r} = \frac{\partial w}{\partial r} = m\dot{r}.$$

Let us find the physical sense of the constants α^1 and α^2. The Hamilton–Jacobi equation (1.3) directly implies that

$$\alpha^1 = \frac{\partial S}{\partial t} = -H = -H_0 = -(T_0 + \Pi_0).$$

Earlier when considering the motion under the action of central forces, the constant

$$h = \frac{2(T_0 + \Pi_0)}{m} = -\frac{2\alpha^1}{m}$$

was introduced instead of the constant $T_0 + \Pi_0$. The second constant corresponding to the area integral $c = r^2\dot{\varphi}$ relates to the constant α^2 as

$$\alpha^2 = \partial S/\partial \varphi = mr^2\dot{\varphi} = mc.$$

Thus, the quantities α^1 and α^2 relate directly to the energy integral and the area integral.

The function $w(r, \alpha^1, \alpha^2)$ is determined up to an additive constant. So, for the sake of simplicity, one can assume that $w = 0$ when $r = r_0$. Integrating Eq. (1.22) under the initial conditions we get

$$w(r, \alpha^1, \alpha^2) = \pm \int_{r_0}^{r} \sqrt{2\mu r - \frac{2\alpha^1 r^2}{m} - \frac{(\alpha^2)^2}{m^2}} \frac{dr}{r}.$$

According to the Jacobi theorem, the general solution with respect to the functions r and φ is defined by the relations

$$\beta_1 = \frac{\partial S}{\partial \alpha^1} = t + \frac{\partial w}{\partial \alpha^1} = t \mp \int_{r_0}^{r} \frac{r \, dr}{\sqrt{2\mu r - \frac{2\alpha^1 r^2}{m} - \frac{(\alpha^2)^2}{m^2}}},$$

$$\beta_2 = \frac{\partial S}{\partial \alpha^2} = \varphi + \frac{\partial w}{\partial \alpha^2} = \varphi \mp \int_{r_0}^{r} \frac{\alpha^2 \, dr}{mr\sqrt{2\mu r - \frac{2\alpha^1 r^2}{m} - \frac{(\alpha^2)^2}{m^2}}},$$

where β_1 and β_2 are arbitrary constants, which can be found from the initial conditions.

Substituting h and c for α^1 and α^2 into these relations and assuming, for the sake of simplicity, that the initial conditions are $t = t_0 = 0, r = r_0, \varphi = \varphi_0 = 0$, we have

$$t = \pm \int_{r_0}^{r} \frac{r \, dr}{\sqrt{hr^2 + 2\mu r - c^2}}, \quad \varphi = \pm \int_{r_0}^{r} \frac{c \, dr}{r\sqrt{hr^2 + 2\mu r - c^2}}. \tag{1.23}$$

If one supposes that $r_0 = r_1$, where r_1 is the smallest root of the equation

$$hr^2 + 2\mu r - c^2 = 0, \tag{1.24}$$

then for $t > 0$ the quantity r increases, and hence,

$$\partial w / \partial r = m\dot{r} > 0, \quad r_1 < r < r_2.$$

Here r_2 is the second positive root of Eq. (1.24). In the case considered in expression (1.22) the radical should be taken with the plus sign. But then we also have the plus sign in the integrals in relations (1.23), i.e.,

$$t = \int_{r_1}^{r} \frac{r \, dr}{\sqrt{hr^2 + 2\mu r - c^2}}, \quad \varphi = \int_{r_1}^{r} \frac{c \, dr}{r\sqrt{hr^2 + 2\mu r - c^2}}.$$

Using the table of integrals, we obtain

$$t = \frac{\sqrt{hr^2 + 2\mu r - c^2}}{h} + \frac{\mu}{h\sqrt{-h}}\left(\arcsin\frac{hr + \mu}{\sqrt{\mu^2 + hc^2}} - \frac{\pi}{2}\right), \quad h < 0,$$

$$\varphi = \arcsin\frac{\mu r - c^2}{r\sqrt{\mu^2 + hc^2}} + \frac{\pi}{2},$$

whence

$$t = \frac{c^2}{\mu + \sqrt{\mu^2 + hc^2}\cos\varphi} = \frac{p}{1 + e\cos\varphi}, \quad p = \frac{c^2}{\mu}, \quad e = \sqrt{1 + \frac{hc^2}{\mu^2}}.$$

2 Integral Invariants

Phase space. Integral invariants. The most general form of differential equations of motion for nonholonomic systems subjected to the action of potential forces is a system of canonical equations in Hamiltonian form

$$\dot{p}_\sigma = -\frac{\partial H(t, q, p)}{\partial q^\sigma}, \quad \dot{q}^\sigma = \frac{\partial H(t, q, p)}{\partial p_\sigma}, \quad \sigma = \overline{1, s}. \tag{2.1}$$

If one introduces continuous numbering by supposing

$$q^1 = \xi_1, \quad q^2 = \xi_2, \dots, q^s = \xi_s,$$

$$p_1 = \xi_{s+1}, \quad p_2 = \xi_{s+2}, \dots, p_s = \xi_{2s},$$

the system can be written as

$$\dot{\xi}_i = \Phi_i(t, \xi), \quad i = \overline{1, 2s}, \tag{2.2}$$

where

$$\Phi_\sigma = \frac{\partial H(t, \xi)}{\partial \xi_{s+\sigma}}, \quad \Phi_{s+\sigma} = -\frac{\partial H(t, \xi)}{\partial \xi_\sigma}, \quad \sigma = \overline{1, s}.$$

We shall characterize the state of a mechanical system at moment t by values of the coordinates $q^\sigma(t)$ and momenta $p_\sigma(t)$. Every set of initial values q_0^σ and $p_{\sigma 0}$ corresponds only to one set of the functions $q^\sigma(t)$ and $p_\sigma(t)$ satisfying the system of Eq. (2.1) and the conditions

$$\left[q^\sigma(t)\right]_{t_0} = q_0^\sigma, \quad \left[p_\sigma(t)\right]_{t_0} = p_{\sigma 0}.$$

In other words, the state of a dynamical system at time t is uniquely specified by its initial state with the help of functions

$$q^\sigma = q^\sigma(t, q_0, p_0), \quad p_\sigma = p_\sigma(t, q_0, p_0).$$ (2.3)

We shall say that a set of variables q^σ, p_σ determines "a point" of the *phase space*. Thus, "the points" of such a space are separate states of the system. From now on we shall assume that this space is constructed is a Euclidean one, and so the chosen frame of axes is orthogonal. Hence, the volume of the space is expressed by the formula

$$\Omega = \int_\Omega dq^1 dq^2 \ldots dq^s dp_1 dp_2 \ldots dp_s = \int_\Omega d\xi_1 d\xi_2 \ldots d\xi_{2s}.$$ (2.4)

In the motion of a system of particles the corresponding point of the phase space moves along its phase trajectory, in so doing through each phase point, in particular, through the initial point $M_0(q_0, p_0)$, only one phase trajectory can pass, as a result of the uniqueness of the solution of the system of differential equations (2.1). Since the quantities q^σ and p_σ are the coordinates of a phase point, it is natural to call the quantities $\dot{q}^\sigma$ and $\dot{p}_\sigma$, as defined by formulas (2.1), the components of the velocity vector of a phase point. We shall denote this vector by $\mathbf{V}^*$.

We shall show that for the Hamiltonian system (2.1) div $\mathbf{V}^* = 0$. Indeed, we have

$$\frac{\partial \dot{q}^\sigma}{\partial q^\sigma} + \frac{\partial \dot{p}_\sigma}{\partial p_\sigma} = \frac{\partial^2 H}{\partial q^\sigma \partial p_\sigma} - \frac{\partial^2 H}{\partial p_\sigma \partial q^\sigma} = 0,$$

and so

$$\text{div } \mathbf{V}^* = \sum_{\sigma=1}^{s} \left(\frac{\partial \dot{q}^\sigma}{\partial q^\sigma} + \frac{\partial \dot{p}_\sigma}{\partial p_\sigma} \right) \equiv \sum_{i=1}^{2s} \frac{\partial V_i^*}{\partial \xi_i} = 0.$$ (2.5)

This condition implies an interesting corollary that the volume of an arbitrary part of the phase space, when taken at an instant t_0, is constant in time. Actually, since the phase points move, the phase volume moves also in time, and in this case, generally speaking, its shape can change.

When moving, the indicated volume could change only if the quantities $\Delta\xi_1$, $\Delta\xi_2$, ..., $\Delta\xi_{2s}$ change, these quantities being the side lengths of $2s$-dimensional parallelepiped. Supposing $\Delta\xi_k = \xi_k' - \xi_k$, we have

$$\frac{d\Delta\xi_k}{dt} = \dot{\xi}_k' - \dot{\xi}_k = V_k^{*'} - V_k^* = \frac{\partial V_k^*}{\partial \xi_k} \Delta\xi_k.$$ (2.6)

The derivative of the elementary volume $\Delta\Omega$ can be written as follows:

$$\frac{d\Delta\Omega}{dt} = \frac{d(\Delta\xi_1\Delta\xi_2\ldots\Delta\xi_{2s})}{dt} = \frac{d\Delta\xi_1}{dt}\Delta\xi_2\Delta\xi_3\ldots\Delta\xi_{2s}+$$

$$+\frac{d\Delta\xi_2}{dt}\Delta\xi_1\Delta\xi_3\ldots\Delta\xi_{2s} + \ldots + \frac{d\Delta\xi_{2s}}{dt}\Delta\xi_1\Delta\xi_2\ldots\Delta\xi_{2s-1}\,.$$

Using formula (2.6), we get

$$\frac{d\Delta\Omega}{dt} = \sum_{i=1}^{2s}\frac{\partial V_i^*}{\partial\xi_i}\Delta\Omega = \operatorname{div}\mathbf{V}^* \cdot \Delta\Omega\,.$$

According to what was proved above, the vector velocity field of a phase point is solenoidal[1] and hence, by (2.5) div $\mathbf{V}^* = 0$. This makes it possible to conclude that the time derivative of the volume $\Delta\Omega$ is also equal to zero, and so the given volume does not change in time. This immediately implies that the total volume Ω remains also constant. This statement is the *Liouville theorem* on conservation of the phase volume. The integral Ω itself (see formula (2.4)) is an example of integral invariants.

Generally speaking, the *integral invariant of the system of differential equations* (2.2) is an integral of the form

$$I = \int\limits_{\Omega} F(t, \xi_1, \xi_2, \ldots, \xi_{2s})\, d\xi_1 d\xi_2 \ldots d\xi_{2s}\,,$$

taken over the domain Ω of a phase space and keeping its value when replacing this domain by any domain into which it goes with time, its points moving along their trajectories. In other words, the integral I does not change under the transformation given by (2.3).

The order of integral invariants is defined by the dimension of integration domain. The integral I, when extended to a $2s$-dimensional domain of the phase space, is called an *integral $2s$-order invariant* (for example, integral (2.4)). The analogous integral that is taken over the l-dimensional manifold, which lies in the phase space, and keeping its value when the points of this manifold move along their trajectories, is called an *integral l-order invariant*. In particular, the curvilinear integral

$$\int\limits_{C} \sum_{i=1}^{2s} F_i(t, \xi_1, \xi_2, \ldots, \xi_{2s})\, d\xi_i \tag{2.7}$$

can be an integral 1-*st*-order invariant, or a linear integral invariant. If the contour C of integration is closed, then a linear invariant is called *relative*, otherwise, *absolute*.[2]

---

[1] The field of vector $\mathbf{u}$ is called *solenoidal*, if it has neither sources nor sinks, i.e., at each point div $\mathbf{u} = 0$ is valid. For example, the velocity field of an incompressible fluid is solenoidal.

[2] The notion of integral invariants was introduced by H. Poincaré. He also analyzed Liouville's theorem on the example of a flow of incompressible fluid for which div $\mathbf{V}^* = 0$ always holds if there are no sources and sinks.

Poincaré linear integral invariant. Suppose, the point M_0, which corresponds to the initial state and lies on a smooth curve C_0, is given in parametric form by the functions

$$q_0^\sigma = q_0^\sigma(\beta), \quad p_{\sigma 0} = p_{\sigma 0}(\beta), \quad 0 \leqslant \beta \leqslant B. \tag{2.8}$$

Let us assume that a curve C_0 is closed, i.e.,

$$q_0^\sigma(0) = q_0^\sigma(B), \quad p_{\sigma 0}(0) = p_{\sigma 0}(B).$$

If one draws the corresponding phase trajectories through the points belonging to the curve of initial states C_0, then we get a *phase space tube*. As follows from expressions (2.3) and (2.8), the phase trajectories forming the phase space tube are specified by the equations

$$q^\sigma = q^\sigma(t, q_0(\beta), p_0(\beta)), \quad p_\sigma = p_\sigma(t, q_0(\beta), p_0(\beta)).$$

These equations make it possible to construct a closed curve C_1 consisting of the points with coordinates

$$q_1^\sigma(\beta) = q^\sigma(t_1, q_0(\beta), p_0(\beta)), \quad p_{\sigma 1}(\beta) = p_\sigma(t_1, q_0(\beta), p_0(\beta))$$

on the phase space tube for any other time instant t_1. Thus, the general solution (2.3) of the system of canonical equations make it possible to establish a correspondence between the points of the closed curves C_0 and C_1.

Let us find out for which functions F_i the curvilinear integrals of kind (2.7), when taken along the curves C_0 and C_1, coincide. To this end, we turn our attention to the action function S which, as was repeatedly mentioned, reflects general properties of integral curves.

In Sect. 10 of Chap. 9 it was shown that if a general solution of a system of canonical equations is known, then the action function S can be regarded as a function of the following kind

$$S = S(t_0, t_1, q_0, q_1). \tag{2.9}$$

It was also noted there that the partial derivatives of the function S with respect to its arguments are calculated by formulas (10.3) and (10.2) of Chap. 9, according to which

$$\frac{\partial S}{\partial t_0} = H_0, \quad \frac{\partial S}{\partial t_1} = -H_1, \quad \frac{\partial S}{\partial q_0^\sigma} = -p_{\sigma 0}, \quad \frac{\partial S}{\partial q_1^\sigma} = p_{\sigma 1}.$$

This implies that the total differential of the function S has the form

$$dS = p_{\sigma 1} dq_1^\sigma - H_1 dt_1 - p_{\sigma 0} dq_0^\sigma + H_0 dt_0. \tag{2.10}$$

The arguments q_0^σ and q_1^σ of the function S of kind (2.9) on the introduced curves C_0 and C_1 are functions of the parameter β. Hence, on these curves the function S

is also the function of β,

$$S(\beta) = S(t_0, t_1, q_0(\beta), q_1(\beta)).$$

In this case, since the time moments t_0 and t_1 do not depend on β, we have by (2.10)

$$dS = \frac{dS}{d\beta} d\beta = p_{\sigma 1} dq_1^{\sigma} - p_{\sigma 0} dq_0^{\sigma}, \tag{2.11}$$

where $dq_k^{\sigma} = \frac{dq_k^{\sigma}}{d\beta} d\beta$, $k = 0, 1$.

The curves C_0 and C_1 are closed, and hence, $S(0) = S(B)$. This implies that the integral of differential form (2.11), when taken over the manifold consisting of the points belonging to the curves C_0 and C_1, must vanish

$$\oint_{C_1} p_{\sigma 1} dq_1^{\sigma} - \oint_{C_0} p_{\sigma 0} dq_0^{\sigma} = \int_0^B dS = S(B) - S(0) = 0.$$

This equality means that the curvilinear integral

$$I_1 = \oint p_{\sigma} dq^{\sigma} = \text{inv}, \tag{2.12}$$

when taken over the contour moving along the tube, remains constant. Integral (2.12) is a sought relative integral invariant of first order and is called *a Poincaré linear relative integral invariant*.

One should distinguish the differentials dq^{σ} that enter the integral I_1 and characterize the vector tangent to the curve C, including phase trajectories, and the differentials dq^{σ} that together with the differentials dp_{σ} are the components of the vector tangent to the phase trajectory.

The phase space introduced before was formed as a manifold with Cartesian coordinates q^{σ} and p_{σ} and had $2s$ dimensions. This space can be extended by joining a coordinate t to it.

In the "extended" phase space , we can also introduce a closed curve C_0^* given in a parametric form by the equations

$$t_0 = t_0(\beta), \quad q_0^{\sigma} = q_0^{\sigma}(\beta), \quad p_{\sigma 0} = p_{\sigma 0}(\beta), \quad 0 \leqslant \beta \leqslant B,$$

$$t_0(0) = t_0(B), \quad q_0^{\sigma}(0) = q_0^{\sigma}(B), \quad p_{\sigma 0} = p_{\sigma 0}(B).$$

Using the curve C_0^* and the general solution (2.3) we construct the "phase (state) tube" consisting of the points with coordinates

$$t = t_0(\beta) + \Delta t, \quad q^{\sigma} = q^{\sigma}(t, q_0(\beta), p_0(\beta)), \quad p_{\sigma} = p_{\sigma}(t, q_0(\beta), p_0(\beta)),$$

where Δt is an arbitrary time interval.

Now consider Δt as a certain continuous function of parameter β subject to the single condition $\Delta t(0) = \Delta t(B)$. Supposing $t_1 = t_0(\beta) + \Delta t(\beta)$, we get on the "phase tube" a closed curve C_1^* set in a parametirc form by the equations

$$t_1(\beta) = t_0(\beta) + \Delta t(\beta), \quad q_1^\sigma(\beta) = q^\sigma(t_1(\beta), q_0(\beta), p_0(\beta)),$$

$$p_{\sigma 1}(\beta) = p_\sigma(t_1(\beta), q_0(\beta), p_0(\beta)).$$

Since we choose the function $\Delta t(\beta)$ in an arbitrary way, a closed curve C_1^* can be regarded as an arbitrary contour including the "phase tube". In this case it is important that every phase "trajectory" passes only through a single point of the contour.

Integrating the differential form (2.10) over the manifold consisting of the points belonging to the curves C_0^* and C_1^*, we obtain

$$\oint_{C_1^*}(p_{\sigma 1}dq_1^\sigma - H_1 dt_1) - \oint_{C_0^*}(p_{\sigma 0}dq_0^\sigma - H_0 dt_0) = \int_0^B dS = S(B) - S(0) = 0.$$

This implies that the integral

$$I_1^* = \oint_{C^*}(p_\sigma dq^\sigma - H dt) = \text{inv} \tag{2.13}$$

does not depend on the choice of the closed contour C^* including the "phase tube". Such an integral is called *a Poincaré–Cartan linear relative integral invariant*.

Since expression (2.10) can be obtained only by using the Hamiltonian equations, the presence of the given equations provides a necessary condition for the existence of the Poincaré–Cartan integral invariant. One can verify the converse: the existence of invariant (2.13) is sufficient for the system of canonical equations in the Poisson form

$$\dot{q}^\sigma = Q_\sigma(t, q, p), \quad \dot{p}_\sigma = P_\sigma(t, q, p) \tag{2.14}$$

to be reducible to the equations in the Hamiltonian form (2.1) with the function H entering the Poincaré–Cartan integral (2.13).

It is obvious that the proof should be based on the independence of the integral I_1^* of the choice of the contour C^*. In order to use this fact, one should represent the coordinates t, q^σ and p_σ of the points of contour C^* in the form

$$t = t_0(\beta) + \Delta t(\beta), \quad q^\sigma = q^\sigma(t, \beta) = q^\sigma(t, q_0(\beta), p_0(\beta)),$$

$$p_\sigma = p_\sigma(t, \beta) = p_\sigma(t, q_0(\beta), p_0(\beta)).$$

Consider a new contour $\widetilde{C}^*$ with the coordinates

$$\widetilde{t} = t + \widetilde{\Delta}t, \quad \widetilde{q}^{\sigma} = q^{\sigma}(\widetilde{t}, \beta), \quad \widetilde{p}_{\sigma} = p_{\sigma}(\widetilde{t}, \beta),$$

where $\widetilde{\Delta}t$ does not depend on β. First of all, we note that in this case the differentials of the variables $\widetilde{t}$ and t as functions of β are equal to each other, i.e., $d\widetilde{t} = dt$.

When $\widetilde{\Delta}t$ is sufficiently small

$$\widetilde{q}^{\sigma} = q^{\sigma} + \dot{q}^{\sigma}\widetilde{\Delta}t, \quad \widetilde{p}_{\sigma} = p_{\sigma} + \dot{p}_{\sigma}\widetilde{\Delta}t, \quad \widetilde{H} = H + \dot{H}\widetilde{\Delta}t,$$

and so, with the accuracy up to the small quantities of first order with respect to $\widetilde{\Delta}t$ we have

$$\widetilde{p}_{\sigma}d\widetilde{q}^{\sigma} - \widetilde{H}d\widetilde{t} = p_{\sigma}dq^{\sigma} - Hdt + \left(\dot{p}_{\sigma}dq^{\sigma} + p_{\sigma}d\dot{q}^{\sigma} - \dot{H}dt\right)\widetilde{\Delta}t. \tag{2.15}$$

Note that the variables $\dot{q}^{\sigma}$, $\dot{p}_{\sigma}$ and $\dot{H}$, entering this expression, are calculated at the points belonging to the contour C^*.

Expression (2.15) implies that the integral I_1^*, when taken along the contour $\widetilde{C}^*$, can be represented as

$$\oint_{\widetilde{C}^*}(\widetilde{p}_{\sigma}d\widetilde{q}^{\sigma} - \widetilde{H}d\widetilde{t}) = \oint_{C^*}(p_{\sigma}dq^{\sigma} - Hdt) + \widetilde{\Delta}t\oint_{C^*}(\dot{p}_{\sigma}dq^{\sigma} + p_{\sigma}d\dot{q}^{\sigma} - \dot{H}dt),$$

when $\widetilde{\Delta}t \neq 0$ is is sufficiently small. So, the equality

$$\oint_{C^*}(\dot{p}_{\sigma}dq^{\sigma} + p_{\sigma}d\dot{q}^{\sigma} - \dot{H}dt) = 0 \tag{2.16}$$

is a corollary of the independence of the integral I_1^* of the choice of the contour.

Calculating the integral $\oint_{C^*} p_{\sigma}d\dot{q}^{\sigma}$ by parts we get

$$\oint_{C^*} p_{\sigma}d\dot{q}^{\sigma} = p_{\sigma}\dot{q}^{\sigma}\Big|_0^B - \oint_{C^*} \dot{q}^{\sigma}dp_{\sigma} = -\oint_{C^*} \dot{q}^{\sigma}dp_{\sigma},$$

as the contour C^* is closed. This implies that expression (2.16) can be rewritten as

$$\oint_{C^*}(\dot{q}^{\sigma}dp_{\sigma} - \dot{p}_{\sigma}dq^{\sigma} + \dot{H}dt) = 0. \tag{2.17}$$

We choose a "phase tube" in the "extended" phase space in an arbitrary way; the contour C^* including it can also be taken arbitrarily. Hence, the integral (2.17) is equal to zero for any closed contour C^*. As is known, this is possible only in the case when the differential form under the integral sign is a total differential of a certain function $\widetilde{H}(t, q, p)$

$$\dot{q}^{\sigma} dp_{\sigma} - \dot{p}_{\sigma} dq^{\sigma} + \dot{H} dt = \frac{\partial \widetilde{H}}{\partial p_{\sigma}} dp_{\sigma} + \frac{\partial \widetilde{H}}{\partial q^{\sigma}} dq^{\sigma} + \frac{\partial \widetilde{H}}{\partial t} dt \,.$$

This implies that

$$\dot{q}^{\sigma} = \frac{\partial \widetilde{H}}{\partial p_{\sigma}}, \quad \dot{p}_{\sigma} = -\frac{\partial \widetilde{H}}{\partial q^{\sigma}}, \quad \dot{H} = \frac{\partial \widetilde{H}}{\partial t} \,. \tag{2.18}$$

Let us calculate the total time derivative of the function $\widetilde{H}(t, q, p)$. Taking into account that the derivatives $\dot{q}^{\sigma}$ and $\dot{p}_{\sigma}$ can be represented in form (2.18) we get

$$\dot{\widetilde{H}} = \frac{\partial \widetilde{H}}{\partial t} + \frac{\partial \widetilde{H}}{\partial q^{\sigma}} \frac{\partial \widetilde{H}}{\partial p_{\sigma}} - \frac{\partial \widetilde{H}}{\partial p_{\sigma}} \frac{\partial \widetilde{H}}{\partial q^{\sigma}} = \frac{\partial \widetilde{H}}{\partial t} \,.$$

Comparing this equality with the third relation from expressions (2.18) one can see that in Eq. (2.18) one should set $\widetilde{H} = H$. Thus, Eq. (2.14) can be indeed reduced to the equations in the Hamiltonian form (2.1).

3   Canonical Transformations

Definition of canonical transformations. In Chap. 6 it was shown that when using the point transformation of the coordinates

$$\widetilde{q}^{\sigma} = \widetilde{q}^{\sigma}(t, q) \tag{3.1}$$

the Lagrange equations of the second kind

$$\frac{d}{dt} \frac{\partial T}{\partial \dot{q}^{\sigma}} - \frac{\partial T}{\partial q^{\sigma}} = Q_{\sigma}$$

do not change and reduce to the equations

$$\frac{d}{dt} \frac{\partial T(t, \widetilde{q}, \dot{\widetilde{q}})}{\partial \dot{\widetilde{q}}^{\tau}} - \frac{\partial T(t, \widetilde{q}, \dot{\widetilde{q}})}{\partial \widetilde{q}^{\tau}} = \widetilde{Q}_{\tau} = Q_{\sigma} \frac{\partial q^{\sigma}}{\partial \widetilde{q}^{\tau}} \,.$$

This property of the above equations is called the *property of their covariance with respect to the point transformations* (3.1).

If the generalized forces Q_{σ} are potential, then the system of Lagrange equations can be written in the canonical form as

$$\dot{p}_{\sigma} = -\frac{\partial H}{\partial q^{\sigma}}, \quad \dot{q}^{\sigma} = \frac{\partial H}{\partial p_{\sigma}}, \quad \sigma = \overline{1, s} \,. \tag{3.2}$$

In this case, the question naturally arises, for what transformations of the kind

$$\widetilde{q}^{\,\sigma} = \widetilde{q}^{\,\sigma}(t, q, p), \quad \widetilde{p}_\sigma = \widetilde{p}_\sigma(t, q, p), \quad \sigma = \overline{1, s}, \tag{3.3}$$

the equations relative to the new variables $\widetilde{q}^{\,\sigma}$, $\widetilde{p}_\sigma$ have the same structure, i.e., have the form

$$\widetilde{\dot{p}}_\sigma = -\frac{\partial \widetilde{H}}{\partial \widetilde{q}^{\,\sigma}}, \quad \widetilde{\dot{q}}^{\,\sigma} = \frac{\partial \widetilde{H}}{\partial \widetilde{p}_\sigma}, \quad \sigma = \overline{1, s}, \tag{3.4}$$

where $\widetilde{H}(t, \widetilde{q}, \widetilde{p})$ is a certain function of the new variables and time that plays the role of the Hamiltonian function. The transformations of this kind are called *canonical*.

The properties of canonicity of transformations. The Lagrange brackets. We need to understand what are the canonical equations from a more general point of view in order to define the conditions under which the transformations (3.3) are canonical. As was already mentioned, such equations can be obtained on the basis of the Legendre equations of the second kind by using the Legendre transformation

$$H = p_\sigma \dot{q}_i^\sigma - L(t, q, \dot{q}).$$

The initial Lagrange equations are necessary conditions for the extremum of the action integral

$$S = \int_{t_0}^{t_1} L(t, q, \dot{q}) \, dt = \int_{t_0}^{t_1} (p_\sigma \dot{q}^\sigma - H) \, dt. \tag{3.5}$$

The equality of the variation of this integral to zero can be taken as an initial principle from which one can obtain the equations of motion of a mechanical system subject to the action of potential forces, and in particular the Hamiltonian equations. The variation of integral (3.5) has the form

$$\delta S = \int_{t_0}^{t_1} \left[\dot{q}^\sigma \delta p_\sigma + p_\sigma \delta \dot{q}^\sigma - \left(\frac{\partial H}{\partial q^\sigma} \delta q^\sigma + \frac{\partial H}{\partial p_\sigma} \delta p_\sigma \right) \right] dt.$$

However,

$$p_\sigma \delta \dot{q}^\sigma = \frac{d}{dt}(p_\sigma \delta q^\sigma) - \dot{p}_\sigma \delta q^\sigma.$$

Since $(\delta q^\sigma)_{t_0} = (\delta q^\sigma)_{t_1} = 0$, we have

$$\int_{t_0}^{t_1} \frac{d}{dt}(p_\sigma \delta q^\sigma) dt = p_\sigma \delta q^\sigma \big|_{t_0}^{t_1} = 0,$$

and hence, the expression for δS can be represented as

$$\delta S = \int_{t_0}^{t_1} \left[\left(\dot{q}^\sigma - \frac{\partial H}{\partial p_\sigma} \right) \delta p_\sigma - \left(\dot{p}_\sigma + \frac{\partial H}{\partial q^\sigma} \right) \delta q^\sigma \right] dt \, . \tag{3.6}$$

First, consider the change of variables by the formulas

$$q^\sigma = q^\sigma(\widetilde{q}, \widetilde{p}) \, , \quad p_\sigma = p_\sigma(\widetilde{q}, \widetilde{p}) \, , \tag{3.7}$$

where $\widetilde{q}$ and $\widetilde{p}$ mean the sets $(\widetilde{q}^{\,1}, \widetilde{q}^{\,2}, \dots, \widetilde{q}^{\,s})$, $(\widetilde{p}_1, \widetilde{p}_2, \dots, \widetilde{p}_s)$, respectively. We have

$$\dot{q}^\sigma = \frac{\partial q^\sigma}{\partial \widetilde{q}^{\,\rho}} \dot{\widetilde{q}}^{\,\rho} + \frac{\partial q^\sigma}{\partial \widetilde{p}_\tau} \dot{\widetilde{p}}_\tau \, , \quad \dot{p}_\sigma = \frac{\partial p_\sigma}{\partial \widetilde{q}^{\,\rho}} \dot{\widetilde{q}}^{\,\rho} + \frac{\partial p_\sigma}{\partial \widetilde{p}_\tau} \dot{\widetilde{p}}_\tau \, , \quad \rho, \tau = \overline{1, s} \, , \tag{3.8}$$

$$\delta q^\sigma = \frac{\partial q^\sigma}{\partial \widetilde{q}^{\,\lambda}} \delta \widetilde{q}^{\,\lambda} + \frac{\partial q^\sigma}{\partial \widetilde{p}_\mu} \delta \widetilde{p}_\mu \, , \quad \delta p_\sigma = \frac{\partial p_\sigma}{\partial \widetilde{q}^{\,\lambda}} \delta \widetilde{q}^{\,\lambda} + \frac{\partial p_\sigma}{\partial \widetilde{p}_\mu} \delta \widetilde{p}_\mu \, , \tag{3.9}$$

$$\sigma, \lambda, \mu = \overline{1, s} \, .$$

Substituting these expressions into formula (3.6) and making simple but cumbersome calculations, we get

$$\delta S = \int_{t_0}^{t_1} \left([\widetilde{q}^{\,\rho}, \widetilde{p}_\mu] \dot{\widetilde{q}}^{\,\rho} \delta \widetilde{p}_\mu + [\widetilde{q}^{\,\rho}, \widetilde{q}^{\,\lambda}] \dot{\widetilde{q}}^{\,\rho} \delta \widetilde{q}^{\,\lambda} + [\widetilde{p}_\tau, \widetilde{q}^{\,\lambda}] \dot{\widetilde{p}}_\tau \delta \widetilde{q}^{\,\lambda} + \right. \tag{3.10}$$

$$\left. + [\widetilde{p}_\tau, \widetilde{p}_\mu] \dot{\widetilde{p}}_\tau \delta \widetilde{p}_\mu - \left(\frac{\partial H}{\partial \widetilde{p}_\mu} \delta \widetilde{p}_\mu + \frac{\partial H}{\partial \widetilde{q}^{\,\lambda}} \delta \widetilde{q}^{\,\lambda} \right) \right) dt \, ,$$

where $[u, v]$ denotes the *Lagrange brackets* with respect to the canonical variables q^σ and p_σ by the parameters u and v given in the form

$$[u, v] = \sum_{\sigma=1}^{s} \left(\frac{\partial q^\sigma}{\partial u} \frac{\partial p_\sigma}{\partial v} - \frac{\partial p_\sigma}{\partial u} \frac{\partial q^\sigma}{\partial v} \right) \, .$$

Taking into account that $[u, v] = -[v, u]$ we represent expression (3.10) as

$$\delta S = \int_{t_0}^{t_1} \left(\left([\widetilde{q}^{\,\rho}, \widetilde{p}_\mu] \dot{\widetilde{q}}^{\,\rho} + [\widetilde{p}_\tau, \widetilde{p}_\mu] \dot{\widetilde{p}}_\tau - \frac{\partial H}{\partial \widetilde{p}_\mu} \right) \delta \widetilde{p}_\mu - \right. \tag{3.11}$$

$$\left. - \left([\widetilde{q}^{\,\lambda}, \widetilde{p}_\tau] \dot{\widetilde{p}}_\tau - [\widetilde{q}^{\,\rho}, \widetilde{q}^{\,\lambda}] \dot{\widetilde{q}}^{\,\rho} + \frac{\partial H}{\partial \widetilde{q}^{\,\lambda}} \right) \delta \widetilde{q}^{\,\lambda} \right) dt \, .$$

This implies that if transformation (3.7) has the following properties:

$$[\widetilde{q}^{\,\rho}, \widetilde{q}^{\,\lambda}] = 0 \, , \quad [\widetilde{p}_\tau, \widetilde{p}_\mu] = 0 \, , \quad \rho, \lambda, \mu, \tau = \overline{1, s} \, , \tag{3.12}$$

$$[\widetilde{q}^{\,\rho}, \widetilde{p}_\mu] = \delta^\rho_\mu = \begin{cases} 1, & \rho = \mu, \\ 0, & \rho \neq \mu, \end{cases} \tag{3.13}$$

then in new variables the variation δS is written as

$$\delta S = \int_{t_0}^{t_1} \left(\left(\dot{\widetilde{q}}^{\,\mu} - \frac{\partial \widetilde{H}}{\partial \widetilde{p}_\mu} \right) \delta \widetilde{p}_\mu - \left(\dot{\widetilde{p}}_\lambda + \frac{\partial \widetilde{H}}{\partial \widetilde{q}^\lambda} \right) \delta \widetilde{q}^\lambda \right) dt, \tag{3.14}$$

where

$$\widetilde{H}(t, \widetilde{q}, \widetilde{p}) = H(t, q(\widetilde{q}, \widetilde{p}), p(\widetilde{q}, \widetilde{p})). \tag{3.15}$$

Formulas (3.6), (3.14) have the same structure, and, hence, making the quantity δS to be zero, we get Eqs. (3.2), (3.4) respectively. And this means that transformations (3.7) are canonical. In other words, conditions (3.12), (3.13) are sufficient for transformations (3.7) to be canonical. Vice versa, if the transformation is canonical, then integral (3.6) can be immediately reduced to integral (3.14), which implies the equality of brackets (3.12) to zero and realization of relations (3.13). Thus, equalities (3.12), (3.13) are necessary and sufficient conditions for the canonicity of transformations (3.7).

Let us now analyze a general case of transformation of canonical variables given by formulas (3.3), which contain the time t explicitly. The quantities δq^σ and δp_σ represent partial differentials of functions (3.3) for fixed t, and in this case formulas (3.9) do not change. The form of formulas (3.8) remains unchanged if one sets $t = \widetilde{q}^{\,0}$, $\dot{\widetilde{q}}^{\,0} = 1$ and assumes that ρ varies from 0 to s. This implies that in the case of transformations of general kind (3.3) the quantity δS can be represented as (3.11) when changing to new variables. Distinguishing the Lagrange brackets containing the time $t = \widetilde{q}^{\,0}$ in this expression and assuming that conditions (3.12), (3.13) are satisfied, we get

$$\delta S = \int_{t_0}^{t_1} \left(\left(\dot{\widetilde{q}}^{\,\mu} + [t, \widetilde{p}_\mu] - \frac{\partial H}{\partial \widetilde{p}_\mu} \right) \delta \widetilde{p}_\mu - \left(\dot{\widetilde{p}}_\lambda - [t, \widetilde{q}^\lambda] + \frac{\partial H}{\partial \widetilde{q}^\lambda} \right) \delta \widetilde{q}^\lambda \right) dt.$$

This expression can be reduced to an integral of form (3.14) if the sum

$$-[t, \widetilde{p}_\mu]\delta \widetilde{p}_\mu - [t, \widetilde{q}^\lambda]\delta \widetilde{q}^\lambda + \frac{\partial H}{\partial \widetilde{p}_\mu} \delta \widetilde{p}_\mu + \frac{\partial H}{\partial \widetilde{q}^\lambda} \delta \widetilde{q}^\lambda$$

is represented as the variation $\delta \widetilde{H}$ of a certain new Hamiltonian function $\widetilde{H}(t, \widetilde{q}, \widetilde{p})$ which cannot be supposed to be given in form (3.15) now. In order to find such a new function $\widetilde{H}$ and simultaneously define possible forms of setting canonical transformations we shall use a relationship between the canonical equations and the Hamilton–Ostrogradskii principle.

The kinds of canonical transformations. The canonical equations (3.2), (3.4) can be regarded as a corollary of the Hamilton–Ostrogradskii principle written in the form

$$\int_{t_0}^{t_1} \delta L\, dt = 0, \qquad \int_{t_0}^{t_1} \delta \widetilde{L}\, dt = 0,$$

where $L = p_\sigma \dot{q}^\sigma - H$, $\widetilde{L} = \widetilde{p}_\sigma \dot{\widetilde{q}}^\sigma - \widetilde{H}$.

This implies that the solution of system (3.2) can be reduced to the solution of system (3.4) by the change of variables according to formulas (3.3), if the Hamilton principle with respect to $\widetilde{L}$ follows from the Hamilton principle with respect to L.

First of all note, that if for the function $L(t, q, \dot{q})$ the condition

$$\delta \int_{t_0}^{t_1} L(t, q, \dot{q})\, dt \equiv \delta \int_{t_0}^{t_1} [p_\sigma \dot{q}^\sigma - H(t, q, p)]\, dt = 0,$$

is satisfied, then the analogous condition is also realized for the function

$$\widetilde{L} = L + \frac{d}{dt} V(t, q, p, \widetilde{q}, \widetilde{p}),$$

where V is an arbitrary differentiable function of $t, q^\sigma, p_\sigma, \widetilde{q}^\sigma, \widetilde{p}_\sigma$. Actually,

$$\delta \int_{t_0}^{t_1} \widetilde{L}\, dt = \delta \int_{t_0}^{t_1} L\, dt + (\delta V)_{t_1} - (\delta V)_{t_0},$$

Since the limits t_0 and t_1 are unchangeable and if the values q^σ and $\widetilde{q}^\sigma$ are unchangeable at moments t_0 and t_1, taking into account relations (3.3) we have $(\delta V)_{t_1} = (\delta V)_{t_0} = 0$, and hence,

$$\delta \int_{t_0}^{t_1} \widetilde{L}\, dt = \delta \int_{t_0}^{t_1} L\, dt.$$

By the above, if one supposes that

$$\widetilde{L} = \widetilde{p}_\sigma \dot{\widetilde{q}}^\sigma - \widetilde{H}(t, \widetilde{q}, \widetilde{p}) = p_\sigma \dot{q}^\sigma - H(t, q, p) + \frac{dV}{dt}, \tag{3.16}$$

then the Hamilton principle is realized for both the variables t, q^σ, p_σ and the variables $t, \widetilde{q}^\sigma, \widetilde{p}_\sigma$. It is convenient to write formula (3.16) as

$$\widetilde{p}\, d\widetilde{q}^\sigma - \widetilde{H}(t, \widetilde{q}, \widetilde{p})\, dt = p_\sigma\, dq^\sigma - H(t, q, p)\, dt + dV. \tag{3.17}$$

Let us use now the arbitrariness of function V relative to the variables t, q^σ, p_σ, $\widetilde{q}^\sigma$, $\widetilde{p}_\sigma$. We shall assume that V is a function of the variables t, q^σ, $\widetilde{q}^\sigma$ only. Denote it by $V_1(t, q, \widetilde{q})$. In this case relation (3.17) appears as

$$\widetilde{p}_\sigma \, d\widetilde{q}^\sigma - \widetilde{H}(t, \widetilde{q}, \widetilde{p}) \, dt = p_\sigma \, dq^\sigma - H(t, q, p) \, dt + \frac{\partial V_1}{\partial t} \, dt + \frac{\partial V_1}{\partial q^\sigma} \, dq^\sigma + \frac{\partial V_1}{\partial \widetilde{q}^\sigma} \, d\widetilde{q}^\sigma \, .$$

This equality becomes an identity if one assumes that

$$\widetilde{p}_\sigma = \frac{\partial V_1(t, q, \widetilde{q})}{\partial \widetilde{q}^\sigma} \, , \tag{3.18}$$

$$p_\sigma = -\frac{\partial V_1(t, q, \widetilde{q})}{\partial q^\sigma} \, , \tag{3.19}$$

$$\widetilde{H}(t, \widetilde{q}, \widetilde{p}) = H(t, q, p) - \frac{\partial V_1}{\partial t} \, . \tag{3.20}$$

Let us assume that the function $V_1(t, q, \widetilde{q})$ is chosen so that

$$\det \left[\frac{\partial^2 V_1}{\partial q^\sigma \partial \widetilde{q}^\tau} \right] \neq 0 \, .$$

In this case, Eq. (3.18) can be solved with respect to q^σ. So, we have

$$q^\sigma = q^\sigma(t, \widetilde{q}, \widetilde{p}) \, .$$

Substituting these expressions into relations (3.19) we have

$$p_\sigma = p_\sigma(t, \widetilde{q}, \widetilde{p}) \, .$$

The transformations given by these formulas are canonical, since in this case equality (3.17) is satisfied identically. And hence, Eqs. (3.2), (3.4) are simultaneously fulfilled.

Formula (3.20) makes it possible to define a new Hamiltonian function $\widetilde{H}(t, \widetilde{q}, \widetilde{p})$. As a result of the fact that the function V is arbitrary, along with canonical transformations given by formulas (3.18), (3.19) there are also canonical transformations of another kind. To find another possible form of setting canonical transformations one should use the identity

$$p_\sigma dq^\sigma = d(p_\sigma q^\sigma) - q^\sigma dp_\sigma \, .$$

In this case relation (3.17) appears as

$$\widetilde{p}_\sigma \, d\widetilde{q}^\sigma - \widetilde{H}(t, \widetilde{q}, \widetilde{p}) \, dt = -q^\sigma \, dp_\sigma - H(t, q, p) \, dt + dV_* \, ,$$

where $V_*(t, q, p, \widetilde{q}, \widetilde{p}) = p_\sigma q^\sigma + V(t, q, p, \widetilde{q}, \widetilde{p})$.

Supposing now that $V_* = V_2(t, \widetilde{q}, p)^{\cdot}$ we obtain

$$\widetilde{p}_\sigma \, d\widetilde{q}^\sigma - \widetilde{H}(t, \widetilde{q}, \widetilde{p}) \, dt = -q^\sigma \, dp_\sigma - H(t, q, p) \, dt +$$
$$+ \frac{\partial V_2}{\partial t} \, dt + \frac{\partial V_2}{\partial \widetilde{q}^\sigma} \, d\widetilde{q}^\sigma + \frac{\partial V_2}{\partial p_\sigma} \, dp_\sigma ,$$

and hence, the canonical transformations can also be set in the form

$$\widetilde{p}_\sigma = \frac{\partial V_2(t, \widetilde{q}, p)}{\partial \widetilde{q}^\sigma} , \quad q^\sigma = -\frac{\partial V_2(t, \widetilde{q}, p)}{\partial p_\sigma} ,$$

$$\widetilde{H}(t, \widetilde{q}, \widetilde{p}) = H(t, q, p) - \frac{\partial V_2}{\partial t} .$$

A similar reasoning shows that the following two partial kinds of a general relation (3.17) are also possible:

$$-\widetilde{q}^\sigma \, d\widetilde{p}_\sigma - \widetilde{H}(t, \widetilde{q}, \widetilde{p}) \, dt = p_\sigma \, dq^\sigma - H(t, q, p) \, dt +$$
$$+ \frac{\partial V_3(t, q, \widetilde{p})}{\partial t} \, dt + \frac{\partial V_3(t, q, \widetilde{p})}{\partial q^\sigma} \, dq^\sigma + \frac{\partial V_3(t, q, \widetilde{p})}{\partial \widetilde{p}_\sigma} \, d\widetilde{p}_\sigma ,$$

$$-\widetilde{q}^\sigma d\widetilde{p}_\sigma - \widetilde{H}(t, \widetilde{q}, \widetilde{p}) \, dt = -q_\sigma \, dp_\sigma - H(t, q, p) \, dt +$$
$$+ \frac{\partial V_4(t, p, \widetilde{p})}{\partial t} \, dt + \frac{\partial V_4(t, p, \widetilde{p})}{\partial p_\sigma} \, dp_\sigma + \frac{\partial V_4(t, p, \widetilde{p})}{\partial \widetilde{p}_\sigma} \, d\widetilde{p}_\sigma .$$

This implies that there exist the following two forms of canonical transformations:

$$\widetilde{q}^\sigma = -\frac{\partial V_3(t, q, \widetilde{p})}{\partial \widetilde{p}_\sigma} , \quad p_\sigma = -\frac{\partial V_3(t, q, \widetilde{p})}{\partial q^\sigma} ,$$

$$\widetilde{H}(t, \widetilde{q}, \widetilde{p}) = H(t, q, p) - \frac{\partial V_3}{\partial t} ,$$

$$\widetilde{q}^\sigma = -\frac{\partial V_4(t, p, \widetilde{p})}{\partial \widetilde{p}_\sigma} , \quad q^\sigma = \frac{\partial V_4(t, p, \widetilde{p})}{\partial p_\sigma} ,$$

$$\widetilde{H}(t, \widetilde{q}, \widetilde{p}) = H(t, q, p) - \frac{\partial V_4}{\partial t} .$$

Thus, depending on the chosen function V the canonical transformations of one of the above kinds can be realized. So, it is only natural to call the function V a *generating function*.

Perturbation method.[3] The perturbation method is perhaps the main method of applying canonical transformations. To describe this method, we consider the system of canonical equations (3.2), in which

$$H(t, q, p) = H_0(t, q, p) + \varepsilon H_1(t, q, p), \quad \varepsilon \ll 1, \quad (3.21)$$

that is, the Hamilton function is the sum of two terms, one of which is small, which is fixed by introducing a small parameter ε. Suppose that for the system of canonical equations with Hamilton function H_0

$$\dot{p}_\sigma = -\frac{\partial H_0}{\partial q^\sigma}, \quad \dot{q}^\sigma = \frac{\partial H_0}{\partial p_\sigma}, \quad \sigma = \overline{1, s}, \quad (3.22)$$

the general solution

$$p_\sigma = p_\sigma(t, \widetilde{p}, \widetilde{q}), \quad q^\sigma = q^\sigma(t, \widetilde{p}, \widetilde{q}), \quad \sigma = \overline{1, s}, \quad (3.23)$$

can be constructed, in which the arbitrary constants are denoted by $\widetilde{p}$, $\widetilde{q}$. Here, as before, for brevity we denote by $\widetilde{p}$, $\widetilde{q}$ the tuples of variables $\widetilde{p} = (\widetilde{p}_1, \dots, \widetilde{p}_s)$, $\widetilde{q} = (\widetilde{q}^1, \dots, \widetilde{q}^s)$.

On the other hand, formulas (3.23) can be considered as the canonical transformation to the new variables $\widetilde{q}$, $\widetilde{p}$, for which the new Hamilton function is $\widetilde{H}_0(t, \widetilde{q}, \widetilde{p}) \equiv 0$, because

$$\dot{\widetilde{p}}_\sigma = -\frac{\partial \widetilde{H}_0}{\partial \widetilde{q}^\sigma} = 0, \quad \dot{\widetilde{q}}^\sigma = \frac{\partial \widetilde{H}_0}{\partial \widetilde{p}_\sigma} = 0, \quad \sigma = \overline{1, s}.$$

In view of (3.20), we have

$$H_0(t, q, p) - \frac{dV_1}{dt} = \widetilde{H}_0(t, \widetilde{q}, \widetilde{p}) = 0.$$

This gives us the generating function V_1 corresponding to the canonical transformation (3.23).

The canonical transformation (3.23) was constructed from the Hamilton function H_0. However, it can be applied for any system of canonical equations with the same number of degrees of freedom s. Indeed, the Lagrange brackets, which appear in the test for the canonicity of transformation (3.12) and (3.13), are independent of the choice of the Hamilton function. Hence we apply the canonical transformation (3.23) to the system with the Hamilton function (3.21). In this way, using (3.20) we get

$$\widetilde{H}(t, \widetilde{q}, \widetilde{p}) = H_0(t, q, p) + \varepsilon H_1(t, q, p) - \frac{dV_1}{dt} = \varepsilon H_1(t, q, p).$$

---

[3] The subsection "Perturbation method" was written by P. E. Tovstik.

and finally,

$$\widetilde{H}(t, \widetilde{q}, \widetilde{p}) = \varepsilon H_1(t, q(t, \widetilde{q}, \widetilde{p}), p(t, \widetilde{q}, \widetilde{p})).$$

As a result of applying the canonical transformation (3.23), the new Hamilton function was found to be small, that is, the new unknowns $\widetilde{p}_\sigma(t)$, $\widetilde{q}^{\,\sigma}(t)$, $\sigma = \overline{1, s}$, by system (3.4) vary slowly with time. There is a possibility of simplifying system (3.4) by replacing the right-hand sides in it by the time-average values. In particular, if the right-hand sides are periodic in time, then one can approximately put

$$f(t, \widetilde{q}, \widetilde{p}) \approx \widehat{f}(\widetilde{q}, \widetilde{p}) = \frac{1}{T} \int_0^T f(t, \widetilde{q}, \widetilde{p})dt,$$

where T is the period and $f(t, \widetilde{q}, \widetilde{p})$ is any of the right-hand sides of system (3.4).

Methods of asymptotic integration have been developed to refine this approximate approach.[4]

Canonical transformations are widely useful in solving problems of celestial mechanics. If we take into account only the attracting force of the Sun, then the planet will describe an elliptic orbit, that is, an analytical solution can be constructed. The perturbation method allows us to take into account the effect of other planets, which leads to a slow evolution of the ellipse parameters.

As an example, consider a system with one degree of freedom with the Hamilton function

$$H(q, p) = H_0(q, p) + \varepsilon H_1(q, p), \quad H_0(q, p) = \frac{1}{2}(q^2 + p^2), \quad H_1(q, p) = F(q),$$

where $F(q)$ is a given function.

The general solution of the system (3.22)

$$\dot{p} = -q, \quad \dot{q} = p$$

has the form

$$q = \widetilde{q}\cos t + \widetilde{p}\sin t, \quad p = \widetilde{p}\cos t - \widetilde{q}\sin t. \tag{3.24}$$

Considering formulas (3.24) as a canonical transformation, we find that the new Hamilton function will be $\widetilde{H} = \varepsilon F(\widetilde{q}\cos t + \widetilde{p}\sin t)$ and the new system of canonical equations will assume the form

$$\overset{\approx}{p} = -\varepsilon f(\widetilde{q}\cos t + \widetilde{p}\sin t)\cos t, \quad \overset{\approx}{q} = \varepsilon f(\widetilde{q}\cos t + \widetilde{p}\sin t)\sin t,$$

$$f(q) = \frac{dF}{dq}. \tag{3.25}$$

---

[4] See the book *N. N. Bogolyubov, Yu. A. Mitropolsky.* Asymptotic methods in theory of nonlinear oscillations. Moscow: Nauka. 1974. 503 p. [in Russian]

Let $F(q) = q^4/4$. Then $f(q) = q^3$ and after averaging over the period $T = 2\pi$ system (3.25) assumes the form

$$\dot{\tilde{p}} = -\frac{3}{8}\varepsilon\tilde{q}(\tilde{q}^2 + \tilde{p}^2), \quad \dot{\tilde{q}} = \frac{3}{8}\varepsilon\tilde{p}(\tilde{q}^2 + \tilde{p}^2). \tag{3.26}$$

Multiplying equations (3.26) by $\tilde{p}$ and by $\tilde{q}$ and adding, we see that this system has the integral $\tilde{q}^2 + \tilde{p}^2 = a^2 = $ const. After this, we find the solution of system (3.26)

$$\tilde{p} = a\cos\left(\frac{3}{8}\varepsilon a^2 t + \alpha\right), \quad \tilde{q} = a\sin\left(\frac{3}{8}\varepsilon a^2 t + \alpha\right), \tag{3.27}$$

where a and α are arbitrary constants. Substituting (3.27) in (3.24), we find the solution of the original system

$$q = a\sin(\omega t + \alpha), \quad p = a\cos(\omega t + \alpha), \quad \omega = 1 + \frac{3}{8}\varepsilon a^2, \tag{3.28}$$

which shows that the frequency of oscillations ø depends on the amplitude a.

In particular, the dimensionless equation of oscillations of a mathematical pendulum has the form $\ddot{\varphi} + \sin\varphi = 0$. For small oscillations, $\ddot{\varphi} + \varphi = 0$, and the frequency of oscillations $\omega_0 = 1$ does not depend on the amplitude a. To investigate oscillations with moderate amplitude, we set $\sin\varphi \approx \varphi - \varphi^3/6$ and use formula (3.28) with $\varepsilon = -1/6$. As a result, we find $\omega = 1 - a^2/16$. For example, for oscillations with amplitude $a = 1/2$ (or 30°) the frequency decreases by 1/64 in comparison with small oscillation frequency.

4 Optical-Mechanical Analogy

The action function S makes it possible, as was shown, to describe the motion of a system on the whole, also allowing one to establish a close relationship between the law of motion of a free particle and the law of propagation of light.

The dependence of the function S on the initial values t_0 and q_0^σ is of no interest in the case under consideration, and so we write

$$S = S(t, q) = \int\limits_{t_0}^{t} L(t, q, \dot{q})\,dt = \int\limits_{t_0}^{t} (T - \Pi)\,dt.$$

In the case of stationary force field when T and Π do not depend on time explicitly, there exists the energy integral $T + \Pi = H = $ const, and hence, $L = 2T - H$. The expression for S in this case appears as

$$S(t, q) = \int\limits_{t_0}^{t} 2T \, dt - \int\limits_{t_0}^{t} H \, dt \,,$$

or, as $T = Mv^2/2$ and $v \, dt = ds$, we have for $t_0 = 0$

$$S(t, q) = \int\limits_{\smile AB} Mv \, ds - Ht = W(q) - Ht \,. \qquad (4.1)$$

For the sake of simplicity, from now on we shall assume that a system consists of one particle and is considered in the Cartesian coordinates. As was already noted, the change to a general case is reduced to the formalization of a mathematical apparatus. So, we assume that

$$S = S(t, x_1, x_2, x_3) \,, \qquad \frac{\partial S}{\partial x_j} = \frac{\partial W}{\partial x_j} = p_j \,, \qquad j = 1, 2, 3,$$

where p_j are the components of impulse which are equal to mv_j. In this case, the total impulse appears as

$$\mathbf{p} = m\mathbf{v} = \nabla S \,.$$

This implies that the vector $m\mathbf{v}$ is orthogonal to the surface $S(t, x) = C = \mathrm{const}$. The equation of this surface, as follows from (4.1), reads as

$$W(x) = Ht + C \,. \qquad (4.2)$$

The function S has the same value C for the instant $t + \Delta t$ on the surface given by the equation

$$W(x) = Ht + C + H \Delta t \,. \qquad (4.3)$$

Comparing the equations of these two surfaces we can see that the surface with equal values of the function S moves in the space.

We define the velocity of displacement of certain points of this surface in such motion. Let $\mathbf{u} = (u_1, u_2, u_3)$ be the velocity vector of interest to us, corresponding to the point with coordinates x_j on the surface (4.2). Then for the sufficiently small Δt this point on the new surface (4.3) has the coordinates $x_j + u_j \Delta t$. The function $W(x + u\Delta t)$ can be represented as

$$W(x + u\Delta t) = W(x) + \frac{\partial W}{\partial x_j} u_j \Delta t \,.$$

with an accuracy up to the terms of the first order of smallness with respect to Δt. Taking into account that points with the coordinates x_j and $x_j + u_j \Delta t$ belong to surfaces (4.2), (4.3), respectively, we have

$$Ht + C + H\Delta t = Ht + C + \frac{\partial W}{\partial x_j} u_j \Delta t \, .$$

Hence, $\frac{\partial W}{\partial x_j} u_j = H$.

Taking into consideration that $\nabla W = \nabla S = m\mathbf{v}$ we get

$$m\mathbf{v} \cdot \mathbf{u} = H \, .$$

The change of coordinates x_j occurs only as a result of the transition from surface (4.2) to surface (4.3). This means that the components $u_j \Delta t$ of the displacement vector are directed along the normal to surface (4.2), i.e., the vectors $\mathbf{u}$ and $\nabla W = m\mathbf{v}$ are collinear. So, we finally have

$$\mathbf{u} = \frac{H}{mv^2}\mathbf{v} = \frac{H}{m^2 v^2}\nabla S \, .$$

This implies that

$$|\mathbf{u}| = u = \frac{H}{mv} = \frac{H}{|\nabla S|} \, . \tag{4.4}$$

We now consider the integral $W = \int_0^t 2T \, dt$ from formula (4.1). It can be written in the Maupertuis–Euler form:

$$W = \int_{\smile AB} mv \, ds \, , \tag{4.5}$$

or, on the basis of formula (4.4), in the form

$$W = \int_{\smile AB} \frac{H ds}{u} = H \int_{\smile AB} \frac{ds}{u} \, . \tag{4.6}$$

Instead of the constant H depending on the initial data, we introduce a new constant H_* of the same dimension, which is assumed to be the same for all the curves. Dividing W by H_* we get

$$\tau = \frac{W}{H_*} = \int_{\smile AB} \frac{ds}{\widetilde{u}} \, , \tag{4.7}$$

where $\widetilde{u} = u H_* / H$.

The quantity τ has the dimension of time. Integral (4.7) is precisely the integral expressing the *Fermat principle*. According to this principle, the light leaving the point A comes to the point B in the shortest time so that integral (4.7) takes the least value for the arc AB.

For the convenience of comparison of integrals (4.5), (4.7) expressing principles of different physical nature we reduce them to a dimensionless kind. Taking c as the velocity standard and l as the length standard, we have

$$W = mcl \int_{\smile AB} \overline{v} \, d\overline{s}, \quad \tau = \frac{l}{c} \int_{\smile AB} \frac{d\overline{s}}{\widetilde{u}},$$

where $d\overline{s} = ds/l$, $\overline{v} = v/c$, $\widetilde{u} = \widetilde{u}/c$, or in the dimensionless kind

$$W = \frac{W}{mcl} = \int_{\smile AB} \overline{v} \, d\overline{s}, \quad \overline{\tau} = \frac{\tau c}{l} = \int_{\smile AB} \frac{d\overline{s}}{\widetilde{u}}.$$

These formulas show that the dimensionless action $\overline{W}$ is equal to the dimensionless time $\overline{\tau}$ if $\overline{v}\widetilde{u} = 1$ or

$$v = c^2/\widetilde{u}. \tag{4.8}$$

Comparing this expression with expression (4.4) $mv = H/u = H_*/\widetilde{u}$, we see that the constant H_* should be chosen in the form

$$H_* = mc^2.$$

Thus, an optical process running with the velocity $\widetilde{u}$ can be put into correspondence with the motion of a particle with velocity v in the potential force field, and vice versa, in this case the velocities $\widetilde{u}$ and v are related by (4.8).

As is known from the course of optics, the propagation of light can be explained by the propagation of disturbance waves that are of oscillatory nature, in this case the phenomenon is subject to the *Huygens principle*. According to this principle, every point of the medium which a disturbance reaches the source of secondary waves with radius $\widetilde{u}\Delta t$, the envelope of which is a new wave surface at the moment $t + \Delta t$.

The Huygens principle can be extended to the case of displacement of a surface with equal values of the function S with the velocity u. A mathematical expression for the Huygens principle is the wave equation

$$\nabla^2 \psi - \frac{1}{\widetilde{u}^2} \frac{\partial^2 \psi}{\partial t^2} = 0. \tag{4.9}$$

Here $\psi(t, x_1, x_2, x_3)$ is the potential, the values of which on the surface of the wave front remain the same; $\widetilde{u}$ is the velocity of propagation of the state ψ. The intrinsic case is the one when the change of function $\psi(t, x_1, x_2, x_3)$ is of oscillatory nature. In the simplest case of light propagation with constant velocity $\widetilde{u}$ along the x-axis, the oscillatory nature of the change of potential ψ satisfying Eq. (4.9) can be specified as

$$\psi(t, x) = A \sin 2\pi \left(\nu t - \frac{\nu x}{\widetilde{u}}\right) \tag{4.10}$$

or in the form

$$\psi(t, x) = A \sin 2\pi\left(\nu t - \frac{x}{\lambda}\right).$$

Here $\lambda = \widetilde{u}/\nu$ is the wave length, ν is the oscillation frequency, A is the amplitude. The quantity $f = \nu t - x/\lambda$ is the oscillation phase. The surfaces with the same phase $\nu t - x/\lambda = $ const represent the planes that are perpendicular to the x-axis and move with the velocity $\widetilde{u} = H_*/(mv) = \nu\lambda$. Such a wave is a called *plane* .

In the case under discussion, the rectilinear light propagation with velocity $\widetilde{u}$ corresponds to the motion of a particle with constant velocity v along the x-axis. As follows from formula (4.1), the function S has the form

$$S = mvx - Ht, \quad s = x.$$

As one can see from this expression, when moving along the x-axis with velocity u, i.e., when $x = x_0 + ut$, the quantity S is constant if the velocity u is given in the form (4.4).

This example shows that there exists the a relationship between the motion of a particle and the wave process. One can establish this relationship due to a more general approach to the mechanical motion of a particle by means of the action function S.

Let us come back to solution (4.10) of the one-dimensional wave equation. The derivative $\partial^2\psi/\partial t^2$ relates to the function ψ as

$$\partial^2\psi/\partial t^2 = -4\pi^2\nu^2\psi.$$

This implies that this function also satisfies the equation

$$\nabla^2\psi + \frac{4\pi^2\nu^2}{\widetilde{u}^2}\,\psi = 0.$$

Invoking the equality $\widetilde{u} = H_*/(mv)$, we get

$$\nabla^2\psi + \frac{8\pi^2\,m\nu^2}{H_*^2}\,\frac{mv^2}{2}\,\psi = 0.$$

Denoting

$$h = \frac{H_*}{\nu} = \frac{mc^2}{\nu} = mv\frac{c^2}{\nu v} = mv\frac{c^2\lambda}{\widetilde{u}v} = p\lambda \qquad (4.11)$$

and taking into account that $mv^2/2 = H - \Pi$ we have

$$\nabla^2\psi + \frac{8\pi^2\,m}{h^2}\,(H - \Pi)\,\psi = 0. \qquad (4.12)$$

The equation with respect to the function ψ of such a kind plays a key role in theoretical physics. Here we have arrived at Eq. (4.12) based on the optical–mechanical analogy and the elementary solution of the wave equation. In this case, we assumed that the velocity $\tilde{u}$, and hence, the potential energy Π are independent of x.

We now consider the properties of Eq. (4.12) supposing that Π is a function of x. First for the analyzed case, it was used by Schrödinger, and is called the *Schrödinger equation*. The total energy H plays a role of the parameter in this equation. It is interesting to note that the solutions that are different from zero and convergent to zero at infinity exist only for some values of the total energy.

Let us study this question in more detail by considering an example of a harmonic oscillator. In this case, the potential energy read as

$$\Pi = c_1 x^2 / 2 .$$

In so doing, equation (4.12) appears as

$$\frac{d^2\psi}{dx^2} + \frac{8\pi^2 m}{h^2} \left(H - \frac{c_1 x^2}{2} \right) \psi = 0 .$$

Setting

$$\alpha = \frac{8\pi^2 m H}{h^2} , \quad \beta = \frac{2\pi}{h} \sqrt{mc_1} , \tag{4.13}$$

we get

$$\frac{d^2\psi}{dx^2} + (\alpha - \beta^2 x^2) \psi = 0 .$$

Let us introduce a new variable $\xi = x \sqrt{\beta}$. Then

$$\frac{d^2\psi}{d\xi^2} + (\overline{H} - \xi^2)\psi = 0 , \tag{4.14}$$

where

$$\overline{H} = \alpha/\beta . \tag{4.15}$$

In this equation the parameter is the quantity $\overline{H}$.

For a plane wave to which corresponds the uniform rectilinear motion of a particle, the function ψ has form (4.10). It is characteristic that this function has the same structure on the whole x-axis. This is explained by the fact that on the whole x-axis the motion is supposed to be uniform and rectilinear. In the case of harmonic oscillator, it occurs inside the finite interval taken on the indicated axis. On this interval this function ψ is not identically zero. Outside the interval the function ψ, which is the solution of equation (4.14), should satisfy the conditions

$$\lim_{x \to +\infty} \psi = 0 , \quad \lim_{x \to -\infty} \psi = 0 . \tag{4.16}$$

In this case, the relationship between a possible motion of the particle and the function ψ is, of course, not so simple as in the uniform rectilinear motion. Here we shall not deal with this special question.

For boundary conditions (4.16) equation (4.14) has the solutions that are not equal to zero identically, but only for some quite definite values of parameter $\overline{H}$, which are called *eigenvalues* or *characteristic numbers*, and the solutions corresponding to them are called *eigenfunctions, proper functions*, or *fundamental functions*. A number of quite definite values of the total energy H corresponds to the eigenvalues of the Schrödinger equation. Hence, the problem of defining the possible discrete values of energy (the problem of energy quantization) reduces to the problem of defining the eigenvalues for equation (4.14) and boundary conditions (4.16).

It is interesting to note that the given mathematical problem had been solved by Hermite long before the Schrödinger equation appeared. He found that the eigenvalues of equation (4.14) are

$$\overline{H}_n = 2n + 1, \quad n = 0, 1, 2, \dots . \tag{4.17}$$

The eigenfunctions ψ_n corresponding to the values $\overline{H}_n$ are called the *Hermite functions* and have the form

$$\psi_n = P_n(\xi)\, e^{-\xi^2/2}, \quad n = 0, 1, 2, \dots ,$$

where $P_n(\xi)$ are the *Hermite polynomials* related by the recurrence relations

$$P_{n+1} = 2\xi P_n - 2n P_{n-1}, \quad n = 1, 2, \dots ,$$

in this case $P_0 = 1$, $P_1 = 2\xi$.

Now let us find to which values of the total energy H_n there correspond the values $\overline{H}_n$.

When a particle oscillates by the law $x = a\,\sin 2\pi\nu t$, its total energy H can be represented as

$$H = \frac{c_1 x^2}{2} + \frac{m\dot{x}^2}{2} = \frac{c_1 x^2}{2} + m \cdot 2\pi^2 a^2 \nu^2 \left(1 - \frac{x^2}{a^2}\right).$$

This implies that for $x = 0$ the total energy $H = m \cdot 2\pi^2 a^2 \nu^2$, and for $x = a$ it is expressed as $H = c_1 a^2/2$, so

$$c_1 a^2/2 = m \cdot 2\pi^2 a^2 \nu^2 .$$

Hence,

$$c_1 = 4\pi^2 \nu^2 m .$$

Substituting this value c_1 into expressions (4.13), we get

$$\beta = \frac{2\pi}{h}\sqrt{mc_1} = \frac{4\pi^2\,m\nu}{h}, \quad \frac{\alpha}{\beta} = \frac{2H}{h\nu},$$

and hence, by formulas (4.15) and (4.17) we have

$$H_n = \overline{H}_n\,\frac{h\nu}{2} = \frac{2n+1}{2}\,h\nu, \quad n = 0, 1, 2, \dots.$$

Thus, the constant H in the Schrödinger equation cannot be taken in a completely arbitrary way—it should be proportional to the oscillation frequency ν and multiple of odd values of the half of a certain elementary initial energy $h\nu$. According to formula (4.11), this energy is expressed by the formula

$$h\nu = mc^2.$$

Chapter 12
Elements of the Special Relativity Theory

N. N. Polyakhov

In this chapter the four-dimensional Poincaré quadratic form was introduced in the study of kinematic relations and the four-dimensional velocity vector was used in the consideration of a composite motion of a point. In the study of dynamics, the generalized Newton law is used to derive the work-energy principle and to write down the Lagrange equations of the second kind and the Hamilton equations. In the study of dynamics, the generalized Newton law is used to derive the work-energy principle and to write down the Lagrange equations of the second kind and the Hamilton equations.

1 Kinematic Relations

Four-dimensional Poincaré quadratic form. The kinematics of a point is concerned with geometric properties of motion in the system of four dimensions, in which the time plays the role of the fourth dimension. However, the time, when considered as the fourth coordinate, appears in kinematic transformations not quite equivalently, as the terms with spatial coordinates. For example, when changing from the $Ox_1x_2x_3$-frame to the $O'x_1'x_2'x_3'$-frame, which moves in a straight line at a constant velocity with respect to the first one, the transformation invariants are, first, the space element length

$$ds^2 = (dx_1)^2 + (dx_2)^2 + (dx_3)^2 = (dx_1')^2 + (dx_2')^2 + (dx_3')^2 \qquad (1.1)$$

and second, the time interval

$$t_2 - t_1 = t_2' - t_1', \qquad (1.2)$$

the latter relation being equivalent to $t = t'$. It is the presence of such invariants that gave Newton an evidence for absolute properties of the space and time.

© Springer Nature Switzerland AG 2021
N. N. Polyakhov et al., *Rational and Applied Mechanics*,
Foundations of Engineering Mechanics,
https://doi.org/10.1007/978-3-030-64061-3_12

The nonequivalence of the spatial and time coordinates follows from formulas (1.1) and (1.2). Clearly, to make these coordinates equivalent and construct a four-dimensional kinematics it is natural to use as a basis the invariant quadratic form

$$d\sigma^2 = (dx_1)^2 + (dx_2)^2 + (dx_3)^2 + (dx_4)^2 =$$
$$= (dx_1')^2 + (dx_2')^2 + (dx_3')^2 + (dx_4')^2 \,, \tag{1.3}$$

where according to Poincaré, $dx_4 = ic\,dt$, $i^2 = -1$, and c is a factor which, clearly, has the dimension of velocity.

The factor i was introduced by Poincaré, because he considered the physical problem involving the reduction of the wave equation

$$\frac{\partial^2 \varphi}{\partial x_1^2} + \frac{\partial^2 \varphi}{\partial x_2^2} + \frac{\partial^2 \varphi}{\partial x_3^2} - \frac{1}{c^2}\frac{\partial^2 \varphi}{\partial t^2} = 0$$

to an equation of the form

$$\frac{\partial^2 \varphi}{\partial x_1^2} + \frac{\partial^2 \varphi}{\partial x_2^2} + \frac{\partial^2 \varphi}{\partial x_3^2} + \frac{\partial^2 \varphi}{\partial x_4^2} = 0 \,,$$

in the latter equation all the four variables are equivalent.

A little later Minkowski proposed to consider the expression

$$dS^2 = c^2(dt)^2 - (dx_1)^2 - (dx_2)^2 - (dx_3)^2$$

as an invariant.

If the form (1.3) is written as

$$d\sigma = ic\,dt(1 - v^2/c^2), \qquad v^2 = \dot{x}_1^2 + \dot{x}_2^2 + \dot{x}_3^2 \,,$$

then we have

$$d\sigma = ic\,dt(1 - v^2/c^2) = ic\,d\tau \,, \tag{1.4}$$

where

$$d\tau = \sqrt{1 - \beta^2}\,dt, \qquad \beta = v/c \,. \tag{1.5}$$

Since the quantity $d\sigma$ was taken for an invariant under the transformation of 4-coordinates, the quantity $d\tau$ is clearly also an invariant.

Lorentz transformations. The requirement that the form (1.3) be invariant means that the space in which the kinematics is now built is the Euclidean 4-space. Since one of the coordinates (namely, x_4) is taken to be purely imaginary, this space will be called the *pseudo-Euclidean space*. The space under consideration is Euclidean, and hence the distance

$$\rho^2 = x_1^2 + x_2^2 + x_3^2 + x_4^2 = (x_1')^2 + (x_2')^2 + (x_3')^2 + (x_4')^2 \tag{1.6}$$

should remain invariant under orthogonal transformations of the space.

For simplicity we shall consider only the case when only the x_1- and x_4-coordinates are transformed. In this case formula (1.6) assumes the form

$$x_1^2 + x_4^2 = (x_1')^2 + (x_4')^2 . \tag{1.7}$$

What are the coefficients α_{ij} of the linear transformation

$$x_1' = \alpha_{11} x_1 + \alpha_{14} x_4, \qquad x_4' = \alpha_{41} x_1 + \alpha_{44} x_4 , \tag{1.8}$$

satisfying condition (1.7)? Substituting (1.8) into (1.7), we find that

$$x_1^2 + x_4^2 = (\alpha_{11}^2 + \alpha_{41}^2) x_1^2 + 2(\alpha_{11}\alpha_{14} + \alpha_{41}\alpha_{44}) x_1 x_4 + (\alpha_{14}^2 + \alpha_{44}^2) x_4^2 ,$$

which shows that the coefficients α_{ij} should obey the equations

$$\begin{gathered} \alpha_{11}^2 + \alpha_{41}^2 = 1, \qquad \alpha_{14}^2 + \alpha_{44}^2 = 1 , \\ \alpha_{11}\alpha_{14} + \alpha_{41}\alpha_{44} = 0 . \end{gathered} \tag{1.9}$$

We also note that since x_1 and x_1' are real and since x_4 and x_4' are purely imaginary, it follows that α_{11} in (1.8) should be real and α_{14} should be purely imaginary.

Let us consider the new variable $i\beta_e = \alpha_{14}/\alpha_{11}$, or what is the same

$$\alpha_{14} = i\beta_e \alpha_{11} . \tag{1.10}$$

Substituting expression (1.10) for α_{14} into Eq. (1.9), we get a system of three equations with the unknowns α_{11}, α_{44} and α_{41}; this system has the parameter β_e. Solving this system, we have

$$\alpha_{11}^2 = \alpha_{44}^2 = \frac{1}{1 - \beta_e^2} , \qquad \alpha_{41}^2 = -\frac{\beta_e^2}{1 - \beta_e^2} .$$

A typical example of an orthogonal transformation of two real variables x and y is the rotation of the Oxy-plane by an angle ψ. It is easily seen that the new coordinates x', y' are related with the old ones by the relations

$$x = x' \cos \psi - y' \sin \psi , \qquad y = x' \sin \psi + y' \cos \psi .$$

Hence we have

$$x' = x \cos \psi + y \sin \psi , \qquad y' = -x \sin \psi + y \cos \psi ,$$

or in the form with subscripts,

$$x_1' = \alpha_{11}x_1 + \alpha_{12}x_2, \qquad x_2' = \alpha_{21}x_1 + \alpha_{22}x_2.$$

The coefficients of this transformation are related as follows:

$$\alpha_{11} = \alpha_{22}, \qquad \alpha_{12} = -\alpha_{21}.$$

The coefficients of transformation (1.8) can also be subject to similar conditions, namely, we may assume that

$$\alpha_{11} = \alpha_{44} = \frac{1}{\sqrt{1 - \beta_e^2}}, \qquad \alpha_{14} = -\alpha_{41} = \frac{i\beta_e}{\sqrt{1 - \beta_e^2}}.$$

Now system (1.8) assumes the form

$$x_1' = \frac{x_1 + i\beta_e x_4}{\sqrt{1 - \beta_e^2}}, \qquad x_4' = \frac{-i\beta_e x_1 + x_4}{\sqrt{1 - \beta_e^2}}.$$

Since $x_4 = ict$, $x_4' = ict'$, we have

$$x_1' = \frac{x_1 - \beta_e ct}{\sqrt{1 - \beta_e^2}}, \qquad t' = \frac{ct - \beta_e x_1}{c\sqrt{1 - \beta_e^2}}. \tag{1.11}$$

From the first equality it follows that if x_1' and the parameter β_e are both assumed to be constant, then

$$\frac{dx_1}{dt} = v_e = \beta_e c, \qquad \beta_e = \frac{v_e}{c}.$$

So, the point fixed in the $O'x_1'x_2'x_3'$-frame has in the $Ox_1x_2x_3$-frame the velocity v_e directed along the x_1-axis. This velocity can be looked upon as the transferred velocity, thereby assuming that the $O'x_1'x_2'x_3'$-frame moves translationally with respect to the $Ox_1x_2x_3$-frame with the velocity v_e directed along the x_1-axis. Expressing in (1.11) the parameter β_e in terms of v_e and taking into account that $x_2' = x_2, x_3' = x_3$, we finally have

$$x_1' = \frac{x_1 - v_e t}{\sqrt{1 - (v_e/c)^2}}, \qquad x_2' = x_2, \qquad x_3' = x_3,$$

$$t' = \frac{t - v_e x_1/c^2}{\sqrt{1 - (v_e/c)^2}}. \tag{1.12}$$

These formulas define the *Lorentz transformations*.

Solving Eqs. (1.12) with respect to the old variables, we find that

$$x_1 = \frac{x_1' + v_e t'}{\sqrt{1 - (v_e/c)^2}}, \qquad x_2 = x_2', \qquad x_3 = x_3',$$

$$t = \frac{t' + v_e x_1'/c^2}{\sqrt{1 - (v_e/c)^2}} . \tag{1.13}$$

This shows that the reverse transition formulas have the same structure as the original formulas (1.12). It remains to find the kinematic sense of the constant c.

As was already pointed out, the invariance of $d\sigma$ corresponds to the invariance of the form of the wave equation

$$\frac{\partial^2 \varphi}{\partial x_1^2} + \frac{\partial^2 \varphi}{\partial x_2^2} + \frac{\partial^2 \varphi}{\partial x_3^2} = \frac{1}{c^2} \frac{\partial^2 \varphi}{\partial t^2} ,$$

underlying the wave optics. Here, c corresponds to the speed of light in a vacuum. Besides, numerous experiments show that this velocity is independent of the velocity v_e of motion of the reference frame and is the same for all systems. This is possible only in the case when, in the new variables x_1', x_2', x_3' and t' (which are related with the old ones by relations (1.12)), the wave equation has the same form[1]:

$$\frac{\partial^2 \varphi}{\partial x_1'^2} + \frac{\partial^2 \varphi}{\partial x_2'^2} + \frac{\partial^2 \varphi}{\partial x_3'^2} = \frac{1}{c^2} \frac{\partial^2 \varphi}{\partial t'^2} .$$

An interesting and important corollary of the Lorentz transforms (1.12) is that the comparison of the distances between points of the 3-space can be made in various frames only with simultaneous comparison of the time intervals. Let us consider this question in more detail.

In the original reference frame, we consider two points with coordinates $(x_{10}, x_{20}, x_{30}, t_0)$ and $(x_{11}, x_{21}, x_{31}, t_1)$, respectively. Assume that the Lorentz transformation sends these points into the points with coordinates $(x_{10}', x_{20}', x_{30}', t_0')$ and $(x_{11}', x_{21}', x_{31}', t_1')$. Hence, as follows from (1.12),

$$x_{11}' - x_{10}' = \frac{x_{11} - x_{10} - v_e(t_1 - t_0)}{\sqrt{1 - (v_e/c)^2}} ,$$

$$t_1' - t_0' = \frac{t_1 - t_0 - (x_{11} - x_{10})v_e/c^2}{\sqrt{1 - (v_e/c)^2}} . \tag{1.14}$$

This shows that with $t_1' = t_0'$ the time interval reads as

$$t_1 - t_0 = (x_{11} - x_{10})v_e/c^2 .$$

Hence, setting $x_{11}' - x_{10}' = l'$, $x_{11} - x_{10} = l$, we get

---

[1] Already in 1887 W. Voigt showed that under transformations (1.12) the form of the wave equation is the same: *W. Voigt.* Über das Dopplersche Prinzip. Göttingen. 1887. S. 41–51.

$$l' = \frac{l(1 - \beta_e^2)}{\sqrt{1 - \beta_e^2}} = l\sqrt{1 - \beta_e^2}, \qquad \beta_e = \frac{v_e}{c}.$$

Assuming that $x'_{11} = x'_{10}$, we find that

$$x_{11} - x_{10} = v_e(t_1 - t_0),$$

and hence,

$$\Delta t' = t'_1 - t'_0 = (t_1 - t_0)\frac{1 - \beta_e^2}{\sqrt{1 - \beta_e^2}} = \Delta t\sqrt{1 - \beta_e^2}.$$

A direct corollary to formulas (1.14) is that

$$l' = \frac{l}{\sqrt{1 - \beta_e^2}}, \qquad \Delta t' = \frac{\Delta t}{1 - \beta_e^2}$$

for $t_1 = t_0$ and for $x_{11} = x_{10}$, respectively.

Composite motion. Let us find the velocity v_1 of the point with respect to the $Ox_1x_2x_3$-frame, assuming that we know, first, the velocity of this point v'_1 with respect the $O'x'_1x'_2x'_3$-frame, and second, the transferred velocity v_e of the system. We shall assume that the point moves parallel to the x'_1-axis, that is $v'_1 = dx'_1/dt'$. Then, using (1.13),

$$v_1 = \frac{dx_1}{dt} = \frac{dx'_1 + v_e dt'}{dt' + v_e dx'_1/c^2} = \frac{v'_1 + v_e}{1 + v_e v'_1/c^2} = \frac{v'_1 + v_e}{1 + \beta_e^2 v'_1/v_e}.$$

This result differs by the factor $(1 + \beta_e^2 v'_1/v_e)^{-1}$ from the formula that corresponds to the invariance of expressions (1.1) and (1.2).

Four-dimensional velocity vector. So far, we have assumed that the four-dimensional quadratic form (1.3) is invariant; this was later written in the form (1.4)

$$d\sigma = \sqrt{\sum_{j=1}^{4}(dx_j)^2} = ic\,dt\sqrt{1 - \beta^2},$$

Then we introduced the invariant quantity (1.5)

$$d\tau = \sqrt{1 - \beta^2}\,dt,$$

which can be looked upon as an invariant time interval.

In the four-dimensional continuum we introduce the four-dimensional velocity vector

$$\mathbf{V} = (V_1, V_2, V_3, V_4),$$

the components of which are defined by the formulas

$$V_j = \frac{dx_j}{d\tau} = \frac{1}{\sqrt{1 - \beta^2}} \frac{dx_j}{dt} = \frac{v_j}{\sqrt{1 - \beta^2}}, \quad j = 1, 2, 3, 4,$$

the time-like component

$$V_4 = \frac{1}{\sqrt{1 - \beta^2}} \frac{dx_4}{dt} = \frac{ic}{\sqrt{1 - \beta^2}}$$

with $\beta < 1$ being purely imaginary.

For convenience in exposition, we shall introduce the unit basis vectors $\mathbf{i}_j = (\delta_j^1, \delta_j^2, \delta_j^3, \delta_j^4)$, where δ_j^k is the Kronecker delta, and the four-dimensional radius-vector $\mathbf{R}$ defined by

$$\mathbf{R} = \sum_{j=1}^{4} x_j \mathbf{i}_j = \mathbf{r} + ict\,\mathbf{i}_4. \tag{1.15}$$

Here $\mathbf{r} = (x_1, x_2, x_3, 0)$ is the vector corresponding to the ordinary radius vector of the point with coordinates (x_1, x_2, x_3). The four-dimensional vector $\mathbf{V}$ can be written in the form

$$\mathbf{V} = \frac{d\mathbf{R}}{d\tau} = \frac{1}{\sqrt{1 - \beta^2}} \frac{d}{dt}(\mathbf{r} + ict\,\mathbf{i}_4) = \frac{\mathbf{v} + ic\,\mathbf{i}_4}{\sqrt{1 - \beta^2}}, \tag{1.16}$$

where $\mathbf{v} = (\dot{x}_1, \dot{x}_2, \dot{x}_3, 0)$ is the vector characterizing the velocity in the ordinary 3-space.

2 The Equations of Dynamics

The generalized Newton law. Let us find out in which sense for a given four-dimensional continuum one should understand the second Newton law, which defines the force. It is naturally to assume that the force, as in the 3-space, should vanish if the velocity vector $\mathbf{V}$ is constant, and to assume that it is proportional to the point mass m. In other words, the generalized Newton law can be put in the form

$$\frac{d(m\mathbf{V})}{d\tau} = \frac{d}{d\tau}\left(m\frac{d\mathbf{R}}{d\tau}\right) = \mathbf{F}, \tag{2.1}$$

where $\mathbf{R}$ and $\mathbf{F}$ are 4-vectors. We wish to direct special attention that the differential $d\tau$ in expression (2.1) is an invariant of the Lorentz transformations (1.12). In other words, it is postulated that there exists a time interval $d\tau$ independent of whether the original reference frame of space coordinates was moving or fixed. Hence the introduction of the differential $d\tau$ in the generalized motion law (2.1) can be looked upon as an extension of Newton's idea on the existence of absolute time.

In analogy with representation (1.15) for an **R**-vector, the quantity **F** is written as

$$\mathbf{F} = \mathbf{f} + i F_4 \mathbf{i}_4,$$

where $\mathbf{f} = (f_1, f_2, f_3, 0)$.

To find a possible connection between the spatial coordinates (f_1, f_2, f_3) of the four-dimensional force vector **F** and the components f_1^*, f_2^*, f_3^* of the ordinary force $\mathbf{f}^*$, we write the generalized Newton law (2.1) as

$$\frac{d}{dt}\left(m_L \frac{d\mathbf{R}}{dt}\right) = \mathbf{F}\sqrt{1-\beta^2} \qquad (2.2)$$

and consider $m_L = m/\sqrt{1-\beta^2}$ as a variable mass. From Eq. (2.2) it follows that the vectors corresponding to the spatial coordinates are related by the expression

$$\frac{d}{dt}(m_L \mathbf{v}) = \mathbf{f}\sqrt{1-\beta^2}, \qquad (2.3)$$

which has the form of the second Newton law in the classical form, and hence we assume that

$$\mathbf{f}\sqrt{1-\beta^2} = \mathbf{f}^* \quad \text{or} \quad f_k = \frac{f_k^*}{\sqrt{1-\beta^2}}, \qquad k = 1, 2, 3. \qquad (2.4)$$

In particular, if the force $\mathbf{f}^*$ has potential Π, then

$$f_k = -\frac{1}{\sqrt{1-\beta^2}}\frac{\partial \Pi}{\partial x_k}, \qquad k = 1, 2, 3. \qquad (2.5)$$

Singling out the imaginary part in Eq. (2.2), we find that

$$\frac{d}{dt}(m_L c) = F_4\sqrt{1-\beta^2} = F_4^*.$$

It should be noted that the introduction of the vectors **r**, **v**, **f**, $\mathbf{f}^*$ with zero fourth coordinate is very convenient, because such vectors can be identified with the ordinary 3-vectors.

Using (1.16) the generalized Newton law (2.1) can be also written in the form

$$\frac{m}{\sqrt{1-\beta^2}}\frac{d}{dt}\frac{\mathbf{v}+ic\,\mathbf{i}_4}{\sqrt{1-\beta^2}} = \mathbf{F}.$$

Expanding the derivatives gives

$$\frac{m\mathbf{w}}{1-\beta^2} + \frac{m(\mathbf{v}+ic\,\mathbf{i}_4)}{\sqrt{1-\beta^2}}\frac{d}{dt}\left(\frac{1}{\sqrt{1-\beta^2}}\right) = \mathbf{F}$$

and further

$$\frac{m\mathbf{w}}{1-\beta^2} + \frac{m(\mathbf{v}+ic\,\mathbf{i}_4)}{(1-\beta^2)^2}\frac{v}{c^2}\frac{dv}{dt} = \mathbf{f} + i\,F_4\,\mathbf{i}_4 . \tag{2.6}$$

Here, $\mathbf{w} = (\ddot{x}_1, \ddot{x}_2, \ddot{x}_3, 0)$.

Separating the real and imaginary parts in (2.6), we find that

$$m\mathbf{w} = \mathbf{f}(1-\beta^2) - \frac{m\beta^2}{1-\beta^2}\mathbf{v}^0\frac{dv}{dt}, \tag{2.7}$$

$$\frac{m\beta}{(1-\beta^2)^2}\frac{dv}{dt} = F_4, \tag{2.8}$$

where $\mathbf{v}^0 = \mathbf{v}/v$ is the unit velocity vector. Equation (2.7), which describes the motion of a point in the ordinary 3-space, is equivalent to the following two scalar equations

$$m\frac{dv}{dt} = f_\varepsilon(1-\beta^2) - \frac{m\beta^2}{1-\beta^2}\frac{dv}{dt}, \tag{2.9}$$

$$m\frac{v^2}{\rho} = f_n(1-\beta^2), \tag{2.10}$$

which can be looked upon as equations in the projections onto the tangent line and the normal line $\mathbf{n}$ to the trajectory. Here ρ is the curvature radius of the trajectory; $f_\varepsilon = \mathbf{f}\cdot\mathbf{v}^0$; $f_n = \mathbf{f}\cdot\mathbf{n}$.

Equations (2.9), (2.10) show that the motion takes place as if the material point was subject, in addition to the force $\mathbf{f}$, to the additional forces

$$f_\varepsilon' = -f_\varepsilon\beta^2 - \frac{m\beta^2}{1-\beta^2}\frac{dv}{dt},$$

$$f_n' = -f_n\beta^2,$$

where f_ε' is the resistance force depending on the velocity squared v^2 and the acceleration dv/dt. Clearly, for a constant force f_ε the acceleration of a point tends to zero with increasing time, because formula (2.9) shows that

$$m\frac{dv}{dt} = f_\varepsilon(1-\beta^2)^2, \tag{2.11}$$

and hence the acceleration w_ε vanishes with $\beta = 1$, that is, when $v = c$. It can be verified that the time requiring to reach such a velocity is infinite.

Let us find the meaning of the component F_4. Using formula (2.11), we write expression (2.8) in the form

$$F_4 = f_\varepsilon \beta = \mathbf{f} \cdot \mathbf{v}/c \,.$$

In terms of the ordinary force f^*, the component F_4 can be expressed as

$$F_4 = \mathbf{f}^* \cdot \mathbf{v}/(c\sqrt{1 - \beta^2}) \,,$$

equality (2.4) being useful. Hence, if the second Newton law, as written in the form (2.1), is considered as the initial one, then the required motion of a material point in the $Ox_1x_2x_3$-system is described by Eq. (2.3) or by (2.7), the force $\mathbf{f}$ from these equations being related by (2.4) with the given force $\mathbf{f}^*$ that controls the motion. Equation (2.3) is called the *relativistic motion equation*.

The work-energy principle. Multiplying formula (2.11) by v, this gives

$$d\left(\frac{mv^2}{2}\right) = f_\varepsilon(1 - \beta^2)^2 \, v \, dt = f_\varepsilon(1 - \beta^2)^2 ds \,, \qquad (2.12)$$

which is the crux of the work-energy principle in the differential form. This formula shows that the role of the force that does the work is played by the force $f_\varepsilon(1 - \beta^2)^2$.

We now assume that the ordinary force $\mathbf{f}^*$ has potential Π. Hence by (2.5) we find that

$$f_\varepsilon = \frac{\mathbf{f} \cdot \mathbf{v}}{v} = -\frac{(1 - \beta^2)^{-1/2}}{v} \sum_{k=1}^{3} \frac{\partial \Pi}{\partial x_k} \frac{dx_k}{dt} = -\frac{(1 - \beta^2)^{-1/2}}{v \, dt} d\Pi \,.$$

Therefore, in this case relation (2.12) can be put in the form

$$d\left(\frac{mv^2}{2}\right) = -(1 - \beta^2)^{3/2} d\Pi \,,$$

or the integral form

$$\frac{mc^2}{2} \int_{\beta_0}^{\beta} \frac{d(\beta^2)}{(1 - \beta^2)^{3/2}} = \Pi_0 - \Pi \,.$$

Assuming for simplicity that the velocity v_0 and, correspondingly, $\beta_o = v_o/c$ are both zero, we find that

$$mc^2 = \left(\frac{1}{\sqrt{1 - \beta^2}} - 1\right) = \Pi_0 - \Pi \,. \qquad (2.13)$$

Setting

$$T = \frac{mv^2}{2} \,, \qquad \varkappa = \frac{2}{\beta^2}\left(\frac{1}{\sqrt{1 - \beta^2}} - 1\right) \,, \qquad (2.14)$$

we have

$$\varkappa T = \Pi_0 - \Pi. \tag{2.15}$$

This is the form which the energy integral assumes in the case of relativistic motion. For $\beta \to 0$, when $\varkappa \to 1$, the energy integral is transformed into the standard energy integral. The quantity $\varkappa T = T_r$ is called the *relativistic kinetic energy*, the energy conservation law assuming the form

$$E_r \equiv T_r + \Pi = \Pi_0 = \text{const}.$$

We write expression (2.13) as

$$\frac{mc^2}{\sqrt{1 - \beta^2}} + \Pi = mc^2 + \Pi_0 = \tilde{\Pi}_0$$

and consider $\tilde{\Pi}_0$ as the initial store of energy. The quantity mc^2 is sometimes called the *self-energy of a point* or the *rest energy*; the quantity $mc^2/\sqrt{1 - \beta^2}$ is called the *"total" energy*. By the above, all variations of the ordinary expressions for the kinetic energy and the conservation law of mechanical energy appear as a consequence of the introduction of the invariant interval $d\tau$ instead of the one that changes when transforming from one system of coordinates to another one in time interval dt. As a result, the ordinary motion law $m\mathbf{w} = \mathbf{f}^*$ assumes the form (2.3). In view of (2.4) Eq. (2.3) can be put in the form

$$\frac{d}{dt}(m_L\mathbf{v}) = \mathbf{f}^*, \qquad m_L = \frac{m}{\sqrt{1 - \beta^2}}.$$

With this form of the principal motion law, the appearance of the factor $\varkappa$, as given by expression (2.14), in the energy integral (2.15) can be explained by the variation of the mass m_L as a function of the velocity.

Lagrange equations. Before we obtained the Lagrange equations of the second kind by writing the vector Newton equation in the projections onto the axes of curvilinear coordinates, which were using the transformation

$$x_i = x_i(q^1, q^2, q^3), \qquad i = 1, 2, 3.$$

In the case of a four-dimensional continuum with invariant (1.3), when the generalized Newton law (2.1) is assumed to hold ab inito, the definite role in the study of motion in the standard three-dimensional space is played by Eq. (2.7), which in view of (2.11) assumes the form

$$m\mathbf{w} = \mathbf{f}(1 - \beta^2) - \frac{f \cdot \mathbf{v}}{c^2}(1 - \beta^2)\mathbf{v}.$$

Let us analyze in the frame of this equation the motion of a material point subject to an ordinary force $\mathbf{f}^*$. Assuming that the vectors $\mathbf{f}^*$ and $\mathbf{f}$ are related via (2.4), we have

$$m\mathbf{w} = \mathbf{f}^*\sqrt{1 - \beta^2} - \frac{\mathbf{f}^* \cdot \mathbf{v}}{c^2}\sqrt{1 - \beta^2}\,\mathbf{v}\,. \tag{2.16}$$

Let us find out the form of this equation in projections onto the axes of the curvilinear frame. For $\beta = 0$ ($c = \infty$) these are the ordinary Lagrange equations of the second kind. For $\beta \neq 0$, multiplying the vector equality (2.16) by the basis vectors $e_\sigma = \partial\mathbf{r}/\partial q^\sigma$ and taking into account that

$$\mathbf{v} = \dot{q}^\rho \mathbf{e}_\rho\,, \qquad \mathbf{f}^* = Q_\tau \mathbf{e}^\tau\,, \qquad \mathbf{e}^\tau \cdot \mathbf{e}_\rho = \delta_\rho^\tau\,,$$

$$g_{\rho\sigma} = \mathbf{e}_\rho \cdot \mathbf{e}_\sigma\,, \qquad \rho, \sigma, \tau = 1, 2, 3,$$

this gives

$$\frac{d}{dt}\frac{\partial T}{\partial \dot{q}^\sigma} - \frac{\partial T}{\partial q^\sigma} = Q_\sigma + \widetilde{Q}_\sigma\,,$$

where

$$\widetilde{Q}_\sigma = Q_\sigma\big(\sqrt{1 - \beta^2} - 1\big) - \frac{Q_\tau \dot{q}^\tau}{c^2}\sqrt{1 - \beta^2}\, g_{\rho\sigma}\dot{q}^\rho\,.$$

If the force $\mathbf{f}^*$ has the potential Π, that is, if $Q_\sigma = -\partial\Pi/\partial q^\sigma$, then introducing the function $L = T - \Pi$ we have

$$\frac{d}{dt}\frac{\partial L}{\partial \dot{q}^\sigma} - \frac{\partial L}{\partial q^\sigma} = \widetilde{Q}_\sigma\,. \tag{2.17}$$

Thus, the relativistic effect is manifested in the fact that on the right of the Lagrange equations there appears a sort of an additional force $\widetilde{Q}_\sigma$, which is not potential. As a result, the energy integral does not exist in the ordinary form, but instead relation (2.15) holds. From Eqs. (2.17) it follows that the Hamilton action is not of "extreme character", inasmuch as

$$\delta \int_{t_0}^{t_1} L\, dt = -\int_{t_0}^{t_1} \big(\widetilde{Q}_\sigma \delta q^\sigma\big)\, dt \neq 0\,.$$

The Hamilton equations in this case can be represented in the form in which they were written for a nonconservative system:

$$\frac{dp_\sigma}{dt} = -\frac{\partial H}{\partial q^\sigma} + \widetilde{Q}_\sigma\,, \qquad \frac{dq^\sigma}{dt} = \frac{\partial H}{\partial p_\sigma}\,.$$

However, it should be recalled that the energy integral in this case exists and reads as

$$\varkappa T + \Pi = \text{const} \quad \text{or} \quad T + \Pi + \varkappa_1 T = \text{const} \,,$$

where

$$\varkappa_1 = \varkappa - 1 = \frac{1}{\beta^2 \sqrt{1 - \beta^2}} \left[2 - (2 + \beta^2) \sqrt{1 - \beta^2} \right] \,.$$

References

General references

1. S.V. Bolotin, A.V. Karapetyan, E.I. Kugushev, D.V. Treshchev, *Theoretical Mechanics (Teoreticheskaya mekhanika)* (Izdat. "Akademiya", Moscow, 2010) 430 p. [in Russian]
2. N.N. Buchholts, *Fundamental Course of Theoretical Mechanics (Osnovnoy kurs teoreticheskoy mekhaniki)* (Izdat. "Nauka", Moscow, 1969). Part I, 468 p.; Part II, 332 p. [in Russian]
3. Yu.F. Golubev, *Fundamentals of Theoretical Mechanics (Osnovy teoreticheskoy mekhaniki)* (Izdat. Moskovskogo universiteta, Moscow, 2019). 728 p. [in Russian]
4. V.F. Zhuravlev, *Fundamentals of Theoretical Mechanics (Osnovy teoreticheskoy mekhaniki)* (Izdat. "Nauka", Moscow, 1997). 320 p. [in Russian]
5. N.A. Kilchevskiy, *Course of Theoretical Mechanics (Kurs teoreticheskoy mekhaniki)*, vol. I (Izdat. "Nauka", Moscow, 1972) 456 p.; 1977. Vol. II, 544 p. [in Russian]
6. A.P. Markeev, *Theoretical Mechanics (Teoreticheskaya mekhanika)* (Izdat. "Nauka", Moscow, 1990). 414 p. [in Russian]
7. G.K. Suslov, *Theoretical Mechanics (Teoreticheskaya mekhanika)* (Izdat. "Gostekhizdat", Moscow-Leningrad, 1946). 655 p. [in Russian]

Supplementary references

8. P. Appell, *Traité de Mécanique Rationnelle*, vols. 1, 2. (Gauthier-Villars, Paris, 1953)
9. V.I. Arnol'd, *Mathematical Methods of Classical Mechanics (Matematicheskie metody klassicheskoy mekhaniki)* (Izdat. "Nauka", Moscow, 1989), 472 p. [in Russian]
10. N.N. Bogolyubov, Yu.A. Mitropol'skiy, *Asymptotic Methods in Theory of Nonlinear Vibration (Asimptoticheskie metody v teorii nelineinykh kolebaniy)*. (Izdat. "Nauka. Fizmatlit", Moscow, 1974), 503 p. [in Russian]
11. V.V. Dobronravov, *Fundamentals of Mechanics of Nonholonomic Systems (Osnovy mekhaniki negolonomnikh system)* (Izdat. "Vysshaya shkola", Moscow, 1970), 272 p. [in Russian]
12. A.S. Galiullin, *Analytical Dynamics (Analiticheskaya dinamika)* (Izdat. "Vysshaya shkola", Moscow, 1989), 264 p. [in Russian]

© Springer Nature Switzerland AG 2021
N. N. Polyakhov et al., *Rational and Applied Mechanics*,
Foundations of Engineering Mechanics,
https://doi.org/10.1007/978-3-030-64061-3

13. N.E. Zhukovskiy, *Theoretical Mechanics (Teoreticheskaya mekhanika)* (Izdat. "Gostekhizdat", Moscow-Leningrad, 1952), 811 p. [in Russian]

14. Sh.Kh. Soltakhanov, M.P. Yushkov, S.A. Zegzhda, *Mechanics of Non-holonomic Systems. A New Class of Control Systems*. (Springer, Berlin, 2009). 329 p

15. S.A. Zegzhda, M.P. Yushkov, Sh.Kh. Soltakhanov, E.A. Shatrov, *Nonholonomic Mechanics and Control Theory (Negolonomnaya mekhanika i teoriya upravleniya)*. Moscow: Izdat. "Nauka. Fizmatlit" (2018). 236 p. [in Russian]

16. A.Yu. Ishlinskiy, *Classical mechanics and inertia forces (Klassicheskaya mekhanika i sily inertsii)* "Nauka" (Izdat., Moscow, 1987). 320 p. [in Russian]

17. J.L. Lagrange, *Analytic Mechanics* (Kluwer Academic Publishers, Dordrecht-Boston-London, 1997), p. 264

18. T. Levi-Civita, U. Amaldi, *Lezioni di Meccanica Razionale*. Vol. 1, 2d Edition, Vol. 2, parts 1 and 2, 1st Edn. (Nicola Zanichelli, Bologna, 1926-30), pp. 807, 526, 684

19. J.W. Leech, *Classical Mechanics* (Methuen and Co Ltd, London, 1958; Wiley, New York 1958)

20. L.G. Loytsyanskiy, A.I. Lur'e, *The Course of Theoretical Mechanics (Kurs teoret008koy mekhaniki)*, "Nauka" (Izd, Moscow, 1982). Vol. I, 352 p.; 1983. Vol. II, 640 p. [in Russian]

21. A.I. Lur'e, *Analytical Mechanics*. "Fizmatgiz" (Izdat, Moscow, 1961). 824 p. [in Russian]

22. A. Liapounoff, *Problème général de la stabilityé du movement (Thèse et articles)*. Seconde édition.

23. D.R. Merkin, *Introduction to the Theory of Stability* (Texts in Applied Mathematics, vol. 24) (Springer Science and Business Media, Berlin, 2012). 320 p

24. Ju.I. Neimark, N.A. Fufaev, *Dynamics of Nonholonomic Systems* (Translations of Mathematical Monographs, vol. 33) (American Mathematical Society, 2004). 518 p

25. I. Newton, *The Mathematical Principles of Natural Philosophy*, vol. I (Benjamin Motte, 1729)

26. N.V. Roze, *Lectures on Analytical Mechanics (Lektsii po analiticheskoi' mekhanike)* (Leingrad University Publishers, Leingrad, 1938). 203 p. [in Russian]

27. E.T. Whittaker, *A Treatise on the Analytical Dynamics of Particles and Rigid Bodies*, 2nd edn. (University Press, Cambridge, 1917). 432 p

28. N.G. Chetayev, *Theoretical Mechanics* (Teoreticheskaya mekhanika). "Nauka" (Izdat, Moscow, 1987). 368 p. [in Russian]

29. N.G. Chetayev. *Stability of motion* (Ustoichivost' dvizheniya). *Works on analytical mechanics* (Raboty po analiticheskoy mekhanike) (AN SSSR Publishers, Moscow, 1962). 536 p. [in Russian]

30. L. Euler. *Fundamentals of the point dynamics* (Osnovy dinamiki tochki) Moscow-Leningrad: Izdat. "ONTI" (1938). 500 p. [in Russian]

31. C. Jacobi, *Vorlesungen über Dynamik* ed. by by A. Clebsch, 2nd edn. (G. Reimer, Berlin 1866); reprint, New York: Chelsea, 1969; English translation by K. Balagangadharan, Jacobi's lectures on dynamics, Biswarup Banerjee, ed., New Delhi: Hindustan Book Agency. 2009

32. J.G. Papastavridis, *Analytical Mechanics* (Oxford, 2002), 1392 p

33. L.A. Pars, *Treatise on Analytical Dynamics* (Heinemann, London, 1965)

34. F.E. Udwadia, R.E. Kalaba, *Analytical Dynamics: A New Approach* (Cambridge, 1996), 262 p

Index

© Springer Nature Switzerland AG 2021
N. N. Polyakhov et al., *Rational and Applied Mechanics*,
Foundations of Engineering Mechanics,
https://doi.org/10.1007/978-3-030-64061-3

Printed in the United States
by Baker & Taylor Publisher Services